Quantum Stochastic Thermodynamics

Philipp Strasberg is a Ramon y Cajal fellow at the Institute of Physics of Cantabria in Santander working on a wide variety of topics in nonequilibrium quantum statistical mechanics. He completed his PhD in 2015 under the supervision of Tobias Brandes at the Technical University of Berlin, before joining Massimiliano Esposito's group at the University of Luxembourg for two years. From 2018 to 2024, he was supported by independent postdoctoral fellowships to work at the Universitat Autònoma de Barcelona in collaboration with Anna Sanpera and Andreas Winter.

Quantum Stochastic Thermodynamics

Foundations and Selected Applications

Philipp Strasberg
Universitat Autònoma de Barcelona

OXFORD
UNIVERSITY PRESS

Great Clarendon Street, Oxford, OX2 6DP,
United Kingdom

Oxford University Press is a department of the University of Oxford.
It furthers the University's objective of excellence in research, scholarship,
and education by publishing worldwide. Oxford is a registered trade mark of
Oxford University Press in the UK and in certain other countries

First published 2022
First published in paperback 2024

Published in the United States of America by Oxford University Press
198 Madison Avenue, New York, NY 10016, United States of America

British Library Cataloguing in Publication Data
Data available

Library of Congress Cataloging in Publication Data
Data available

ISBN 978–0–19–289558–5 (Hbk.)
ISBN 978–0–19–893158–4 (Pbk.)

DOI:10.1093/oso/9780192895585.001.0001

Printed and bound by
CPI Group (UK) Ltd, Croydon, CR0 4YY

Cover image: Xanya69/Shutterstock.com

This book is dedicated to the memory of my PhD supervisor, Tobias Brandes

Preface

Recent decades have seen much progress in our understanding of thermodynamic processes at the nanoscale. But nanoscale systems are very small, in most applications far from equilibrium, often subject to strong fluctuations and sometimes even characterized by exotic quantum properties—so it seems that these features rule out any possibility of finding a consistent thermodynamic description for them.

It is the primary objective of this book to show that this is not the case. There is a thermodynamic framework, characterized by a remarkable internal consistency, that is able to describe nanoscale systems even under extreme conditions. Moreover, this framework not only reaffirms common folklore around thermodynamics (*there is no perpetual motion machine, etc.*), but also provides a wealth of beautiful results beyond the traditional scope of thermodynamics—opening up the possibility of understanding a plethora of different physical situations from a unified perspective.

The main title *Quantum Stochastic Thermodynamics* suggests that the present book is about a synthesis of two research fields: classical stochastic thermodynamics and quantum thermodynamics. Both have pushed the boundaries of the applicability of the laws of thermodynamics by explaining and supplementing them with microscopic considerations. For a considerable large class of nanoscale systems and processes, I believe that most foundational questions have been settled by now. The present book is intended to fill a gap in the literature by justifying this claim in detail for a large variety of situations. I am also convinced that its content will prove important in exploring new territories at the rapidly evolving frontiers of this field.

The subtitle *Foundations and Selected Applications* emphasizes that the reader can mainly expect explanations about the basic theoretical pillars. These explanations are intended to be pedagogically accessible. However, the book is also driven by the desire to introduce a general and versatile framework characterized by conceptual clarity—in complete awareness of the fact that this poses additional technical obstacles for the beginner. To remedy this, a considerable effort has been made to transparently explain common 'jargon' in the community (non-equilibrium entropies, local detailed balance, Landauer's principle, entropy production, time-reversal symmetry, the arrow of time, etc.), which often appears unnecessarily mystifying (a problem that seems to have a tradition in thermodynamics and statistical mechanics). The reader will often find the same (or closely related) results derived in different ways in order to generate confidence and trust in the framework.

The field of quantum stochastic thermodynamics is, however, fascinating, not only because it allows us to addresss foundational questions about the nature of heat, entropy or the second law, but also because it might have direct practical applications in a world with increasing nanotechnological abilities. These applications could come in the form of efficient thermoelectric devices, powerful energy harvesters, fast cooling strategies, or energy-efficient computers, among other more exotic applications. To form a close connection between foundational and practical problems, this book treats a few selected applications in detail.

Unfortunately, because of a lack of time and understanding on the author's side, this book cannot cover all possible directions that are currently under investigation. The selected material is obviously biased and I ask for forbearance from my many colleagues who feel that the present book misses some important ideas.

How to Read this Book

This book is written for graduate students who know the basics of quantum mechanics and equilibrium statistical mechanics and who cannot wait to combine these fields to understand non-equilibrium phenomena. As a rule of thumb, you are ready to delve into the book if the following equations do not scare you off:

$$\begin{aligned}
\frac{\partial}{\partial t}\rho(t) &= -\frac{i}{\hbar}[H,\rho(t)], \\
\rho &= \sum_n \lambda_n |n\rangle\langle n|, \\
\mathcal{Z}(\beta) &= \mathrm{tr}\{e^{-\beta H}\}, \\
\mathcal{U}(\beta) &= -\frac{\partial}{\partial\beta}\ln \mathcal{Z}(\beta), \\
S_B &= k_B \ln W, \\
f(\epsilon) &= \frac{1}{e^{\beta(\epsilon-\mu)}+1}.
\end{aligned}$$

Of course, this book is also written for more experienced researchers. I hope that they will have no trouble in jumping between different sections of this book (though it might help to first look at the Basic Notation section below). To become further acquainted with the book, I summarize here its most important features.

STRUCTURE OF THE BOOK: To get a complete picture, I believe one should sooner or later read the entire book and, then, it would perhaps be most beneficial to go through it in linear order. However, I also believe that the impatient reader should not be afraid to skip sections or chapters. For instance, Chapter 1 (Quantum Stochastic Processes) appears to be the most abstract one, in particular its second half. While I believe that this more abstract point of view helps to view the entire field in a clear and unified way, it is certainly not necessary to reach an understanding of Chapter 2 (Classical Stochastic Thermodynamics), which solely requires some basic background knowledge of the theory of classical stochastic processes. Likewise, readers with some familiarity with open quantum system theory can directly start reading Chapter 3 (Quantum Thermodynamics Without Measurements). To understand Chapter 4 (Quantum Fluctuation Theorems), some background information from previous chapters is required. For some sections of Chapter 5 (Operational Quantum Stochastic Thermodynamics) it is necessary to have also read and understood the end of Chapter 1. Finally, three appendices complement this book by providing information about topics that appear at various points in the main text, but whose detailed exposition requires a longer detour, which would blur the main narrative.

STRUCTURE OF THE SECTIONS: Typically, I have tried to start each section with a small paragraph motivating its content and to end each section with a small summary or outlook. To facilitate orientation, some sections (in particular longer ones) are divided into subsections using unnumbered subtitles. Furthermore, important statements are distinguished by longer italic text and boxed equations highlight important definitions or results.

INDEX: I have tried to make a long and informative index list. Words or phrases appearing in this list are printed in a **boldface** font in the main text at the point where they are first introduced or explained. However, as I said above, the book is characterized by presenting similar concepts in different contexts and from different perspectives. Thus, the book is not written as an encyclopaedia, but tries to maintain a narrative that is most beneficial for pedagogical purposes.

EXERCISES: Various exercises are scattered throughout the text. These exercises, sometimes supplemented by (hopefully) helpful hints on how to solve them, should be rather simple because I believe there is no benefit in torturing the reader. Exercises fall, however, into two categories. The first category of exercises are the short ones. They are supposed to supply simple cross-checks for the reader or, by asking for derivations of some equations, to acquaint the reader with standard mathematical manipulations in the field. Then, there are also various longer exercises, which might even require some simple symbolic programming. These longer exercises are typically meant to introduce ideas, concepts or results whose detailed exposition would probably bore the more experienced reader in the field. Thus, the exercises help to keep the book more concise, while allowing me at the same time to cover a wider range of topics. I remark that all exercises appear during the text at the point where they best fit the overall narrative.

REFERENCES: In contrast to the exercises, all references are relegated to a special Further reading section at the end of each chapter. Having the overall pedagogical purpose of the book in mind, I indeed believe that there is little benefit from mentioning references during the main exposition of the material. Moreover, it ought to be clear that, given the breadth and scope of the present book, it is impossible to give credit to all contributions and, seen again from a pedagogical perspective, I believe there is little benefit from trying to do so here. Hence, this book should not be confused with a conventional 'review article'. Most citations are given to the latest work, where the main concepts or equations I rely on were first introduced or derived. Other citations typically refer to expositions that go beyond the material presented here in a wider sense (i.e. books, reviews or introductory articles about fields related to but not part of quantum stochastic thermodynamics). Finally, some citations are added for historical clarity. Thus, the overall idea is that the list of references provides a first *orientation* for the newcomer, not an exhaustive list of contributions to the field.

Basic Notation

I tried to keep abbreviations to a minimum and used only those that are widely used in the literature. The three most important abbreviations, which are scattered throughout the text, are CP (completely positive), CPTP (completely positive and trace-preserving) and BMS (Born–Markov secular). In some equations, I write h.c. to denote the Hermitian conjugate. Also the abbreviation POVM (positive operator-valued measure) is used occasionally.

Below, I further provide a non-exhaustive list of the most important notation used throughout the book.

GENERAL MATHEMATICS: Matrices are denoted by capital letters (such as M). Boldface letters are used for vectors (e.g. a vector of probabilities $\boldsymbol{p}$) and sequences (e.g. a sequence of measurement results $\mathbf{r}$), but their elements are written as, for example, p_j or r_n. Multiplication of vectors and matrices is written without a dot (e.g. $M\boldsymbol{p}$). $[A, B] = AB - BA$ and $\{A, B\} = AB + BA$ denote the commutator and anti-commutator, respectively. A superscript $*$, T or $\dagger$ denotes the complex conjugate, transpose or the conjugate transpose, respectively. The imaginary unit is denoted i and the symbol $\mathcal{O}(x)$ denotes that $\lim_{x\to 0} |\mathcal{O}(x)/x| < \infty$. Total and partial derivatives (e.g. with respect to time t) appearing in in-line equations are written as d_t and ∂_t, respectively. Frequently used functions include the Heaviside step function $\Theta(x)$, the Kronecker delta $\delta_{m,n}$ and the Dirac delta function $\delta(x - y)$.

QUANTUM DYNAMICS: I use Dirac notation for states $|\psi\rangle$ and their conjugate transpose $\langle\psi|$ that live in a Hilbert space $\mathcal{H}$ and its dual, respectively. The scalar product is written as $\langle\phi|\psi\rangle$. To avoid unwanted technicalities, I assume Hilbert spaces are (or can be approximated to be) finite dimensional: $d = \dim \mathcal{H} < \infty$. The density matrix is typically denoted by ρ, whereas most other operators are denoted by capital letters such as $H, P, X, \ldots$ and I is the identity. Exceptions are the familiar bosonic and fermionic creation and annihilation operators (denoted typically by $a^{(\dagger)}, b^{(\dagger)}, c^{(\dagger)}, d^{(\dagger)}$) and the Pauli matrices:

$$\sigma_x = \begin{pmatrix} 0 & 1 \\ 1 & 0 \end{pmatrix}, \quad \sigma_y = \begin{pmatrix} 0 & -i \\ i & 0 \end{pmatrix}, \quad \sigma_z = \begin{pmatrix} 1 & 0 \\ 0 & -1 \end{pmatrix}.$$

The trace of some operator O is denoted $\mathrm{tr}\{O\}$. Superoperators (also simply called maps), which map operators onto operators, are always denoted by calligraphic letters such as $\mathcal{C}, \mathcal{D}, \mathcal{L}, \ldots$. The tensor product is denoted $\otimes$ and additional subscripts $A, B, S \ldots$ are used to indicate on which subspace some operator is acting (e.g. ρ_S) or to denote the partial trace (e.g. $\mathrm{tr}_B\{\ldots\}$).

STATISTICAL MECHANICS: Equilibrium concepts are denoted by calligraphic letters such as the equilibrium internal energy $\mathcal{U}$ or equilibrium entropy $\mathcal{S}$. Exceptions are parameters such as temperature T or chemical potential μ. Out-of-equilibrium quantities are denoted by, for example, U and S if they refer to some expectation value or ensemble average. Thermodynamic quantities defined along single stochastic trajectories are denoted by small Latin letters (e.g. u and s). The canonical ensemble (or

Gibbs state) is denoted by the Greek letter π. For instance, $\pi_S(\beta)$ denotes the Gibbs state of some system S at inverse temperature β.

FINALLY, every process starts at the initial time $t_0 = 0$. Moreover, I do *not* set Boltzmann's and Planck's constants k_B and $\hbar$ to 1. I believe that this makes the physical content of many equations more insightful. For practical manipulations, this choice is, of course, not the most convenient one, but since the majority of the literature sets $k_B \equiv 1$ and $\hbar \equiv 1$, I thought it would be good to keep them explicit here.

Acknowledgements

This book is dedicated to the memory of Tobias Brandes because—long before I enjoyed being his PhD student—his inspiring and unprecedented lectures were the reason why I actually turned towards theoretical physics. In my opinion, Tobias's approach to physics was dual, tackling deep and conceptual problems while having at the same time an eye on experimentally well-grounded approaches and models, which also work 'in practice'. I view my own research, and in particular this book, in the tradition of his philosophy, perhaps with some bias more towards conceptual and general ideas. I hope Tobias would have enjoyed reading it.

Concerning the topics exposed here I further owe much of my detailed knowledge to Massimiliano Esposito and Gernot Schaller. Neither ever hesitated to share their insights and ideas with me about various topics in statistical mechanics, stochastic thermodynamics, non-equilibrium physics and open quantum systems. Much of their knowledge is reflected here; yet, I believe I succeeded in also adding my own twist.

A special thanks goes to Kavan Modi. A short, but in retrospect important, discussion at the Kavli Institute in Santa Barbara in 2018 inspired me to look at the problem from a different angle and part of the material presented in Chapter 5, and in some sense the motivation to write this book, is a consequence of that discussion.

Many more colleagues have shared their insights and thoughts with me. Those on whom I could particularly rely concerning the topics presented in this book include Robert Alicki, Janet Anders, Felipe Barra, Victor Bastidas, Javier Cerrillo, Luis Correa, María García Díaz, David Gelbwaser-Klimovsky, John Goold, Giacomo Guarnieri, Géraldine Haack, Christopher Jarzynski, Matteo Lostaglio, Mark Mitchison, Kavan Modi, Wolfgang Muschik, Juan Parrondo, Martí Perarnau-Llobet, Matteo Polettini, Andreu Riera-Campeny, Felix Ritort, Àngel Rivas, Dominik Šafránek, Rafael Sánchez, Anna Sanpera, Udo Seifert, Michalis Skotiniotis, Christopher Wächtler and Andreas Winter. It makes me a bit sad to know that I probably forgot to correctly acknowledge all the people who contributed to my actual understanding of this topic. There were clearly many more people who influenced my thinking.

Furthermore, since writing a book is quite a 'mammoth project', I am grateful to Javier Cerrillo, Marco Merkli, Kavan Modi, Andreu Riera-Campeny, Àngel Rivas and Rafael Sánchez, who read (parts of) early drafts of the book and provided valuable comments. In particular, a special thanks goes to Teresa Reinhard for making sure that various parts of the book do not sound too cryptic to the outsider.

I also want to thank the team at Oxford University Press. In particular, I much appreciated the uncomplicated manner of communicating with them, which clearly contributed to the fact that the writing of this book was an overall very joyful experience.

Of course, the constant support I receive from my family and friends, including the little lion, are invaluable.

Most parts of the work reported here were financially supported by the DFG (project STR 1505/2-1). For further support I thank the Spanish Agencia Estatal de Investigación, projects IJC2019-040883-I and PID2019-107609GB-I00, the Spanish MINECO FIS2016-80681-P (AEI/FEDER, UE) and the Generalitat de Catalunya CIRIT 2017-SGR-1127.

Finally and almost needless to say, all mistakes and opinions expressed here are entirely my own. I am always grateful to receive further comments, questions and feedback (to get in contact with me I suggest you type my name into your preferred search engine and look up my up-to-date email address). Errata will be announced on my homepage.

Contents

1
Quantum Stochastic Processes

Summary. This chapter describes the basic features of open quantum systems, i.e. quantum systems that are affected by noise due to uncontrollable degrees of freedom of an environment or bath. This noise is responsible for effects such as dissipation, decoherence and irreversibility. We study the equilibrium states of open quantum systems and review tools from quantum measurement theory, which describe how to extract information from an (open) quantum system. We generalize these tools to multi-time statistics and define the notion of a quantum stochastic process and a quantum Markov process. Finally, we study in which cases a quantum stochastic process looks classical.

1.1 Isolated Quantum Systems

Time-independent case

A quantum system is described by a Hilbert space $\mathcal{H}$ and a state ρ called the density matrix, which acts on that space. The density matrix is characterized by the facts that it is Hermitian, $\rho^\dagger = \rho$ (with $\dagger$ denoting the Hermitian conjugate), has unit trace, $\text{tr}\{\rho\} = 1$, and is positive, $\rho \geq 0$. Here, the notation $\rho \geq 0$ is shorthand for $\langle\psi|\rho|\psi\rangle \geq 0$ for all $|\psi\rangle \in \mathcal{H}$.

The state of an isolated quantum system, i.e. a quantum system which is not in contact with any other part of the world, obeys the Liouville–von Neumann equation

$$\frac{\partial}{\partial t}\rho(t) = -\frac{i}{\hbar}[H, \rho(t)]. \tag{1.1}$$

Here, H is the Hamiltonian operator characterizing the total energy of the system and $[A, B] \equiv AB - BA$ is the commutator. Furthermore, i and $\hbar$ are the familiar imaginary unit and Planck's constant, respectively. Note that we will also use the notation $\partial_t\rho(t)$ to denote a partial derivative with respect to time.

Isolated quantum systems have some important characteristics:

(i) There exists a unitary time evolution operator $U(t) \equiv \exp(-iHt/\hbar)$, which means that $U(t)U(t)^\dagger = U(t)^\dagger U(t) = I$, where I denotes the identity matrix. This time evolution operator propagates the system state according to

$$\rho(t) = U(t)\rho(0)U(t)^\dagger, \tag{1.2}$$

where $\rho(0)$ denotes the initial state of the quantum system.

Quantum Stochastic Thermodynamics. Philipp Strasberg, Oxford University Press.
 DOI: 10.1093/oso/9780192895585.003.0001

(ii) The spectrum of the state $\rho(t)$, i.e. its eigenvalues λ_k, do not change in time:

$$\rho(t) = \sum_k \lambda_k |\psi_k(t)\rangle\langle\psi_k(t)|. \tag{1.3}$$

Here, λ_k is time independent and $|\psi_k(t)\rangle$ belongs to an orthonormal basis of wave functions, i.e. $\langle\psi_k(t)|\psi_\ell(t)\rangle = \delta_{k,\ell}$ with the Kronecker delta $\delta_{k,\ell}$. Note that the spectrum can be degenerate, i.e. it is possible that $\lambda_k = \lambda_\ell$ for some $k \neq \ell$. Equation (1.3) follows from the fact that any unitary transformation of the form (1.2) leaves the spectrum invariant since the characteristic polynomial does not change:

$$\det\{\rho(t) - \lambda I\} = \det\{U(t)[\rho(0) - \lambda I]U(t)^\dagger\} = \det\{\rho(0) - \lambda I\}. \tag{1.4}$$

Here, $\det\{\dots\}$ denotes the determinant of a matrix.

(iii) The purity of $\rho(t)$, which measures the 'mixedness' of a state and is defined as $\mathrm{tr}\{\rho^2\}$, is conserved. This follows immediately from point (ii) above or, alternatively, from eqn (1.1) and the fact that the trace is cyclic, i.e. $\mathrm{tr}\{ABC\} = \mathrm{tr}\{CAB\}$. In particular, if the initial state is pure, i.e. $\mathrm{tr}\{\rho(0)^2\} = 1$, we have $\rho(t) = |\psi(t)\rangle\langle\psi(t)|$ for some wave function $|\psi(t)\rangle$. This wave function evolves in time according to the familiar Schrödinger equation:

$$\frac{\partial}{\partial t}|\psi(t)\rangle = -\frac{i}{\hbar}H|\psi(t)\rangle. \tag{1.5}$$

Since the Schrödinger equation can be applied to compute the time evolution of any of the $|\psi_k(t)\rangle$ in eqn (1.3), it is equivalent to the Liouville–von Neumann equation (1.1).

(iv) The **von Neumann entropy** of the system, which is defined as

$$S_{\mathrm{vN}}(\rho) \equiv -\mathrm{tr}\{\rho \ln \rho\}, \tag{1.6}$$

is constant in time. This follows again from point (ii) above because $S_{\mathrm{vN}}[\rho(t)] = -\sum_k \lambda_k \ln \lambda_k$, which implies that the von Neumann entropy quantifies the classical uncertainty about the state of the system. Alternatively, the conservation of von Neumann entropy follows by using

$$\frac{d}{dt}S_{\mathrm{vN}}[\rho(t)] = -\mathrm{tr}\left\{\frac{d\rho(t)}{dt}\ln\rho(t)\right\} \tag{1.7}$$

and eqn (1.1). Note that eqn (1.7) holds only if the rank of the density operator does not change during the evolution, which is guaranteed by point (ii) above. If the rank changes in time, the von Neumann entropy is not differentiable. Note that we use a subscript 'vN' for the von Neumann entropy throughout the book because we want to distinguish it from the notion of *thermodynamic* entropy introduced later on. Readers unfamiliar with information-theoretic concepts of entropy can find an overview in Appendix A.

Time-dependent case

Quantum systems are often subjected to time-dependent fields in a laboratory, and these fields can be treated semiclassically (e.g. laser light). In this case, the Hamiltonian becomes time dependent and is denoted by $H(\lambda_t)$, where the time-dependent parameter λ_t specifies the external fields. We prefer the notation $H(\lambda_t)$ instead of $H(t)$ to avoid any possible confusion with the Heisenberg picture. In the following, λ_t is called a *driving* or *control protocol* and the state of such a *driven system* evolves in time according to the Liouville–von Neumann equation (1.1) with H replaced by $H(\lambda_t)$.

The evolution $\rho(t) = U(t,0)\rho(0)U^\dagger(t,0)$ is still described by a unitary operator $U(t,0)$, but its explicit computation is now more complicated. Formally, we have

$$
\begin{aligned}
U(t,0) &= \sum_{n=0}^{\infty}\left(-\frac{i}{\hbar}\right)^n \int_0^t dt_1 H(\lambda_1)\int_0^{t_1} dt_2 H(\lambda_2)\cdots\int_0^{t_{n-1}} dt_n H(\lambda_n) \\
&= \sum_{n=0}^{\infty}\frac{1}{n!}\left(-\frac{i}{\hbar}\right)^n \int_0^t dt_1\cdots\int_0^t dt_n [H(\lambda_1)\ldots H(\lambda_n)]_+ \\
&\equiv \exp_+\left[-\frac{i}{\hbar}\int_0^t ds H(\lambda_s)\right],
\end{aligned} \tag{1.8}
$$

where we abbreviated $\lambda_{t_j} \equiv \lambda_j$ and the subscript $+$ denotes time ordering. By dividing the time interval $[0,t]$ into $n = t/\delta t$ small steps δt, we can also write

$$
U(t,0) \approx \prod_{j=0}^{n-1} e^{-iH(\lambda_j)\delta t/\hbar}, \tag{1.9}
$$

which becomes exact in the limit $n \to \infty$.

Strictly speaking, a driven quantum system is not isolated as it is in contact with the external driving field. However, it is easy to check that points (i)–(iv) above are still satisfied for such a driven system. In contrast, the time evolution of a quantum system which interacts with another quantum system is markedly different, as we will see in the rest of this chapter. Therefore, it seems wise to call a quantum system 'isolated' as long as it evolves in time according to the Liouville–von Neumann equation (1.1), whether the Hamiltonian is time independent or not.

1.2 System–Bath Theories and the Origin of Noise

In reality, nobody has ever observed an isolated quantum system at the very end because the mere act of 'observing' a quantum system requires it to be coupled with an external detector. We will, however, postpone the discussion of quantum measurement theory to Section 1.4 and are here rather concerned with the fact that many quantum systems interact with *uncontrollable* degrees of freedom of a so-called *environment* or *bath*. One example is an atom (the 'system') interacting with the many electromagnetic modes of the surrounding space (the 'environment'). Another example is a single spin, e.g. an impurity in a metal or crystal, interacting with many remaining spins and

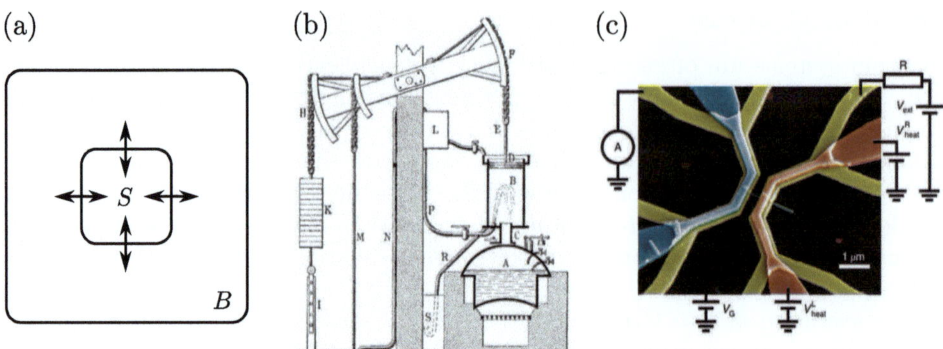

Fig. 1.1 Open quantum systems. (a) A rough sketch of a system S in contact with a bath B with which it can exchange energies, particles, entropy, etc. (b) Diagram of the Newcomen atmospheric steam engine as an example of an ancient 'open quantum system': thermodynamics had already divided the universe into a system part and reservoirs. (c) (False-coloured) scanning electron microscope image of a modern open quantum system: a quantum dot formed by a nanowire (thin green line) in contact with electron reservoirs (yellow) and heaters (blue and red). Such setups will be treated in further detail in Section 3.10. Reprinted by permission from Springer Nature: Springer Nature, *Nature Nanotechnology* (A quantum-dot heat engine operating close to the thermodynamic efficiency limits, M. Josefsson et al.), Copyright by Springer Nature (2018).

phonons (lattice vibrations) in the surroundings. Finally, the historical origin of the theory of thermodynamics is rooted in the desire to understand, for example, steam in a container in contact with hot air produced by burning coal on the one side and cold water on the other side. Here, the system is defined by the container, which contains the so-called working medium or working fluid, whereas the outside hot air and cold water are called a *heat bath* or *reservoir*. We will also use these words from Chapter 2 onwards. In this chapter, however, we use the broader term 'environment' or 'bath' to refer to any external, uncontrollable part of the world, not necessarily described by a thermodynamic variable such as temperature. It will become clear throughout this book that the predictive power of the second law comes from an efficient description of these uncontrollable degrees of freedom about which we have only very little information.

Quantum systems which are not isolated but in contact with a bath or environment are called *open* quantum systems; see Fig. 1.1 for sketches. Conceptually, many different approaches exist to describe them theoretically and a very powerful one is to model the bath itself as another quantum system such that the system *and* the bath (which we sometimes also call the *universe*) constitute one big isolated quantum system. This is the origin of system–bath theory and we will see below that this is indeed not an assumption: any open quantum system can be seen as being part of a larger isolated quantum system.

Mathematically, the system–bath composite is a *bipartite* quantum system described by the tensor product of the system and bath Hilbert space: $\mathcal{H}_S \otimes \mathcal{H}_B$. The dimension of that space is $\dim(\mathcal{H}_S \otimes \mathcal{H}_B) = \dim \mathcal{H}_S \cdot \dim \mathcal{H}_B$. For many applications the dimension of $\mathcal{H}_S$ is very small, e.g. $\dim \mathcal{H}_S = 2$ for a single spin, whereas the dimension of $\mathcal{H}_B$ is often very large, e.g. $\dim \mathcal{H}_B = 2^{N_A}$, where the Avogadro number N_A is of the order of 10^{23}. The dynamics of the system and the bath is governed by the Hamiltonian $H_{SB} = H_S \otimes I_B + I_S \otimes H_B + V_{SB}$. Here, H_S and H_B denote the Hamiltonian of the isolated system or bath, respectively, and V_{SB} denotes their interaction. They sum up to the total Hamiltonian H_{SB}. Whereas the system and bath Hamiltonian commute (since they live on different Hilbert spaces) we have in general $[V_{SB}, H_S] \neq 0$ and $[V_{SB}, H_B] \neq 0$. In the following, we suppress tensor products with the identity for notational simplicity and write the system–bath Hamiltonian as

$$H_{SB} = H_S + H_B + V_{SB}. \tag{1.10}$$

For simplicity, we assumed no driving λ_t here, but this will change in later chapters. We remark that the tensor product structure of the system–bath composite assumes the system and bath to be distinguishable objects and implies a particular choice of gauge made when identifying what is the 'system' and what is the 'bath'.

Since the system–bath composite is isolated, its global state described by the density operator $\rho_{SB}(t)$ evolves according to the Liouville–von Neumann equation (1.1) with H replaced by H_{SB}. The system evolution is obtained by taking the partial trace:

$$\rho_S(t) = \mathrm{tr}_B\{\rho_{SB}(t)\}. \tag{1.11}$$

In contrast to the isolated case, the evolution of an open quantum system is significantly different. None of the four points mentioned in Section 1.1 remain true:

(i′) The time evolution of $\rho_S(t)$ is not described by a unitary operator.

(ii′) The eigenvalues of the system state change in time, i.e. eqn (1.3) is replaced by

$$\rho_S(t) = \sum_k \lambda_k(t) |\psi_k(t)\rangle_S \langle \psi_k(t)|. \tag{1.12}$$

(iii′) Since the eigenvalues of $\rho_S(t)$ change, the purity of the state can change too. In particular, an initially pure state becomes mixed in general.

(iv′) The von Neumann entropy is no longer conserved. Note that the von Neumann entropy of the system can become larger or smaller during the evolution, without violating the second law of thermodynamics. The second law states only that the thermodynamic entropy of the universe, i.e. the system and the bath, cannot become smaller in time. It does not imply that the von Neumann entropy of the system state cannot decrease.

Illustrative example

We illustrate the above arguments by considering an assembly of n interacting spins. Readers unfamiliar with open quantum systems are invited to explicitly follow this example by doing their own numerics for it. We assume that the spins are described by the following global Hamiltonian:

$$H_{SB} = \frac{\hbar\Omega}{2} \sum_{i=1}^{n} \sigma_z^{(i)} + \frac{\hbar}{2} \sum_{i=1}^{n} \sum_{j>i} g_{ij} \sigma_x^{(i)} \sigma_x^{(j)}, \tag{1.13}$$

where $\sigma_\alpha^{(i)}$ $(\alpha = x, y, z)$ are the familiar Pauli matrices acting on spin i. The first term describes n isolated spins with energy gap $\hbar\Omega$. The second term describes the interaction between two spins with coupling strength g_{ij}. As our system we now choose one of the spins, say the first, such that $H_S = \hbar\Omega\sigma_z^{(1)}/2$. The bath Hamiltonian consequently becomes $H_B = \hbar\Omega\sum_{i=2}^n \sigma_z^{(i)}/2 + \hbar\sum_{i=2}^n\sum_{j>i} g_{ij}\sigma_x^{(i)}\sigma_x^{(j)}/2$ and the remaining part defines the interaction Hamiltonian V_{SB}. The initial state of the system and bath is assumed to be decorrelated:

$$\rho_{SB}(0) = \rho_S(0) \otimes \rho_B(0). \tag{1.14}$$

In the simulations, we take $\rho_S(0) = |+\rangle\langle+|_S$ to be a pure state. Here, $|+\rangle \equiv (|0\rangle + |1\rangle)/\sqrt{2}$ denotes a coherent superposition of the eigenstates of σ_z (with $\sigma_z|0\rangle = -|0\rangle$ and $\sigma_z|1\rangle = +|1\rangle$), which coincide with the energy eigenstates of H_S. The bath instead is taken to be a canonical equilibrium (Gibbs) ensemble with respect to the inverse temperature β, denoted by

$$\rho_B(0) = \pi_B \equiv \frac{e^{-\beta H_B}}{\mathcal{Z}_B}, \quad \mathcal{Z}_B \equiv \mathrm{tr}_B\{e^{-\beta H_B}\}, \tag{1.15}$$

where $\mathcal{Z}_B$ is the partition function of the bath. Therefore, we assume the following situation: previous to the initial time the first spin is decoupled from the others and prepared in a superposition of energy eigenstates. Then, at time $t = 0$ we suddenly switch on the interaction with the bath, which is assumed to be thermalized.

We aim at a numerical exact simulation of the system–bath dynamics based on the Liouville–von Neumann equation (1.1). For n spins the total density matrix as well as the unitary time evolution operator are $2^n \times 2^n$ matrices. This exponential growth in size necessarily limits us to consider only a few spins, precisely we consider seven spins in total (i.e. the bath consists of six spins). Note that one is often interested in a bath that is *much* larger in size. Computing the exact dynamics of an open quantum system then quickly becomes impossible, which motivates the need for efficient and reliable approximation schemes. For the moment, however, a small bath suffices to illustrate our points above. To continue with the discussion of our model, let us choose the spin–spin interactions g_{ij} random from a uniform distribution over $[0, 1]$. We do not explicitly write down the values for g_{ij} here because the dynamics are qualitatively similar for most choices.

Numerical results for an initial (dimensionless) bath temperature of $\beta\hbar\Omega = 10$ are shown in Fig. 1.2. Note that this bath temperature is relatively cold, i.e. the energy gap $\hbar\Omega$ of each single spin is 10 times larger than the typical energy $k_BT = \beta^{-1}$ of a thermal excitation. Figure 1.2a shows the time evolution of the expectation value $\langle\sigma_x^{(1)}\rangle(t) = \mathrm{tr}_1\{\sigma_x^{(1)}\rho_S(t)\}$. For better comparison we also plot its time evolution in the case that the spin was isolated, i.e. for $V_{SB} = 0$ (thin grey line). Their difference is quite striking; in particular, it is difficult to recognize any structure or pattern for the open quantum system case. Figure 1.2b shows the purity, which decreases as expected. Note that the minimal value of the purity for a two-level system is 1/2. Figure 1.2c shows the evolution of the von Neumann entropy, which is no longer conserved. Note that the maximum value for the von Neumann entropy of a two-level

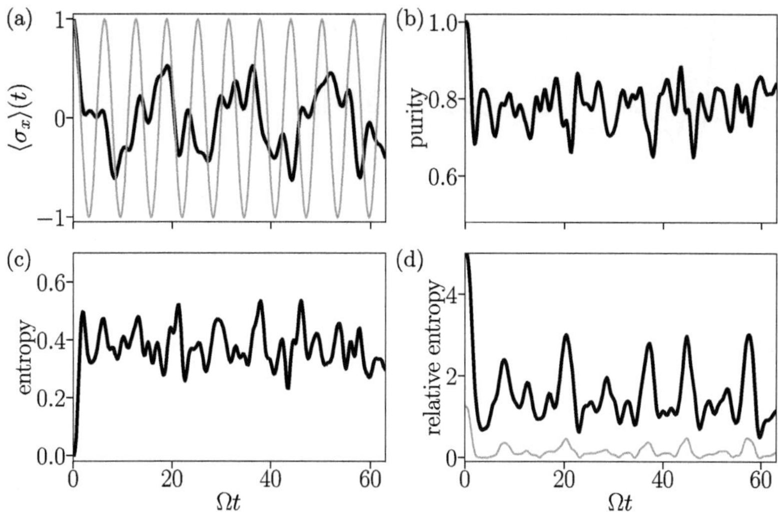

Fig. 1.2 Exemplary time evolution of an open quantum system for a 'cold' bath. The thin grey line in (a) shows the time evolution of an isolated system with an identical system Hamiltonian. The thin grey line in (d) shows the time evolution of the quantum relative entropy with respect to a reference state, which describes a refined equilibrium state introduced in Section 1.3.

system is $\ln 2 \approx 0.7$. In Fig. 1.2d we show the time evolution of a quantity that plays an important role throughout the book. It is known as the **quantum relative entropy**, which is defined in general as

$$D(\rho|\sigma) \equiv \text{tr}\{\rho(\ln\rho - \ln\sigma)\} \geq 0 \tag{1.16}$$

for two arbitrary density matrices ρ and σ. The relative entropy is non-negative and only zero if $\rho = \sigma$. It can be regarded as a measure of statistical 'distance' between ρ and σ; further information is provided in Appendix A. In particular, we plot

$$D\left[\rho_S(t)\middle|\frac{e^{-\beta H_S}}{\mathcal{Z}_S}\right], \tag{1.17}$$

i.e. the distance between the reduced system state and its associated Gibbs state at the inverse temperature of the bath. One could naively expect that this difference should become very small for long times in accordance with equilibrium statistical mechanics. This is not the case for two important reasons. First, a bath of only six spins is still too small to induce equilibration of a small quantum system. Second, even if the bath were larger, equilibrium statistical mechanics in fact predicts a state *different* from the Gibbs state, as we will explain in detail in the next section. The time evolution of the relative entropy with respect to this different state is shown by the thin grey line, which is much closer to zero than eqn (1.17).

Finally, Fig. 1.3 shows the same plots as Fig. 1.2, but for a different initial temperature of the bath. This time we chose $\beta\hbar\Omega = 1$, such that the thermal excitation energies

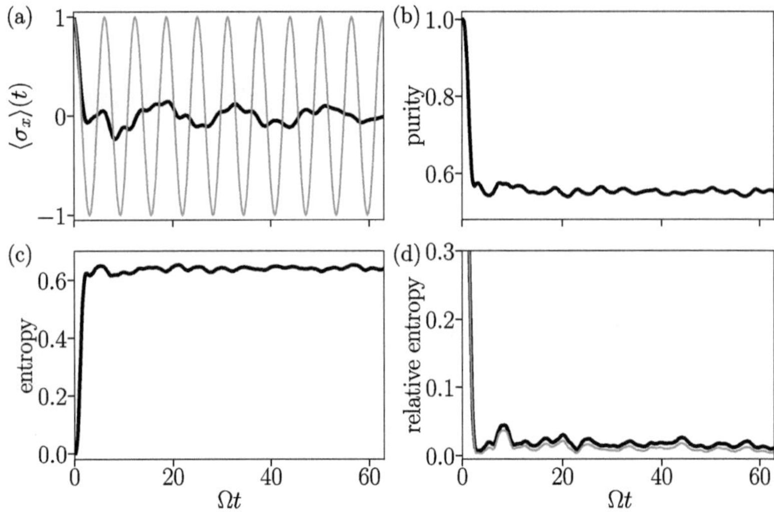

Fig. 1.3 Exemplary time evolution of an open quantum system for a 'hot' bath.

are comparable with the energy gap of each single spin. Equivalently, one could say that we have more 'noise' in the bath. The difference compared with the cold case $\beta\hbar\Omega = 10$ is quite striking. Oscillations are much less pronounced and eqn (1.17) becomes quite small after some transient time. This means that the standard equilibrium Gibbs ensemble describes the system state quite well. In fact, the system is well described by a completely mixed state $\rho_S(t) \approx I_S/2$, which follows by considering the time evolution of the purity.

To conclude, an open quantum system behaves very differently from an isolated quantum system. Even for a very small bath of only six spins effects such as equilibration and thermalization start to become visible. Next, we consider equilibrium states of open quantum systems in the case that the bath is much larger.

1.3 Equilibrium States of Open Quantum Systems

Before we analyse equilibrium states of open quantum systems, it might be worth asking whether an open quantum system can actually reach an equilibrium state. In fact, we have introduced open quantum systems via the system–bath paradigm, i.e. a composite system that evolves unitarily in time. This implies that, as long as the system–bath composite can be described by a finite-dimensional Hilbert space, the dynamics is **quasi-periodic**. By this we mean the following. Let $|\psi(0)\rangle$ denote the initial state of an arbitrary isolated system, which is here assumed to be pure for simplicity. We denote the eigenvalues and eigenstates of its Hamiltonian H by E_n and $|n\rangle$, respectively. Then, if we expand the initial state as $|\psi(0)\rangle = \sum_n c_n|n\rangle$ with complex coefficients c_n obeying $\sum_n |c_n|^2 = 1$, we can write the state at time t as

$$|\psi(t)\rangle = \sum_n c_n e^{-iE_n t/\hbar}|n\rangle, \tag{1.18}$$

i.e. it is a sum of periodic functions with frequencies $\omega_n \equiv E_n/\hbar$. If we wait long enough, we might wonder whether there exists some time t and a set of natural numbers $\{k_n\}$ such that $\omega_n t = 2\pi k_n$ for all n. This would then imply that $\langle\psi(0)|\psi(t)\rangle = 1$, i.e. the system returned back to its initial state. If all frequencies are rational numbers, we can straightforwardly find such a time t. By assumption we can then write $\omega_n = q_n/r_n$ for some $q_n \in \mathbb{N}$ and $r_n \in \mathbb{N}$. In particular, if we define $N \equiv \prod_n r_n$, we can write $\omega_n = m_n/N$ for all n and some $m_n \in \mathbb{N}$. Thus, by choosing $t = 2\pi kN$, we obtain $\langle\psi(0)|\psi(t)\rangle = 1$ for all $k \in \mathbb{N}$, i.e. the quantum system returns infinitely often to its initial state.

Things start to become more subtle if the ratio $\omega_n/\omega_{n'}$ of some pair of frequencies becomes *irrational*. Indeed, it then never happens that $\langle\psi(0)|\psi(t)\rangle = 1$. However, irrational numbers can be approximated arbitrarily well by rational numbers. Hence, it seems reasonable to expect from the foregoing argument that the system returns after a sufficiently long time to its initial state up to a *very small error*. Indeed, one can prove that an isolated finite-dimensional quantum system returns arbitrarily close to its initial state infinitely often, i.e. for any $\epsilon > 0$ there exist infinitely many times t such that $|\langle\psi(0)|\psi(t)\rangle| = 1-\epsilon$. This is known as quasi-periodicity. This result is also known from classical mechanics as **Poincaré's recurrence theorem**. Obviously, since an open quantum system is a part of a larger isolated system, this also implies that any open quantum system state returns infinitely often arbitrarily close to its initial state. This seems to imply that we should not expect an open quantum system to equilibrate at all and Poincaré's recurrence theorem has played an important historical role in the debate on whether it is possible to derive the laws of thermodynamics from an underlying microscopic perspective: If all states return arbitrarily close to their initial state, why do we observe any irreversibility?

The key insight to resolve this question lies in the fact that the next time t for which $|\langle\psi(0)|\psi(t)\rangle| = 1-\epsilon$ happens grows incredibly fast with the number of coefficients c_n in eqn (1.18). This number is expected to scale with the number of particles in the system because it is experimentally extremely unlikely to prepare a many-body system in a state $|\psi(0)\rangle$, which contains only a few energy eigenstates $|n\rangle$. Then, if the isolated system is composed of N particles, e.g. the N spins in the previous example, one generically expects the average recurrence time to scale *double exponentially* with N, i.e. $t = \mathcal{O}\{\exp[\exp(N)]\}$. If one assumes that the number of particles is typically of the order of 10^{23}, the problem of Poincaré recurrences becomes irrelevant for all human time scales.

Therefore, we assume for now that the system equilibrates and reaches a stationary state $\lim_{t\to\infty} \rho_S(t)$, where the notation $t \to \infty$ means that we consider times much larger than typical time scales of the open system evolution, but still, of course, smaller than the Poincaré recurrence time. The next question is then: What is the equilibrium state of the open quantum system? Finding a complete answer to that question turns out to be difficult and is still the subject of intense research. Since this is not the topic of the book, we here restrict ourselves to invoking arguments from equilibrium statistical mechanics, which work well in many cases.

For this purpose, we consider the familiar **Gibbs state** or **canonical ensemble**, defined for any system with Hamiltonian H by

$$\pi \equiv \frac{e^{-\beta H}}{\mathcal{Z}}, \qquad \mathcal{Z} \equiv \mathrm{tr}\{e^{-\beta H}\}. \tag{1.19}$$

Here, $\beta = (k_B T)^{-1}$ is the inverse temperature and $\mathcal{Z}$ is the partition function. If the internal energy $\mathcal{U} = \mathrm{tr}\{H\pi\}$ is fixed, the Gibbs state is characterized by the fact that it maximizes the von Neumann entropy with the inverse temperature implicitly defined through the relation $\mathcal{U} = -\partial_\beta \ln \mathcal{Z}$. Vice versa, for a fixed von Neumann entropy $S_{\mathrm{vN}}(\pi)$, the Gibbs state minimizes the energy with the inverse temperature implicitly defined through the relation $S_{\mathrm{vN}}(\pi) = -\beta^2 \partial_\beta(\beta^{-1} \ln \mathcal{Z})$.

Now, suppose that the system–bath composite is well described by a global Gibbs state $\pi_{SB} = e^{-\beta H_{SB}}/\mathcal{Z}_{SB}$ with $\mathcal{Z}_{SB} \equiv \mathrm{tr}_{SB}\{e^{-\beta H_{SB}}\}$. What does the reduced system state $\mathrm{tr}_B\{\pi_{SB}\}$ look like? Perhaps surprisingly, it turns out that in general $\mathrm{tr}_B\{\pi_{SB}\} \neq \pi_S = e^{-\beta H_S}/\mathcal{Z}_S$. Only in the case of very weak coupling between the system and the bath, i.e. if V_{SB} is negligible compared with H_S and H_B, does the reduced system state equal the conventional canonical ensemble. For non-negligible coupling V_{SB}, however, we define $\pi_S^* \equiv \mathrm{tr}_B\{\pi_{SB}\}$ and we have $\pi_S^* \neq \pi_S$.

Nevertheless, it is still possible to write π_S^* in an *apparent* Gibbs form with respect to an *effective* Hamiltonian $\tilde{H}_S$:

$$\pi_S^* = \mathrm{tr}_B\{\pi_{SB}\} = \frac{e^{-\beta \tilde{H}_S}}{\tilde{\mathcal{Z}}_S}. \tag{1.20}$$

In fact, this is possible for every density matrix ρ_S by defining $\tilde{H}_S \equiv -\beta^{-1} \ln(\tilde{\mathcal{Z}}_S \rho_S)$, and not only for π_S^*. Notice that $\tilde{H}_S$ is fixed only up to an arbitrary choice of the positive normalization constant $\tilde{\mathcal{Z}}_S$. For the reduced state π_S^* of a canonical equilibrium ensemble there is one convenient choice for the effective partition function, which we denote by $\mathcal{Z}_S^* = \tilde{\mathcal{Z}}_S$ and which is obtained by setting

$$\boxed{\mathcal{Z}_S^* \equiv \frac{\mathcal{Z}_{SB}}{\mathcal{Z}_B} = \frac{\mathrm{tr}_{SB}\{e^{-\beta H_{SB}}\}}{\mathrm{tr}_B\{e^{-\beta H_B}\}}.} \tag{1.21}$$

That is, the effective partition function $\mathcal{Z}_S^*$ is the ratio of the partition functions of the system–bath composite and of the bath *alone*. The corresponding effective Hamiltonian is known as the **Hamiltonian of mean force** and reads explicitly

$$\boxed{H_S^* = -\frac{1}{\beta} \ln(\mathcal{Z}_S^* \pi_S^*) = -\frac{1}{\beta} \ln \frac{\mathrm{tr}_B\{e^{-\beta H_{SB}}\}}{\mathcal{Z}_B}.} \tag{1.22}$$

One quickly verifies that $\mathrm{tr}_B\{\pi_{SB}\} = e^{-\beta H_S^*}/\mathcal{Z}_S^*$. The Hamiltonian of mean force provides a neat concept used various times in the following chapters. Physically speaking, it can be seen as an effective free energy landscape for the system, a claim that becomes clearer in Chapter 2. Notice that the Hamiltonian of mean force depends explicitly on the inverse temperature β. For weak coupling V_{SB}, H_S^* reduces to H_S. Thus, the familiar canonical ensemble only emerges in the weak coupling regime.

It is natural to ask: Does the effective Gibbs state (1.20) represent a good approximation of the equilibrated open quantum system state, i.e. is $\pi_S^* \approx \lim_{t\to\infty} \rho_S(t)$?

If that is the case, we say that the system *thermalizes*, which is a stronger requirement than equilibration alone, which only assumes the system state to become time independent for long times. Paralleling the objections raised by Poincaré's recurrence theorem above that equilibration can never strictly happen, similar doubts may rise for the case of thermalization. In particular, no initial system–bath state $\rho_{SB}(0)$ different from π_{SB} will ever reach the latter state during the evolution.

Exercise 1.1 Show that, if $\rho_{SB}(0) \neq \pi_{SB}$ and if H_{SB} is time independent, $\rho_{SB}(t) \neq \pi_{SB}$ for all t.

However, similar to the case of equilibration, thermalization should also be regarded as a convenient *illusion* by recalling that our experimental capabilities are limited. For a bath with its prosaic 10^{23} degrees of freedom, the full system–bath state $\rho_{SB}(t)$, which is defined on a Hilbert space of dimension of the order of $10^{10^{23}}$, remains experimentally inaccessible. Instead, one typically has access only to a restricted set of observables, for instance those determined by looking at the open quantum system. Thermalization to eqn (1.20) is then very likely to happen as long as one can meaningfully associate some macroscopic temperature T with the bath. Exceptions are, for instance, a bath consisting of two parts kept at different temperatures (a scenario that becomes relevant in later chapters when studying transport processes) or a bath prepared in a giant Schrödinger cat state as a superposition of two very different energies (an unlikely scenario). Furthermore, also in the presence of additional conservation laws, i.e. if some observables commute with the global Hamiltonian, thermalization to eqn (1.20) will typically not happen. However, if these scenarios can be excluded, many open quantum systems thermalize to eqn (1.20) regardless of the initial state $\rho_S(0)$. In fact, this observation is confirmed by our example at the end of Section 1.2. The grey line in Figs. 1.2d and 1.3d shows the time evolution of $D[\rho_S(t)|\pi_S^*]$. We see that, even for a bath of only six spins, π_S^* can represent a good approximation to the open system state, even at low temperatures, i.e. in the regime where quantum fluctuations dominate thermal fluctuations.

The last observation raises the question of whether the deviation of π_S^* from the standard Gibbs state π_S is mainly caused by quantum effects. In general, this does not need to be the case, but we will return to this question more rigorously in Section 3.6 when we discuss the zeroth law of thermodynamics. For now, we conclude this section by studying an important class of models for which there is a clear difference between π_S^* and π_S in the quantum and classical regime.

Exercise 1.2 Consider a harmonic oscillator with frequency ω and Hamiltonian (in mass-weighted coordinates) $H_S = \frac{1}{2}(p_S^2 + \omega^2 x_S^2)$ coupled to a 'bath' of one other harmonic oscillator with the same frequency ω. The global Hamiltonian is

$$H_{SB} = H_S + \frac{1}{2}\left[p_B^2 + \omega^2\left(x_B - \frac{c}{\omega^2}x_S\right)^2\right]. \tag{1.23}$$

Here, the coupling strength is c and we have written the Hamiltonian in a manifestly positive form such that $H_{SB} \geq 0$ for any choice of ω and c. Now, show first that, if treated as a classical system, there is no correction to the thermal state of the system: $\pi_S^* = \pi_S = e^{-\beta H_S}/\mathcal{Z}_S$. *Hint:*

Remember that the trace over the bath degrees of freedom becomes classically an integral over the phase-space coordinates (x_B, p_B) of the bath.

Then, show that quantum mechanically this is no longer true. *Hint:* You are not asked to directly compute H_S^*, but only to show that $\pi_S^* \neq \pi_S$. This can be done in various ways. One way is to compute the equilibrium variance of the system coordinate $\mathrm{tr}\{x_S^2 \pi_{SB}\}$ by changing to normal modes. Denote the (squared) eigenfrequencies of H_{SB} by $\Omega_\pm^2 \equiv [c^2 + 2\omega^4 \pm \sqrt{c^4 + 4c^2\omega^4}]/2\omega^2$ and show that

$$\mathrm{tr}\{x_S^2 \pi_{SB}\} = \hbar \frac{e^{\beta\hbar(\Omega_+ + \Omega_-)} - 1}{(\Omega_+ + \Omega_-)(e^{\beta\hbar\Omega_+} - 1)(e^{\beta\hbar\Omega_-} - 1)}. \tag{1.24}$$

Confirm that in the classical limit ($\hbar \to 0$) this result is independent of the coupling strength c. Also confirm that for $c \to 0$ you recover the result $\mathrm{tr}_S\{x_S^2 \pi_S\} = \hbar \coth(\beta\hbar\omega/2)/2\omega$, showing that even at zero temperature quantum fluctuations give rise to a non-zero variance.

The above result can be generalized to a system–bath Hamiltonian of the form

$$H_{SB} = H_S + \frac{1}{2}\sum_k \left[p_k^2 + \omega_k^2 \left(x_k - \frac{c_k}{\omega_k^2} S\right)^2\right]. \tag{1.25}$$

This describes a system with arbitrary Hamiltonian H_S coupled via an arbitrary system coupling operator S to a bunch of harmonic oscillators. Such system–bath models, which are well justified whenever the statistics of the bath are (approximately) Gaussian, are known as **Caldeira–Leggett models** and they play an important role in the theory of open quantum systems. Convince yourself of the fact that for a classical system the Hamiltonian of mean force is identical to the system Hamiltonian: $H_S^* = H_S$. Obviously, this no longer holds for quantum systems. Computing the exact reduced equilibrium state of the open quantum system is possible, but it is no longer a triviality.

1.4 Quantum Measurement Theory

We now start to focus on the question of how to obtain information about a quantum system by considering measurements at a *single* point in time. For that purpose it is sufficient to simply consider an arbitrary system state ρ_S, neglecting for a moment the possible presence of a bath. We will come back to the question of how to treat multi-time measurements under the influence of a bath in Section 1.6 onwards.

Projective measurements

In quantum mechanics the measurement of a system observable X_S is described by the projection postulate. Let $X_S = \sum_{x=1}^n \lambda(x)\Pi_S(x)$ be the spectral decomposition of the observable with n different eigenvalues $\lambda(x)$, which are labelled by the index x. Furthermore, $\{\Pi_S(x)\}$ is a set of projection operators satisfying $\Pi_S(x)\Pi_S(x') = \delta_{x,x'}\Pi_S(x)$ and $\sum_{x=1}^n \Pi_S(x) = I_S$. If the state of the system is ρ_S, then the probability of observing result x is $p(x) = \mathrm{tr}_S\{\Pi_S(x)\rho_S\}$. The *post*-measurement state $\rho_S'(x)$ *conditional* on observing x becomes

$$\rho_S'(x) = \frac{\Pi_S(x)\rho_S\Pi_S(x)}{p(x)}. \tag{1.26}$$

The *average* or *unconditional* post-measurement state ρ_S' of the system follows as

$$\rho'_S = \sum_{x=1}^{n} p(x)\rho'_S(x) = \sum_{x=1}^{n} \Pi_S(x)\rho_S\Pi_S(x). \tag{1.27}$$

Notice that this state is in general different from the pre-measurement state, $\rho'_S \neq \rho_S$, which reflects the well-known fact that quantum measurements are disturbing. This measurement backaction is absent only if the initial state commutes with the observable: $[\rho_S, X_S] = 0$. Even in this case, however, the conditional state (1.26) is in general different from the pre-measurement state, i.e. $\rho'_S(x) \neq \rho_S$. This is not a quantum effect but a consequence of the fact that the observer updates its state of knowledge about the system. Readers who feel unfamiliar with this description are advised to do the following exercise.

Exercise 1.3 Consider the case where all projectors $\Pi_S(x) = |x\rangle\langle x|_S$ are of rank 1 and the system state is pure: $\rho_S = |\psi\rangle\langle\psi|_S$. Convince yourself of the fact that the above framework then reduces to the conventional picture known from introductory textbooks in quantum mechanics. What corresponds to the Born rule? Which equation describes the 'collapse' of the wave function? When does the measurement reveal no information, i.e. when do we have $\rho'_S(x) = \rho_S$? Can you think about physical examples where the projectors are not of rank 1?

Remarkably, it is possible to derive eqn (1.27), which describes the average effect of a quantum measurement, using only unitary dynamics and *no* projection postulate. This works as follows. Consider a second n-dimensional 'auxiliary' quantum system A, called the *ancilla* in the following, which interacts with our system S of interest. Imagine that the initial system–ancilla state is decorrelated, $\rho_{SA} = \rho_S \otimes \rho_A$, and the ancilla is prepared in some fixed pure state $\rho_A = |1\rangle\langle 1|_A$. Furthermore, assume that the overall system–ancilla interaction is described by the following unitary operator:

$$V = \sum_{x=1}^{n} \Pi_S(x) \otimes \sum_{r=1}^{n} |r + x - 1\rangle\langle r|_A, \tag{1.28}$$

where we interpret $r + x - 1$ modulo n whenever $r + x - 1 > n$. It is easy to verify that V is unitary, satisfying, $VV^\dagger = V^\dagger V = I_{SA}$. The joint system–ancilla state after the interaction reads

$$\rho'_{SA} = V\rho_S \otimes \rho_A V^\dagger. \tag{1.29}$$

The reduced state of the system is obtained by tracing out the ancilla,

$$\rho'_S = \mathrm{tr}_A\{V\rho_S \otimes \rho_A V^\dagger\} = \sum_{x=1}^{n} \Pi_S(x)\rho_S\Pi_S(x), \tag{1.30}$$

which is identical to eqn (1.27). Thus, the average effect of a system measurement arises in this picture from interactions of the system with an outside ancilla. By tracing out the ancilla, we lose information and the reduced dynamics is no longer unitary.

To derive the conditional post-measurement state (1.26) in this picture, assume that we subject the ancilla to a projective measurement of the observable $R_A =$

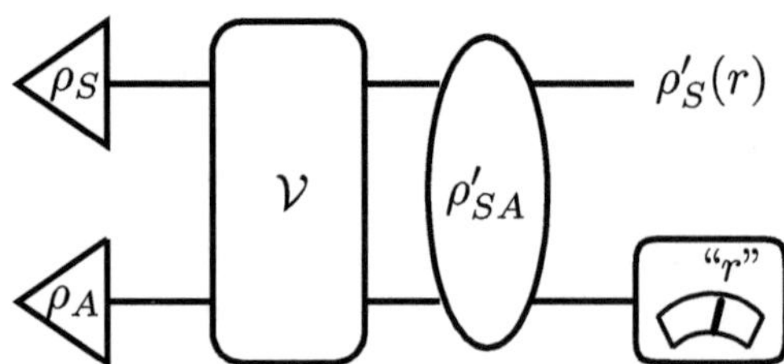

Fig. 1.4 Circuit representation of an ancilla-assisted measurement with time running from left to right. The input states to the process (left, triangles) are a system and ancilla state ρ_S and ρ_A. Both interact via a unitary transformation, denoted here with a calligraphic symbol $\mathcal{V}$. The joint state ρ'_{SA} after the interaction is in general correlated. To finally infer something about the system, the experimentalist measures the state of the ancilla, here with projector $|r\rangle\langle r|_A$. The conditional post-measurement state of the system is denoted $\rho'_S(r)$.

$\sum_{r=1}^{n} r|r\rangle\langle r|_A$. Suppose this projective measurement of R_A reveals outcome r, then the conditional post-measurement state reads $\rho'_{SA}(r) = |r\rangle\langle r|_A V\rho_S \otimes \rho_A V^\dagger |r\rangle\langle r|_A$. Keeping the information about r but tracing out the ancilla degrees of freedom reveals

$$\tilde{\rho}'_S(r) \equiv \mathrm{tr}_A\{|r\rangle\langle r|_A V\rho_S \otimes \rho_A V^\dagger\} = \Pi_S(r)\rho_S\Pi_S(r). \tag{1.31}$$

After identifying $x \equiv r$, we obtain eqn (1.26) up to normalization. In fact, the trace of the non-normalized state $\tilde{\rho}'_S(r)$ equals the probability of obtaining the measurement result r: $p(r) = \mathrm{tr}_S\{\tilde{\rho}'_S(r)\} = \mathrm{tr}_S\{\Pi_S(r)\rho_S\}$. Hence, the normalized state reads $\rho'_S(r) = \tilde{\rho}'_S(r)/p(r)$, in agreement with eqn (1.26).

As we will see throughout the remainder of this book, ancillas provide us with a flexible mathematical tool to think about quantum measurements and much more. They can be seen as comprising the essential features of an environment in an abstract and minimal way. Physically, the use of ancillas can be motivated as follows. In order to find out something about a system, an experimentalist prepares an external *probe* and puts it into contact with the system. Then, both start to interact and, depending on the state of the system, the system imprints some information onto the probe. Finally, the experimentalist reads out the state of the probe in order to infer something about the system. This picture is often justified by recognizing that an experimentalist has precise control about the detectors in the laboratory—how they are prepared, how they intervene with the system and how they are read off—whereas the system itself is given by nature and can be only indirectly accessed via the detectors found in the laboratory. An experiment in which this picture becomes particularly transparent will be treated in detail at the end of this book in Section 5.7. A pictorial representation of this process is provided in Fig. 1.4. For the rest of this section we focus on investigating the process in Fig. 1.4 in greater detail, finding that it can describe *generalized* quantum measurements beyond the projection postulate.

Generalized measurements

We generalize the picture above by starting with an arbitary initial ancilla state, written in its eigenbasis as $\rho_A = \sum_j p_j|j\rangle\langle j|_A$, an arbitrary unitary V, not necessarily the one specified in eqn (1.28), and a final projective measurement of some ancilla observable $R_A = \sum_{r=1}^{n} r|r\rangle\langle r|_A$. As above, we now take a look at the conditional state of the system given the measurement result r of R_A. Tracing out the ancilla, we get the non-normalized system state

$$\tilde{\rho}'_S(r) = \mathrm{tr}_A\{|r\rangle\langle r|_A V \rho_S \otimes \rho_A V^\dagger\} = \sum_j p_j \langle r|_A (V|j\rangle_A) \rho_S \langle j|_A (V^\dagger |r\rangle_A). \tag{1.32}$$

Next, we introduce the operators

$$K_j(r) \equiv \sqrt{p_j} \langle r|_A \,(V|j\rangle_A)\,. \tag{1.33}$$

Notice that the $K_j(r)$ are still operators acting on the *system* Hilbert space. Using them, we can express eqn (1.32) more compactly as

$$\tilde{\rho}'_S(r) = \sum_j K_j(r) \rho_S K_j^\dagger(r). \tag{1.34}$$

This equation makes it more transparent how the system state actually gets transformed by computing the operators $K_j(r)$ via eqn (1.33). In terms of them, we can also express the probability of obtaining measurement result r as

$$p(r) = \mathrm{tr}_S\{\tilde{\rho}'_S(r)\} = \sum_j \mathrm{tr}_S \left\{ K_j^\dagger(r) K_j(r) \rho_S \right\}. \tag{1.35}$$

This equation suggests introducing the following *probability operators*:

$$M(r) \equiv \sum_j K_j^\dagger(r) K_j(r), \tag{1.36}$$

which are always positive, $M(r) \geq 0$, which follows from the fact that any operator of the form $A^\dagger A$ is positive. Furthermore, they satisfy the *completeness relation*

$$\sum_r M(r) = I_S. \tag{1.37}$$

Moreover, the set of operators $\{M(r)\}$ completely fixes the measurement *statistics* and therefore plays an important role in quantum measurement theory, where they are known as a **positive operator-valued measure** or, in short, a **POVM**.

Exercise 1.4 Show that any set of operators $\{M(r)\}_r$ satisfying the completeness relation and $M(r) \geq 0$ for all r gives rise to a set of well-defined probabilities $p(r) = \mathrm{tr}_S\{M(r)\rho_S\}$ for any state ρ_S. *Hint:* By 'well defined' we mean probabilities that are positive and sum up to one.

We now study two examples to illustrate how the projection postulate can be generalized. The first example describes projective measurement of the initially studied observable $X_S = \sum_{x=1}^{n} \lambda(x)\Pi_S(x)$, but it includes classical measurement errors. For this purpose we relabel $j \equiv x$ in eqn (1.33) and model the measurement with operators $K_x(r) = \sqrt{p(r|x)}\Pi_S(x)$, giving rise to the probability operator

$$M(r) = \sum_x p(r|x)\Pi_S(x). \tag{1.38}$$

For $\{M(r)\}$ to be a POVM, we demand that the $p(r|x)$ are non-negative numbers normalized according to $\sum_r p(r|x) = 1$. Using eqn (1.34), the non-normalized post-measurement state of the system becomes

$$\tilde{\rho}'_S(r) = \sum_x p(r|x)\Pi_S(x)\rho_S\Pi_S(x). \tag{1.39}$$

Its norm equals the probablity of obtaining result r, $p(r) = \sum_x p(r|x)\mathrm{tr}_S\{\Pi_S(x)\rho_S\}$, which has a transparent interpretation: since $\mathrm{tr}_S\{\Pi_S(x)\rho_S\}$ equals the probability of obtaining result x in an error-free projective measurement, $p(r|x)$ is the conditional probability of obtaining result r given that an error-free measurement had given result x. Thus, owing to some errors in the classical data processing, erroneous detection events of $r \neq x$ can occur. The case of error-free measurements is recovered if $p(r|x) = \delta_{r,x}$, where eqn (1.39), after normalization, reduces to eqn (1.26).

The second example also describes a measurement of X_S with errors, but this time the errors cannot be explained classically. For this purpose, let us consider operators $K(r) = \sum_x \sqrt{p(r|x)}\Pi_S(x)$ (without any label j), which gives rise to

$$M(r) = \sum_x p(r|x)\Pi_S(x), \tag{1.40}$$

which *coincides* with eqn (1.38). Thus, they give rise to the same measurement statistics. However, the post-measurement state looks very different:

$$\tilde{\rho}'_S(r) = K(r)\rho_S K(r) = \sum_x \sqrt{p(r|x)}\Pi_S(x)\rho_S \sum_{x'} \sqrt{p(r|x')}\Pi_S(x'). \tag{1.41}$$

This teaches us an important lesson, namely that the probablity operators $M(r)$ do not uniquely fix how the state changes as a result of the measurement. The next exercise elucidates the difference between eqns (1.39) and (1.41) further.

Exercise 1.5 Consider an initially pure state $\rho_S = |\psi\rangle\langle\psi|_S$. Then, show that the post-measurement state given by eqn (1.41) is always pure, i.e. $[\rho'_S(r)]^2 = \rho'_S(r)$, whereas this is in general not the case for eqn (1.39). Thus, eqn (1.41) preserves the 'quantum character' of the state, whereas eqn (1.39) inevitably introduces classical noise.

Summary

We summarize what we have found out in this chapter. By shifting the description of the projective measurement from the system itself to an external ancilla, with which the system interacts in a unitary way, we obtained a more general description of quantum measurements expressed by system state changes of the form (1.34). It includes the case of projective measurements, but as eqns (1.39) and (1.41) have shown also more general transformations. By summing over r, we obtain the change of the unconditional system state as

$$\rho'_S = \sum_r p(r)\rho'_S(r) = \sum_r \tilde{\rho}'_S(r) = \sum_r \sum_j K_j(r)\rho_S K_j^\dagger(r). \tag{1.42}$$

To simplify the notation, we introduce the multi-index $\alpha = (r, j)$ and write

$$\rho'_S = \sum_\alpha K_\alpha \rho_S K_\alpha^\dagger. \tag{1.43}$$

Owing to trace conservation, we know that the operators K_α must satisfy the completeness relation

$$\sum_\alpha K_\alpha^\dagger K_\alpha = 1. \tag{1.44}$$

Apart from this relation, it turns out that the operators K_α are indeed arbitrary. We will learn more about them from an abstract perspective, which is not necessarily related to quantum measurements, in the next section.

It is instructive to point out the similarities and differences between the system–ancilla picture and the system–bath picture of Section 1.2. In both cases the system is *open* because of its coupling to some external degrees of freedom. However, we did not call the ancilla a 'bath' here because its interpretation is quite different in the present context. Whereas the bath was introduced in Section 1.2 as an uncontrollable and typically large object, the small and controllable ancilla of this section was introduced to probe the state of a quantum system. The physical interpretation of this situation is different, but—as the next sections will show in greater detail—the mathematical description is essentially the same. Before proceeding, we conclude this section with a final exercise to elucidate the difference between classical and quantum measurements.

Exercise 1.6 Construct the classical counterpart of the theory above. To this end, replace the notion of the density matrix ρ_S by a vector $\boldsymbol{p}$ with elements $p(x)$ denoting the probability of finding the classical system in state x. Let $p(r|x)$ be the conditional probability of obtaining the measurement result r given that the system is in state x and define the matrix $M(r)$ with elements $M_{xx'}(r) = \delta_{x,x'}p(r|x)$. What do $\{M(r)\}$ and a POVM have in common? What is the quantum counterpart of the classical expression $M(r)\boldsymbol{p}$? Relate the probability $p(r) = \sum_x p(r|x)p(x)$ to obtain outcome r to the expression $M(r)\boldsymbol{p}$. Show that the (normalized) post-measurement state of the system given result r obeys **Bayes' rule** $p'(x|r) = p(r|x)p(x)/p(r)$. Finally, verify that the average post-measurement state does not change: $\boldsymbol{p}' = \boldsymbol{p}$. Thus, Bayes' rule describes the most general *non-disturbing* classical measurement.

1.5 Operations, Interventions and Instruments

The goal of this section is to review the formalism of quantum measurement theory from an abstract perspective and to introduce some terminology and powerful mathematical results. From now on, we call the map (1.43) in general a (*control*) *operation* or *intervention*. These notions emphasize that an experimentalist has a large amount of freedom to manipulate a quantum system and the intention might not always be to 'measure' the system in a literal sense. We further abbreviate eqn (1.43) as

$$\boxed{\rho_S' = \mathcal{C}\rho_S \equiv \sum_\alpha K_\alpha \rho_S K_\alpha^\dagger.} \tag{1.45}$$

The map $\mathcal{C}$, which takes an operator and maps it to another operator, is also known as a **superoperator** to distinguish it from the notion of a 'usual' operator, which 'only' maps vectors onto vectors. Mathematically speaking, this distinction is superfluous: the space of matrices itself forms a vector space and, therefore, one can represent any superoperator $\mathcal{C}$ also by a (large) matrix. For physics applications it is, however, advantageous to use a terminology and notation that distinguishes whether an operator acts on states $|\psi\rangle$ or density matrices ρ. From now on, we frequently use superoperators as an efficient bookkeeping tool. By convention, superoperators are always written with a calligraphic letter such as $\mathcal{C}$ and they act on all operators to their right, i.e. $\mathcal{C}_2\mathcal{C}_1\rho = \mathcal{C}_2[\mathcal{C}_1(\rho)]$. Note that the action of different superoperators does not commute in general, i.e. $\mathcal{C}_2\mathcal{C}_1\rho \neq \mathcal{C}_1\mathcal{C}_2\rho$. Readers who are exposed to this for the first time are advised to consult Appendix B before proceeding.

We also introduce the superoperator $\mathcal{C}(r)$ corresponding to eqn (1.34):

$$\boxed{\tilde{\rho}_S'(r) = \mathcal{C}(r)\rho_S \equiv \sum_{\alpha_r} K_{\alpha_r} \rho_S K_{\alpha_r}^\dagger,} \tag{1.46}$$

which is obtained from eqn (1.45) by using only a subset of the operators K_α. The subsets of indices $\{\alpha_r\}$ are obtained from the total set $\{\alpha\}$ by a disjoint decomposition such that $\sum_r \mathcal{C}(r) = \mathcal{C}$. In the following we call a set of maps $\{\mathcal{C}(r)\}$ that decomposes an operation $\mathcal{C}$ of the form (1.45) such that $\sum_r \mathcal{C}(r) = \mathcal{C}$ an **instrument**. Notice that, given only eqn (1.45), there are many instruments that decompose $\mathcal{C}$.

Now, our claim is that eqn (1.45) describes the *most general* control operation or intervention we can implement in a laboratory on a quantum system at a single time. To verify this, let us find the minimial conditions we would like to have satisfied by an instrument $\{\mathcal{C}(r)\}$ such that it describes a physically allowed control operation in a laboratory. Clearly, if the state of the system was previously described by a valid density matrix ρ_S, the minimal requirement is that also the final set of states $\{\tilde{\rho}_S'(r) = \mathcal{C}(r)\rho_S\}$ can be interpreted in a legitimate way as the state of a physical system. Recall that any density matrix ρ is positive, has unit trace and if ρ_1 and ρ_2 are two valid density matrices, then their *convex combination* $\lambda\rho_1 + (1-\lambda)\rho_2$ for any $\lambda \in [0,1]$ is another valid density matrix. Therefore, we postulate that any physically legitimate instrument $\{\mathcal{C}(r)\}_r$ should at least satisfy the following:

(i) $\mathcal{C}(r)$ is positive: if $\rho_S \geq 0$, then also $\mathcal{C}(r)\rho_S \geq 0$.

(ii) $\mathcal{C}(r)$ is trace non-increasing: if $\text{tr}\{\rho_S\} = 1$, then $p(r) = \text{tr}\{\mathcal{C}(r)\rho_S\}$ is the probability of applying the map $\mathcal{C}(r)$ (in view of a quantum measurement, it is the probability of obtaining result r).

(iii) $\mathcal{C}(r)$ is convex linear: if $\sum_i \lambda_i \rho_S^{(i)}$ is a statistical mixture of quantum states with $\lambda_i \geq 0$ satisfying $\sum_i \lambda_i = 1$, then

$$\mathcal{C}(r)\sum_i \lambda_i\rho_S^{(i)} = \sum_i \lambda_i\mathcal{C}(r)\rho_S^{(i)}. \tag{1.47}$$

Now, to make things even a little more complicated, it turns out that desideratum (i) is not enough. Remember that $\mathcal{C}(r)$ could describe, for instance, a measurement of a system S coupled to a bath B. Prior to the measurement the system–bath state ρ_{SB} can be arbitrary and we would like to ensure that the post-measurement state also describes a legitimate quantum state. This means that we actually want the global state $\tilde{\rho}'_{SB}(r) = [\mathcal{C}(r)\otimes\mathcal{I}_B]\rho_{SB} \geq 0$ to be positive, where $\mathcal{I}_B$ denotes the identity operation acting on the bath defined via $\mathcal{I}_B\rho_B \equiv \rho_B$ for any state ρ_B. This is *not* necessarily guaranteed by the notion of a positive map. Therefore, we replace (i) by the stronger condition:

(i′) $\mathcal{C}(r)$ is **completely positive**: consider the composite system $\mathcal{H}_S\otimes\mathcal{H}_B$ with $\mathcal{H}_B$ an arbitrary Hilbert space. If $\rho_{SB} \geq 0$ is an arbitrary positive state of the composite system, then $[\mathcal{C}(r)\otimes\mathcal{I}_B]\rho_{SB} \geq 0$ is also positive.

Exercise 1.7 An example for a positive but not completely positive map is the transpose operation $\mathcal{T}$. Let $\rho = \rho_{00}|0\rangle\langle 0| + \rho_{01}|0\rangle\langle 1| + \rho_{10}|1\rangle\langle 0| + \rho_{11}|1\rangle\langle 1|$ be an arbitrary density matrix of a qubit, then the transpose operation with respect to the basis $\{|0\rangle, |1\rangle\}$ is defined via $\mathcal{T}\rho \equiv \rho_{00}|0\rangle\langle 0| + \rho_{10}|0\rangle\langle 1| + \rho_{01}|1\rangle\langle 0| + \rho_{11}|1\rangle\langle 1|$. Show that this map is positive, but not completely positive. *Hint:* As a counterexample consider the transpose operation acting on the first qubit of the pure and maximally entangled bipartite state $(|00\rangle + |11\rangle)/\sqrt{2}$.

We remark that the distinction between positivity and complete positivity is a quantum phenomenon: classically, all postive maps are also completely positive. Furthermore, the next exercise shows that any convex linear map acting on density matrices can be extended to a linear map acting on arbitrary matrices. We therefore use the notions convex linearity and linearity in the following synonymously.

Exercise 1.8 Let $\mathcal{C}$ be a convex linear map acting on density matrices ρ of a Hilbert space with dimension $\dim\mathcal{H} = d$. Show that every complex $d\times d$ matrix A can be written as a linear combination of density matrices ρ_i with complex coefficients $c_i \in \mathbb{C}$, i.e. $A = \sum_i c_i\rho_i$. *Hint:* To do so, you can use the two fundamental results that, first, every A can be written as $A = A_1 + iA_2$ with $A_1 = A_1^\dagger$ and $A_2 = A_2^\dagger$ being Hermitian and, second, every Hermitian A can be decomposed as $A = A_+ - A_-$ with $A_\pm \geq 0$ being positive.

The extension $\tilde{\mathcal{C}}$ of $\mathcal{C}$ is then defined via $\tilde{\mathcal{C}}A \equiv \sum_i c_i\mathcal{C}\rho_i$ and in the following, tacitly assuming this extension, we identify $\mathcal{C} \equiv \tilde{\mathcal{C}}$.

To conclude, physically legitimate operations are mathematically described by an instrument that is a collection of completely positive, trace non-increasing and linear maps, which we call **CP maps**. All CP maps of an instrument have to add up to a completely positive and trace-preserving map, which we call a **CPTP map.** There are two important theorems to characterize CP(TP) maps mathematically.

Operator–sum representation. *A map $\mathcal{C}(r)$ satisfies desiderata (i′), (ii) and (iii) if and only if it can be written as in eqn (1.46) for some operators K_{α_r} satisfying*

$\sum_{\alpha_r} K^\dagger_{\alpha_r} K_{\alpha_r} \leq I_S$. *If* $\sum_{\alpha_r} K^\dagger_{\alpha_r} K_{\alpha_r} = I_S$, *then* $\mathcal{C}(r) = \mathcal{C}$ *is trace-preserving. Equations (1.45) or (1.46) are called the operator–sum representation of a CP(TP) map.*

It is not too hard to show that a map of the form (1.46) with $\sum_{\alpha_r} K^\dagger_{\alpha_r} K_{\alpha_r} \leq I_S$ satisfies points (i′), (ii) and (iii). The other direction is harder to show and not done here. Instead, we only note that the operator–sum representation (1.46) is *not* unique, as explicitly shown by the next exercise.

Exercise 1.9 Consider the map $\mathcal{C}\rho = \sum_{\alpha=1}^{d} K_\alpha \rho K^\dagger_\alpha$ and define $K_\alpha \equiv \sum_\alpha U_{\alpha\beta} \tilde{K}_\beta$ for an arbitrary $d \times d$ unitary matrix U. Show that $\mathcal{C}\rho = \sum_{\alpha=1}^{d} \tilde{K}_\alpha \rho \tilde{K}^\dagger_\alpha$.

It turns out that there is another important representation theorem for an instrument, which brings us back to our ancilla construction used in the previous section.

Unitary dilation theorem. *A set of maps* $\{\mathcal{C}(r)\}$ *acting on a d-dimensional system Hilbert space* $\mathcal{H}_S$ *forms an instrument if and only if there exists a* d^2*-dimensional 'ancilla' Hilbert space* $\mathcal{H}_A$*, a unitary* U_{SA} *acting on* $\mathcal{H}_S \otimes \mathcal{H}_A$*, a pure ancilla state* $|\phi\rangle_A \in \mathcal{H}_A$ *and a set of projectors* $\{\Pi_A(r)\}$ *acting on the ancilla space such that*

$$\boxed{\mathcal{C}(r)\rho_S = \mathrm{tr}_A\{\Pi_A(r) U_{SA}(\rho_S \otimes |\phi\rangle\langle\phi|_A) U^\dagger_{SA}\}} \tag{1.48}$$

for all r. *In particular, a map* $\mathcal{C}$ *is CPTP if and only if it can be written as*

$$\boxed{\mathcal{C}\rho_S = \mathrm{tr}_A\{U_{SA}(\rho_S \otimes |\phi\rangle\langle\phi|_A) U^\dagger_{SA}\}.} \tag{1.49}$$

We are not going to prove the unitary dilation theorem here, but only discuss its consequences. For instance, the next exercise shows that the number of operators K_α appearing in the operator–sum representation is at most d^2 with $d = \dim \mathcal{H}_S$.

Exercise 1.10 Use the unitary dilation theorem to derive eqns (1.45) and (1.46) by giving explicit expressions for the operators K_{α_r} and K_α in terms of U_{SA}, $|\phi\rangle_A$ and $\Pi_A(r)$. Confirm that the maximum number of operators K_α is d^2. *Hint:* Remember eqn (1.33).

We remark that it is also possible to use the operator–sum representation to derive the unitary dilation theorem. Hence, both are equivalent statements. The physical significance of the unitary dilation theorem is that it teaches us that every allowed state transformation in quantum mechanics satisfying desiderata (i′), (ii) and (iii) can be constructed by more primitive transformations, namely unitary evolution and projective measurements. We therefore do not need to add any axiom to the standard textbook framework of quantum mechanics in order to describe generalized measurements or other general state transformations.

The unitary dilation theorem also fits well into our system–bath paradigm from Section 1.2 if we regard the ancilla as the bath coupled to the system. From that perspective, it tells us that every state change of a quantum system can be explained through the interaction with an external environment, which we might call an ancilla or a bath depending on the context. Care is, however, required here because the unitary dilation theorem is an *abstract* statement and does not tell us for a given transformation of the system what *is* the actual environment causing this transformation. This is related to the non-uniqueness of the operator–sum representation: there are infinitely many Hilbert spaces $\mathcal{H}_A$, unitary matrices U_{SA}, states $|\phi\rangle_A$ (possibly also mixed states) and projectors $\Pi_A(r)$ satisfying eqns (1.48) and (1.49).

Before concluding this section, we point out a subtle but important observation. What the unitary dilation theorem also teaches us is that satisfying desiderata (i′), (ii) and (iii) is equivalent to representing the interaction of the system with some environment that is initially *decorrelated* from the system, i.e. of the form $\rho_S \otimes |\phi\rangle\langle\phi|_A$. However, the evolution of an open quantum system (1.11) could also result from a system initially *correlated* with the environment. In this case, the operator–sum representation and the unitary dilation theorem *break down*. The reason is that desideratum (iii) (linearity) is then no longer satisfied. This is exemplified in the next exercise and a way out of this 'dilemma' is presented in Section 1.7.

Exercise 1.11 Consider the interaction of two qubits. The first qubit is called the system and the second the ancilla. We further introduce the maximally entangled states $|\pm\rangle = (|00\rangle + |11\rangle)/\sqrt{2}$, where by convention we set $|ij\rangle \equiv |i\rangle_S \otimes |j\rangle_A$ for $i, j \in \{0, 1\}$. Let the system–ancilla interaction be modelled by the unitary $U_{SA} = |+\rangle\langle+| + |-\rangle\langle 10| + |10\rangle\langle -| + |01\rangle\langle 01|$ (check that this is a unitary matrix). The evolution of the composite system is therefore modelled via $\rho'_{SA} = U_{SA}\rho_{SA}U_{SA}^\dagger$, which is clearly linear with respect to the *joint* input state ρ_{SA} and CPTP.

Next, consider the reduced dynamics of S assuming that the initial system–ancilla state $\rho_{SA} = |+\rangle\langle+|$ is entangled. Verify the following results: (1) the initial reduced system state is $\rho_S = (|0\rangle\langle 0| + |1\rangle\langle 1|)/2$; (2) the final system state is identical to the initial system state, $\rho'_S = \mathrm{tr}_A\{U_{SA}|+\rangle\langle+|U_{SA}^\dagger\} = \rho_S$; (3) if we first perform a measurement of the initial system in its eigenbasis, the initial *system* state does not change on average: $|0\rangle\langle 0|\rho_S|0\rangle\langle 0| + |1\rangle\langle 1|\rho_S|1\rangle\langle 1| = \rho_S$; but (4) the final system state after such an initial measurement

$$\mathrm{tr}_A\{U_{SA}(|0\rangle\langle 0|\rho_{SA}|0\rangle\langle 0| + |1\rangle\langle 1|\rho_{SA}|1\rangle\langle 1|)U_{SA}^\dagger\} = \frac{1}{4}|0\rangle\langle 0| + \frac{3}{4}|1\rangle\langle 1| \tag{1.50}$$

is *different* from ρ_S. Hence, we have a 'paradox' because the same reduced initial system state gives rise to two different final states. Thus, we cannot associate any map $\mathcal{C}$, which only acts on the system part, with this input–output relation.

How can this be resolved? Clearly, from a global point of view there is no paradox: the two initial states $\rho_{SA} = |+\rangle\langle+|$ and $|0\rangle\langle 0|_S\rho_{SA}|0\rangle\langle 0|_S + |1\rangle\langle 1|_S\rho_{SA}|1\rangle\langle 1|_S$ are different (albeit they give rise to the same reduced system state), and hence there can be two different output states. With respect to our construction above, the key insight is to realize that *it is no longer meaningful to speak about different initial system states if the system is initially entangled with its environment* (here, the ancilla). In particular, we assumed that it is possible to mix (or convex combine) different system states without influencing the dynamics, i.e. the definition of the map. This assumption is incompatible with having an initial system state entangled with its environment.

We summarize the content of the last two sections, which play an important role in the following. First, quantum systems can undergo more general state transformations than unitary evolutions and projective measurements. However, even general state transformations can be constructed using only unitary evolutions and projective measurements on a bigger system–ancilla (or system–bath) space. If the desiderata (i′), (ii) and (iii) are satisfied, where point (iii) assumes that the preparation of the initial system state can be disentangled from the effect of the state transformation, then we can use either the operator–sum representation or the unitary dilation theorem. If there is no post selection (i.e. conditioning on a measurement result r), the state transformation is described by a CPTP map represented by either eqn (1.45) or eqn (1.49). If there is conditioning, the state transformation is described by a CP map represented by either eqn (1.46) or eqn (1.48). All CP maps of an instrument have to add up to a CPTP map. Vice versa, every CPTP map can be decomposed into a set of CP maps. Such a set is called an instrument.

Finally and for completeness, we mention one extension of the framework introduced here. So far, we have assumed that the dimension d of the system space does not change during the control operation, but for some applications it makes sense to relax this requirement. These applications play a minor role in quantum stochastic thermodynamics and the mathematical modifications we have to add only amount to correct bookkeeping of the different input and output Hilbert spaces in the notation. Nevertheless, the final exercise shows a couple of neat examples of such control operations, which are useful to keep in mind.

Exercise 1.12 Verify that the following maps are CPTP by finding an operator–sum representation for them. *Example 1:* The 'trace map' is defined for any input state ρ by $\mathcal{C}\rho \equiv \text{tr}\{\rho\}$. This map is also often associated with 'discarding a system' as it destroys all the information contained in a system and replaces it by a 'trivial' system living in the Hilbert space $\mathcal{H} = \mathbb{C}$ with the only possible density matrix $\rho_{\mathbb{C}} = 1$. *Example 2:* Consider a bipartite system with Hilbert space $\mathcal{H}_1 \otimes \mathcal{H}_2$. Show that the 'partial trace map' is CPTP: $\mathcal{C}\rho_{12} \equiv \text{tr}_2\{\rho_{12}\} = \rho_1$. *Example 3:* In contrast to Example 1, one can also 'create' a system using a CPTP map, which takes input states from the Hilbert space $\mathcal{H} = \mathbb{C}$, by defining $\mathcal{C}_\rho c \equiv \rho$ for any $c \in \mathbb{C}$ and some fixed density matrix ρ. Note that, according to this example, 'states' in quantum mechanics are CPTP maps. *Example 4:* An extension of Example 3 and in some sense the opposite of Example 2 is the map which 'adds' a fixed state ρ_2 to an input state ρ_1: $\mathcal{C}_{\rho_2}\rho_1 \equiv \rho_1 \otimes \rho_2$.

1.6 Classical Stochastic Processes

In the previous sections we have learned about some fundamental aspects of open quantum systems and quantum measurement theory. Before we put these two ingredients together to define a quantum stochastic process, it is worthwhile recapitulating the definition of a classical stochastic process. Classical stochastic processes have found widespread application in the natural and social sciences. They are commonly used to describe a process in time that is characterized by a random variable whose observation can be safely assumed not to change the process. It is obvious that this assumption is no longer satisfied for quantum systems. Even classically, this assumption can be violated. This leads to the much richer framework of classical causal models—a topic which we briefly touch on at the end of this section.

Let R denote some random variable characterizing the state of a system evolving in time. Simple examples for R with relevance for (quantum) stochastic thermodynamics include the position of a colloidal particle suspended in water (which yields to *Brownian motion*), the number of electrons in a nanostructure such as a quantum dot or the number of photons in a cavity and conformational states of a macromolecule (e.g. folded or unfolded), among many others. In order to infer something about the time evolution of the system, the experimenter measures the value of R at an arbitrary set of times $\{t_\ell\}_{\ell=0}^{n}$, where here and in the following we assume the order $0 = t_0 < t_1 < \cdots < t_n$. The results of the measurement at time t_ℓ are denoted by r_ℓ. By repeating the experiment many times, the experimenter can gather enough statistics to construct (approximately) the joint probability distribution $p(r_n, t_n; \ldots; r_1, t_1; r_0, t_0)$ to observe r_0 at time t_0, r_1 at time t_1, and so on and so forth up to the last measurement giving result r_n at time t_n. Since the subscript at the measurement result r_ℓ is in one-to-one correspondence with the time t_ℓ of the measurement, we write for brevity

$$p(r_n, \ldots, r_1, r_0) \equiv p(r_n, t_n; \ldots; r_1, t_1; r_0, t_0). \tag{1.51}$$

To be a valid probability distribution, $p(r_n, \ldots, r_1, r_0)$ must satisfy

$$p(r_n, \ldots, r_1, r_0) \geq 0, \quad \sum_{r_n} \cdots \sum_{r_1} \sum_{r_0} p(r_n, \ldots, r_1, r_0) = 1. \tag{1.52}$$

To save further space in the notation, we write the sequence of measurement results in boldface as $\mathbf{r}_n \equiv (r_n, \ldots, r_1, r_0)$. In this notation the content of eqn (1.52) reduces to $p(\mathbf{r}_n) \geq 0$ and $\sum_{\mathbf{r}_n} p(\mathbf{r}_n) = 1$. Note that the sequence $\mathbf{r}_n$ has $n+1$ entries (and not n) since our first measurement happens by convention at time t_0 (and not at t_1).

The conditions (1.52) of positivity and normalization have to be satisfied by any probability distribution. Therefore, $p(\mathbf{r}_n)$ does not yet describe a stochastic process that has one important additional structure. To uncover this additional structure, we need to look at the joint probability distribution $p(r_n, \ldots, r_{\ell+1}, r_{\ell-1}, \ldots, r_0)$ to obtain the results $r_n, \ldots, r_{\ell+1}$, $r_{\ell-1}, \ldots, r_0$ by measuring the system at times $t_n, \ldots, t_{\ell+1}$, $t_{\ell-1}, \ldots, t_0$, i.e. at all times *except* for time t_ℓ, where we perform *no* measurement. To emphasize this fact, we denote this probability by $p(r_n, \ldots, \not{r}_\ell, \ldots, r_0)$. Then, we define a classical stochastic process by the requirement that

$$\boxed{p(r_n, \ldots, \not{r}_\ell, \ldots, r_0) = \sum_{r_\ell} p(r_n, \ldots, r_\ell, \ldots, r_0)} \tag{1.53}$$

for all $\ell \in \{0, 1, \ldots, n\}$. This is known as the **Kolmogorov consistency condition** and it tells us that not measuring the observable R at some point in time is equivalent to measuring it followed by marginalizing about the measurement result. Therefore, the joint probabilities of a classical stochastic process form a *hierarchy*, where the $(n+1)$-time joint probability $p(r_n, \ldots, r_1, r_0)$ contains *all* the information about the k-time probabilities (with $k < n+1$) obtained from measuring the system at any subset of the times $\{t_n, \ldots, t_1, t_0\}$. Furthermore, an important mathematical result known as the *Daniell–Kolmogorov extension theorem* says that we can also go the reverse way: whenever we have a hierarchy of n-time joint probabilities satisfying the Kolmogorov consistency condition, then they can be constructed as marginals of N-time joint probabilities with $N > n$, which also satisfy the consistency condition. The important point here is that this holds even in the limit of $n \to \infty$. The Daniell–Kolmogorov extension theorem therefore provides a bridge between experimental reality (where any measurement statistics are always finite and described by $n < \infty$) and its theoretical description (which often uses continuous-time dynamics in the form of, for example, stochastic *differential* equations, which assume $n \to \infty$).

It is clear that the Kolmogorov consistency condition is in general not satisfied for quantum systems. It is, however, noteworthy that even for classical systems eqn (1.53) can be broken. This can happen simply for the reason that the measurement of a classical system can be disturbing as well. Another reason is an experimenter choosing to *actively break* the consistency condition. Examples of the latter kind include the use of feedback control, where an external agent manipulates the dynamics of a system based on the so far available information (say, based on $\mathbf{r}_k$ up to some time t_k). The future probabilities of observing r_n ($n > k$) are then in general different from the situation where the external agent had not observed the system previously (and, consequently, had not applied any feedback control). Another example is a clinical trial where the health of patients is monitored at regular intervals and where the process can be actively changed by giving drugs. The health of the patients in the future then depends on the question of whether their health was monitored previously or not (and, consequently, whether they received drugs or not).

These examples show that the theory of classical stochastic processes breaks down if an external agent intervenes in the process. The mathematical language appropriate for such situations is the theory of *classical causal models.* Without going into its details, we point out that causal models can *distinguish* between causation and correlation, whereas classical stochastic processes *cannot.* Indeed, classical stochastic processes allow correlations to be quantified between two random variables, say X and Y, but the consistency condition forbids us to infer whether X is a *cause* of Y, Y is a cause of X, or whether there is no causal relation between X and Y and their correlation is the result of another common cause Z. In order to infer causal relationships, it is important that we can change the process and thus violate the consistency condition, for instance by looking at the behaviour of Y after forcing X to take on a certain value. The following example makes this qualitative reasoning quantitative.

Exercise 1.13 We consider three binary random variables S, B and C with values s, b and c. On a given day S describes whether the sun is shining ($s = 1$) or not ($s = 0$), B describes whether the number of sunburns is high ($b = 1$) or low ($b = 0$) and C describes whether the number of ice cream sales is high ($c = 1$) or low ($c = 0$). We assume the notion of 'high'/'low' to be chosen according to some reasonable threshold. We further set the conditional probabilities to be $p(b = 1|s = 1) = p(c = 1|s = 1) = p(b = 0|s = 0) = p(c = 0|s = 0) = \lambda$, with $\lambda \in [1/2, 1]$; for example, $\lambda = 1$ implies that the number of sunburns is always high if the sun is shining. By conservation of probability, $p(b = 0|s = 1) = p(c = 0|s = 1) = p(b = 1|s = 0) = p(c = 1|s = 0) = 1 - \lambda$. Furthermore, we assume the probability for a sunny day to be $p(s = 1) = 1/2$, which implies that the sun is not shining with the same probability $p(s = 0) = 1/2$. Now, first confirm that B and C are correlated unless $\lambda = 1/2$. This can be done in several ways, for instance by computing the mutual information, which we introduce in Appendix A in detail, between B and C:

$$I_{B:C} \equiv \sum_{b,c} p(b,c) \ln \frac{p(b,c)}{p(b)p(c)} = \ln 2 - S_{\text{Sh}}(2\lambda - 2\lambda^2). \tag{1.54}$$

Here, $S_{\text{Sh}}(p) \equiv -p \ln p - (1-p) \ln(1-p)$ denotes the *binary* Shannon entropy. Verify eqn (1.54) by using $p(b,c) = \sum_s p(b,c,s) = \sum_s p(b|s)p(c|s)p(s)$. Show that $I_{B:C} = 0$ (no correlations) implies $\lambda = 1/2$ and $I_{B:C} = \ln 2$ (maximal correlations) implies $\lambda = 1$. Thus, while B and C are in general correlated, they are not causally related as demonstrated below. Does this sound plausible to you?

Next, we turn to the correlations between S and B. Confirm that $I_{B:S} = \ln 2 - S_{\text{Sh}}(\lambda)$ and find the values of λ for which S and B are maximally correlated/uncorrelated.

Finally, we strongly believe that S is the cause of B (and also of C), i.e. the shining sun triggers sunburn (and ice cream sales). But how can we make this intuition rigorous?

Assume that we have an external mechanism that can change whether the sun is shining or not (which seems unrealistic at first sight, but we come back to it below). We therefore introduce an additional *intervention variable* I_S (not to be mixed up with the mutual information), which labels the following three actions i: do nothing, i.e. leave the sun as it is ($i = \text{idle}$), make the sun shining ($i = 1$) or block sun shine ($i = 0$). Now, consider the conditional probability $p(b|s,i)$ for sunburns given sunshine s and intervention i. For $i = \text{idle}$ we set $p(b|s, \text{idle}) = p(b|s)$ as defined above. Furthermore, we assume $p(b|s, i = 1) = \lambda$ and $p(b|s, i = 0) = 1 - \lambda$ *independent* of s because $i \in \{0, 1\}$ overwrites the natural value of s to be identical to i. Next, confirm that the mutual information between B and I_S is

$$I_{B:I_S} = \sum_{b,i} p(b,i) \ln \frac{p(b,i)}{p(b)p(i)} = S_{\text{Sh}}[p(b)] - [1 - p(i = \text{idle})] S_{\text{Sh}}(\lambda) - p(i = \text{idle}) \ln 2, \quad (1.55)$$

where the marginal probability $p(i)$ that we perform a certain intervention is assumed to be controllable in an experiment. We now define that *S is a* ***cause*** *of B if there are correlations between I_S and B.* Confirm that there are no correlations between B and I_S, i.e. S is *not* the cause of B, if one of the following two cases happen: either $p(i = \text{idle}) = 1$, which corresponds to the case that we do not perform any intervention and, hence, cannot test for causality, or $\lambda = 1/2$, which implies that there are no correlations between B and S in the first place. Furthermore, confirm that in general $p(b,s) \neq \sum_i p(b,s,i)$, where $p(b,s) = p(b|s)p(s)$ is the joint probability from the beginning obtained *without* interventions and $p(b,s,i) = p(b|s,i)p(s)p(i)$ is the joint probability with interventions. Thus, the Kolmogorov consistency condition (1.53) is broken in general. Show that the consistency condition is obeyed if and only if $p(i = \text{idle}) = 1$ or $p(i = 0) = p(i = 1)$. Can you see why? Of course, we could also replace B by C above and find that sunshine *causes* a high number of ice cream sales.

Finally, let us return to our assumption that we can change the sunlight by an external intervention. Indeed, such a mechanism is not easy to construct for a human. But to distinguish causation and correlation, it is not necessary that humans perform the intervention: it could be also done by nature, for instance, due to a solar eclipse. Important is only that we can fix the intervention variable independent of the other variables in the model.

1.7 Quantum Stochastic Processes

In this section we introduce the notion of a quantum stochastic process describing the response of an open quantum system to the most general interventions we can perform on it. This generality comes at the price of a high level of abstractness, even though it might help to keep in mind that much of what follows below 'merely' presents convenient terminology and notation. We proceed by first applying the notion of a classical stochastic process directly to quantum systems. Afterwards, we show that the idea can be fruitfully generalized to define a quantum stochastic process. Then, we show how to reconstruct a quantum stochastic process experimentally and we finish by exploring further mathematical consequences of this formalism.

Quantum statistics of a projectively measured system

Even though quantum measurements are in general disturbing, the joint probability $p(\mathbf{r}_n)$ in eqn (1.51) remains a well-defined object in quantum mechanics. Experimentally, it describes the probability of obtaining the results $\mathbf{r}_n = (r_n, \ldots, r_1, r_0)$ after performing projective measurements of the quantum mechanical observable R with eigenvalues labelled by r at times $0 = t_0 < t_1 < \cdots < t_n$. Theoretically, the probability is computed for an isolated quantum system as

$$\begin{aligned} p(\mathbf{r}_n) = \mathrm{tr}\{&\Pi(r_n)U(t_n,t_{n-1})\cdots\Pi(r_1)U(t_1,0)\Pi(r_0)\rho(0) \\ &\times \Pi(r_0)U^\dagger(t_1,0)\Pi(r_1)\cdots U^\dagger(t_n,t_{n-1})\Pi(r_n)\}. \end{aligned} \tag{1.56}$$

Here, $\rho(0)$ denotes the initial state of the quantum system, followed by a projective measurement $\Pi(r_0)\rho(0)\Pi(r_0)$ with result r_0, followed by a unitary time evolution $U(t_1,0)\Pi(r_0)\rho(0)\Pi(r_0)U^\dagger(t_1,0)$ from 0 to t_1, and so on and so forth up to time t_n. Since the expression (1.56) looks quite cumbersome, we introduce the superoperators $\mathcal{P}(r_\ell)$ and $\mathcal{U}(t_\ell,t_k)$, defined via their action $\mathcal{P}(r_\ell)\rho \equiv \Pi(r_\ell)\rho\Pi(r_\ell)$ and $\mathcal{U}(t_\ell,t_k)\rho = U(t_\ell,t_k)\rho U^\dagger(t_\ell,t_k)$ on an arbitrary state ρ. Then, eqn (1.56) can be expressed more compactly as

$$p(\mathbf{r}_n) = \mathrm{tr}\{\mathcal{P}(r_n)\mathcal{U}(t_n,t_{n-1})\cdots\mathcal{P}(r_1)\mathcal{U}(t_1,0)\mathcal{P}(r_0)\rho(0)\}. \tag{1.57}$$

The following exercise shows that these probabilities are well defined, but in general break the Kolmogorov consistency condition (1.53).

Exercise 1.14 Confirm that $p(\mathbf{r}_n)$ satisfies eqn (1.52) but in general not eqn (1.53).

Next, we return to the system–bath paradigm of Section 1.2. The spirit of system–bath theories was to view the open system S as the *accessible* part of the 'universe', which can be easily measured and manipulated. In contrast, the bath B represents the *inaccessible* part of the universe, about which we have only limited information. Following this spirit, we adapt eqn (1.56) to the case of open quantum systems by restricting R to be a system observable R_S. Then, the joint probability $p(\mathbf{r}_n)$ is

$$\begin{aligned} p(\mathbf{r}_n) = \mathrm{tr}_{SB}\{&[\mathcal{P}_S(r_n)\otimes\mathcal{I}_B]\mathcal{U}_{SB}(t_n,t_{n-1})\cdots \\ &\times [\mathcal{P}_S(r_1)\otimes\mathcal{I}_B]\mathcal{U}_{SB}(t_1,0)[\mathcal{P}_S(r_0)\otimes\mathcal{I}_B]\rho_{SB}(0)\}. \end{aligned} \tag{1.58}$$

Here, $\rho_{SB}(0)$ denotes the initial system–bath state, $\mathcal{U}_{SB}(t_\ell,t_k)$ the unitary time evolution from t_k to t_ℓ generated by some system–bath Hamiltonian and $[\mathcal{P}_S(r_\ell)\otimes\mathcal{I}_B]\rho_{SB} = [\Pi_S(r_\ell)\otimes I_B]\rho_{SB}[\Pi_S(r_\ell)\otimes I_B]$ the local projective measurement of the system. To save space, we drop all identity operations in the following. In addition, we understand that all superoperators *without* a subscript act only on the system, e.g. we write $\mathcal{P}(r_\ell)$ instead of $\mathcal{P}_S(r_\ell)\otimes\mathcal{I}_B$. Then, eqn (1.58) becomes

$$p(\mathbf{r}_n) = \mathrm{tr}_{SB}\{\mathcal{P}(r_n)\mathcal{U}_{SB}(t_n,t_{n-1})\cdots\mathcal{P}(r_1)\mathcal{U}_{SB}(t_1,0)\mathcal{P}(r_0)\rho_{SB}(0)\}. \tag{1.59}$$

In fact, this notation is still not short enough for our purposes. Instead, we are aiming at a notation that explicitly distinguishes between the experimentally controllable part that is here represented by the projective measurements $\mathcal{P}(r_\ell)$ we are performing on the system, and the uncontrollable or inaccessible part of the evolution encoded in the unitary system–bath evolution $\mathcal{U}_{SB}(t_\ell, t_k)$, which is fixed in a given experimental setup. Therefore, we define

$$\mathfrak{T}[\mathcal{P}(r_n), \ldots, \mathcal{P}(r_1), \mathcal{P}(r_0)] \equiv \\ \mathrm{tr}_B\{\mathcal{P}(r_n)\mathcal{U}_{SB}(t_n, t_{n-1}) \cdots \mathcal{P}(r_1)\mathcal{U}_{SB}(t_1, 0)\mathcal{P}(r_0)\rho_{SB}(0)\} \tag{1.60}$$

such that eqn (1.59) simply becomes $p(\mathbf{r}_n) = \mathrm{tr}_S\{\mathfrak{T}[\mathcal{P}(r_n), \ldots, \mathcal{P}(r_1), \mathcal{P}(r_0)]\}$.

The process tensor and quantum stochastic processes

In contrast to the classical case, projective measurements of a quantum system generically disturb the dynamics. Furthermore, in Sections 1.4 and 1.5 we have learned that there is a much larger class of control operations that we can implement in a laboratory, and not only projective measurements. This motivates us to extend the definition (1.60) to include arbitrary instruments applied at times t_ℓ, which are characterized by a set of CP maps $\{\mathcal{C}(r_\ell)\}$ adding up to a CPTP map $\mathcal{C}_\ell \equiv \sum_{r_\ell} \mathcal{C}(r_\ell)$. Thus, we define

$$\boxed{\mathfrak{T}[\mathcal{C}(r_n), \ldots, \mathcal{C}(r_1), \mathcal{C}(r_0)] \equiv \\ \mathrm{tr}_B\{\mathcal{C}(r_n)\mathcal{U}_{SB}(t_n, t_{n-1}) \cdots \mathcal{C}(r_1)\mathcal{U}_{SB}(t_1, t_0)\mathcal{C}(r_0)\rho_{SB}(0)\},} \tag{1.61}$$

and call $\mathfrak{T}$ the **process tensor**, which forms the central object for our studies in the remainder of this chapter. The probability of getting the results $\mathbf{r}_n = (r_n, \ldots, r_1, r_0)$, i.e. the probability of realizing the CP maps $\mathcal{C}(r_n), \ldots, \mathcal{C}(r_1), \mathcal{C}(r_0)$, is

$$\boxed{p(\mathbf{r}_n) = \mathrm{tr}_S\{\mathfrak{T}[\mathcal{C}(r_n), \ldots, \mathcal{C}(r_1), \mathcal{C}(r_0)]\},} \tag{1.62}$$

which generalizes the result for a single measurement, eqn (1.35), to multiple times. The output of the process tensor is the non-normalized system state at time t_n after $n+1$ control operations conditioned on the results $\mathbf{r}_n$. In agreement with our previous notation for a single control operation, we denote this state as

$$\boxed{\tilde{\rho}_S(t_n|\mathbf{r}_n) \equiv \mathfrak{T}[\mathcal{C}(r_n), \ldots, \mathcal{C}(r_1), \mathcal{C}(r_0)].} \tag{1.63}$$

We have $p(\mathbf{r}_n) = \mathrm{tr}_S\{\tilde{\rho}_S(t_n|\mathbf{r}_n)\}$.

The process tensor is the most abstract and complex mathematical object we meet in this book. If one calls $\mathcal{C}(r_\ell)$ a superoperator, then $\mathfrak{T}$ is a *super-superoperator* and, although we seldomly attempt to compute it explicitly, it is our guiding theoretical framework in the following. Once we have appreciated its vast generality, it allows us to view the structure of many upcoming problems with greater clarity.

The process tensor encodes in an abstract way all what can happen to an open quantum system by 'interrogating' it at an arbitrary set of discrete times with instruments $\{\mathcal{C}(r_\ell)\}$, which act only on the system degrees of freedom. Thus, all the information we can locally obtain about the evolution of an open quantum system is contained in the process tensor. This neatly extends the idea of a classical stochastic process, where one also assumes the ability to perfectly measure the system at arbitrary times, to the quantum regime. Hence, we make the following postulate.

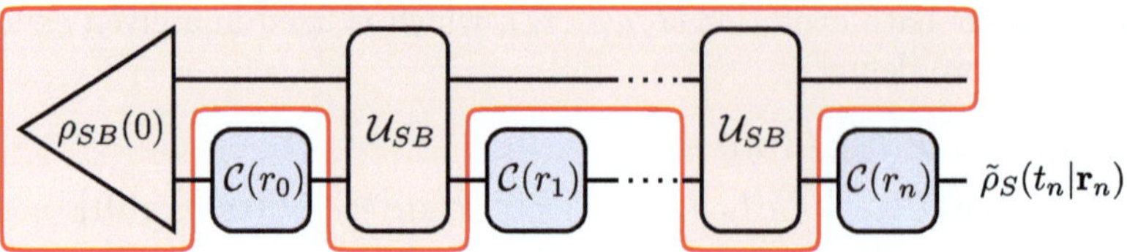

Fig. 1.5 Sketch of the process tensor with time running from left to right. Quantum systems (such as S or B) are marked with a line. The initial state is drawn as a triangle and state transformations are portrayed with boxes. Thus, the shaded red region presents all the information encoded in the process tensor, whereas the inputs or arguments of the process tensor are the interventions $\mathcal{C}(r_\ell)$ shaded in blue. For notational simplicity we dropped the time dependence of the system–bath unitary $\mathcal{U}_{SB}$.

Quantum stochastic process. *Every process tensor describes a quantum stochastic process. Vice versa, every quantum stochastic process can be described by a process tensor.*

Therefore, both classical and quantum stochastic processes have in common the idea of explicitly including the control operations applied in an experiment into the theoretical description, but a quantum stochastic process allows for a larger class of interventions beyond non-disturbing projective measurements. Throughout this book, we call a theoretical description that explicitly accounts for any interventions performed in an experiment **operational**. The rest of this section explains further properties of the process tensor (or a quantum stochastic process), which is pictorially represented in Fig. 1.5. Indeed, after having emphasized its operational character, it seems worth first wondering how to experimentally access the process tensor.

Experimental reconstruction of the process tensor

To experimentally reconstruct the process tensor, it is useful to note first that it really is a *tensor*, i.e. a multi-linear object with respect to its arguments. If this is not yet evident, the next exercise asks you to verify it directly. Explicit representations of the process tensor are constructed in Appendix B.3.

Exercise 1.15 Show based on definition (1.61) that for every $\ell \in \{0, 1, \dots, n\}$

$$\begin{aligned}\mathfrak{T}[\mathcal{A}_n, \dots, a_\ell \mathcal{A}_\ell + b_\ell \mathcal{B}_\ell, \dots, \mathcal{A}_0] = {}& a_\ell \mathfrak{T}[\mathcal{A}_n, \dots, \mathcal{A}_\ell, \dots, \mathcal{A}_0] \\ &+ b_\ell \mathfrak{T}[\mathcal{A}_n, \dots, \mathcal{B}_\ell, \dots, \mathcal{A}_0],\end{aligned} \tag{1.64}$$

where $\mathcal{A}_0, \dots, \mathcal{A}_\ell, \dots, \mathcal{A}_n$ and $\mathcal{B}_\ell$ are arbitrary maps and $a_\ell, b_\ell \in \mathbb{C}$ are two arbitrary complex numbers. Therefore, the process tensor is linear in all its entries. But on which space is the process tensor acting?

For this purpose consider a quantum system with Hilbert space $\mathcal{H}_S$ of dimension $d = \dim \mathcal{H}_S$. We denote the space of all linear maps (or operators) acting on $\mathcal{H}_S$ by $\mathcal{L}(\mathcal{H}_S)$. $\mathcal{L}(\mathcal{H}_S)$ is itself a vector (or Hilbert) space of dimension d^2. Examples for objects living in $\mathcal{L}(\mathcal{H}_S)$ are the density matrix ρ_S and the Hamiltonian H_S. Furthermore, we denote the space of all linear maps (or superoperators) acting on $\mathcal{L}(\mathcal{H}_S)$ by $\mathcal{L}(\mathcal{L}(\mathcal{H}_S))$. This again forms a Hilbert space, now of dimension d^4, and examples include all the CP and CPTP maps we have studied above such as $\mathcal{C}(r_\ell)$. The process tensor therefore maps objects from the $(n+1)$-fold tensor product space of $\mathcal{L}(\mathcal{L}(\mathcal{H}_S))$ to the state space $\mathcal{L}(\mathcal{H}_S)$, symbolically:

$$\mathfrak{T} : \underbrace{\mathcal{L}(\mathcal{L}(\mathcal{H}_S)) \otimes \cdots \otimes \mathcal{L}(\mathcal{L}(\mathcal{H}_S))}_{(n+1) \text{ times}} \to \mathcal{L}(\mathcal{H}_S). \tag{1.65}$$

Deduce that the dimension of the input space of $\mathfrak{T}$ is $d^{4(n+1)}$.

The basic idea to reconstruct the process tensor experimentally is similar to the case of a classical stochastic process, where one constructs a histogram of joint probabilities with respect to all possible measurement results. Now, however, we have to construct a 'histogram' of density matrices with respect to all kinds of possible interventions. As a preliminary step the next exercise first reviews the basics of quantum *state* tomography before we turn to the more general case of quantum *process* tomography.

Exercise 1.16 Quantum state tomography describes the experimental procedure to determine an unknown quantum state ρ. To approach it, we start with an unknown state of a classical two-level system (a 'bit'), here denoted as $\rho_{\text{bit}} = p|0\rangle\langle 0| + (1-p)|1\rangle\langle 1|$ with $p \in [0,1]$ unknown. Clearly, to determine p we simply need sufficiently many copies of ρ_{bit} and measure its state many times in the basis $\{|0\rangle, |1\rangle\}$. The frequency of measurement outcomes m_0/M and m_1/M (with M the total number of measurements) then approximates p and $1-p$.

Now, suppose that you have sufficiently many copies of a qubit in the state $\rho = p|0\rangle\langle 0| + (1-p)|1\rangle\langle 1| + c|0\rangle\langle 1| + c^*|1\rangle\langle 0|$ at your disposal with unknown $p \in [0,1]$ and $c \in \mathbb{C}$, which obeys $|c|^2 \leq p(1-p)$ (show that this follows from the positivity of ρ). Now, devise a measurement strategy to determine ρ. *Hint:* There are many solutions. One convenient approach is offered by the Bloch sphere representation $\rho = (I + \boldsymbol{r} \cdot \boldsymbol{\sigma})/2$ with the vector of Pauli matrices $\boldsymbol{\sigma} = (\sigma_x, \sigma_y, \sigma_z)$ and the Bloch vector $\boldsymbol{r} = \text{tr}\{\boldsymbol{\sigma}\rho\}$ obeying $r_x^2 + r_y^2 + r_z^2 \leq 1$. Show also that it is sufficient to know the three probabilities $p_{x=1} = \text{tr}\{|1\rangle\langle 1|_x \rho\}$, $p_{y=1} = \text{tr}\{|1\rangle\langle 1|_y \rho\}$ and $p_{z=1} = \text{tr}\{|1\rangle\langle 1|_z \rho\}$, where $|1\rangle_{x,y,z}$ is the eigenvector of $\sigma_{x,y,z}$ with eigenvalue $+1$, to determine ρ completely. A set of projectors such as $\{|1\rangle\langle 1|_x, |1\rangle\langle 1|_y, |1\rangle\langle 1|_z\}$ whose outcome probabilities determine the entire state of a quantum system are also known as **informationally complete.**

Despite some similarity in the quantum and classical case, there are also differences. Confirm that, given that you know that the classical bit is in a pure state, there is only a *single* measurement needed to determine its state. Convince yourself of the fact that this no longer holds true for a qubit.

Do you see how to generalize the above procedure to arbitrary d dimensional quantum systems? In this case, a convenient parametrization of ρ could be the generalized Bloch representation

$$\rho = \frac{1}{d}\left(I + \sqrt{\frac{d(d-1)}{2}}\boldsymbol{r} \cdot \boldsymbol{\Lambda}\right). \tag{1.66}$$

Here, the generalized Bloch vector $\boldsymbol{r} \in \mathbb{R}^{d^2-1}$ obeys $\sum_i r_i^2 \leq 1$ and $\boldsymbol{\Lambda}$ denotes a vector of d^2-1 traceless Hermitian matrices Λ_i obeying $\text{tr}\{\Lambda_i \Lambda_j\} = 2\delta_{ij}$.

Given that we now know how to tomographically reconstruct a quantum state, how can we perform **quantum process tomography** of the full process tensor? Luckily, since the process tensor is multi-linear, it suffices to know its elements with respect to a basis of quantum operation performed at different times. For a single time, one such basis is offered by the quantum operations

$$\mathcal{B}_{\alpha\beta}\rho_S \equiv P_\alpha \text{tr}\{\Pi_\beta \rho_S\}. \tag{1.67}$$

Here, the set $\{\Pi_\beta\}_\beta$ presents an informationally complete set of projectors (see last exercise). Thus, $\{\Pi_\beta\}_\beta$ has d^2-1 elements to which we add one more element such that we have a basis for the vector space of complex $d \times d$ matrices. Furthermore, $\{P_\alpha\}_\alpha$ denotes a set of d^2 independent quantum states, which span the space $\mathcal{L}(\mathcal{H}_S)$. If we fix a given basis $\{|n\rangle\}$ of $\mathcal{H}_S$, then one convenient parametrization is given by the double index $\alpha \equiv (m,n)$, such that $P_\alpha = |\psi_{m,n}\rangle\langle\psi_{m,n}|$ with $|\psi_{n,n}\rangle = |n\rangle$, $|\psi_{m,n}\rangle = (|m\rangle + |n\rangle)/\sqrt{2}$ for $m > n$ and $|\psi_{m,n}\rangle = (i|m\rangle + |n\rangle)/\sqrt{2}$ for $m < n$.

Exercise 1.17 Confirm that the above defined set of states P_α linearly spans the entire space of $d \times d$ matrices.

Physically, the control operations (1.67) correspond to measuring the quantum state and obtaining (probabilistically) the result β followed by repreparing the system in the state P_α. Every map $\mathcal{C}$ can be linearly expanded in that basis

$$\mathcal{C} = \sum_{\alpha,\beta} c_{\alpha\beta} \mathcal{B}_{\alpha\beta} \tag{1.68}$$

with in general complex coefficients $c_{\alpha\beta} \in \mathbb{C}$. If $\mathcal{C}$ is a CP or CPTP map, even more can be said about the coefficients, as summarized in the next exercise.

Exercise 1.18 First, show that, if $\mathcal{C}$ preserves Hermiticity, then the coefficients $c_{\alpha\beta} \in \mathbb{R}$ are real. Next, show that, if the map $\mathcal{C}$ is trace-preserving, then $\sum_\alpha c_{\alpha\beta} = 1$.

Now, all that we have to do in order to reconstruct the full process tensor is to measure the quantum system at times t_ℓ, $\ell \in \{0, 1, \ldots, n\}$, in the informationally complete basis $\{\Pi_{\beta_\ell}\}$, followed by randomly repreparing the state in one of the P_{α_ℓ}'s after the measurement. By repeating this procedure many times, we obtain the elementary process tensors $\mathfrak{T}[\mathcal{B}_{\alpha_n\beta_n}, \ldots, \mathcal{B}_{\alpha_1\beta_1}, \mathcal{B}_{\alpha_0\beta_0}]$. Given them, the action of the process tensor on any set of instruments can be reconstructed by linear combination

$$\begin{aligned} &\mathfrak{T}[\mathcal{C}(r_n), \ldots, \mathcal{C}(r_1), \mathcal{C}(r_0)] \\ &\quad = \sum_{\alpha_n,\beta_n} \cdots \sum_{\alpha_1,\beta_1} \sum_{\alpha_0,\beta_0} c_{\alpha_n\beta_n} \cdots c_{\alpha_1\beta_1} c_{\alpha_0\beta_0} \mathfrak{T}[\mathcal{B}_{\alpha_n\beta_n}, \ldots, \mathcal{B}_{\alpha_1\beta_1}, \mathcal{B}_{\alpha_0\beta_0}] \end{aligned} \tag{1.69}$$

for some coefficients $c_{\alpha_\ell\beta_\ell} \in \mathbb{R}$. In fact, we can now recognize that the sequence of interventions $\mathcal{C}(r_n), \ldots, \mathcal{C}(r_1), \mathcal{C}(r_0)$ forms only a special subclass of all conceivable interventions, namely those that are *decorrelated*. The next exercise explains this.

Exercise 1.19 Recall eqn (1.65), which shows that the process tensor can be seen as a map acting on $\bigotimes_{i=0}^n \mathcal{L}(\mathcal{L}(\mathcal{H}_S))$. Using the basis (1.67), an arbitrary element $\boldsymbol{C}_{n:0}$ of that space can be written as

$$\boldsymbol{C}_{n:0} = \sum_{\alpha_n,\beta_n} \cdots \sum_{\alpha_1,\beta_1} \sum_{\alpha_0,\beta_0} c_{\alpha_n\beta_n,\ldots,\alpha_1\beta_1,\alpha_0\beta_0} \mathcal{B}_{\alpha_n\beta_n} \otimes \cdots \otimes \mathcal{B}_{\alpha_1\beta_1} \otimes \mathcal{B}_{\alpha_0\beta_0}, \tag{1.70}$$

where the coefficients $c_{\alpha_n\beta_n,\ldots,\alpha_1\beta_1,\alpha_0\beta_0}$ need not factor as in eqn (1.69). Confirm that the element $\boldsymbol{C}_{n:0}(\mathbf{r}_n)$ corresponding to applying a sequence of control operations $\mathcal{C}(r_n), \ldots, \mathcal{C}(r_0)$ can be written as a tensor product in that space: $\boldsymbol{C}_{n:0}(\mathbf{r}_n) = \mathcal{C}(r_n) \otimes \cdots \otimes \mathcal{C}(r_0)$. This can be called a decorrelated control operation or intervention.

Equation (1.70) therefore tells us that more general, correlated control operations $\boldsymbol{C}_{n:0}(\mathbf{r}_n)$ are possible. In this case we denote the process tensor as $\mathfrak{T}[\boldsymbol{C}_{n:0}(\mathbf{r}_n)] = \tilde{\rho}_S(t_n|\mathbf{r}_n)$, where $\tilde{\rho}_S(t_n|\mathbf{r}_n)$ still denotes the non-normalized system state at time t_n resulting from the action of some correlated control operation happening with probability $p(\mathbf{r}_n) = \mathrm{tr}_S\{\tilde{\rho}_S(t_n|\mathbf{r}_n)\}$. Examples for correlated control operations are, for instance, classical feedback control or conditioning of the dynamics. To be specific, let $\mathcal{P}(r_0)$ denote a projective measurement of some system observable at time t_0 giving result r_0. Let $\mathcal{C}_1(r_0)$ denote some CPTP map at time t_1, which depends on the previously recorded measurement result r_0. The corresponding control operation is $\boldsymbol{C}_{1:0}(r_0) = \mathcal{C}_1(r_0) \otimes \mathcal{P}(r_0)$, which is still decorrelated. However, the average effect of the control operation is given by $\boldsymbol{C}_{1:0} = \sum_{r_0} \mathcal{C}_1(r_0) \otimes \mathcal{P}(r_0)$, which is correlated. Correlated control operations therefore allow us to treat all conceivable scenarios of feedback control within the process tensor framework.

To conclude, the above procedure shows how to experimentally reconstruct the process tensor. This reconstruction is based on two assumptions. First, the experimentalist must be able to locally measure and manipulate the open quantum system. This does not necessarily mean that one must be able to implement the specific control operations (1.67), any other basis of maps suffices as well. Second, the control operations were assumed to happen *instantaneously*, similar to the measurements in a classical stochastic process. In practice, this statement translates into the requirement that the time scales of the environmentally induced system evolution, determined by $\mathcal{U}_{SB}$, must be long compared with the time it takes to implement the control operation. This is required in order to disentangle the effect of the environment from the effect of the external controller. Even if these requirements are satisfied, experimentally reconstructing the process tensor remains a formidable challenge since the number of parameters we have to estimate grows exponentially as d^{4n} with the number n of time steps (for a classical stochastic process, it grows as d^n). Luckily, for one theoretically and experimentally important case of applications, the process tensor can be constructed in a simple manner. This case is described by quantum *Markov* processes, which we will introduce in the next section. Before turning to them, we list three further relevant properties of the process tensor.

Further properties of the process tensor

In Section 1.6 we have seen that the joint probabilities of a classical stochastic process form a hierarchy. The same is also true for the process tensor, which satisfies a *containment property*, as explored in the next exercise. We remark that this containment property is the basis to show that there exists a generalized Daniell–Kolmogorov extension theorem also for quantum stochastic processes.

Exercise 1.20 The process tensor $\mathfrak{T}$ was defined on the set of times $\{t_0, t_1, \ldots, t_n\}$. Consider any subset of times $T \subset \{t_0, t_1, \ldots, t_n\}$. Show that the process tensor $\mathfrak{T}_T$ defined on this subset of times is *contained* in the original $\mathfrak{T}$. By this we mean that all probabilities predictable from $\mathfrak{T}_T$ can also be recovered from $\mathfrak{T}$. Thus, process tensors $\mathfrak{T}_{T_1}, \ldots, \mathfrak{T}_{T_N}$ for discrete sets of times $T_1 \subset \cdots \subset T_N$ form a hierarchy with $\mathfrak{T}_{T_\ell}$ containing $\mathfrak{T}_{T_k}$ for $\ell \geq k$.

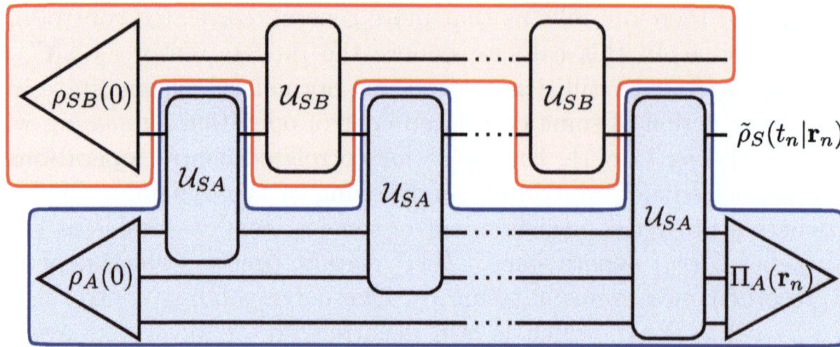

Fig. 1.6 Similar to the unitary dilation theorem, an arbitrary correlated control operation $\boldsymbol{C}_{n:0}(\mathbf{r}_n)$ can also be simulated by introducing a sufficiently large ancilla state $\rho_A(0)$ that is initially decorrelated from the system and usually contains multiple ancillas (represented by the different lines in the figure) and that interacts unitarily with the system through $\mathcal{U}_{SA}$. Note that we are a bit sloppy here and use the same symbol $\mathcal{U}_{SA}$ for in general different system–ancilla unitaries. Finally, the joint ancilla state is measured using some projector $\Pi_A(\mathbf{r}_n)$. The blue and red objects together represent the unitary dilation of the process.

Furthermore, one can extend the operator–sum representation and the unitary dilation theorem to the process tensor, but we are not going to write down the corresponding operator–sum representation theorem here and the unitary dilation theorem of the process tensor is only pictorially presented in Fig. 1.6. Mathematically, this is possible because the process tensor satisfies requirements similar to those imposed on a single-time control operation in Section 1.5 (requirements (i′), (ii) and (iii)).

First, the process tensor preserves complete positivity (requirement (i′)) in the following sense. Let ρ_{SA} be an arbitrary initial system–ancilla state and let $\boldsymbol{C}_{n:0}^{SA}(\mathbf{r}_n)$ be any $(n+1)$-time control operation acting on the system–ancilla state. We can then define an extended process tensor $\mathfrak{T}_S \otimes \mathfrak{I}_A$ as in eqn (1.61) by replacing $\rho_{SB}(0)$ by $\rho_{SBA}(0)$ and $\mathcal{U}_{SB}(t_\ell, t_k)$ by $\mathcal{U}_{SB}(t_\ell, t_k) \otimes \mathcal{I}_A$. Thus, $\mathfrak{T}_S$ equals $\mathfrak{T}$ as defined in eqn (1.69) and $\mathfrak{I}_A$ is the 'identity process'. It then follows that the output state $\tilde{\rho}_{SA}(t_n|\mathbf{r}_n) = \mathfrak{T}_S \otimes \mathfrak{I}_A(\boldsymbol{C}_{n:0}^{SA}) \geq 0$ is always positive.

Second, the trace non-increasing property (ii) is clearly satisfied for the process tensor and it also satisfies a modified requirement (iii) of linearity by recalling that the process tensor is linear with respect to the applied interventions. In fact, those interventions are the objects that we can freely control in an experiment, in contrast to the initial system state $\rho_S(0)$. This resolves the conundrum of Exercise 1.11.

Exercise 1.21 The process tensor does not depend linearly on the initial system state $\rho_S(0)$ in general. However, show that it does for an initial state of the form $\rho_{SB}(0) = \rho_S(0) \otimes \rho_B(0)$. Then, convince yourself that in this case the first control operation $\mathcal{C}(r_0)$ becomes redundant and one can define the process tensor as $\mathfrak{T}[\mathcal{C}(r_n), \ldots, \mathcal{C}(r_1), \rho_S(0)]$, where $\rho_S(0)$ is now taken as arbitrary. Beyond that case, the present approach provides an operational resolution to the conundrum of Exercise 1.11 by realizing that *quantum dynamics is linear with respect to the state preparation* $\mathcal{C}(r_0)$ *of an experiment*, but not with respect to the initial system state.

The final point we want to make is that the process tensor can be used to compute correlation functions of the form

$$\langle A(t)B(0)\rangle = \mathrm{tr}_{SB}\{AU_{SB}(t,0)B\rho_{SB}(0)U_{SB}^{\dagger}(t,0)\}. \tag{1.71}$$

Here, A and B are two arbitrary system observables. The reader is asked to explicitly verify this in the next exercise, which concludes this section.

Exercise 1.22 We write the spectral decomposition of A as $A = \sum_a a|a\rangle\langle a|$. Since

$$\langle A(t)B(0)\rangle = \sum_a a\mathrm{tr}_{SB}\{|a\rangle\langle a|U_{SB}(t,0)B\rho_{SB}(0)U_{SB}^{\dagger}(t,0)\}, \tag{1.72}$$

it is clear that the final control operation at time t_1 has to be a projective measurement of the observable A. We therefore write $\langle A(t)B(0)\rangle = \sum_a a\mathrm{tr}_S\{\mathfrak{T}[\mathcal{P}(a),\mathcal{B}_0]\}$, where $\mathfrak{T}[\cdot,\cdot]$ is the two-time process tensor defined at times 0 and t and the control operation $\mathcal{B}_0$ is defined by $\mathcal{B}_0\rho_{SB}(0) = B\rho_{SB}(0)$. Unfortunately, $\mathcal{B}_0$ is not CP and does not even preserve Hermiticity. Therefore, it cannot directly be implemented in a laboratory using a suitably tailored system–ancilla interaction. Nevertheless, by linearity we can find a set of experimentally implementable CP maps $\mathcal{B}_{\alpha\beta}$ with coefficients $c_{\alpha\beta}$ such that $\mathcal{B}_0 = \sum_{\alpha,\beta} c_{\alpha\beta}\mathcal{B}_{\alpha\beta}$; see eqn (1.68). To see this explicitly, we use $B = \sum_b b|b\rangle\langle b|$ and $I = \sum_b |b\rangle\langle b|$ to write

$$\mathcal{B}_0\rho_{SB}(t_0) = B\rho_{SB}(t_0) = \sum_b b|b\rangle\langle b|\rho_{SB}\sum_{b'}|b'\rangle\langle b'| = \sum_{b,b'} b|b\rangle\langle b'|\mathrm{tr}_S\{|b'\rangle\langle b|\rho_{SB}(0)\}. \tag{1.73}$$

Now, this already looks close to a linear combination of the elementary control operations (1.67) with the difference that the elements $|b\rangle\langle b'|$ in general do not belong to the sets of $\{P_\alpha\}_\alpha$ and $\{\Pi_\beta\}_\beta$. However, as Exercise 1.17 has shown, we can always write

$$|b\rangle\langle b'| = \sum_\alpha c_\alpha^{bb'}P_\alpha, \quad |b'\rangle\langle b| = \sum_\beta \tilde{c}_\beta^{b'b}\Pi_\beta \tag{1.74}$$

for some (in general complex) coefficients $c_\alpha^{bb'}$ and $\tilde{c}_\beta^{b'b}$. Thus, in terms of the elementary control operations (1.67) we have

$$\langle A(t)B(0)\rangle = \sum_{a,b,b',\alpha,\beta} abc_\alpha^{bb'}\tilde{c}_\beta^{b'b}\mathrm{tr}_S\{\mathfrak{T}[\mathcal{P}(a),\mathcal{B}_{\alpha\beta}]\}. \tag{1.75}$$

Now, the process $\mathfrak{T}[\mathcal{P}(a),\mathcal{B}_{\alpha\beta}]$ is implementable in a laboratory using only local measurements and state preparations. Furthermore, the coefficient $abc_\alpha^{bb'}\tilde{c}_\beta^{b'b}$ is known. Hence, the correlation function $\langle A(t)B(0)\rangle$ is measurable and its value is encoded in the process tensor.

If you had problems following the steps above, you should explicitly verify them for a qubit and the linear map $\mathcal{B}_0\rho = \sigma_z\rho$ using a basis of your choice for the elementary control operations $\mathcal{B}_{\alpha\beta}\rho = P_\alpha\mathrm{tr}\{\Pi_\beta\rho\}$. Furthermore, convince yourself of the fact that the above statement can be extended to arbitary multi-time correlation functions $\langle A_n(t_n)\dots A_1(t_1)A_0(t_0)\rangle$ as long as all A_ℓ are system observables.

1.8 Quantum Markov Processes and Dynamical Maps

Markov processes, whether they are quantum or classical in nature, play an important role in physics. In part, this seems to be caused by the fact that they considerably simplify the life of a theoretician and typically allow for at least some analytical progress in the description. Also in this book we will encounter Markov processes, albeit a considerable part of this book is devoted to showing that all our main results continue to hold for non-Markovian processes. Colloquially, Markov processes are associated with *memoryless* processes. This means that the environment quickly *forgets* the state of the system in the past such that the system's future evolution is only influenced by its current state. We now make this reasoning rigorous.

To approach the problem, we first of all introduce the notion of a **causal break**. A causal break is an intervention that reprepares the system in a state that is independent of all past interventions. They are in general written as

$$\mathcal{B}_{\alpha\beta}\rho_S \equiv \sigma_S^{(\alpha)}\mathrm{tr}_S\{P_\beta \rho_S\}. \tag{1.76}$$

Here, P_β is an arbitrary element of a POVM and $\sigma_S^{(\alpha)}$ an arbitrary system density matrix. Thus, a causal break describes an intervention in which we read out the state of the system in an arbitary way (remember that $P_\beta = I_S$ is a legitimate choice too), then discard the system state and finally prepare a fresh state $\sigma_S^{(\alpha)}$ that is *independent* of the previous system state or the result of the measurement P_β. This ensures that $\sigma_S^{(\alpha)}$ can have no memory about the past. Note that the operations introduced in eqn (1.67) are causal breaks and, hence, causal breaks can be used as a basis to span the space of linear maps.

Furthermore, we introduce the following notation to denote the *normalized* system state after an arbitrary sequence $\mathcal{C}(r_n), \dots, \mathcal{C}(r_1), \mathcal{C}(r_0)$ of interventions:

$$\rho_n[\mathcal{C}(r_n), \dots, \mathcal{C}(r_1), \mathcal{C}(r_0)] \equiv \frac{\tilde{\rho}_S(t_n|\mathbf{r}_n)}{p(\mathbf{r}_n)} = \frac{\mathfrak{T}[\mathcal{C}(r_n), \dots, \mathcal{C}(r_1), \mathcal{C}(r_0)]}{\mathrm{tr}_S\{\mathfrak{T}[\mathcal{C}(r_n), \dots, \mathcal{C}(r_1), \mathcal{C}(r_0)]\}}. \tag{1.77}$$

We now define a quantum Markov process.

Quantum Markov process. *A quantum stochastic process is Markovian if the normalized system state at time t_ℓ after a causal break (1.76) at time $t_k < t_\ell$ depends only on the input state $\sigma_S^{(\alpha_k)}$ for any set of previous interventions $\mathcal{C}(r_{k-1})$, ..., $\mathcal{C}(r_0)$. In equation format,*

$$\rho_\ell[\mathcal{B}_{\alpha_k\beta_k}, \mathcal{C}(r_{k-1}), \dots, \mathcal{C}(r_0)] = \rho_\ell[\sigma_S^{(\alpha_k)}]. \tag{1.78}$$

This definition captures the idea that everything which has happened to the system in the past does not influence its future after any causal connection at the system level is cut. Thus, if eqn (1.78) is violated, this must be due to the fact that the *environment* kept some memory about what has happend to the system in the past. To become familiar with this definition, it is best to confirm an intuitve result such as the following.

Exercise 1.23 Verify that an isolated (i.e. unitarily evolving) system is Markovian.

The following important result, which the reader is asked to prove in an exercise below, shows that quantum Markov processes have a particularly simple mathematical structure.

Factorization of the process tensor. *For a Markovian process the process tensor 'factorizes':*

$$\boxed{\mathfrak{T}[\mathcal{C}(r_n),\ldots,\mathcal{C}(r_0)] = \mathcal{C}(r_n)\mathcal{E}(t_n,t_{n-1})\ldots\mathcal{E}(t_1,0)\mathcal{C}(r_0)\rho_S(0).} \tag{1.79}$$

Here, $\rho_S(0) = \text{tr}_B\{\rho_{SB}(0)\}$ *is the initial system state and* $\mathcal{E}(t_\ell,t_k)$ *are CPTP maps independent of* $\mathcal{C}(r_n),\ldots,\mathcal{C}(r_0)$.

The physical meaning of the CPTP maps $\mathcal{E}(t_\ell,t_k)$ is to propagate the system state forwards in time. This becomes transparent by considering the process tensor obtained by applying the identity maps restricted to the subset of times $\{t_k,0\}$ with $k>0$:

$$\rho_S(t_k) = \mathfrak{T}_{k,0}[\mathcal{I}_k,\mathcal{I}_0] = \mathcal{E}(t_k,0)\rho_S(0). \tag{1.80}$$

Furthermore, we can use the interventions $\mathcal{C}(r_k)$ to prepare arbitrary system states at any time t_k. Since eqn (1.79) holds for all interventions and all subsets of times $t_\ell > t_k > t_j > t_0$, we infer that the CPTP maps obey the composition rule

$$\mathcal{E}(t_\ell,t_j) = \mathcal{E}(t_\ell,t_k)\mathcal{E}(t_k,t_j). \tag{1.81}$$

In fact, whenever it is possible to define a set of cptp maps $\{\mathcal{E}(t_k,t_j)\}$ for all times $t_k > t_j$ that propagate the system state $\rho(t_k) = \mathcal{E}(t_k,t_j)\rho(t_j)$ and obey eqn (1.81) one calls the dynamics **CP divisible**. Note that a dynamics can be non-Markovian but CP divisible because the notion of CP divisibility does not rely on interventions.

Note that a CPTP map $\mathcal{E}(t_k,0)$ propagating the system state forwards in time can *always* be defined if the system–bath state at time t_0 factorizes. Then, we can write

$$\rho_S(t_k) = \text{tr}_B\{U_{SB}(t_k,0)\rho_S(0)\otimes\rho_B(0)U_{SB}^\dagger(t_k,0)\} \equiv \mathcal{E}(t_k,0)\rho_S(0) \tag{1.82}$$

and by virtue of the unitary dilation theorem the map $\mathcal{E}(t_k,0)$ is CPTP. Markovianity precisely says that we can always write $\rho_S(t_\ell) = \mathcal{E}(t_\ell,t_k)\rho_S(t_k)$ for all pairs of times $t_\ell > t_k$ with $\mathcal{E}(t_\ell,t_k)$ CPTP and independent of what has happened in the past. Since the maps $\mathcal{E}(t_\ell,t_k)$ play such an important role in the evolution of open quantum systems, they are called **dynamical maps**. In quantum information theory they are also often called **channels** to signify that they can be used to *communicate* a quantum state from one laboratory to another. That is to say, the CPTP maps $\mathcal{E}$ cannot only be used to propagate the system state forwards in time, but also forwards in *space*–time. In non-relativistic quantum mechanics $\mathcal{E}$ describes therefore the most general time evolution obeying the properties (i′), (ii) and (iii) of Section 1.5. The proof that Markov processes factorize is relegated to an exercise.

Exercise 1.24 First, show that if the process tensor factorizes as in eqn (1.79) then the process is Markovian. Conversely, let us assume that the process is Markovian. Consider first an elementary process tensor obtained by applying only causal breaks. Note that the action of a causal breaks implies that the system–bath state becomes decorrelated: $\mathcal{B}_{\alpha\beta}\rho_{SB} = \sigma_S^{(\alpha)}\otimes\text{tr}_S\{P_\beta\rho_{SB}\}$. We can therefore always write

$$\begin{aligned}&\mathfrak{T}[\mathcal{B}_{\alpha_n\beta_n},\ldots,\mathcal{B}_{\alpha_1\beta_1},\mathcal{B}_{\alpha_0\beta_0}] = \\ &\qquad \sigma_S^{(\alpha_n)}\text{tr}_S\{P_{\beta_n}\tilde{\mathcal{E}}(t_n,t_{n-1})\sigma_S^{(\alpha_{n-1})}\}\ldots\text{tr}_S\{P_{\beta_1}\tilde{\mathcal{E}}(t_1,t_0)\sigma_S^{(\alpha_0)}\}\text{tr}_S\{P_{\beta_0}\rho_S(t_0)\}.\end{aligned} \tag{1.83}$$

Here, the $\tilde{\mathcal{E}}(t_\ell, t_k)$ are CPTP maps which can be reconstructed by quantum process tomography by preparing a set of basis states $\{\rho_S^{(\alpha_k)}\}$ and by measuring an informationally complete set of POVMs $\{P_{\beta_\ell}\}$. In general, however, the maps $\tilde{\mathcal{E}}(t_\ell, t_k)$ depend on all previous interventions $\beta_k, \alpha_{k-1}, \beta_{k-1}, \dots, \alpha_0, \beta_0$ because the reduced state of the bath at time t_k carries information about the history of the system. Therefore, the maps $\tilde{\mathcal{E}}(t_\ell, t_k)$ are usually not very helpful. However, for Markovian processes the claim is precisely that these maps are *independent* of the history of the system. Prove this! *Hint:* It might be useful to try a proof by contradiction, i.e. assume that there are two different histories $\mathbf{h} \equiv (\beta_k, \alpha_{k-1}, \beta_{k-1}, \dots, \alpha_0, \beta_0)$ and $\mathbf{h}' \equiv (\beta_k', \alpha_{k-1}', \beta_{k-1}', \dots, \alpha_0', \beta_0')$ of causal breaks with $\mathbf{h} \neq \mathbf{h}'$ such that $\tilde{\mathcal{E}}(t_\ell, t_k) \equiv \tilde{\mathcal{E}}(t_\ell, t_k|\mathbf{h}) \neq \tilde{\mathcal{E}}(t_\ell, t_k|\mathbf{h}') \equiv \tilde{\mathcal{E}}'(t_\ell, t_k)$. Show that this leads to a contradiction for a Markov process.

It is important to remark that the Markov property is purely defined in terms of the process tensor. Hence, it is a property *inherent* to the process itself. In particular, the question of whether the process is Markovian or not is *independent* of the control operations applied to the system. This holds true even for correlated control operations, for instance if we apply feedback control at time t_n depending on previous measurement results $\mathbf{r}_k$ $(k < n)$. The fact that the environment for a Markov process does not keep any memory about previous interventions is not changed by the decision of an external agent to keep a memory about them. The connection between classical and quantum Markov processes is explored in the next section.

1.9 Classical Quantum Stochastic Processes

In the final section of this chapter we connect the notion of quantum stochastic processes to classical stochastic processes. We unravel that quantum stochastic processes look classical in some measurement basis whenever coherences do not influence the dynamics in that basis. Later on in this book, we will indeed see that there is a preferred basis in which at least weakly coupled open quantum systems behave 'classical'. This basis is the energy eigenbasis, which is related to the fact that the thermal equilibrium state $\pi_S = e^{-\beta H_S}/\mathcal{Z}_S$ is diagonal in that basis. In the second part of this section we connect the notion of quantum Markovianity to classical Markovianity.

Non-classicality and coherence

To investigate the question of when a quantum stochastic process mimics a classical stochastic process, we return to the case of a projectively measured system observable R_S with a joint probability $p(\mathbf{r}_n)$ as defined in eqns (1.59) and (1.60). It turns out that our answer depends on the question of whether the observable R_S is degenerate or not. For the moment we assume that it is non-degenerate, i.e. all projectors are of rank 1, which means that $R_S = \sum_r r|r\rangle\langle r|_S$. To approach the problem, let us first consider a case in which the quantum stochastic process evidently looks classical. This case is described by assuming that at each time t_ℓ, $\ell \in \{0, 1, \dots, n\}$, the joint system–bath state can be written as

$$\rho_{SB}(t_\ell) = \sum_{r_\ell} p(r_\ell, t_\ell)|r_\ell\rangle\langle r_\ell|_S \otimes \rho_B(t_\ell|r_\ell). \tag{1.84}$$

This describes a state with classical but no quantum correlations. With probability $p(r_\ell, t_\ell)$ the system is in state $|r_\ell\rangle$ and the bath is in state $\rho_B(t_\ell|r_\ell)$, on which we put no further restriction. A unique characteristic of the state (1.84) is that a projective measurement of R_S does not change it *on average*, i.e.

$$\rho_{SB}(t_\ell) = \sum_{r_\ell} |r_\ell\rangle\langle r_\ell|_S \rho_{SB}(t_\ell)|r_\ell\rangle\langle r_\ell|_S. \tag{1.85}$$

Put differently, $\rho_{SB}(t_\ell)$ is characterized by the fact that we can find a local measurement basis (in this case $\{|r_\ell\rangle_S\}$) that has no quantum backaction, i.e. it does not disturb the state. Bipartite states, which have this property, are also said to possess *zero quantum discord.* It is interesting to remark that such states allow definition (1.82) of a dynamical map, originally given only for decorrelated system–bath states of the form $\rho_{SB}(0) = \rho_S(0) \otimes \rho_B(0)$, to be extended to classically correlated cases, but in general not beyond. This is shown in the next exercise.

Exercise 1.25 Consider a system–bath state with zero quantum discord with respect to a complete set of system projectors $|j\rangle\langle j|_S$ such that

$$\rho_{SB}(0) = \sum_j |j\rangle\langle j|_S \rho_{SB}(0)|j\rangle\langle j|_S = \sum_j p_j |j\rangle\langle j|_S \otimes \rho_B(j). \tag{1.86}$$

Here, p_j is a probablity distribution and $\rho_B(j)$ is the state of the bath given the state $|j\rangle\langle j|_S$ of the system. Show that the reduced system dynamics $\rho_S(t) = \mathrm{tr}_B\{U_{SB}\rho_{SB}(0)U_{SB}^\dagger\}$ is CPTP for any unitary U_{SB}. *Hint:* Show that you can write

$$\rho_S(t) = \sum_{b,b'} \left[\sum_j \left\langle b \middle| U_{SB}\sqrt{\rho_B(j)} \middle| b' \right\rangle |j\rangle\langle j|_S\right] \rho_S(0) \left[\sum_k |k\rangle\langle k|_S \left\langle b' \middle| \sqrt{\rho_B(k)} U_{SB}^\dagger \middle| b \right\rangle\right], \tag{1.87}$$

where $\{|b\rangle\}$ and $\{|b'\rangle\}$ are arbitrary bases in the bath Hilbert space and the square root of $\rho_B(j)$ is well defined since $\rho_B(j)$ is a positive operator. Then, put eqn (1.87) into operator–sum representation.

Another consequence of the assumption (1.84) is that the projective measurement statistics of $R_S = \sum_r r|r\rangle\langle r|_S$ are classical because they satisfy the Kolmogorov consistency condition. The reader is asked to verify this in the next exercise.

Exercise 1.26 Show, assuming eqn (1.84) for all t_ℓ, that the joint probabilities (1.59) satisfy the Kolmogorov consistency condition (1.53).

While states of the form (1.84) provide a sufficient criterion to decide that $p(\mathbf{r}_n)$ cannot be distinguished from a classical stochastic process, the answer is not yet fully satisfactory for two reasons. First, deciding whether eqn (1.84) applies requires explicit knowledge about the system–bath correlations, which is usually not available in an experiment. Second, even if the system–bath state is not of the form (1.84), the resulting measurement statistics $p(\mathbf{r}_n)$ might nevertheless be indistinguishable from a classical stochastic process. We now show that this happens whenever *coherences* in the system state with respect to the basis $\{|r_\ell\rangle_S\}_{r_\ell}$ do not influence the dynamics of the *populations*.

Recall that for a given basis $\{|r\rangle\}_r$ the density matrix can always be written as

$$\rho = \sum_r \rho_{rr}|r\rangle\langle r| + \sum_{r'>r}(\rho_{rr'}|r\rangle\langle r'| + \text{h.c.}) = \begin{pmatrix} \rho_{11} & \rho_{12} & \cdots & \rho_{1n} \\ \rho_{12}^* & \rho_{22} & & \vdots \\ \vdots & & \ddots & \rho_{n-1n} \\ \rho_{1n}^* & \cdots & \rho_{n-1n}^* & \rho_{dd} \end{pmatrix}, \tag{1.88}$$

where the diagonal elements of the matrix are called populations and the off-diagonal elements coherences. The average effect of a projective measurement of an observable R is called a **dephasing operation**, denoted by $\mathcal{D}_R$ in the following. Its action on the previous state ρ is to delete all coherences:

$$\mathcal{D}_R\rho \equiv \sum_r |r\rangle\langle r|\rho|r\rangle\langle r| = \sum_r \rho_{rr}|r\rangle\langle r| = \begin{pmatrix} \rho_{11} & 0 & \dots & 0 \\ 0 & \rho_{22} & & \vdots \\ \vdots & & \ddots & 0 \\ 0 & \dots & 0 & \rho_{dd} \end{pmatrix}. \tag{1.89}$$

We now introduce the notion of an **incoherent process**, which is supposed to capture the absence of observable effects of the coherences on the population dynamics. Namely, we call a quantum stochastic process n-incoherent (with respect to the system observable R_S) if all process tensors

$$\mathfrak{T}\left[\mathcal{D}_{R_n}, \begin{Bmatrix} \mathcal{D}_{R_{n-1}} \\ \mathcal{I}_{n-1} \end{Bmatrix}, \dots, \begin{Bmatrix} \mathcal{D}_{R_1} \\ \mathcal{I}_1 \end{Bmatrix}, \mathcal{C}_0\right] \tag{1.90}$$

are equal, where the curly bracked notation $\{\cdot\}$ signifies that we are free to choose to apply either a dephasing operation $\mathcal{D}_{R_\ell}$ or nothing (i.e. the identity operation $\mathcal{I}_\ell$) at time t_ℓ. Remember our convention to leave out the subscript S on operations that act only on the system space. Furthermore, $\mathcal{C}_0$ describes an arbitrary but fixed initial state preparation (recall Exercise 1.21). Therefore, an n-incoherent process is characterized by the fact that the populations of its output state $\rho_S(t_n|\mathcal{C}_0)$ given any initial state preparation at time t_0 are insensitive to the question of whether we erased all coherences at any intermediate time step t_ℓ, $\ell \in \{1, \dots, n-1\}$, or not. Put differently, the coherences do not affect the dynamics of the populations. In the following we are only interested in processes that are ℓ-incoherent for all $\ell \in \{1, 2, \dots, n\}$, and call them *incoherent*.

Finally, we recall that the probability of obtaining the measurement results $r_n, \dots,$ r_1 for a given state preparation $\mathcal{C}_0$ is $p(r_n, \dots, r_1|\mathcal{C}_0) = \text{tr}_S\{\mathfrak{T}[\mathcal{P}(r_n), \dots, \mathcal{P}(r_1), \mathcal{C}_0]\}$. After these preliminary considerations, we can formulate the first main result of this section, the proof of which is left as an exercise.

Classicality implies incoherence. *If the hierarchy of probabilities $p(r_n, \dots, r_1|\mathcal{C}_0)$ obeys the Kolmogorov consistency condition (1.53), then the process is incoherent.*

Exercise 1.27 Prove this statement. *Hint:* Since $R_S = \sum_r r|r\rangle\langle r|_S$ is assumed to be non-degenerate, it follows that $\mathfrak{T}[\mathcal{P}(r_n), \dots, \mathcal{P}(r_1), \mathcal{C}_0] = p(r_n, \dots, r_1|\mathcal{C}_0)|r_n\rangle\langle r_n|$.

Thus, we see that the requirement for classicality as characterized by the Kolmogorov consistency condition is quite strong as it immediately implies that coherences can have no detectable effect on the population dynamics. However, experimentally confirming the Kolmogorov consistency condition can become quite involved, whereas confirming the notion of incoherent dynamics only requires the response of a system to one or multiple dephasing operations to be probed. The second main result of this section specifies under which condition incoherent dynamics is sufficient to conclude that the measurement statistics are classical.

Classicality from incoherence. *If the dynamics is Markovian, invertible and incoherent for all preparations $\mathcal{C}_0$, then the statistics are classical for any preparation.*

To understand the notion of invertibility, recall eqn (1.79), which shows that Markovian dynamics is characterized by a set of dynamical maps $\{\mathcal{E}(t_\ell, t_k)|n \geq \ell \geq k \geq 0\}$. *Invertible* Markovian dynamics is then defined by the requirement that the dynamical maps $\mathcal{E}_{\ell,k}$ are invertible (in the usual sense of an invertible matrix), where here and in the rest of this chapter we use the abbreviation $\mathcal{E}_{\ell,k} \equiv \mathcal{E}(t_\ell, t_k)$. This puts us in a position to prove the above statement.

Using that the process is incoherent and Markovian, we confirm that

$$\mathcal{D}_{R_{\ell+1}}\mathcal{E}_{\ell+1,\ell}\mathcal{D}_{R_\ell}\mathcal{E}_{\ell,0}\mathcal{C}_0\rho_S(t_0) = \mathcal{D}_{R_{\ell+1}}\mathcal{E}_{\ell+1,\ell}\mathcal{I}_\ell\mathcal{E}_{\ell,0}\mathcal{C}_0\rho_S(t_0) \tag{1.91}$$

must hold for all preparations $\mathcal{C}_0$. Since the dynamics is invertible, this becomes a superoperator identity: $\mathcal{D}_{R_{\ell+1}}\mathcal{E}_{\ell+1,\ell}\mathcal{D}_{R_\ell} = \mathcal{D}_{R_{\ell+1}}\mathcal{E}_{\ell+1,\ell}$. By multiplying this equation by $\mathcal{P}(r_{\ell+1})$, we obtain the relation

$$\sum_{r_\ell}\mathcal{P}(r_{\ell+1})\mathcal{E}_{\ell+1,\ell}\mathcal{P}(r_\ell) = \mathcal{P}(r_{\ell+1})\mathcal{E}_{\ell+1,\ell}, \tag{1.92}$$

where we used that $\mathcal{D}_{R_\ell} = \sum_{r_\ell}\mathcal{P}(r_\ell)$. Based on this, it is easy to show that the Kolmogorov consistency condition is satisfied:

$$\begin{aligned}\sum_{r_\ell} p(r_n, \dots, r_1|\mathcal{C}_0) &= \sum_{r_\ell} \mathrm{tr}_S\{\mathcal{P}(r_n)\mathcal{E}_{n,n-1}\dots\mathcal{P}(r_{\ell+1})\mathcal{E}_{\ell+1,\ell}\mathcal{P}(r_\ell)\dots\mathcal{E}_{1,0}\mathcal{C}_0\rho_S(t_0)\} \\ &= \mathrm{tr}_S\{\mathcal{P}(r_n)\mathcal{E}_{n,n-1}\dots\mathcal{P}(r_{\ell+1})\mathcal{E}_{\ell+1,\ell}\mathcal{I}_\ell\dots\mathcal{E}_{1,0}\mathcal{C}_0\rho_S(t_0)\} \\ &= p(r_n, \dots \not{r_\ell}, \dots, r_1|\mathcal{C}_0).\end{aligned} \tag{1.93}$$

This concludes the proof.

The question remains of whether it is possible to relax some conditions above, for instance by looking at Markovian dynamics, which is incoherent only for some preparations $\mathcal{C}_0$ or not invertible. The following exercise excludes this possibility.

Exercise 1.28 Find examples of non-classical processes that are Markovian and incoherent with respect to a restricted set of preparations $\mathcal{C}_0$ or not invertible. *Hint:* It suffices to consider the unitary evolution of a two-level system prepared in the maximally mixed state $\rho_S(t_0) = (|0\rangle\langle 0| + |1\rangle\langle 1|)/2$.

Finally, we briefly discuss the case of degenerate observables $R_S = \sum_r r\Pi_S(r)$, where some projectors $\Pi_S(r)$ have a rank larger than 1. One easily confirms that the proof above made no use of the fact that the observable is non-degenerate. Thus, classicality continues to follow from incoherence under the assumptions spelled out above, even for degenerate observables. On the other hand, our first main result that classicality implies incoherence can break down. The reason for this is that coherence could affect the system state in the degenerate subspaces with respect to R_S, whereas the probabilities $p(r_n, \ldots, r_1|\mathcal{C}_0)$ do not reveal any information about these subspaces.

Classical Markov processes

Our exposition of quantum stochastic and Markov processes would be quite incomplete without mentioning the definition of a classical Markov processes.

Classical Markov process. *We call a classical stochastic process described by probabilities $p(\mathbf{r}_n)$ on a set of times $\{t_0, t_1, \ldots, t_n\}$ Markovian if*

$$\boxed{p(r_n|\mathbf{r}_{n-1}) \equiv \frac{p(\mathbf{r}_n)}{p(\mathbf{r}_{n-1})} = p(r_n|r_{n-1})} \tag{1.94}$$

holds for all joint probabilities in the hierarchy, i.e. for all subsets of times $T \subset \{t_0, t_1, \ldots, t_n\}$.

Equation (1.94) encodes the fact that the conditional probability $p(r_n|\mathbf{r}_{n-1})$ of a Markov process depends only on the *last* measurement result r_{n-1}, but not on previous ones $r_{n-2}, \ldots, r_0$. It is important to note that we assumed in our definition that the probabilities $p(\mathbf{r}_n)$ actually describe a *classical* stochastic process, i.e. we assumed that the Kolmogorov consistency condition (1.53) holds. The description of processes satisfying only eqn (1.94) but *not* the consistency condition is more involved and differs significantly from the standard theory of classical Markov processes. An example of such a process is an isolated, unitarily evolving system interrupted by rank-1 projective measurements. This process satisfies eqn (1.94), and it is also a *quantum Markov* process (recall Exercise 1.23), but the probabilities $p(\mathbf{r}_n)$ in general violate the consistency condition because the projective measurements disturb the dynamics.

The connection between a quantum Markov process and a classical Markov process is indeed subtle. At least, however, a statement analogous to eqn (1.79) holds also for classical Markov processes. Namely, using the definition of conditional probabilities, we can always write the joint probability $p(\mathbf{r}_n)$ as

$$\begin{aligned} p(\mathbf{r}_n) &= \frac{p(\mathbf{r}_n)}{p(\mathbf{r}_{n-1})}\frac{p(\mathbf{r}_{n-1})}{p(\mathbf{r}_{n-2})}\cdots\frac{p(\mathbf{r}_1)}{p(r_0)}p(r_0) \\ &= p(r_n|\mathbf{r}_{n-1})p(r_{n-1}|\mathbf{r}_{n-2})\ldots p(r_1|r_0)p(r_0). \end{aligned} \tag{1.95}$$

Now, for a classical Markov process we can simplify that to

$$p(\mathbf{r}_n) = p(r_n|r_{n-1})p(r_{n-1}|r_{n-2})\dots p(r_1|r_0)p(r_0), \tag{1.96}$$

where the conditional probabilities $p(r_\ell|r_k)$ play an analogous role to the dynamical maps $\mathcal{E}_{\ell,k}$: they also propagate the system state, now described by a vector of probabilities $p(r_k)$, forwards in time. In the next chapter we will often use these conditional probabilities $p(r_\ell|r_k)$ and call them *transition matrices.*

Now, the third and last main result of this section establishes the connection among quantum stochastic processes, classical stochastic processes and the Markov property (1.94).

Classical from quantum Markovianity. *Consider a quantum stochastic process that yields for a fixed set of interventions the probabilities $p(\mathbf{r}_n)$. It is true that:*

1. *If the quantum stochastic process is Markovian and if all interventions are causal breaks, then the probabilities $p(\mathbf{r}_n)$ satisfy the Markov property (1.94).*
2. *If we add to the assumptions of point (1) that the probabilities $p(\mathbf{r}_n)$ also satisfy the Kolmogorov consistency condition, then these probabilities describe a classical Markov process.*

If these statements are clear and understandable, you have successfully mastered the entire first chapter of this book. If there is some scepticism left, it helps to try to prove them.

Exercise 1.29 Prove the statement above. *Hint:* For the first part it is helpful to note the following equivalent statement: *If for some set of interventions, which are causal breaks, the probabilities $p(\mathbf{r}_n)$ do not satisfy the Markov property (1.94), then the quantum stochastic process is non-Markovian.*

Further reading

1.1. The first section should be part of the standard quantum mechanics curriculum.

1.2. Introducing open quantum systems by partitioning an isolated 'universe' into a system and an environment is also fairly standard. Indeed, this idea is already inherent to thermodynamics itself and, therefore, it was used long before the advent of microscopic system–bath theories. Which particular assumption and gauge is implied by splitting the Hilbert space of the universe into a *tensor product* of a system and a bath Hilbert space is discussed by Stokes, A. and Nazir, A. (2022).

1.3. In our discussion about equilibrium states of open quantum systems we started from the conventional Gibbs ensemble. We did not address, however, the question of how this ensemble can be justified for an isolated quantum system. Readers interested in this non-trivial question can find a large body of recent reviews on this topic (Gemmer *et al.*, 2004; Borgonovi *et al.*, 2016; D'Alessio *et al.*, 2016; Gogolin and Eisert, 2016; Goold *et al.*, 2016; Deutsch, 2018; Mori *et al.*, 2018). Furthermore, readers interested in a proof of Poincaré's recurrence theorem can look, for example, at the work of Wallace (2015). Calculating the exact Poincaré recurrence time for a given model is very hard, but estimates can be found at various places (Reimann, 2008; Venuti, 2015). The Hamiltonian of mean force was first discussed almost 100 years ago by Onsager, Kirkwood and others, and the concept is widely used in chemical physics (Roux and Simonson, 1999), but it gathered only recently attention in quantum thermodynamics (Trushechkin *et al.*, 2022). The Caldeira–Leggett Hamiltonian, which we introduced in Exercise 1.2 and which we will meet again from time to time, became a standard model to describe the dynamics of open quantum systems; for in-depth treatments see, for example, the books of Nitzan (2006) and Weiss (2008).

1.4. Section 1.4 reviewed well-known results covered in greater detail elsewhere (Nielsen and Chuang, 2000; Holevo, 2001; Wiseman and Milburn, 2010; Jacobs, 2014).

1.5. The content of Section 1.5 is reviewed at many places in greater detail (Nielsen and Chuang, 2000; Holevo, 2001; Wiseman and Milburn, 2010; Jacobs, 2014). An early influential book was written by Kraus (1983). For this reason, the operators K_α appearing in eqn (1.45) are sometimes called *Kraus operators* and the entire operation is sometimes called a *Kraus map.* We did not follow this terminology here as those maps and operators have been studied already earlier. In general, the historical development of the field is quite intertwined, involving many mathematicians and physicists (re)discovering the same results from different perspectives. The proof of the operator–sum representation can be found, for example, in the book of Nielsen and Chuang (2000) (Theorem 8.1 therein). This proof makes use of the Choi–Jamiołkowski isomorphism, an elegant mathematical construction reviewed also in Appendix B. For a proof of the unitary dilation theorem see Nielsen and Chuang (2000) or Holevo (2001).

1.6. The field of classical stochastic processes is covered in many textbooks, but only a very few prove the extension theorem mentioned in Section 1.6; see, for example, the original work of Kolmogorov (2018). A standard book covering many aspects of classical causal modelling was written by Pearl (2009).

1.7. In contrast to the previous sections, the treatment of quantum stochastic processes as done in Section 1.7 has not yet become standard material, although the basic idea is quite old (Lindblad, 1979). However, it is only recently that this approach regained independent attention from many different directions; see Milz and Modi (2021) for a detailed introduction citing many references. The generalized extension theorem for quantum stochastic processes is discussed by Milz *et al.* (2020*b*). The generalized Bloch representation (1.66) is due to Byrd and Khaneja (2003).

1.8. The definition of a quantum Markov process has caused some debate; see Li *et al.* (2018) for the latest review of many different perspectives. Our definition coincides with the one of Pollock *et al.* (2018a). It has the advantage that it reduces to the conventional Markov definition for a classical stochastic process in its respective limit. In fact, for classical systems it coincides with the *causal Markov condition* of classical causal models (Pearl, 2009).

1.9. In the first part of the last section we followed Strasberg and Díaz (2019) and Milz *et al.* (2020*a*). Putting multi-time statistics aside, the influence of initial system–bath correlations as quantified by quantum discord on the definition of a dynamical map (Exercise 1.25) was investigated by Rodríguez-Rosario *et al.* (2008).

2
Classical Stochastic Thermodynamics

Summary. After giving an introduction to the phenomenological theory of non-equilibrium thermodynamics, it is shown how this theory can be derived and extended for small systems described by a classical Markov process. Thermodynamic definitions for internal energy, heat, work, entropy and entropy production are provided along a single stochastic trajectory. It is shown that the fluctuations in work and entropy production satisfy universal constraints, which are known as fluctuation theorems. By providing an independent derivation of them starting from microscopically reversible Hamiltonian dynamics in the full system–bath phase space, it is demonstrated that fluctuation theorems also hold in the non-Markovian regime. The theoretical framework established here is called (classical) stochastic thermodynamics and it has found widespread applications in biology and biochemistry, in soft condensed matter physics as well as in the study of various artificial nanostructures down to the regime where quantum effects start to matter. The chapter finishes with a discussion of the particularly relevant setting of single-molecule-pulling experiments.

2.1 Phenomenological Non-Equilibrium Thermodynamics

The theory of thermodynamics arose out of the desire to understand macroscopic transformations of matter in chemistry and engineering in the 19th century. It is an independent physical theory, *a priori* not relying on the validity of classical or quantum mechanics. In fact, its most important message is perhaps that there are universal laws in nature that are independent of the microscopic details. Nevertheless, within the desire to reduce all known laws in physics to more 'fundamental' laws, a central quest of statistical mechanics is to derive the laws of thermodynamics from classical or quantum mechanics. This is a non-trivial and not yet completed endeavour. In order to actually know what we want to derive, it seems advisable to briefly review the theory of thermodynamics.

Back in the 19th century the systems under investigation were macroscopic and the enormous simplifications arising in thermodynamics are due to the observation that the state of such systems can be well described by a very few variables. Typical examples of these variables include the (absolute) temperature T, pressure p and volume V, but different or additional variables (chemical potentials, polarization, magnetization, shear stress, etc.) can also appear. These macroscopic bodies are then allowed to exchange heat Q with their surroundings and external work W can be supplied to (or

Quantum Stochastic Thermodynamics. Philipp Strasberg, Oxford University Press.
 DOI: 10.1093/oso/9780192895585.003.0002

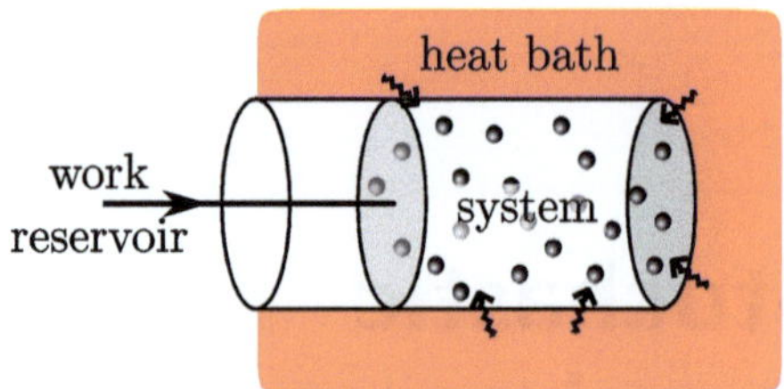

Fig. 2.1 Thermodynamic set-up where the system is a gas in a container. By pushing a piston, the thermodynamic variables (such as T, p or V) can be changed in a mechanically controlled way, which is abstracted as the action of a 'work reservoir'. Furthermore, through the walls of the container the gas is in simultaneous contact with a heat bath, with which it can exchange energy. This exchange of energy is accompanied by an exchange of entropy, which is the defining property to call this energy exchange 'heat'.

extracted from) them. A prototypical example of a thermodynamic set-up is sketched in Fig. 2.1.

To such a macroscopic system one attributes a *state function* U_S, which depends only on the state variables (such as T, p or V) and which is called the *internal energy*. The **first law of thermodynamics** then asserts that the change ΔU_S in internal energy of the system is balanced by heat and work:

$$\Delta U_S = Q + W. \tag{2.1}$$

Note that we define heat and work to be positive whenever they increase the internal energy of the system. The first law links the three quantities ΔU_S, Q and W, but it does not tell us how to compute or measure them. Historically, determining the work W was the simplest problem as it can be controlled in a mechanical way and it is typically given by the integral $\int pdV$. Thus, if we know the pressure of the gas in the container sketched in Fig. 2.1, which can also be expressed as a function $p = p(T, V)$, we can compute the work supplied since the change in volume is easily measurable and controllable. The correct determination of heat Q is more subtle and requires the knowledge of the *heat capacity* C of the system, which allows us to compute $Q = \int CdT$. In practice, these measurements can be involved as the heat capacity C can depend in a complicated way on the state variables describing the system. Here, the *ideal gas* played an important historical role as a reference system because the relations between the state variables T, p and V as well as the heat capacity and internal energy were well known (we do not go into further details here as these can be looked up in any standard textbook). Then, given enough patience, it was possible to determine the change in internal energy ΔU_S as a function of the initial and final values of the state variables. Alternatively, of course, given that we know the internal energy of a system (such as the internal energy of an ideal gas), we can also use the first law to compute the heat flow (provided we know W). Moreover, the heat flowing into the system equals the heat flowing out of the heat bath, which offers an alternative way to infer its value given that one knows the heat capacity of the bath and its temperature.

Furthermore, an important simplification arises by assuming that the internal energy is an **extensive quantity**, which is justified whenever bulk (or volume) properties dominate the surface (or boundary) properties of the substance. For many (but not all) macroscopic systems, this is the case. Then, extensivity means that, when expressing the internal energy as a function of the volume $U_S = U_S(V)$, we have

$$U_S(V_1 + V_2) = U_S(V_1) + U_S(V_2) \tag{2.2}$$

if the temperature and pressure are held fixed. Thus, if we succeeded in determining the internal energy of 1 litre of water for some temperature and pressure, we know its internal energy for any other volume at that temperature and pressure as well.

As we will see later, the first law follows from energy conservation applied to the system, the heat bath and the work reservoir. However, the fundamental distinction between heat and work only becomes transparent by considering the **second law of thermodynamics**. The second law in its most general form asserts that the entropy of the universe tends to a maximum. In equations, for any process,

$$\Delta S_{\text{univ}} \geq 0, \tag{2.3}$$

where S_{univ} denotes the *thermodynamic* entropy of the universe, which should be clearly distinguished from any information-theoretic notion of entropy at this point. The change in entropy of the universe is called the **entropy production**, denoted by $\Sigma \equiv \Delta S_{\text{univ}}$. If $\Sigma = 0$, the process is called **reversible**; otherwise it is **irreversible**. We remark that the terminology 'universe' does not necessarily refer to the cosmological universe as a whole, but to any part of the world which can be approximated as sufficiently isolated from its surroundings during the experiment.

For a system–bath set-up, for example, as sketched in Fig. 2.1, the entropy of the universe can be typically split into the entropy of the system and that of the environment: $S_{\text{univ}} = S_S + S_{\text{env}}$. This additive splitting assumes that the entropy is dominated by bulk and not by surface properties. Then, the second law becomes

$$\Sigma = \Delta S_S + \Delta S_{\text{env}} \geq 0. \tag{2.4}$$

For our setting in Fig. 2.1 the environment is composed of a work source and a heat bath. Now, the defining property of a work source is that its entropy does not change although it exchanges energy with a system. Hence, ΔS_{env} reduces to the change in entropy of the heat bath, which is typically assumed to be well described by an equilibrium state at a (in general varying) temperature T_B, which can, in principle, be different from the system's temperature T. Its change in entropy is then equated to be $\Delta S_{\text{env}} = -\int đQ/T_B$. Here, $đQ$ denotes an infinitesimal heat flow into the system satisfying $Q = \int đQ$. Then, the second law takes on the traditional form

$$\Sigma = \Delta S_S - \int \frac{đQ}{T_B} \geq 0. \tag{2.5}$$

Equation (2.5) is also known as **Clausius's inequality**. Finally, if the bath gets only slightly perturbed away from its initial temperature, here denoted by $T_B(0)$, then eqn (2.5) reduces to

$$\Sigma = \Delta S_S - \frac{Q}{T_B(0)} \geq 0. \tag{2.6}$$

The first law together with the second law, both of which can be extended to the presence of multiple heat baths and situations involving particle exchanges (we here considered only energy exchanges), constitute the basic building blocks of phenomenological non-equilibrium thermodynamics. They dictate the efficiency of any heat engine, refrigerator, heat pump or, more generally, any thermodynamic process. Namely, *the larger the entropy production* Σ, *the smaller the efficiency of a heat engine* (or any other thermodynamic process). The following exercise shows this explicitly.

Exercise 2.1 The first law for a system in contact with a work source and a hot and a cold heat bath reads $\Delta U_S = W + Q_H + Q_C$, where $Q_{H(C)}$ is the heat flow from the hot (cold) bath. Assuming that the baths are described throughout the process by constant temperatures $T_H > T_C$, the second law (2.6) generalizes to $\Sigma = \Delta S_S - Q_H/T_H - Q_C/T_C \geq 0$. Now, assume that we want to use this set-up as a heat engine, i.e. we want to extract work from it: $W < 0$. We further consider a *cyclically* working heat engine, which has eventually reached a steady state characterized by $\Delta U_S = 0$ and $\Delta S_S = 0$ per cycle. First, show in this case that $W < 0$ implies $Q_H > 0$. Hence, the following number is positive and called the heat engine's **efficiency** per cycle:

$$\eta \equiv \frac{-W}{Q_H}. \tag{2.7}$$

Next, use $\Delta U_S = 0$ and $\Delta S_S = 0$ and the first and second laws of thermodynamics to show that the following relations hold:

$$\eta = 1 - \frac{T_C}{T_H} - \frac{T_C \Sigma}{Q_H} \leq 1 - \frac{T_C}{T_H} \equiv \eta_C. \tag{2.8}$$

Here, η_C denotes the **Carnot efficiency**, which is the maximum efficiency of any engine working between two heat baths with fixed temperatures. Thus, any excess in the entropy production Σ diminishes the efficiency of the engine since $\eta_C - \eta = T_C\Sigma/Q_H \geq 0$.

In contrast to the determination of internal energy changes ΔU_S, measuring the change in entropy ΔS_S is quite subtle. Clearly, for a reversible process we can set

$$\Delta S_S = \int \frac{đQ}{T_B} = \int \frac{đQ}{T}, \tag{2.9}$$

where we used that a reversible process is characterized by the fact that the system and bath temperature are equal at all times. For an irreversible process, where the state variables change from, say, (T_0, p_0, V_0) to (T_1, p_1, V_1), Clausius suggested computing $\Delta S_S = S_S(T_1, p_1, V_1) - S_S(T_0, p_0, V_0)$ by using another process, which *reversibly* transforms (T_1, p_1, V_1) back to (T_0, p_0, V_0), such that

$$-\Delta S_S = S_S(T_0, p_0, V_0) - S_S(T_1, p_1, V_1) = \int_r \frac{đQ_r}{T}. \tag{2.10}$$

Here, the minus sign appears because we transform the system *back* to its initial state and the subscript '*r*' shall remind us that this is done reversibly. Note that the heat flows in the reversible and irreversible processes are different in general: $Q_r \neq Q$.

We have now reached the point where it is good to pause a moment and reflect on what we have just said. On the one hand, we found out that Clausius and his contemporaries were well aware of the fact that thermodynamics is *not* restricted to reversible equilibrium processes (*contrary* to some frequently made statements). On the other hand, Clausius assumes that state variables such as temperature T, pressure p and volume V remain well defined even out of equilibrium and sufficiently describe the state of the system.

It is the last point that appears questionable in general. In particular, it is not clear whether a process as assumed in eqn (2.10) actually always exists. The question of what are the correct state variables to describe non-equilibrium systems and how should a thermodynamic entropy be defined for such systems has thus become a major research problem. If the system is still macroscopic, then the *local equilibrium assumption* is a useful approach, i.e. the idea that temperature and pressure can still be well defined locally on smaller scales (despite a missing *single* global temperature and pressure for the entire system). Unfortunately, in this book we are interested in systems that are so small that it might not even be possible to divide them into local regions. In this regime, another fruitful approach is based on the assumption that the thermodynamic forces applied to the system are very small (e.g. small temperature gradients or slow variations of other external parameters such as the volume). This is the *linear response regime*, where a variety of theoretical tools is known for a long time. But unfortunately again, in this book we are interested in small systems, which can be driven *far* from equilibrium by applying large thermodynamic forces to them.

Thus, our goal is to find a thermodynamic framework that is valid beyond commonly used assumptions in thermodynamics and statistical physics. We want to find the 'state variables' describing such small far-from-equilibrium systems and definitions for internal energy, heat, work and entropy, which give rise to the same laws of thermodynamics as introduced above and in unison with previously known results. This non-trivial endeavour is clearly not yet completed, but we have reached a point where it seems important to summarize the significant and definite achievements made so far.

We end this exposition on phenomenological non-equilibrium thermodynamics by reminding the reader that there is also a *zeroth law of thermodynamics* and a *third law of thermodynamics*, albeit that—compared with the first and second laws—both play a minor role in applications. The zeroth law of thermodynamics has implications for the foundations of thermodynamics, in particular for the equilibrium concept of temperature, and we will discuss its role (and its violation) in small systems in Section 3.6. The third law of thermodynamics in its traditional form poses restictions on the behaviour of thermodynamic systems in the low-temperature (i.e. deep quantum) regime. We will discuss this in Section 4.8.

2.2 From Equilibrium Entropies to Landauer's Principle

As argued above, finding a definition of thermodynamic entropy out of equilibrium is challenging based on purely phenomenological considerations. The goal of this section is to make plausible the idea that Shannon entropy is a good candidate for thermodynamic entropy even out of equilibrium *provided* that certain conditions are met.

To approach the problem, we start by briefly reviewing two entropy concepts from statistical mechanics, which work well at equilibrium. Then, after explaining the idea that the notion of equilibrium is always *relative* with respect to our time scales of perception, we study the example of an idealized computer memory to explain how and when Shannon entropy is related to thermodynamic entropy. This identification is the essence of Landauer's principle, which we use to motivate subsequent definitions for thermodynamic entropy of a small *open* (classical or quantum) system. A general discussion about non-equilibrium entropies is given in Section 3.8.

Equilibrium entropies

There are two standard candidates to compute the entropy of a system at equilibrium in statistical mechanics: the Boltzmann and the Gibbs–Shannon–von Neumann entropies. To introduce them, we use here a quantum mechanical notation for simplicity; their classical counterpart is conceptually analogous.

We start with Boltzmann's entropy, which is based on a counting argument. Let X denote the collection of relevant variables or observables describing the state of a thermodynamic system. For instance, X could include the energy, the particle number or the polarization and, in case that inhomogeneity plays a role, these variables could depend on the region (e.g. the number of gas molecules in the *left* half of a container). Then, let $V(X)$ denote the number of microstates compatible with the constraint of having some fixed value for X. Classically, $V(X)$ can be defined by the *volume* of microscopic states in phase space (described by the collection of positions and momenta of all particles) giving rise to the macrostate X (which should not to be confused with the geometric volume appearing in Section 2.1). Quantum mechanically, $V(X)$ is defined by the dimension of a subspace of the total Hilbert space, which contains all wave functions giving rise to the value X. Then, **Boltzmann entropy** is defined as

$$S_B(X) \equiv k_B \ln V(X), \tag{2.11}$$

where k_B is Boltzmann's constant.

Writing down the precise definition of Boltzmann's entropy for a general X is not easy. Thus, we focus on a simple but relevant example. Consider a quantum system with Hamiltonian $H = \sum_k \epsilon_k |\epsilon_k\rangle\langle\epsilon_k|$ and let $X = E$ denote its *coarse-grained* energy, i.e. its energy known up to some uncertainty δ. The corresponding projector is

$$\Pi(E) \equiv \sum_{\epsilon_k \in [E, E+\delta)} |\epsilon_k\rangle\langle\epsilon_k| \tag{2.12}$$

and the related volume term, i.e. the number of microstates in the energy shell $[E, E+\delta)$, is $V(E) = \text{tr}\{\Pi_E\}$. In this case, Boltzmann's entropy reduces to

$$S_B(E) = k_B \ln V(E). \tag{2.13}$$

We emphasize that Boltzmann's entropy is a measure about the *available* phase or Hilbert space for a given constraint, which associates the *same value* with all states compatible with that constraint. In particular, also a pure state, say $\rho = |\epsilon_k\rangle\langle\epsilon_k|$ with $\epsilon_k \in [E, E+\delta)$, has a (non-zero) Boltzmann entropy equal to eqn (2.13).

In contrast, the second entropy concept requires the notion of an *ensemble* of states. Sticking to the example of an isolated system described by its coarse-grained energy, the relevant equilibrium ensemble is the microcanonical ensemble

$$\omega(E) \equiv \frac{\Pi(E)}{V(E)}. \tag{2.14}$$

It is built on the postulate of equal *a priori* probabilities, i.e. all microstates compatible with the constraint E are regarded as equally likely. The entropy of that state can then be computed by using the **Gibbs–Shannon–von Neumann entropy**

$$k_B S_{\rm vN}[\omega(E)] = -k_B \text{tr}\{\omega(E) \ln \omega(E)\} = k_B \ln V(E), \tag{2.15}$$

which agrees with Boltzmann's entropy. Classically, the trace is replaced by an integral over phase space. For brevity, we typically use below the terminology von Neumann *or* Shannon entropy if it is clear that we are referring to a quantum or classical system.

Related to the microcanonical ensemble is the canonical ensemble $\pi(\beta) = e^{-\beta H}/\mathcal{Z}$ at inverse temperature $\beta = (k_B T)^{-1}$, which we have already met in Chapter 1. For typical macroscopic systems, one finds

$$k_B S_{\rm vN}[\pi(\beta)] \approx S_B(E) \tag{2.16}$$

if one fixes the temperature of the canonical ensemble by the definition

$$\frac{1}{T} \equiv \frac{\partial S_B(E)}{\partial E} \tag{2.17}$$

and if the *equivalence of ensembles* applies. The equivalence of ensembles is an important result in equilibrium statistical mechanics; it asserts that $\omega(E) \approx \pi(\beta)$ for macroscopic systems if β is fixed via eqn (2.17) and if the Boltzmann entropy $S_B(E)$ is a concave function of energy. We do not prove this statement here. Moreover, to avoid ambiguities and controversies, we assume throughout this book that the heat bath in our description satisfies the equivalence of ensembles.

We remark that the Gibbs–Shannon–von Neumann entropy can be defined for any ensemble, in particular also for a pure state. Then, however, we always have $S_{\rm vN}(|\psi\rangle\langle\psi|) = 0$ in strong contrast to Boltzmann's entropy. In general, the Gibbs–Shannon–von Neumann entropy therefore reflects well thermodynamic entropy only in conjuction with the use of an appropriate statistical ensemble. In the following, our goal is to show that Shannon entropy also captures important aspects of thermodynamic entropy for ensembles that describe an open system coupled to an ideal heat bath, even if the open system is *out of equilibrium*. A more general discussion of out-of-equilibrium entropies is relegated to Section 3.8.

Before we proceed, we emphasize an important point to counteract sceptical voices claiming that thermodynamic entropy is only defined at equilibrium. This argument can be reduced *ad absurdum* by pointing out that, strictly speaking, there exist no equilibrium states in nature. All equilibrium states are always *constrained* equilibrium states. To illustrate this point, consider a cup of hot coffee, whose properties are well described by an equilibrium ensemble with respect to some (hot) temperature.

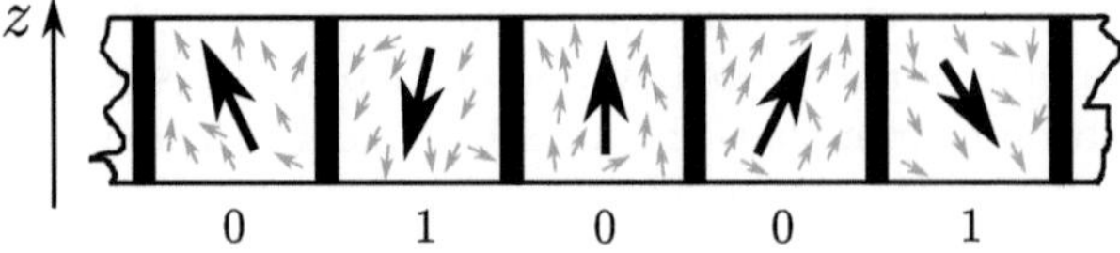

Fig. 2.2 A magnetic memory used to encode a string of bits can be thought of as a tape divided into non-interacting domains consisting of many single spins (small grey arrows). Collectively they give rise to some net magnetization (thick black arrow). Depending on some rule, for example whether the net magnetization is pointing up or down in the z-direction, we say that the domain encodes a bit in state '0' or '1'.

However, this cup of coffee stands on a table in some room whose temperature is much colder than that of the coffee. Strictly speaking we therefore have to deal with a *non-equilibrium* situation. The reason why equilibrium statistical mechanics applies well to this problem is that our time scales of perception are very slow compared with the individual motions of the coffee molecules (and, hence, are incapable of sensing non-equilibrium deviations in the actual state of the coffee), but our perception is fine enough to sense the hotness of the coffee when drinking it (instead of a very coarse perception, which would only sense the average coffee and room temperature). Thus, equilibrium is always defined *relative* to certain time scales. The observation that many processes in nature are characterized by vastly different time scales is known as *time-scale separation*, which plays an important role in the following. Thus, we should not be surprised to see that the realm in which we define meaningful thermodynamic entropies grows together with our increased nanotechnological abilities to resolve shorter time scales.

The symmetric memory and Landauer's principle

To illustrate the use of Shannon entropy as non-equilibrium thermodynamic entropy, we consider an example inspired by the thermodynamics of computation. Imagine a memory where we use small magnetic domains to encode information on it. Each domain consists of many single spins, which collectively add up to some net magnetization $\mathbf{m} \in \mathbb{R}^3$. Depending on some rule, we encode one bit of information on each domain: say, if $m_z \equiv \mathbf{m} \cdot \mathbf{e}_z \geq 0$, we give it value '0', and '1' otherwise (here, $\mathbf{e}_z$ denotes the unit vector in the z-direction). A sketch is shown in Fig. 2.2. We assume that without further perturbation the values of 0 and 1 are relatively stable and have a long lifetime (as one would wish for a *memory*) relative to the dynamics of a single spin in each domain, which is prone to thermal fluctuations. Those thermal fluctuations arise by assuming that the entire memory is in contact with a heat bath at temperature T. This encodes the idea of time-scale separation: the state of the collective degree of freedom of the bit is long-lived and observable compared with the state of a single microscopic spin.

Mathematically, we denote by $x \in \{0, 1\}$ the value of one bit and by i_x a particular microscopic configuration of all the single spins, which gives rise to the collective state x. We call x a *mesostate* and i_x a *microstate*. Using time-scale separation and the fact that the memory is in contact with a heat bath, we approximate the conditional probability $p(i_x|x)$ to find some microstate i_x given a mesostate x to be in thermal equilibrium. Thus, we denote by $p(i_x|x) = \pi(i_x|x)$ an equilibrium state *constrained* to the value x (we leave the exact microscopic spin Hamiltonian unspecified as it does not play any role in our argument). Hence, given x, we can apply equilibrium statistical mechanics to compute the entropy of a single domain, which we denote by $\mathcal{S}_x \equiv k_B S_{\text{Sh}}[\pi(i_x|x)]$ and for simplicity we set $\mathcal{S}_0 = \mathcal{S}_1 \equiv \mathcal{S}_{\text{mic}}$. In fact, the assumption $\mathcal{S}_0 = \mathcal{S}_1$ is justified by symmetry if we assume that the energy of the system is symmetric under flipping the sign of magnetization in the z-direction: $E(m_z) = E(-m_z)$.

Suppose that we have an ensemble of $N \gg 1$ non-interacting bits, assumed to be all in state 0. The entropy of a single system, i.e. the entropy per bit, is

$$S_S(p_x = \delta_{x,0}) = \mathcal{S}_{\text{mic}}, \tag{2.18}$$

where $p_x = \delta_{x,0}$ describes the probability distribution of the bit mesostate. Even though we said that the mesostates x are relatively stable, after a very long time they will also eventually thermalize (unless we additionally protect them). Assuming $E(m_z) = E(-m_z)$, the thermal distribution for them is $\pi_x = 1/2$. Thus, the entropy of the *global* equilibrium state per bit is

$$\begin{aligned} S_S(\pi_x) &= k_B S_{\text{Sh}}[\pi(i_x|x)\pi_x] = -k_B \sum_x \sum_{i_x} \pi(i_x|x)\pi_x \ln[\pi(i_x|x)\pi_x] \\ &= k_B S_{\text{Sh}}(\pi_x) + k_B \sum_x \pi_x S_{\text{Sh}}[\pi(i_x|x)] = k_B S_{\text{Sh}}(\pi_x) + \mathcal{S}_{\text{mic}}, \end{aligned} \tag{2.19}$$

where we used $k_B S_{\text{Sh}}[\pi(i_x|x)] = \mathcal{S}_{\text{mic}}$ and $\sum_x \pi_x = 1$ at the end. The change in equilibrium thermodynamic entropy when going from the constrained equilibrium state of N bits in state 0 to the global equilibrium state is therefore per bit

$$\Delta S_S = S_S(\pi_x) - S_S(p_x = \delta_{x,0}) = k_B S_{\text{Sh}}(\pi_x) = k_B \ln 2 > 0, \tag{2.20}$$

where we used $\pi_x = 1/2$. As expected, the process of full equilibration increases the thermodynamic entropy of the universe in accordance with the second law of thermodynamics. Interestingly, for the present process we can express the change in thermodynamic entropy solely in terms of the Shannon entropy of the mesostates x: $\Delta S_S = k_B\{S_{\text{Sh}}(\pi_x) - S_{\text{Sh}}[p_x(0)]\}$, where $p_x(0) = \delta_{x,0}$ denotes the initial distribution of the mesostates. Note that, despite saying that the states x are relatively stable, the initial distribution is, nevertheless, globally seen a *non-equilibrium* state. Is it possible to generalize this argument further to arbitrary initial distributions $p_x(0)$?

Suppose we have a particular initial state with a fraction of $p_0 = N_0/N$ bits in state 0 and $p_1 = N_1/N = 1 - p_0$ bits in state 1. In addition, we assume for the moment that we know exactly the state of each single bit, for instance, by knowing that we have a string of bits in the special state

$$(\underbrace{0, 0, \dots, 0}_{N_0}, \underbrace{1, 1, \dots, 1}_{N_1}). \tag{2.21}$$

We also assume again that the bits are non-interacting (for instance, by assuming that they are separated by thick domain walls, as indicated in Fig. 2.2). Since the value of each bit is known, the joint entropy is additive and simply $N\mathcal{S}_{\text{mic}}$, i.e. the entropy per bit equals $\mathcal{S}_{\text{mic}}$ as in eqn (2.18).

Next, assume that we know only p_0 and p_1, and that we know *nothing more* about the string of bits. Such a disordered string of bits, in which each sequence with N_0 bits in state 0 and N_1 bits in state 1 is equally probable, can be obtained from randomly *mixing* a string of bits such as the one in eqn (2.21). Since each bit is characterized by the same equilibrium properties and does not interact with the others, bits with state 0 or state 1 behave like two distinguishable ideal gases at the same temperature and pressure. But the entropy change due to mixing two ideal gases labelled '0' and '1', each initially occupying a volume V_0 and V_1, is well known and becomes $\Delta S_{\text{mix}} = -Nk_B(p_0 \ln p_0 + p_1 \ln p_1)$, where $N = N_0 + N_1$ is the total number of gas molecules and we identified $p_0 = V_0/V = N_0/N$ and $p_1 = V_1/V = N_1/N$. Thus, the entropy per bit of an ensemble of bits defined by a probability distribution p_x equals the entropy $\mathcal{S}_{\text{mic}}$ associated with the constrained equilibrium state of each bit *plus* the mixing entropy:

$$S_S(p_x) = k_B S_{\text{Sh}}(p_x) + \mathcal{S}_{\text{mic}}. \tag{2.22}$$

Then, given an initial state $p_x(0)$, the change in thermodynamic entropy associated with an equilibration process $p_x(0) \to \pi_x$ as investigated above becomes

$$\Delta S_S = k_B\{S_{\text{Sh}}(\pi_x) - S_{\text{Sh}}[p_x(0)]\} \geq 0. \tag{2.23}$$

This result is now valid for any non-equilibrium distribution $p_x(0)$. Thus, at least if we keep in mind the assumption that the microscopic configuration of the single spins is always in a conditional equilibrium state $\pi(i_x|x)$, whose entropy $\mathcal{S}_{\text{mic}}$ is independent of x, it is justified to equate $k_B S_{\text{Sh}}(p_x)$ with the thermodynamic entropy of the bits.

An interesting consequence of the aforementioned treatment appears if we consider the reverse process of equilibration by resetting the state π_x back to some fixed reference state p_x. This process cannot happen spontaneously as it decreases the entropy of the universe. It therefore needs some energy investment from the outside in the form of work. To elucidate this, remember that the second law (2.6) stipulates that

$$\Sigma = k_B[S_{\text{Sh}}(p_x) - S_{\text{Sh}}(\pi_x)] - \frac{Q}{T} \geq 0. \tag{2.24}$$

To proceed, we have to specify the internal energy of the system, which is characterized by some intrinsic energy $\mathcal{U}_x$ for each bit x, which we could in principle compute using the conditional equilibrium state $\pi(i_x|x)$ and some microscopic Hamiltonian. However, as for the entropy we will assume only that, owing to symmetry, $\mathcal{U}_x = \mathcal{U}_{\text{mic}}$ for all x and leave $\mathcal{U}_{\text{mic}}$ unspecified. This implies for the change in internal energy that

$$\Delta U_S = \sum_x p_x U_x - \sum_x \pi_x \mathcal{U}_x = \mathcal{U}_{\text{mic}} \sum_x p_x - \mathcal{U}_{\text{mic}} \sum_x \pi_x = 0. \tag{2.25}$$

Thus, the second law becomes, after taking into account the first law $0 = \Delta U_S = W + Q$,

$$\Sigma = k_B[S_{\text{Sh}}(p_x) - S_{\text{Sh}}(\pi_x)] + \frac{W}{T} \geq 0. \tag{2.26}$$

If we consider the special situation where we reset the bits to the state 0 described by the distribution $p_x = \delta_{x,0}$, and if we use $\pi_x = 1/2$, we obtain after rearrangement

$$\boxed{W \geq k_B T \ln 2.} \tag{2.27}$$

This equation describes the minimum work cost associated with the *erasure* of one bit of information, where 'erasure' means that we reset some unknown state to some definite and known state (here, 0). Equation (2.27) is often referred to as Landauer's principle.

In retrospect, one could object that Landauer's principle is a mere tautology. If one accepts the notion of Shannon entropy as non-equilibrium thermodynamic entropy, then *Landauer's principle is nothing but the second law (2.24) of non-equilibrium thermodynamics applied to a special situation.* Making any distinction between the second law and Landauer's principle hence seems misleading and inappropriate because the latter is a special case of the former. One could therefore reformulate Landauer's principle by realizing that Landauer's basic insight actually was to find the conditions under which Shannon entropy provides a legitimate candidate for thermodynamic entropy:

Landauer's principle. *Shannon entropy (times k_B) equals the non-equilibrium thermodynamic entropy for a system that is in contact with a fast thermal bath and whose states are characterized by equal intrinsic entropies.*

This is the main message of this section. We remark that a single domain in a magnetic memory is considerably smaller than the size of traditional thermodynamic systems from the 19th century, but it is still somewhat large compared with the size of a single molecule or a single atom, which is what we are interested in in this book.[1] However, as long as we can measure the system fast enough (time-scale separation) and as long as the system is coupled to a thermal bath, the above arguments can also be used for systems as small as a single electron. Note that the states of a single electron (spin up or down) are by construction characterized by equal intrinsic entropies because those states are already the microstates, i.e. it is not possible to partition them further.

The next exercise illustrates what changes when different intrinsic entropies come into play and it serves as a warning that one should not blindly use the Shannon entropy as thermodynamic entropy, even when time-scale separation applies.

Exercise 2.2 In our exposition above we have used that the two mesostates $x \in \{0,1\}$ have the same thermodynamic properties, i.e. the same intrinsic entropy $\mathcal{S}_x = \mathcal{S}_{\text{mic}}$ and internal energy $\mathcal{U}_x = \mathcal{U}_{\text{mic}}$. This is a symmetric memory. The task is now to generalize the above

[1]For comparison, according to Wikipedia's article on the 'Hard disk drive' (accessed on 20 April 2020) current magnetic storage systems reach a density of 1 Tbit per square inch, which means 10^{12} bits per approximately 6.5 cm^2 in ordinary units. Using a thickness of 20 nm of the magnetic material and boldly assuming the magnetic material to be iron with a density of 7.87 g/cm^3 at room temperature (one indeed uses a ferromagnetic material, but not a simple one as iron but a more sophisticated alloy), one arrives at the conclusion that 1 Tbit is stored on a mass of 10^{-4} g or one bit on 10^{-16} g. While this number is incredibly small, the number of iron atoms per bit is nevertheless approximately *one million*, which gives an estimate for the number of spins involved in the data storage.

expressions to the case where $\mathcal{S}_0 \neq \mathcal{S}_1$ and $\mathcal{U}_0 \neq \mathcal{U}_1$. Show that the second law for an equilibration process (no external work supplied) starting with some p_x and ending with π_x becomes

$$\Sigma = k_B S_{\rm Sh}(\pi_x) - k_B S_{\rm Sh}(p_x) + \sum_x (\pi_x - p_x)\mathcal{S}_x - \frac{1}{T}\sum_x \mathcal{U}_x(\pi_x - p_x) \geq 0. \tag{2.28}$$

Note that for an asymmetric memory we have in general $\pi_0 \neq \pi_1$. In fact, it does not even need to be the case that $S_{\rm Sh}(\pi_x) \geq S_{\rm Sh}(p_x)$, i.e. the Shannon entropy of the bits can decrease provided that the microstates take up enough entropy. For instance, set $\pi_0 = 3/4$ and $p_0 = 1/2$, which implies $S_{\rm Sh}(\pi_x) < S_{\rm Sh}(p_x)$. Find explicit numerical values for $\mathcal{U}_x$ and $\mathcal{S}_x$ such that $\Sigma \geq 0$ for this choice, i.e. the process $p_x \to \pi_x$ can happen spontaneously.

2.3 Classical Markov Processes and Local Detailed Balance

In this section, we consider the dynamical description of systems described by a finite set of discrete states $\{x\}$ in detail. These states could denote the value of some bits in a small memory (as in the previous section), the state of an enzyme in your body (where $x \in \{0, 1\}$ could denote whether the enzyme is bound to some substrate or not), the conformational states of a macromolecule (see Section 2.9) or the number of electrons on a quantum dot (see Section 3.10), among many other and often more complicated examples. For simplicity, we assume that the system obeys a Markovian classical stochastic process. Whereas the assumption of a classical stochastic process is well justified for the examples above, the question of whether the process is Markovian is more intricate and far from universally answered. For a couple of experimentally relevant situations the Markovian approximation is, however, well justified and we therefore use it here and return to the treatment of non-Markovian systems from Section 2.7 onwards.

Rate master equation

A classical Markov process is completely characterized by the conditional probabilities $p(x_\ell|x_k)$ to reach state x_ℓ at time t_ℓ given that the state was x_k at time t_k. In this context $p(x_\ell|x_k)$ is also called a *transition matrix* because, if we describe the states at time t_k (t_ℓ) by probability vectors $\boldsymbol{p}(t_k)$ $[\boldsymbol{p}(t_\ell)]$ with elements $p_{x_k}(t_k)$ $[p_{x_\ell}(t_\ell)]$, the state can be propagated forwards in time using ordinary matrix multiplication $\boldsymbol{p}(t_\ell) = T_{\ell,k}\boldsymbol{p}(t_k)$. Here, we introduced the matrix $T_{\ell,k}$ with elements $(T_{\ell,k})_{x_\ell,x_k} = p(x_\ell|x_k)$, which obeys $(T_{\ell,k})_{x_\ell,x_k} \geq 0$ for all x_ℓ and x_k and $\sum_{x_\ell}(T_{\ell,k})_{x_\ell,x_k} = 1$ for all x_k. Matrices with these properties are called **stochastic matrices**.

Even though the transition matrices are sufficient to describe the dynamics, in physics one is rather used to working with differential equations describing the *continuous* time evolution of a system. In order to derive such a differential equation, we need to assume that the Markov assumption holds for a sufficiently small time step δt and we write the time evolution starting at t as $\boldsymbol{p}(t+\delta t) = T_{t+\delta t,t}\boldsymbol{p}(t)$. Now, we expand the elements of the transition matrix to go from x' to x to first order in δt:

$$(T_{t+\delta t,t})_{x,x'} = \delta_{x,x'} + R_{x,x'}(t)\delta t + \mathcal{O}(\delta t^2). \tag{2.29}$$

Here, the lowest-order term $\delta_{x,x'}$ arises from the fact that the system does not change if $\delta t = 0$. The first-order term was summarized in the *rate matrix* $R_{x,x'}(t)$, which has the dimension of a rate (i.e. an inverse time). From the conservation of probability $\sum_x (T_{t+\delta t,t})_{x,x'} = 1$ we deduce the requirement that

$$\sum_x R_{x,x'}(t) = 0 \tag{2.30}$$

for all x' and t. Furthermore, since $(T_{t+\delta t,t})_{x,x'} \geq 0$ for all x, x', we infer that $R_{x,x'}(t) \geq 0$ for $x \neq x'$ and consequently that $R_{x,x}(t) = -\sum_{x' \neq x} R_{x,x'}(t) < 0$ because of eqn (2.30). If it is justified to neglect terms of order $\mathcal{O}(\delta t^2)$, we can write the time evolution of the system in terms of a differential equation,

$$\boxed{\frac{d}{dt} p_x(t) \approx \frac{p_x(t+\delta t) - p_x(t)}{\delta t} = \sum_{x'} R_{x,x'}(t) p_{x'}(t),} \tag{2.31}$$

or in matrix form simply $d_t \boldsymbol{p}(t) = R(t)\boldsymbol{p}(t)$. Equation (2.31) is called a **rate master equation.** If the stochastic process is Markovian, the rate master equation completely describes the process, as explained in the next exercise.

Exercise 2.3 Derive that the transition matrix $T_{\ell,k}$ for a finite time step from t_k to t_ℓ follows from eqn (2.31) as the time-ordered exponential (defined in eqn (1.8)):

$$T_{\ell,k} = \exp_+ \left[\int_{t_k}^{t_\ell} R(t) dt \right]. \tag{2.32}$$

If the rate matrix does not depend on time, show that $T_{\ell,k} = e^{(t_\ell - t_k)R}$. Then, under the assumption that the dynamics is Markovian, show that for any $n \in \mathbb{N}$

$$p(x_n, \ldots, x_1, x_0) = (T_{n,n-1})_{x_n,x_{n-1}} \cdots (T_{1,0})_{x_1,x_0} p_{x_0}(0), \tag{2.33}$$

where the joint probability $p(x_n, \ldots, x_1, x_0)$ completely characterizes the stochastic process.

We remark that master equations can also be derived for *non-Markovian* dynamics. In this case, however, the master equation can be used only to propagate the state vector $\boldsymbol{p}(t)$ forwards in time; it cannot be used to compute the joint probability (2.33), which was only possible because we assumed $p(x_n|x_{n-1}, \ldots, x_1, x_0) = p(x_n|x_{n-1})$.

We note that in any realistic application δt describes a *finite* time step, but, as long as we can approximate $(T_{t+\delta t,t})_{x,x'} \approx \delta_{x,x'} + R_{x,x'}(t)\delta t$, it is justified to use the differential master equation. Focusing on rate matrices R instead of transition matrices T can turn out to be advantageous. For instance, deriving a master equation for a specific physical situation is often easier than deriving the transition matrices directly, and the rate matrix R contains all the physically relevant information about the problem and fully determines the transition matrices. Equation (2.31) is also our starting point for the next three sections, which establish the framework of classical stochastic thermodynamics. But in order to achieve this, we first of all have to equip the rate matrix with a bit more physical substance—up to now eqn (2.31) has been a purely mathematical tool without any connection to thermodynamics and not even to physics.

First of all, and almost needless to say, we need to associate with each state x some physical properties. Below and in the next section, we assume this property to be only its energy, denoted E_x. Afterwards, we will also associate an intrinsic entropy with each state, and in later chapters also a particle number. In principle, even more attributes are conceivable, depending on the situation.

Next, we address the time dependence of the rate matrix $R(t)$, which can in principle have two different physical origins. One source of time dependence arises from an environment whose own dynamics influences the system, i.e. from the system's point of view the environment does not appear static. We neglect this possibility here and assume that *time-scale separation* applies, i.e. for all *practical* purposes the (slow) dynamics of the system does not 'see' the (fast) dynamics of the environment, which appears equilibrated instead. We met the idea of time-scale separation in the previous section. We discuss it further and make it more quantitative below and in Sections 2.5, 2.8 and 3.2.

The second source of time dependence arises from the possibility that an external agent controls some physical parameter of the system in time. An example could be an electromagnetic field (e.g. a laser) used to manipulate the system dynamics. The important assumption here is that this sort of time dependence is known to the external agent and can be precisely controlled. We denote this time dependence by λ_t. By specifying its value as a function of time t from some initial time to some final time one defines a so-called *control protocol* or *driving protocol.* We will see soon that this time dependence is responsible for injecting or extracting work into or from the system as it resembles the action of the piston in Fig. 2.1, which changes the volume V in a controlled way. In the following, we therefore write $R(t) = R(\lambda_t)$.

Local detailed balance

Another crucial property of the rate matrices arises from the assumption that they are derived from an underlying microscopic Hamiltonian model, for which time-scale separation applies. This has profound consequences for the structure of the rate matrix, whose elements $R_{x,x'}(\lambda_t)$ have to obey the symmetry relation

$$\boxed{\frac{R_{x,x'}(\lambda_t)}{R_{x',x}(\lambda_t)} = \exp\left[\frac{E_{x'}(\lambda_t) - E_x(\lambda_t)}{k_B T}\right].} \tag{2.34}$$

Here, T denotes the temperature of the environment and $E_x(\lambda_t)$ the energy of the microstate x, which can depend—according to what we said above—on the control protocol λ_t. The relation (2.34) is called **local detailed balance** and it plays a crucial role in the following three sections and also in some sections of later chapters. We postpone a derivation of it, which also reveals that eqn (2.34) is linked to the fact that the dynamics tends to maximize the thermodynamic entropy, to the end of this section. First, we discuss some simple but important consequences.

A first important consequence of the local detailed balance condition arises in the *absence* of any driving, i.e. when λ_t = constant such that we can write $R_{x,x'}(\lambda_t) = R_{x,x'}$. Then, the equilibrium Gibbs ensemble

$$\pi_x = \frac{e^{-\beta E_x}}{\mathcal{Z}_S}, \quad \mathcal{Z}_S = \sum_x e^{-\beta E_x} \tag{2.35}$$

is a steady state of the master equation, as expected from arguments of equilibrium statistical mechanics. The reader is asked to confirm this in the following exercise.

Exercise 2.4 Use the rate master equation (2.31) together with the condition of local detailed balance to show that eqn (2.35) is a steady-state, i.e. $R\boldsymbol{\pi} = 0$, where $\boldsymbol{\pi}$ denotes the vector with elements π_x. An important question is whether this is the *unique* steady-state solution or whether there exists another probability vector $\bar{\boldsymbol{p}} \neq \boldsymbol{\pi}$, which also satisfies $R\bar{\boldsymbol{p}} = 0$? It turns out that every rate master equation has a unique steady-state solution if it is *fully connected* or *irreducible*. By this we mean that for any two states x and x' it is always possible to construct a path $x \to x_1 \to x_2 \to \cdots \to x_n \to x'$ using other states x_i, $i \in \{1, \dots, n\}$, such that the product $R_{x',x_n} \cdots R_{x_2,x_1} R_{x_1,x}$ does not vanish. We are not going to prove this theorem here, but the reader is asked to construct a rate matrix with multiple steady states and to confirm that this matrix is not irreducible. What kind of physical situations could be described by rate master equations with multiple steady states?

We consider the time-independent case (λ_t = constant) a bit further. It follows from the local detailed balance condition and the form of the Gibbs state (2.35) that

$$R_{x,x'}\pi_{x'} - R_{x',x}\pi_x = 0. \tag{2.36}$$

Notice that the term on the left-hand side of this equation is nothing but the *net probability current* from x' to x: the term $R_{x,x'}\pi_{x'}$ tells us how much probability is transferred from state x' to x per time step δt and from this we substract the probability flow $R_{x',x}\pi_x$ in the opposite direction. The thermal equilibrium state of a rate master equation obeying local detailed balance is therefore characterized by the fact that no net current is flowing in the system: each transition is carefully balanced by its reversed transition. Clearly, this is expected: an equilibrium state should be *time independent* and the presence of a non-vanishing current implies that the system cannot be time independent from a *macroscopic* point of view.

At this point, it is useful to distinguish the notion of *local* detailed balance with the related but different notion of (global) **detailed balance.** A rate master equation is said to be detailed balanced if there exists a probability vector $\boldsymbol{p}$ satisfying

$$R_{x,x'}p_{x'} - R_{x',x}p_x = 0 \quad \text{for all } x, x'. \tag{2.37}$$

If such a vector $\boldsymbol{p}$ exists, it is a steady state of the master equation, $R\boldsymbol{p} = 0$, which is easy to check, but the converse is not true: there are master equations whose steady state does not obey eqn (2.37). In fact, we will soon encounter such master equations when we treat systems in contact with *multiple* heat baths. For a single heat bath, we note that local detailed balance implies detailed balance. The converse is true if one assumes that $\boldsymbol{p} = \boldsymbol{\pi}$ is a steady state of the rate master equation.

We now come back to the time-dependent case. Clearly, the result of Exercise 2.4 implies that the time-dependent Gibbs state $\pi_x(\lambda_t) = e^{-\beta E_x(\lambda_t)}/\mathcal{Z}_S(\lambda_t)$ satisfies at each single time $R(\lambda_t)\boldsymbol{\pi}(\lambda_t) = 0$. This, however, does *not* imply that the system

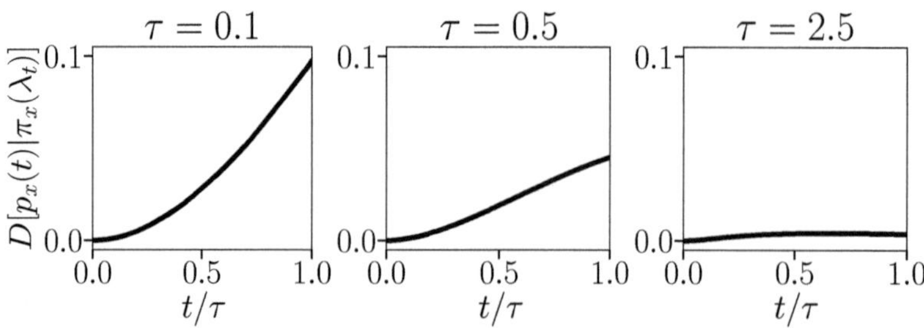

Fig. 2.3 Relative entropy $D[p_x(t)|\pi_x(\lambda_t)]$ between the time-dependent solution $p_x(t)$ of eqn (2.38) and the instantaneous equilibrium state $\pi_x(\lambda_t)$. The time-dependent energy gap was parametrized as $\Delta(\lambda_t) = t/\tau$. Thus, the driving gets slower from left to right, implying smaller deviations from equilibrium. Further numerical parameters are $\beta = 1$ and $\Gamma = 1$.

remains all the time in thermal equilibrium, even if the initial condition is $\boldsymbol{p}(0) = \boldsymbol{\pi}(\lambda_0)$. The reason for this is that for a driving protocol λ_t, which changes faster than the time needed to relax back to equilibrium, the system state $\boldsymbol{p}(t)$ 'lags behind' the instantaneous equilibrium state $\boldsymbol{\pi}(\lambda_t)$. As a consequence, the system gets *driven out of equilibrium.* The next exercise illustrates this behaviour for a simple example.

Exercise 2.5 Consider a two-level system with states labelled '0' and '1' described by the rate master equation

$$\frac{d}{dt}\begin{pmatrix} p_0(t) \\ p_1(t) \end{pmatrix} = \Gamma \begin{pmatrix} -e^{-\beta\Delta(\lambda_t)/2} & e^{\beta\Delta(\lambda_t)/2} \\ e^{-\beta\Delta(\lambda_t)/2} & -e^{\beta\Delta(\lambda_t)/2} \end{pmatrix} \begin{pmatrix} p_0(t) \\ p_1(t) \end{pmatrix}. \tag{2.38}$$

Here, Γ denotes some overall relaxation rate and $\Delta(\lambda_t)$ the time-dependent energy gap between 0 and 1. Assume the system starts in equilibrium, $\boldsymbol{p}(0) = \boldsymbol{\pi}(\lambda_0)$. By considering a numerical parametrization of your choice, convince yourself of the fact that the system follows the instantaneous equilibrium state, i.e. $\boldsymbol{p}(t) \approx \boldsymbol{\pi}(\lambda_t)$, whenever the driving is *slow*, which means that $\dot{\lambda}_t \ll \Gamma$. *Hint:* One simple example is shown in Fig. 2.3.

Derivation of local detailed balance

We return to the derivation of eqn (2.34) and afterwards finish this section with a short summary. Two ingredients enter the derivation of the local detailed balance condition. First, the assumption of time-scale separation, which we already met in Section 2.2: it allows us to model the bath state as a constrained or conditional equilibrium state during the dynamics. Second, we assume that the system states and the global Hamiltonian of the underlying microscopic dynamics are invariant with respect to *time reversal.* This is the case for many interesting situations, but we note that time-reversal invariance can be broken, for instance, by an external magnetic field. In this case, the local detailed balance condition needs to be generalized. Since time reversal is an essential concept in this book, a detailed discussion of it is given in Appendix C. In this section, we only use an intuitive result of it, but readers unfamiliar with this concept should sooner or later consult the appendix. Finally, we note that our derivation is quantum mechanical. This is, of course, ultimatively always justified, but in addition it also simplifies the notation.

We start by looking at the rates $R_{x,x'}$ from a microscopic perspective, assuming that the universe is a large isolated Hamiltonian system. We also neglect any driving for the moment: λ_t = constant. We define our set of system states $\{x\}$ by a complete set of orthogonal projectors $\{\Pi_x\}$. In spirit of system–bath theories, these projectors could describe a projective measurement of the system, but the present treatment is in fact more general and also includes measurements of some collective degree of freedom such as the net magnetization of an assembly of spins, as in Section 2.2. If we denote the unitary time evolution operator from t to $t + \delta t$ by $U(\delta t)$, the joint probability of obtaining outcome x' at t and x at $t + \delta t$ reads

$$p(x, t + \delta t; x', t) = \mathrm{tr}\left\{\Pi_x U(\delta t)\Pi_{x'}\rho(t)\Pi_{x'}U(\delta t)^\dagger\right\}, \tag{2.39}$$

where $\rho(t)$ describes the state of the universe at time t. Dividing by the probability $p_{x'}(t)$ to obtain measurement result x', the elements of the rate matrix for $x \neq x'$ become

$$R_{x,x'} = \frac{1}{\delta t}\mathrm{tr}\left\{\Pi_x U(\delta t)\frac{\Pi_{x'}\rho(t)\Pi_{x'}}{p_{x'}(t)}U(\delta t)^\dagger\right\} \tag{2.40}$$

according to eqn (2.29).

So far, we have not used any assumptions. This changes now. First, since the universe is isolated, we assume that it has a well-defined initial energy E in a macroscopic sense, i.e. the entire dynamics happens in an energy shell defined by a projector of the form (2.12). Furthermore, we use the assumption of time-scale separation, which means that δt—albeit small in comparison with characteristic time scales of the system evolution—is large enough such that the unobserved microscopic degrees of freedom effectively equilibrate. Thus, the universe is assumed to be in all states compatible with the constraint x with equal probability, i.e. we set $\rho(t) = \sum_x p_x(t)\omega(E, x)$, where $\omega(E, x) = \Pi_{E,x}/V_{E,x}$ denotes a microcanonical state compatible with the constraint (E, x); compare with eqn (2.14). Since the total macroscopic energy E of the universe is conserved, the dynamics remains within the energy shell around E and we can replace the projector Π_x by $\Pi_{E,x}$ in eqn (2.40). Hence, we obtain

$$R_{x,x'} = \frac{\mathrm{tr}\{\Pi_{E,x}U(\delta t)\omega(E, x')U(\delta t)^\dagger\}}{\delta t} = \frac{\mathrm{tr}\{\Pi_{E,x}U(\delta t)\Pi_{E,x'}U(\delta t)^\dagger\}}{\delta t V_{E,x'}}. \tag{2.41}$$

We now use time-reversal symmetry. Namely, we assume that both the observed states x and the Hamiltonian of the universe are invariant with respect to time reversal. Examples for such observables are, for example, the position of some object or the number of particles in some region. Furthermore, the energy is also invariant if the momenta enter the microscopic Hamiltonian only quadratically and if we can neglect the spin. Taken together, this implies the relation

$$\mathrm{tr}\{\Pi_{E,x}U(\delta t)\Pi_{E,x'}U(\delta t)^\dagger\} = \mathrm{tr}\{\Pi_{E,x'}U(\delta t)\Pi_{E,x}U(\delta t)^\dagger\}, \tag{2.42}$$

which we prove in Appendix C (Exercise C.7), where we also discuss the case of what happens beyond the present assumptions.

Exercise 2.6 If $\Pi_{E,x} = |x\rangle\langle x|$ are projectors of rank 1, then eqn (2.42) reduces to $|\langle x|Ux'\rangle|^2 = |\langle x'|Ux\rangle|^2$, i.e. the transition probability from x' to x equals the one from x to x'. Consider a single spin 1/2 particle with Hamiltonian $H = B\sigma_z$ (which is *not* invariant under time reversal) and construct an example where $|\langle x|Ux'\rangle|^2 \neq |\langle x'|Ux\rangle|^2$.

From eqn (2.42) we can conclude, without the need to ever evaluate eqn (2.41) explicitly, that

$$\boxed{\frac{R_{x,x'}}{R_{x',x}} = \frac{V_{E,x}}{V_{E,x'}} = \exp\left[\frac{S_B(E,x) - S_B(E,x')}{k_B}\right]}, \tag{2.43}$$

where $S_B(E,x)$ denotes the Boltzmann entropy introduced in eqn (2.11). This relation is not yet of the form (2.34), but it can in fact be seen as a more general form of local detailed balance. It predicts that the rate $R_{x,x'}$ is larger than $R_{x',x}$ whenever the entropy $S_B(E,x)$ is larger than the entropy $S_B(E,x')$, i.e. *the dynamics maximizes the entropy at each time step.*

The more restricted form (2.34) of local detailed balance follows from eqn (2.43) when the projectors $\{\Pi_x\}$ describe energy measurements of a small subpart of the entire universe. Let us denote the outcomes by E_x. Then, since the global energy E is fixed, $S_B(E,x') = S_B(E-E_x)$ is the Boltzmann entropy of the remaining microstates, which have energy $E - E_x$. Assuming that $E \gg E_x$, we can write $S_B(E - E_x) \approx S_B(E) - E_x/T$, where we used the definition (2.17) of temperature. Then, eqn (2.43) reduces to

$$\frac{R_{x,x'}}{R_{x',x}} = \frac{V_{E,x}}{V_{E,x'}} = \exp\left(\frac{E_{x'} - E_x}{k_B T}\right). \tag{2.44}$$

While we assumed the Hamiltonian of the universe to be time independent to arrive at this result, the derivation can be extended to include an external driving protocol λ_t if it changes *slowly* enough such that the assumption of time-scale separation continues to hold. In this case, the previous equation generalizes to eqn (2.34), provided that we remain close to the initial energy shell around E; otherwise, the temperature T in eqn (2.34) would become time dependent too. This concludes our derivation of the local detailed balance condition; an alternative derivation of it will also be presented in the context of quantum master equations in Section 3.3.

Summary

Classical Markov systems described by a discrete set of states are conveniently described by a rate master equation of the form (2.31). If the bath can be approximated as being in conditional equilibrium during the time-scales of the system evolution, we say that time-scale separation applies. This implies that the only time dependence of the rate matrix $R(\lambda_t)$ results from an externally controlled driving protocol λ_t and that the ratio of the rates satisfies the local detailed balance condition (2.34) or the more general form (2.43). Investigating the thermodynamic consequences of these assumptions forms the central part of the theory of *stochastic thermodynamics*, as summarized in Fig. 2.4.

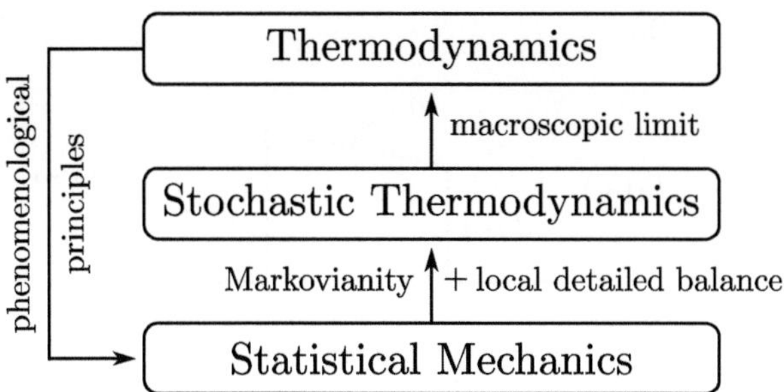

Fig. 2.4 Stochastic thermodynamics provides a bridge between statistical mechanics and thermodynamics: it reproduces the thermodynamics of Sec. 2.1 in the macroscopic limit and arises from a microscopic statistical description by assuming Markovianity and local detailed balance. Of course, these assumptions are made in agreement with the phenomenological principles of thermodynamics: we will see in Sec. 2.4 that local detailed balance is equivalent to the second law at each time step. Also, we remark that there is no univocally accepted definition of what stochastic thermodynamics *is*, and in Sections 2.7 to 2.9 we will turn to the thermodynamic description of small systems beyond the Markov assumption.

We remark that this book leaves out the treatment of classical Markov systems described by *continuous* degrees of freedom (e.g. Brownian motion of a colloidal particle in water). In this context the master equation is called a *Fokker–Planck equation.* Using the same assumptions, a relation similar to local detailed balance appears, linking the diffusion and friction coefficient of the Fokker–Planck equation via the temperature of the surrounding heat bath. We treat, however, non-Markovian classical systems with continuous degrees of freedom from Section 2.7 onwards.

2.4 Fluctuating Internal Energy, Heat, Work and Entropy

This section establishs a thermodynamic framework for a system with states described by time-dependent microscopic energies $E_x(\lambda_t)$. The system is supposed to obey a rate master equation satisfying the local detailed balance condition (2.34). We first consider a system in contact with a single heat bath and we later generalize to multiple baths. Moreover, we establish this framework even along a single stochastic trajectory.

To define a stochastic trajectory, consider a fictitious experiment in which we measure the state of the system in time steps δt. The time step is assumed to be small enough such that $\delta t R_{x,x'}(\lambda_t) \ll 1$ for all $x \neq x'$. A trajectory from the initial time $t_0 = 0$ to some final time t_n is then characterized by a sequence of states $(x_n, \ldots, x_1, x_0)$, where x_ℓ denotes the system state at time $t_\ell = \ell \delta t$. In short, we denote this trajectory by a boldface notation $\mathbf{x}_n$. A simple example for a two-state system is sketched in Fig. 2.5. The stochasic trajectory is characterized by sudden, unpredictable jumps between the two states, which is also known as a *random telegraph signal.* Note that the probability of two jumps in a time step δt is negligible because $\delta t R_{x,x'}(\lambda_t) \ll 1$.

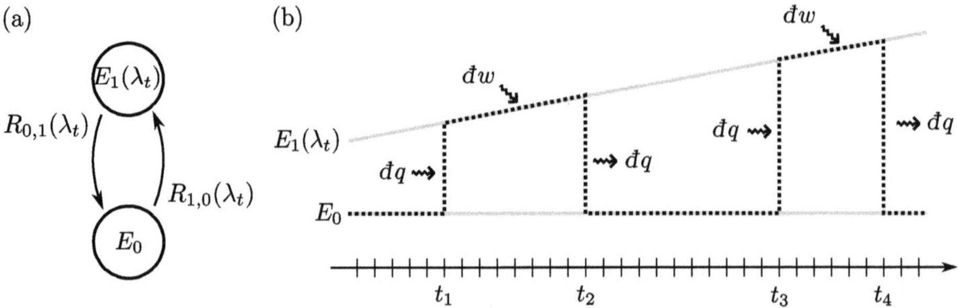

Fig. 2.5 (a) Sketch of a two-state system described by a rate master equation. (b) Example of a stochastic trajectory. The grey solid lines depict the ground state energy E_0 (here assumed to be time independent) and the excited state energy $E_1(\lambda_t)$ (here sketched as a linear ramp). The system is assumed to start in the ground state and it is measured in small time steps δt. Most of the times 'nothing' happens. But at some times (here at t_1, t_2, t_3 and t_4) the system jumps from the ground to the excited state or vice versa. This jump is accompanied by an exchange of heat $đq$. During the time that the system spends in the excited state, work $đw$ is done on it.

Stochastic energetics and first law

Our goal is to define notions of internal energy, heat and work, which satisfy the first law even along a single stochastic trajectory. To emphasize that these definitions refer to a single stochastic trajectory $\mathbf{x}_n$, we call them *stochastic* internal energy, heat and work, and denote them by small letters u_S, q and w, respectively. Later on, we also consider their averages, which we denote by capital letters (i.e. U_S, Q and W).

Clearly, the obvious choice to define **stochastic internal energy** at time t_ℓ is

$$\boxed{u_S(x_\ell, t_\ell) \equiv E_{x_\ell}(\lambda_\ell),} \tag{2.45}$$

where λ_ℓ denotes the value of the control protocol λ_t at time $t = t_\ell$. If the states x are not microstates but mesostates (i.e. collections of microstates as studied in Section 2.2), definition (2.45) needs to be corrected as discussed in the next section. The **stochastic first law** for a single time step δt reads

$$\boxed{du_S(t_\ell) \equiv u_S(x_{\ell+1}, t_{\ell+1}) - u_S(x_\ell, t_\ell) = đq(t_\ell) + đw(t_\ell),} \tag{2.46}$$

where we left the dependence on the states x_ℓ and $x_{\ell+1}$ implicit in the notation. It remains to find meaningful definitions for the increments in stochastic heat $đq(t_\ell)$ and work $đw(t_\ell)$. In anticipation of the fact that heat and work are path-dependent quantities and not a change in a state function, we used the notation $đ$ here.

Intuitively, we expect the **stochastic heat** to be linked to energy exchanges with the bath caused by jumps of the system from one state to another. Hence, we define

$$\boxed{đq(t_\ell) \equiv E_{x_{\ell+1}}(\lambda_{\ell+1}) - E_{x_\ell}(\lambda_{\ell+1}).} \tag{2.47}$$

Thus, $đq(t_\ell) = 0$ whenever the system stays in the same state $x_{\ell+1} = x_\ell$. Definition (2.47) can be justified by recalling the local detailed balance condition. From eqns (2.34) and (2.43), we find that

$$đq(t_\ell) = k_BT \ln \frac{R_{x_\ell,x_{\ell+1}}(\lambda_{\ell+1})}{R_{x_{\ell+1},x_\ell}(\lambda_{\ell+1})} = -T\left[S_B(E - E_{x_{\ell+1}}) - S_B(E - E_{x_\ell})\right]. \tag{2.48}$$

Thus, whenever the system jumps from x_ℓ to $x_{\ell+1}$, the associated heat is proportional to *minus* the entropy change in the bath, with the proportionality factor being the temperature T, i.e. $đq = -TdS_B$. This coincides with the phenomenological Clausius relation defining an ideal heat bath (keeping in mind that $-đq$ is positive if it increases the bath energy). Relation (2.48) underlines the importance of the local detailed balance condition for the foundations of classical stochastic thermodynamics.

The definition of **stochastic work** can now be inferred from the stochastic first law and reads

$$\boxed{đw(t_\ell) \equiv E_{x_\ell}(\lambda_{\ell+1}) - E_{x_\ell}(\lambda_\ell).} \tag{2.49}$$

It quantifies how the energy for a given state of the system x_ℓ changes by varying the control protocol λ_t. This is intuitively appealing, but more microscopic justifications for the definition of mechanical work are provided in Sections 2.7 and 3.1.

We end the discussion about the stochastic first law for a single heat bath by noting that there are alternative definitions for stochastic heat and work, namely $đq(t_\ell) = E_{x_{\ell+1}}(\lambda_\ell) - E_{x_\ell}(\lambda_\ell)$ and $đw(t_\ell) = E_{x_{\ell+1}}(\lambda_{\ell+1}) - E_{x_{\ell+1}}(\lambda_\ell)$. Those are merely a matter of convention and their difference vanishes as $\delta t \to 0$.

Finally, we turn to the average description that is valid along an arbitrary solution $\boldsymbol{p}(t)$ of the rate master equation (2.31). We define and denote averages by

$$F \equiv \langle f(\mathbf{x}_n)\rangle_{\mathbf{x}_n} \equiv \sum_{\mathbf{x}_n} p(\mathbf{x}_n) f(\mathbf{x}_n), \tag{2.50}$$

where $f(\mathbf{x}_n)$ is some quantity defined in terms of a stochastic trajectory $\mathbf{x}_n$ and $p(\mathbf{x}_n)$ is the joint probability of observing this trajectory; see eqn (2.33).

Since the state $p_x(t_\ell)$ at time t_ℓ follows from maginalizing $p(\mathbf{x}_n)$ over all times except time t_ℓ, we immediately see that the **average internal energy** becomes

$$\boxed{U_S(t) = \sum_x E_x(\lambda_t) p_x(t).} \tag{2.51}$$

Since t_ℓ was arbitrary, we set $t_\ell \equiv t$ here. It also requires only a little more effort to confirm that the average increments in heat and work are

$$đQ(t) = \sum_x E_x(\lambda_t)[p_x(t+\delta t) - p_x(t)], \tag{2.52}$$

$$đW(t) = \sum_x [E_x(\lambda_{t+\delta t}) - E_x(\lambda_t)] p_x(t). \tag{2.53}$$

In particular, after dividing by δt, we find for the **heat flux** $\dot{Q}(t)$ and **power** $\dot{W}(t)$

$$\boxed{\dot{Q}(t) \equiv \frac{đQ(t)}{\delta t} = \sum_x E_x(\lambda_t)\frac{dp_x(t)}{dt}, \quad \dot{W}(t) \equiv \frac{đW(t)}{\delta t} = \sum_x \frac{dE_x(\lambda_t)}{dt} p_x(t).} \tag{2.54}$$

Obviously, these definitions give rise to the **average first law**:

$$\boxed{\frac{d}{dt}U_S(t) = \dot{Q}(t) + \dot{W}(t).} \tag{2.55}$$

Fluctuating entropy and second law

If our definition of stochastic heat above was correct, it has to appear in the second law at the right place. However, before we can address the second law, we need a definition of **stochastic entropy**, which we take to be

$$\boxed{s_S(x_\ell, t_\ell) \equiv -k_B \ln p_{x_\ell}(t_\ell).} \tag{2.56}$$

This definition takes the term $-k_B \ln p_x(t)$, where $p_x(t)$ is the solution of the rate master equation (2.31) for a given initial condition $p_x(0)$, and evaluates it along the stochastic trajectory $\mathbf{x}_n$. In contrast to internal energy, heat and work, the evaluation of eqn (2.56) therefore needs, first, to specify an initial condition $p_x(0)$ and, second, to solve the master equation.

Now, the **stochastic entropy production** or **stochastic second law** follows as

$$\boxed{đ\sigma(t_\ell) = ds_S(t_\ell) - \frac{đq(t_\ell)}{T}.} \tag{2.57}$$

From eqn (2.48) and definition (2.56) it follows that the stochastic entropy production can be expressed as

$$đ\sigma(t_\ell) = [-k_B \ln p_{x_{\ell+1}}(t_{\ell+1}) + S_B(E - E_{x_{\ell+1}})] - [-k_B \ln p_{x_\ell}(t_\ell) + S_B(E - E_{x_\ell})], \tag{2.58}$$

which makes it transparent that the stochastic entropy production quantifies a change in (stochastic) thermodynamic entropy.

In contrast to the phenomenological second law, we do not demand that $đ\sigma(t_\ell) \geq 0$ along a single trajectory. In fact, the phenomenological second law was established for macroscopic systems, which behave deterministically with negligible fluctuations. In statistical mechanics, fluctuations are prevalent and the second law acquires a statistical character. Hence, along a single trajectory it can happen that $đ\sigma(t_\ell) < 0$ and, as we will see in Section 2.6, there is a remarkable identity that restricts the fluctuation in stochastic entropy production.

Nevertheless, we demand, of course, that the entropy production is *on average positive*. The average **system entropy** follows from eqn (2.56) as (setting again $t_\ell = t$)

$$\boxed{S_S(t) = k_B S_{\text{Sh}}[\boldsymbol{p}(t)] = -k_B \sum_x p_x(t) \ln p_x(t).} \tag{2.59}$$

The average **entropy production rate** $\dot{\Sigma}(t) \equiv đ\Sigma/\delta t$ with $đ\Sigma = \langle đ\sigma \rangle_{\mathbf{x}_n}$ becomes

$$\boxed{\dot{\Sigma}(t) = k_B \frac{d}{dt} S_{\text{Sh}}[\boldsymbol{p}(t)] - \frac{\dot{Q}(t)}{T} \geq 0,} \tag{2.60}$$

which we claim to be always positive—for any initial state $\boldsymbol{p}(0)$ and any driving protocol λ_t. The reader is asked to verify this statement in the next exercise.

Exercise 2.7 There are surprisingly many ways to show that $\dot{\Sigma}(t) \geq 0$ and the reader is asked to confirm two of them here. First, show that the entropy production rate can be alternatively expressed in terms of the relative entropy as follows:

$$\dot{\Sigma}(t) = -k_B \left.\frac{\partial}{\partial t}\right|_{\lambda_t} D[\boldsymbol{p}(t)|\boldsymbol{\pi}(\lambda_t)], \tag{2.61}$$

where the derivative is taken with respect to a *fixed* λ_t. Use this expression to show that the entropy production rate is positive. *Hint:* Use monotonicity of relative entropy (Theorem A.4) and discretize time by considering a small time step δt.

Alternatively, confirm that the entropy production rate can be expressed as

$$\dot{\Sigma}(t) = \frac{k_B}{2} \sum_{x,x'} [R_{x,x'}(\lambda_t)p_{x'}(t) - R_{x',x}(\lambda_t)p_x(t)] \ln \frac{R_{x,x'}(\lambda_t)p_{x'}(t)}{R_{x',x}(\lambda_t)p_x(t)}. \tag{2.62}$$

Prove $\dot{\Sigma}(t) \geq 0$ by using that $(a - b)\ln(a/b) \geq 0$ for all positive numbers a and b.

We remark that both forms not only are mere mathematical identities, but also convey useful physical information. The form (2.61) teaches us that the entropy production rate describes the tendency of the bath to draw the state $\boldsymbol{p}(t)$ at each time closer to the equilibrium state $\boldsymbol{\pi}(\lambda_t)$. At equilibrium, when $\boldsymbol{p}(t) = \boldsymbol{\pi}(\lambda_t)$, we have $\dot{\Sigma}(t) = 0$. Equation (2.62) teaches us that a non-zero entropy production rate is related to the existence of non-zero *currents* $R_{x,x'}(\lambda_t)p_{x'}(t) - R_{x',x}(\lambda_t)p_x(t) \neq 0$ from x' to x. Each non-zero current produces entropy and the sum of all contributions equals the entropy production rate.

We remark that positivity of the entropy production *rate* is a stronger result than positivity of the total entropy production (i.e. the *integrated* entropy production rate)

$$\Sigma(t) = \int_0^t ds\dot{\Sigma}(s) = \Delta S_S(t) - \frac{Q(t)}{T} \geq 0. \tag{2.63}$$

Here, we denote the change in any state function X as $\Delta X(t) = X(t) - X(0)$ and $Q(t) \equiv \int_0^t ds\dot{Q}(s)$. Later, we will indeed encounter situations where it is temporarily possible that the entropy production rate becomes negative, i.e. $\dot{\Sigma}(s) < 0$ for some $s \in [0, t]$, whereas $\Sigma(t) \geq 0$ always.

Multiple baths

We turn to the treatment of multiple baths, which we label by $\nu \in \{1, 2, \dots\}$ and assume to be described by different fixed temperatures T_ν. In general, the treatment of this situation is considerably more involved than the single bath case. However, at least for one important class of situations the treatment becomes much simpler and basically follows from the single bath case. This class is described by assuming that the different baths enter *additively* in the rate master equation, by which we mean that the rate matrix can be decomposed as

$$R(\lambda_t) = \sum_\nu R^{(\nu)}(\lambda_t). \tag{2.64}$$

Here, the rate matrices $R^{(\nu)}(\lambda_t)$ associated with each bath satisfy separately the local detailed balance condition

$$\frac{R^{(\nu)}_{x,x'}(\lambda_t)}{R^{(\nu)}_{x',x}(\lambda_t)} = \exp\left(\frac{E_{x'}(\lambda_t) - E_x(\lambda_t)}{k_B T_\nu}\right). \tag{2.65}$$

In Section 3.2, we will microscopically justify the decomposition (2.64) for a system weakly coupled to several independent baths. By 'independent baths' we mean that the baths influence each other only via the system and any direct correlation between them turns out to be a negligible higher-order effect. Then, the local detailed balance condition (2.65) again arises by assuming time-scale separation for each bath.

It follows from eqn (2.65) that the rate matrix associated with each bath satisfies $R^{(\nu)}(\lambda_t)\boldsymbol{\pi}(\beta_\nu, \lambda_t) = 0$, where the Gibbs state with respect to bath ν is defined as $\pi_x(\beta_\nu, \lambda_t) = e^{-\beta_\nu E_x(\lambda_t)}/\mathcal{Z}_S(\beta_\nu, \lambda_t)$ with $\mathcal{Z}_S(\beta_\nu, \lambda_t) = \sum_x e^{-\beta_\nu E_x(\lambda_t)}$. Consequently, each bath tries to thermalize the system according to its own temperature, but—unless all T_ν are equal—the system can no longer be described by a canonical ensemble with respect to a single temperature. Thus, even in absence of driving, $\lambda_t =$ constant, the system needs to find a 'compromise' between the different temperatures involved. A **non-equilibrium steady state** arises, which is typically characterized, as we see below, by non-vanishing currents and a positive entropy production rate. Non-equilibrium steady states are ubiquitous in nature as they are intimately linked with the phenomenon of *transport*. They play an important role in this book.

To describe this situation thermodynamically, we first of all note that the definitions of internal energy, system entropy and work remain unchanged. Hence, we only need to identify with each bath ν a respective heat flux. Along a single stochastic trajectory this means that we have to identify, whenever the system jumps from state x' to x, which of the baths has caused the jump. Thus, we decompose the stochastic heat as $đq(t_\ell) = \sum_\nu đq_\nu(t_\ell)$ and *symbolically* define the stochastic heat from bath ν as

$$đq_\nu(t_\ell) = [E_{x_{\ell+1}}(\lambda_{\ell+1}) - E_{x_\ell}(\lambda_{\ell+1})]_\nu, \tag{2.66}$$

which is non-zero only if $x_{\ell+1} \neq x_\ell$ *and* if the jump was caused by bath ν. We note that, whenever $x_{\ell+1} \neq x_\ell$, there can only be one bath ν that caused the jump according to eqn (2.64). Thus, the stochastic first law becomes

$$du_S(t_\ell) = \sum_\nu đq_\nu(t_\ell) + đw(t_\ell). \tag{2.67}$$

As for the single bath case, eqn (2.48), local detailed balance relates the heat flow $đq_\nu$ from each bath ν to a respective entropy decrease $đq_\nu/T_\nu = -dS_\nu$ in that bath.

We now take a look at the average heat flux from bath ν:

$$đQ_\nu(t_\ell) = \sum_{x_\ell, x_{\ell+1}} [E_{x_{\ell+1}}(\lambda_{\ell+1}) - E_{x_\ell}(\lambda_{\ell+1})]_\nu p(x_{\ell+1}, x_\ell). \tag{2.68}$$

When evaluating it, we have to keep in mind that $đq_\nu(t_\ell)$ is by definition zero whenever the jump from x_ℓ to $x_{\ell+1}$ is not caused by bath ν. We therefore write

$$đQ_\nu(t_\ell) = \sum_{x_\ell, x_{\ell+1}} [E_{x_{\ell+1}}(\lambda_{\ell+1}) p_\nu(x_{\ell+1}|x_\ell) - E_{x_\ell}(\lambda_{\ell+1})] p_{x_\ell}(t_\ell) \tag{2.69}$$

and need to find the conditional probability $p_\nu(x_{\ell+1}|x_\ell)$ that bath ν causes a jump from x_ℓ to $x_{\ell+1}$. Owing to the additive structure (2.64) this is $p_\nu(x_{\ell+1}|x_\ell) = \delta_{x_{\ell+1},x_\ell} + \delta t R^{(\nu)}_{x_{\ell+1},x_\ell}(\lambda_\ell)$. Hence,

$$đQ_\nu(t_\ell) = \delta t \sum_{x_\ell, x_{\ell+1}} E_{x_{\ell+1}}(\lambda_{\ell+1}) R^{(\nu)}_{x_{\ell+1},x_\ell}(\lambda_\ell) p_{x_\ell}(t_\ell). \tag{2.70}$$

Thus, after some relabelling, we can write the heat flux from bath ν at time t as

$$\dot{Q}_\nu(t) = \sum_{x,x'} E_x(\lambda_t) R^{(\nu)}_{x,x'}(\lambda_t) p_{x'}(t). \tag{2.71}$$

The average first law for multiple baths follows from the stochastic first law (2.67):

$$\boxed{\frac{d}{dt} U_S(t) = \sum_\nu \dot{Q}_\nu(t) + \dot{W}(t).} \tag{2.72}$$

Finally, the stochastic entropy production reads

$$đ\sigma(t_\ell) = ds(t_\ell) - \sum_\nu \frac{đq_\nu(t_\ell)}{T_\nu}, \tag{2.73}$$

which can have either sign, but the average entropy production rate

$$\boxed{\dot{\Sigma}(t) = k_B \frac{d}{dt} S_{\text{Sh}}[\boldsymbol{p}(t)] - \sum_\nu \frac{\dot{Q}_\nu(t)}{T_\nu} \geq 0} \tag{2.74}$$

is always positive. The reader is asked to verify this in the next exercise.

Exercise 2.8 Prove that the entropy production rate is positive for multiple baths by showing that eqns (2.61) and (2.62) can both be generalized as follows:

$$\begin{aligned} \dot{\Sigma}(t) &= -k_B \sum_\nu \left.\frac{\partial}{\partial t}\right|_{\nu,\lambda_t} D[\boldsymbol{p}(t)|\boldsymbol{\pi}(\beta_\nu, \lambda_t)] \\ &= \frac{k_B}{2} \sum_\nu \sum_{x,x'} \left[R^{(\nu)}_{x,x'}(\lambda_t) p_{x'}(t) - R^{(\nu)}_{x',x}(\lambda_t) p_x(t)\right] \ln \frac{R^{(\nu)}_{x,x'}(\lambda_t) p_{x'}(t)}{R^{(\nu)}_{x',x}(\lambda_t) p_x(t)}. \end{aligned} \tag{2.75}$$

In the first line, we introduced the notation $\partial_t|_\nu \boldsymbol{p}(t) \equiv R^{(\nu)}(\lambda_t)\boldsymbol{p}(t)$.

The physics of multiple baths and non-equilibrium steady states is very rich compared with the single-bath case. Some subtleties are explored in the following exercise.

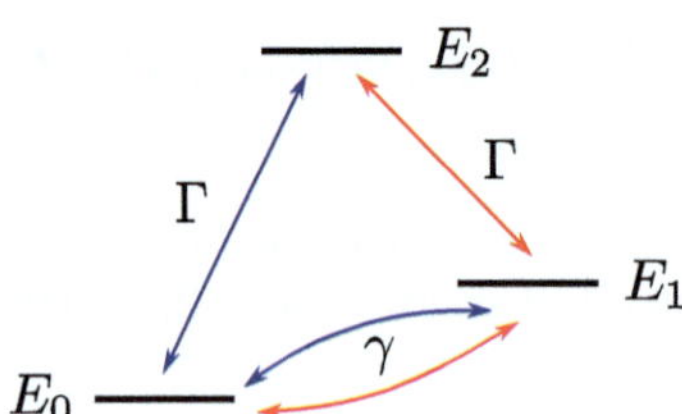

Fig. 2.6 Sketch of a system with three different energy levels $E_0 < E_1 < E_2$. The transitions $0 \leftrightarrow 2$ ($1 \leftrightarrow 2$) are mediated by a cold (hot) bath and happen at a time scale $1/\Gamma$, whereas the transition $0 \leftrightarrow 1$ is mediated by both baths and happens at a time scale $1/\gamma$.

Exercise 2.9 Consider a three–state system with energies $E_2 > E_1 > E_0$ coupled to a hot (H) and a cold (C) bath (see Fig. 2.6). The following rate matrix with respect to the basis $\boldsymbol{p}(t) = [p_0(t), p_1(t), p_2(t)]^T$ fixes the dynamics:

$$R = \begin{pmatrix} -(\dots) & \gamma[e^{\beta_C \Delta_{10}/2} + e^{\beta_H \Delta_{10}/2}] & \Gamma e^{\beta_C \Delta_{20}/2} \\ \gamma[e^{-\beta_C \Delta_{10}/2} + e^{-\beta_H \Delta_{10}/2}] & -(\dots) & \Gamma e^{\beta_H \Delta_{21}/2} \\ \Gamma e^{-\beta_C \Delta_{20}/2} & \Gamma e^{-\beta_H \Delta_{21}/2} & -(\dots) \end{pmatrix}.$$

Here, $\Delta_{ij} \equiv E_i - E_j$ is the energy difference and Γ and γ are rates. Furthermore, we did not write down the diagonal elements of the rate matrix as they are fixed by conservation of probability. First, confirm that this rate matrix has the desired additive structure (2.64) and satisfies local detailed balance. Next, compute the steady state as well as the average heat flows $\dot{Q}_H$ and $\dot{Q}_C$ at steady state. Convince yourself of the following facts.

(1) If $\gamma = 0$, the steady state is a non-equilibrium steady state, i.e. it cannot be described by an equilibrium distribution $\pi_x(\beta)$ for any β, but the heat flows are zero: $\dot{Q}_H = \dot{Q}_C = 0$. Consequently, the entropy production rate is also zero at steady state.

(2) In contrast, if $\Gamma = 0$ and if we set $p_2(0) = 0$, which implies $p_2(t) = 0$ for all t, the steady state is out of equilibrium, but it can be effectively described by an equilibrium distribution $\pi_x(\beta)$ for some effective β (find it!). That the system is out of equilibrium follows from the positive entropy production rate.

Can you find a general argument for why the heat fluxes are zero in the first case and non-zero in the second case? *Hint:* Try to identify *non-equilibrium cycles* in the graph shown in Fig. 2.6, i.e. sequences of states $x_1 \to x_2 \to \cdots \to x_1$ starting and ending at the same state whose transitions are not triggered solely by a single bath, and link the existence of such cycles to the existence of non-zero currents.

We end this section with three remarks. First, we have here considered heat baths, which are only characterized by their temperature. In principle, they can also be characterized by additional thermodynamic variables such as a chemical potential, which we consider later in this book. Multiple baths with different chemical potentials indeed play an important role in applications, but the modifications are rather straightforward once one has understood how to treat different temperatures.

Second, for most experimental situations in classical stochastic thermodynamics, the trajectories $\mathbf{x}_n$ can be observed without disturbing the dynamics, which makes the theory easily verifiable (or falsifiable) in an experiment. Indeed, it is remarkable that the non-equilibrium thermodynamics of a small system can be completely described by the local dynamics of the system only. However, there is an important operational distinction between the single-bath case and the multiple baths case. In the latter case, the heat is only determined if we additonally know *from which bath* the energy came. This information cannot be inferred by only knowing $\mathbf{x}_n$, at least not in general.

Third, the assumption that measurements do not disturb the dynamics is clearly unjustified for quantum systems in general, but—as we saw in Chapter 1—also for classical systems the Kolmogorov consistency condition can be broken, for instance, in the case of feedback control. Then, the framework above needs to be modified to account for the thermodynamic resources invested in an experiment to observe the system. In fact, as explained in Section 2.2 in the context of Landauer's principle, the storage of information has an entropic cost. These topics are the content of Chapter 5.

2.5 Coarse-Graining and Time-Scale Separation

In the previous section we considered microstates x without internal structure and associated a stochastic internal energy $E_x(\lambda_t)$ and stochastic entropy $-k_B \ln p_x(t)$ with them. In contrast, in Section 2.2 we considered states x, which described collections of microstates i_x. States x with such an internal structure are called *mesostates* and it turns out that one has to associate an additional intrinsic internal energy and system entropy with them, as done in Section 2.2. Consequently, the particularly simple local detailed balance condition (2.34) needs to be replaced by the more general condition (2.43). Here, we present an alternative approach to the problem, which starts from the description in the previous section and automatically equips us with the sought-after definitions of the intrinsic internal energy and entropy. Moreover, the crucial role of time-scale separation becomes particularly transparent in this approach.

Derivation of the coarse-grained master equation

We start with a set of microstates i, which could label, for example, all the microscopic spin configurations in the example treated in Section 2.2; see Fig. 2.2. We assume that the probability distribution $p_i(t)$ of the microstates follows a Markovian rate master equation with rates here denoted by $L_{i,i'} = L_{i,i'}(\lambda_t)$ satisfying the local detailed balance condition (2.34). For notational simplicity, we drop the dependence on λ_t for a moment. Then, with each microstate i we associate a unique mesostate x, e.g. the state of the bit from the example of Section 2.2, and consequently we label the microstate $i = i_x$. In the spirit of the example from Section 2.2, we want to encode into the rate matrix L_{i_x,i'_y} the fact that the microstates for a given mesostate evolve *fast* whereas the transitions between different mesostates are *slow*. This is achieved by writing

$$L_{i_x,i'_y} = \begin{cases} F_{i_x,i'_x} & \text{if } x = y, \text{ but } i_x \neq i'_x, \\ S_{i_x,i'_y} & \text{if } x \neq y, \\ -\sum_{i'_x \neq i_x} F_{i'_x,i_x} - \sum_{y \neq x, i'_y} S_{i'_y,i_x} & \text{if } i_x = i'_y, \end{cases} \tag{2.76}$$

where the last line arises from the normalization condition $\sum_{x,i_x} L_{i_x,i'_y} = 0$. The splitting (2.76) does not yet entail any assumption, but we now demand that the transitions between different microstates within the same mesostate are much faster than the transitions between microstates belonging to different mesotates, i.e.

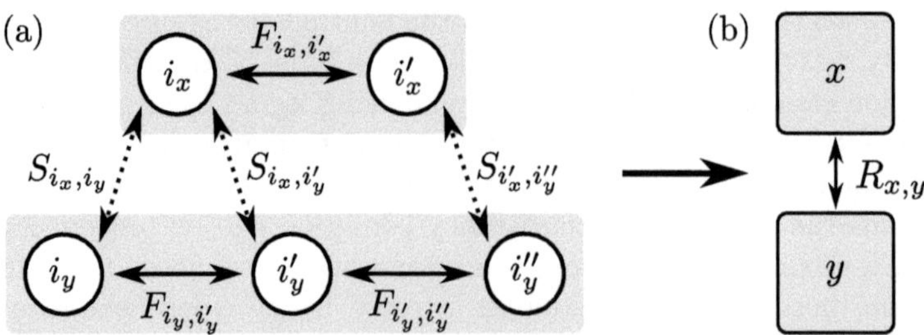

Fig. 2.7 (a) Sketch of a network of microstates i. The states i_x and i'_x are grouped as mesostate x, while i_y, i'_y and i''_y belong to mesostate y (shaded grey rectangles). The dynamics within each mesostate is given by fast rates F (solid arrows), whereas the dynamics between different mesostates is given by slow rates S (dotted arrows). (b) After coarse-graining and assuming time-scale separation, the mesostates x and y obey a rate master equation with effective rates $R_{x,y}$ defined in eqn (2.82).

$$F_{i_x,i'_x} \gg S_{i_x,i'_y}. \tag{2.77}$$

This can be seen as the central assumption behind **time-scale separation.** A sketch of the idea is shown in Fig. 2.7.

Our goal is to derive an effective master equation for the mesostates. To this end, we split the microstate probabilities as $p_{i_x}(t) = p_{i|x}(t)P_x(t)$, where $p_{i|x}(t)$ denotes the conditional probability for a particular microstate i given the mesostate x. From the definition of the mesostate probabilities $P_x(t) = \sum_{i_x} p_{i_x}(t)$ we get

$$\frac{d}{dt}P_x(t) = \sum_{i_x}\sum_{y,i'_y} L_{i_x,i'_y}p_{i'|y}(t)P_y(t) = \sum_{i_x}\sum_{y,i'_y} S_{i_x,i'_y}p_{i'|y}(t)P_y(t), \tag{2.78}$$

where the second equality follows from $\sum_{i_x} F_{i_x,i'_x} = 0$. So far, eqn (2.78) does not present any simplification as its solution requires explicit knowledge about the time evolution of the conditional probabilities $p_{i|x}(t)$.

Now, we make use of our time-scale separation assumption (2.77). In particular, observe that the evolution time scale of the mesostates is proportional to the slow rates S_{i_x,i'_y} according to eqn (2.78). This allows us to use the following argumentation. First, we return to the master equation describing the microstate evolution. From $p_{i_x}(t) = p_{i|x}P_x(t)$ we get after multiplication with dt

$$\begin{aligned}[dP_x(t)]p_{i|x} + P_x(t)dp_{i|x}(t) = dt \sum_{i'_x \neq i_x} \left[F_{i_x,i'_x}p_{i'|x}(t) - F_{i'_x,i_x}p_{i|x}(t)\right] P_x(t) \\ + dt\sum_{y\neq x}\sum_{i'_y}\left[S_{i_x,i'_y}p_{i'_y}(t) - S_{i'_y,i_x}p_{i_x}(t)\right].\end{aligned} \tag{2.79}$$

Next, consider a time step δt such that $\delta t S_{i_x,i'_y} \ll 1$ and $\delta t F_{i_x,i'_x} \lesssim 1$. During such a time step, condition (2.77) allows us to approximate $dP_x(t) = P_x(t+\delta t) - P_x(t) \approx 0$ and to neglect terms proportional to the slow rates S_{i_x,i'_y}. Then, eqn (2.79) reduces to

$$dp_{i|x}(t) \approx \delta t \sum_{i'_x \neq i_x} \left[F_{i_x,i'_x} p_{i'|x}(t) - F_{i'_x,i_x} p_{i|x}(t)\right]. \tag{2.80}$$

Now, we consider a coarser time scale $\delta\tau \gg \delta t$ such that $\delta\tau S_{i_x,i'_y} \lesssim 1$. On this time scale we can use eqn (2.78) for the mesostates, where the conditional micro-probabilities $p_{i|x}(t)$ are computed from eqn (2.80). In the strict limit of time-scale separation, where $F_{i_x,i'_x}/S_{i_x,i'_y} \to \infty$, we can let $\delta\tau/\delta t \to \infty$ and replace $p_{i|x}(t)$ in eqn (2.78) with the *steady-state solution* of eqn (2.80). Assuming that the system is coupled to a single heat bath and that all microstates i_x within a mesostate x are *connected* (see Exercise 2.4 for the definition), the steady state describes a constrained equilibrium state

$$p_{i|x}(t) \approx \pi_{i|x} \equiv \frac{e^{-\beta E_{i_x}}}{\mathcal{Z}_x}, \quad \mathcal{Z}_x \equiv \sum_{i_x} e^{-\beta E_{i_x}}. \tag{2.81}$$

Finally, we introduce the driving protocol λ_t again, which is assumed to change slowly compared with the microstate dynamics, i.e. $(\lambda_{t+\delta t} - \lambda_t)/\delta t \approx 0$. Then, within time-scale separation, our first central result is that the mesostate dynamics obeys

$$\boxed{\frac{d}{dt}P_x(t) = \sum_y R_{x,y}(\lambda_t)P_y(t), \quad R_{x,y}(\lambda_t) \equiv \sum_{i_x,i'_y} S_{i_x,i'_y}(\lambda_t)\pi_{i'|y}(\lambda_t).} \tag{2.82}$$

This coarse-grained master equation follows from eqn (2.78) after replacing $p_{i|x}(t)$ with $\pi_{i|x}(\lambda_t)$, which follows from eqn (2.81) after replacing E_{i_x} and $\mathcal{Z}_x$ by $E_{i_x}(\lambda_t)$ and $\mathcal{Z}_x(\lambda_t)$. The next exercise explicitly confirms that the dynamics of the mesostates is Markovian in the strict limit of time-scale separation. This seems obvious, but remember that this requires us to check whether the classical Markov property holds for the entire hierarchy of joint probability distributions.

Exercise 2.10 The microstate dynamics is Markovian and completely described by the conditional probablities (or transition matrix elements) $p(i_x|i'_y) \equiv p(i_x, t+\delta t|i'_y, t)$ to jump from i'_y at t to i_x at $t+\delta t$ (below we suppress any explicit dependence on time for simplicity). Thus, the joint probablity of observing a sequence of mesostates $\mathbf{x}_n = (x_n, \dots, x_1, x_0)$ is

$$P(\mathbf{x}_n) = \sum_{i_{x_n}} \cdots \sum_{i_{x_0}} p(i_{x_n}|i_{x_{n-1}}) \dots p(i_{x_1}|i_{x_0}) p_{i_0|x_0}(t_0) P_{x_0}(t_0). \tag{2.83}$$

Show that this describes a Markov process if the probability of ending up in a mesostate is independent of the precise initial microstate. In equations,

$$\sum_{i_{x_\ell}} p(i_{x_\ell}|i'_{x_k}) = \sum_{i_{x_\ell}} p(i_{x_\ell}|i''_{x_k}) \quad \text{for any} \quad i'_{x_k} \neq i''_{x_k}. \tag{2.84}$$

Remarkably, the converse is also true, although we do not prove this result here.

Next, consider the transition matrix $p(i_x|i'_y)$ within the limit of time-scale separation and up to first order in $S_{i_x,i'_y}\delta\tau$ satisfying $F^{-1}_{i_x,i_{x'}} \ll \delta\tau \ll S^{-1}_{i_x,i'_y}$. Convince yourself of the fact that

$$p(i_x|i'_y) = \delta_{x,y}\pi_{i|x}\left(1 - \delta\tau \sum_{z \neq x}\sum_{k_z} S_{k_z,i_x}\right) + (1-\delta_{x,y})\delta\tau \sum_{k_y} S_{i_x,k_y}\pi_{k|y}. \tag{2.85}$$

This transition matrix obeys an even stronger condition than eqn (2.84), namely $p(i_x|i'_y) = p(i_x|i''_y)$ for all $i'_y \neq i''_y$. Therefore, the coarse-grained process is Markovian. The converse is not true: Markovianity does not, in general, imply the property of time-scale separation.

Stochastic thermodynamics after coarse-graining

We now turn to the thermodynamic description of eqn (2.82). For this purpose we first of all associate with each mesostate an intrinsic free energy $\mathcal{F}_x(\lambda_t) \equiv -k_B T \ln \mathcal{Z}_x(\lambda_t)$ and note the following remarkable property:

$$\boxed{\frac{R_{x,y}(\lambda_t)}{R_{y,x}(\lambda_t)} = \exp\left[\frac{\mathcal{F}_y(\lambda_t) - \mathcal{F}_x(\lambda_t)}{k_B T}\right].} \tag{2.86}$$

This follows directly from the definition of the rates $R_{x,y}(\lambda_t)$—see eqn (2.82)—together with the assumption that the original rates S_{i_x,i'_y} satisfy the local detailed balance condition (2.34). Thus, eqn (2.86) describes nothing other than local detailed balance again, but this time in unison with the more general condition (2.43)

To see this last point, notice that the steady state of the mesostates for a *fixed* λ_t becomes, as a result of eqn (2.86),

$$\pi^*_x(\lambda_t) = \frac{e^{-\beta\mathcal{F}_x(\lambda_t)}}{\mathcal{Z}(\lambda_t)}, \quad \mathcal{Z}(\lambda_t) = \sum_x e^{-\beta\mathcal{F}_x(\lambda_t)}. \tag{2.87}$$

Here, we used the notation $\pi^*_x(\lambda_t)$ to emphasize that this steady state arises from moving in an *effective free energy landscape* $\mathcal{F}_x$ in contrast to the steady state (2.35), where the E_x denoted microscopic energies. In particular, $\mathcal{F}_x = \mathcal{F}_x(\beta)$ depends on the temperature of the environment whereas E_x did not. Now, using arguments from equilibrium statistical mechanics, we also expect that the probability $\pi^*_x(\lambda_t)$ should be proportional to the number of microstates $V_{E,x}(\lambda_t)$ of the *universe* (i.e. the system including the micro- and mesostates as well as all the bath degrees of freedom) compatible with the constraint x. Thus, $\pi^*_x(\lambda_t) = V_{E,x}(\lambda_t)/V_E$, where the normalization constant $V_E = \sum_x V_{E,x}(\lambda_t)$ denotes the *total* number of states. The exact value of this (huge) number is in fact unimportant as it cancels out in the following expression:

$$\frac{\pi^*_x(\lambda_t)}{\pi^*_y(\lambda_t)} = e^{\beta[\mathcal{F}_y(\lambda_t) - \mathcal{F}_x(\lambda_t)]} = \frac{V_{E,x}(\lambda_t)}{V_{E,y}(\lambda_t)}. \tag{2.88}$$

Using Boltzmann's entropy formula $S_B(E,x) = k_B \ln V_{E,x}$, we see that eqn (2.86) is *consistent* with our earlier result (2.43). Of course, all this was to be expected as the assumption of time-scale separation also played a crucial role in Section 2.3.

Exercise 2.11 Perhaps the reader has recognized that we used the superscript $*$ to denote the equilibrium state π^* of the coarse-grained degrees of freedom, as we also did to denote the reduced equilibrium state of a strongly coupled open system in Section 1.3. Show that these two states indeed agree by verifying that eqn (2.87) coincides with

$$\pi_x = \sum_{i_x} \frac{e^{-\beta E_{i_x}}}{\mathcal{Z}}, \quad \mathcal{Z} = \sum_x \sum_{i_x} e^{-\beta E_{i_x}}. \tag{2.89}$$

This justifies the claim made in Section 1.3 that the Hamiltonian of mean force $[H_S^*]$ can be seen as an effective free energy landscape $[\mathcal{F}_x]$ for the system.

Next, we turn to the thermodynamics of the mesostates. Since we have associated with each mesostate x an intrinsic free energy $\mathcal{F}_x(\lambda_t)$, it seems reasonable to also introduce intrinsic internal energies and entropies according to

$$\boxed{\mathcal{U}_x(\lambda_t) \equiv \sum_{i_x} E_{i_x}(\lambda_t)\pi_{i|x}(\lambda_t), \quad \mathcal{S}_x(\lambda_t) \equiv -k_B \sum_{i_x} \pi_{i|x}(\lambda_t) \ln \pi_{i|x}(\lambda_t).} \tag{2.90}$$

Note that these definitions satisfy the relation $\mathcal{F}_x(\lambda_t) = \mathcal{U}_x(\lambda_t) - T\mathcal{S}_x(\lambda_t)$. These intrinsic energies and entropies naturally suggest generalizing the definitions of internal energy (2.51) and system entropy (2.59) as follows:

$$\boxed{U_S(t) \equiv \sum_x \mathcal{U}_x(\lambda_t)P_x(t), \quad S_S(t) \equiv \sum_x P_x(t)\left[\mathcal{S}_x(\lambda_t) - k_B \ln P_x(t)\right].} \tag{2.91}$$

One quickly verifies that $U_S(t)$ and $S_S(t)$ coincide with the microscopic internal energy $\sum_i E_i p_i(t)$ and entropy $-k_B \sum_i p_i(t) \ln p_i(t)$ within time-scale separation.

The correct definition of heat and work, however, turns out to be more subtle. At first sight, in analogy to eqn (2.54), it is tempting to define

$$\dot{Q}_{\text{cg}}(t) \equiv \sum_x \mathcal{U}_x(\lambda_t)\frac{dP_x(t)}{dt}, \quad \dot{W}_{\text{cg}}(t) \equiv \sum_x \frac{d\mathcal{U}_x(\lambda_t)}{dt}P_x(t), \tag{2.92}$$

where we used a subscript 'cg' for coarse-graining. Clearly, these definitions satisfy the first law $d_t U_S(t) = \dot{Q}_{\text{cq}}(t) + \dot{W}_{\text{cg}}(t)$. However, if we compare them with the microscopic definition $\dot{Q}(t) = \sum_i E_i(\lambda_t)[\partial_t p_i(t)]$ and $\dot{W}(t) = \sum_i [\partial_t E_i(\lambda_t)]p_i(t)$ in the limit of time-scale separation, differences appear. Specifically,

$$\dot{Q}(t) - \dot{Q}_{\text{cg}}(t) = T\sum_x \frac{d\mathcal{S}_x(\lambda_t)}{dt}P_x(t) = \dot{W}_{\text{cg}}(t) - \dot{W}(t), \tag{2.93}$$

which is in general non-zero. The reader is asked to verify this equation as an exercise.

Exercise 2.12 Derive the last equation by using $p_{i_x}(t) = \pi_{i|x}(\lambda_t)P_x(t)$.

Thus, the definitions (2.92) are *blind* to the fact that there is in general a hidden heat flow and work cost associated with the intrinsic dynamics of the microstates within a given mesostate. This is a novel feature of our coarse-grained dynamics in comparison with the microscopic dynamics of the previous section. Now, the driving protocol λ_t not only can change the energetics of the system, but also can affect the *entropy* of the mesostates. As a consequence, we need to formulate the second law as follows:

$$\dot{\Sigma}(t) = \frac{d}{dt}S_S(t) - \frac{\dot{Q}(t)}{T} = \frac{d}{dt}S_S(t) - \frac{\dot{Q}_{\text{cg}}(t)}{T} - \sum_x \frac{d\mathcal{S}_x(\lambda_t)}{dt}P_x(t) \geq 0. \tag{2.94}$$

Exercise 2.13 Assume λ_t = constant and consider a relaxation process from some non-equilibrium initial state $P_x(0)$ to the final equilibrium state π_x^*. Show that the entropy production coincides with the entropy production computed in Exercise 2.2.

Furthermore, the foregoing average definitions have a clear counterpart at the trajectory level, as the reader is asked to verify in the next exercise.

Exercise 2.14 Show that the average (with respect to the coarse-grained dynamics) of the following stochastic definitions gives rise to the thermodynamic quantities introduced above:

$$u_S(x_\ell, t_\ell) \equiv \mathcal{U}_{x_\ell}(\lambda_\ell), \tag{2.95}$$

$$s_S(x_\ell, t_\ell) \equiv \mathcal{S}_{x_\ell}(\lambda_\ell) - k_B \ln p_{x_\ell}(t_\ell), \tag{2.96}$$

$$đq_{\text{cq}}(t_\ell) \equiv \mathcal{U}_{x_{\ell+1}}(\lambda_{\ell+1}) - \mathcal{U}_{x_\ell}(\lambda_{\ell+1}), \tag{2.97}$$

$$đw_{\text{cq}}(t_\ell) \equiv \mathcal{U}_{x_\ell}(\lambda_{\ell+1}) - \mathcal{U}_{x_\ell}(\lambda_\ell), \tag{2.98}$$

where the mesostates are assumed to be x_ℓ and $x_{\ell+1}$ at times t_ℓ and $t_{\ell+1}$, respectively. Verify that the first law also holds at the trajectory level.

We end this section with two observations. First, in practice the assumption of time-scale separation is not always satisfied. Generalizing stochastic thermodynamics beyond the regime, where eqn (2.82) is approximately justified, turns out to be difficult. Nevertheless, we will soon introduce useful tools to deal with this situation.

The second observation concerns the extension of the above concept to multiple heat baths. If the microstates i_x belonging to a mesostate x are all coupled to the same heat bath, i.e. if the different baths are coupled separately to different mesostates, then the theory above can be extended in a straightforward way. However, in general the microstates i_x for a given mesostate x can couple to multiple baths. Then, even under the assumption of time-scale separation, the thermodynamic description at the mesolevel becomes non-trivial. The reason for this is that it is no longer possible to characterize the mesostates in a simple way by using an intrinsic internal energy $\mathcal{U}_x(\lambda_t)$ and entropy $\mathcal{S}_x(\lambda_t)$ only. Instead, the mesostates themselves have an intrinsic entropy production assosciated with them, which can depend in a complicated way on the microstates.

2.6 Time-Reversed Trajectories and Fluctuation Theorems

In the previous two sections we have introduced definitions of internal energy, heat, work and entropy along a single trajectory, but we have not yet studied their fluctuations in detail—except for finding that the first law also holds at the trajectory level and that the second law must be obeyed on average. An important feature of stochastic thermodynamics is that it allows us to study the fluctuations of thermodynamic quantities. In the spirit of the motto 'the noise is the signal' we have good reasons to expect that these fluctuations contain valuable information and offer additional insights into the physics. In this section we derive an important result, which severely constrains the possible fluctuations in the stochastic entropy production σ and which, in some sense, promotes the status of the second law from an inequality to an *exact equality*. This relation is our first encounter with an entire family of similar relations known in general as *fluctuation theorems*.

We focus on the setting studied in Section 2.4 and shift the treatment of coarse-grained master equations to an exercise later on. We start with the case of a single heat bath and consider a trajectory $\mathbf{x}_n$. The entropy production along such a trajectory follows from eqn (2.57) and reads

$$\sigma(\mathbf{x}_n) = \sum_\ell đ\sigma(t_\ell) = \Delta s_S(t_n) - \frac{q(\mathbf{x}_n)}{T}. \tag{2.99}$$

Here, $q(\mathbf{x}_n) = \sum_\ell đq(t_\ell)$ is the net heat absorbed along the trajectory and $\Delta s_S(t) = -k_B \ln[p(x_n;t_n)/p(x_0;0)]$ is the total change in the stochastic system entropy. We note that we typically write in this section $p(x;t) = p_x(t)$ for the marginal single-time probabilities. The entire trajectory is described by the joint probability

$$p(\mathbf{x}_n) = p(x_n|x_{n-1};\lambda_n)\dots p(x_2|x_1;\lambda_2)p(x_1|x_0;\lambda_1)p(x_0;0), \tag{2.100}$$

where we explicitly included in the notation the dependence on the control protocol λ_t, i.e. $p(x_\ell|x_{\ell-1};\lambda_\ell) = \delta_{x_\ell,x_{\ell-1}} + \delta t R_{x_\ell,x_{\ell-1}}(\lambda_\ell) + \mathcal{O}(\delta t^2)$.

Next, we use a tool that we will encounter multiple times in this book and that is based on considering the **time-reversed process** (a general introduction to time-reversal *symmetry*, which is not obeyed by a Markovian master equation, is given in Appendix C). For this purpose consider the same set-up—a system coupled to a single heat bath described by a rate master equation obeying local detailed balance—but now we change the driving protocol in the *reverse* order from λ_t 'back' to λ_0, which we denote as $\lambda_t^\dagger$. Using discrete time steps, this means that $\lambda_\ell^\dagger \equiv \lambda_{n-\ell}$. Note that this process can be realized in an experiment as usual with time running 'forwards' by implementing a driving protocol in the reverse order. We therefore do not really 'reverse time' and we also use the terminology *forward* and *backward* process to distinguish between the process driven by λ_t and $\lambda_t^\dagger$, respectively. Furthermore, the probability of observing a trajectory $\mathbf{x}_n = (x_n,\dots,x_1,x_0)$ starting at x_0 and ending in x_n in the backward process reads

$$p_{\text{tr}}(\mathbf{x}_n) = p(x_n|x_{n-1};\lambda_1)\dots p(x_2|x_1;\lambda_{n-1})p(x_1|x_0;\lambda_n)p_{\text{tr}}(x_0;0), \tag{2.101}$$

with the 'initial' condition $p_{\text{tr}}(x_0;0)$ for the backward dynamics. Here, the subscript 'tr' signifies 'time reversed'. Finally, we introduce the concept of a time-reversed trajectory

$\mathbf{x}_n^\dagger \equiv (x_0, x_1, \ldots, x_n)$, which starts at x_n and ends in x_0, i.e. it traverses $\mathbf{x}_n$ in the reverse order.

We now state an important and astonishing fact that links the heat exchange statistics along a particular forward trajectory to its backward trajectory.

Crooks' lemma. *Consider a classical Markov process obeying local detailed balance. Let $\mathbf{x}_n$ be a stochastic trajectory of the system and $q(\mathbf{x}_n)$ the stochastic heat exchanged along that trajectory. Then,*

$$\boxed{\ln \frac{p(\mathbf{x}_n|x_0)}{p_{\text{tr}}(\mathbf{x}_n^\dagger|x_n)} = -\frac{q(\mathbf{x}_n)}{k_B T},} \tag{2.102}$$

where $p(\mathbf{x}_n|x_0) = p(\mathbf{x}_n)/p(x_0;0)$ is the conditional probability of observing the trajectory $\mathbf{x}_n$ in the forward process given the initial condition x_0. Similarly, $p_{\text{tr}}(\mathbf{x}_n^\dagger|x_n) = p_{\text{tr}}(\mathbf{x}_n^\dagger)/p_{\text{tr}}(x_n;0)$.

The proof goes as follows. By definition we have

$$\frac{p(\mathbf{x}_n|x_0)}{p_{\text{tr}}(\mathbf{x}_n^\dagger|x_n)} = \prod_{\ell=0}^{n-1} \frac{p(x_{\ell+1}|x_\ell;\lambda_{\ell+1})}{p(x_\ell|x_{\ell+1};\lambda_{\ell+1})}. \tag{2.103}$$

We focus on a single time step:

$$\frac{p(x_{\ell+1}|x_\ell;\lambda_{\ell+1})}{p(x_\ell|x_{\ell+1};\lambda_{\ell+1})} = \frac{\delta_{x_{\ell+1},x_\ell} + \delta t R_{x_{\ell+1},x_\ell}(\lambda_{\ell+1})}{\delta_{x_{\ell+1},x_\ell} + \delta t R_{x_\ell,x_{\ell+1}}(\lambda_{\ell+1})}. \tag{2.104}$$

There are two cases to distinguish. First, we can have $x_\ell = x_{\ell+1}$ and then

$$\frac{p(x_\ell|x_\ell;\lambda_{\ell+1})}{p(x_\ell|x_\ell;\lambda_{\ell+1})} = \frac{1 + \delta t R_{x_\ell,x_\ell}(\lambda_{\ell+1})}{1 + \delta t R_{x_\ell,x_\ell}(\lambda_{\ell+1})} = 1. \tag{2.105}$$

Second, if $x_\ell \neq x_{\ell+1}$, we can simplify the expression using local detailed balance:

$$\frac{p(x_{\ell+1}|x_\ell;\lambda_{\ell+1})}{p(x_\ell|x_{\ell+1};\lambda_{\ell+1})} = \exp\left[\frac{E_{x_\ell}(\lambda_{\ell+1}) - E_{x_{\ell+1}}(\lambda_{\ell+1})}{k_B T}\right]. \tag{2.106}$$

In fact, the previous equation agrees with eqn (2.105) for $x_\ell = x_{\ell+1}$. Hence, we can write in general

$$\frac{p(\mathbf{x}_n|x_0)}{p_{\text{tr}}(\mathbf{x}_n^\dagger|x_n)} = \prod_{\ell=0}^{n-1} \exp\left[\frac{E_{x_\ell}(\lambda_{\ell+1}) - E_{x_{\ell+1}}(\lambda_{\ell+1})}{k_B T}\right] = \exp\left[-\frac{q(\mathbf{x}_n)}{k_B T}\right]. \tag{2.107}$$

This proves Crooks' lemma.

A consequence of Crooks' lemma is the following remarkable result.

Integral fluctuation theorem for entropy production. *For a classical Markov process obeying local detailed balance, the fluctuations in stochastic entropy production satisfy*

$$\boxed{\left\langle e^{-\sigma/k_B} \right\rangle_{\mathbf{x}_n} \equiv \sum_{\mathbf{x}_n} p(\mathbf{x}_n) \exp\left[-\frac{\sigma(\mathbf{x}_n)}{k_B}\right] = 1.} \tag{2.108}$$

The proof proceeds as follows. From Crooks' lemma we obtain the general identity

$$\frac{p(\mathbf{x}_n)}{p_{\mathrm{tr}}(\mathbf{x}_n^\dagger)} = \frac{p(\mathbf{x}_n|x_0)p(x_0;0)}{p_{\mathrm{tr}}(\mathbf{x}_n^\dagger|x_n;0)p_{\mathrm{tr}}(x_n)} = \exp\left[-\ln\frac{p_{\mathrm{tr}}(x_n;0)}{p(x_0;0)} - \frac{q(\mathbf{x}_n)}{k_BT}\right]. \tag{2.109}$$

Note that $p_{\mathrm{tr}}(\mathbf{x}_n^\dagger)$ is the probability of observing the trajectory $\mathbf{x}_n^\dagger = (x_0, \ldots, x_n)$ in the backward process starting from the *initial condition* x_n, which occurs with probability $p_{\mathrm{tr}}(x_n;0)$. Since we are free to choose the initial condition of the backward process, we choose it to match the final probablity distribution of the forward process:

$$p_{\mathrm{tr}}(x_n;0) \equiv p(x_n;t_n). \tag{2.110}$$

This choice implies

$$\frac{p(\mathbf{x}_n)}{p_{\mathrm{tr}}(\mathbf{x}_n^\dagger)} = \exp\left[\frac{\Delta s_S(t_n)}{k_B} - \frac{q(\mathbf{x}_n)}{k_BT}\right] = \exp\left[\frac{\sigma(\mathbf{x}_n)}{k_B}\right]. \tag{2.111}$$

Consequently, we obtain the chain of equalities

$$\sum_{\mathbf{x}_n} p(\mathbf{x}_n)\exp\left[-\frac{\sigma(\mathbf{x}_n)}{k_B}\right] = \sum_{\mathbf{x}_n} p(\mathbf{x}_n)\frac{p_{\mathrm{tr}}(\mathbf{x}_n^\dagger)}{p(\mathbf{x}_n)} = \sum_{\mathbf{x}_n} p_{\mathrm{tr}}(\mathbf{x}_n^\dagger) = 1. \tag{2.112}$$

In the last step we used that the operation '†' is an *involution*, meaning that $(\mathbf{x}_n^\dagger)^\dagger = \mathbf{x}_n$, which is easy to check. Hence, it is a bijection such that we can replace $\sum_{\mathbf{x}_n} = \sum_{\mathbf{x}_n^\dagger}$. Since $p_{\mathrm{tr}}(\mathbf{x}_n^\dagger)$ is normalized, this completes the proof.

The integral fluctuation theorem has some immediate consequences, which the reader is asked to derive in the exercise below. First, the integral fluctuation theorem implies the second law, i.e. the non-negativity of entropy production on average. Therefore, it promotes the status of the second law from an inequality to an *exact* relation constraining the likelihood of observing 'violations' of the second law along a single trajectory. In fact, the integral fluctuation theorem proves that there must exist trajectories with negative stochastic entropy production $\sigma(\mathbf{x}_n) < 0$, i.e. the second law is a statement that really refers to *averages* only.

Exercise 2.15 Prove that eqn (2.108) implies $\Sigma = \langle\sigma(\mathbf{x}_n)\rangle_{\mathbf{x}_n} \geq 0$. *Hint:* Use $e^x \geq 1 + x$.

Convince yourself that eqn (2.108) implies the existence of trajectories $\mathbf{x}_n$ with $\sigma(\mathbf{x}_n) < 0$ unless $\sigma(\mathbf{x}_n) = 0$ for all $\mathbf{x}_n$.

Furthermore, to become acquainted with the techniques used in the proofs above, the next exercise asks the reader to verify that these results can be extended to the coarse-grained level under the assumption of time-scale separation.

Exercise 2.16 Return to Section 2.5 and consider a stochastic trajectory of mesostates $\mathbf{x}_n$. Use time-scale separation to show that eqn (2.102) generalizes to the mesolevel as

$$\ln\frac{p(\mathbf{x}_n|x_0)}{p_{\mathrm{tr}}(\mathbf{x}_n^\dagger|x_n)} = -\frac{q_{\mathrm{cg}}(\mathbf{x}_n)}{k_BT} - \sum_{\ell=0}^{n-1}\frac{\mathcal{S}_{x_\ell}(\lambda_{\ell+1}) - \mathcal{S}_{x_\ell}(\lambda_\ell)}{k_B} + \frac{\mathcal{S}_{x_n}(\lambda_n) - \mathcal{S}_{x_0}(\lambda_0)}{k_B}. \tag{2.113}$$

With the choice made in eqn (2.110) show that this implies $\ln[p(\mathbf{x}_n)/p_{\mathrm{tr}}(\mathbf{x}_n^\dagger)] = \sigma(\mathbf{x}_n)/k_B$ in agreement with eqn (2.111). Here, $\sigma(\mathbf{x}_n) \equiv \Delta s_S(t_n) - q_{\mathrm{cg}}(\mathbf{x}_n)/T - \sum_{\ell=0}^{n-1}[\mathcal{S}_{x_\ell}(\lambda_{\ell+1}) - \mathcal{S}_{x_\ell}(\lambda_\ell)]$

is the stochastic entropy production at the mesolevel with the stochastic entropy s_S defined in eqn (2.96).

Although we used the backward process to prove eqn (2.108), it is important to note that eqn (2.108) can be checked experimentally *without* implementing any backward process. However, if statistics about the backward process are available too, we can refine the integral fluctuation theorem (2.108) further. To this end, we write the stochastic entropy production associated to the forward and time–reversed process as

$$\frac{\sigma_{\text{fw}}(\mathbf{x}_n)}{k_B} = \ln \frac{p(\mathbf{x}_n)}{p_{\text{tr}}(\mathbf{x}_n^\dagger)}, \quad \frac{\sigma_{\text{tr}}(\mathbf{x}_n)}{k_B} = \ln \frac{p'_{\text{tr}}(\mathbf{x}_n)}{p'(\mathbf{x}_n^\dagger)}. \tag{2.114}$$

Here, the probabilities $p(\mathbf{x}_n)$ and $p_{\text{tr}}(\mathbf{x}_n)$ are linked by the choice (2.110) that the initial state of the time–reversed process coincides with the final state of the forward process. In analogy, $p'(\mathbf{x}_n)$ and $p'_{\text{tr}}(\mathbf{x}_n)$ are linked by demanding $p'(x_n;0) = p'_{\text{tr}}(x_n;t_n)$, i.e., the initial state of the forward process is chosen to coincide with the final state of the time–reversed process. For later use we note the identity

$$\frac{\sigma_{\text{fw}}(\mathbf{x}_n) + \sigma_{\text{tr}}(\mathbf{x}_n^\dagger)}{k_B} = \ln \left[\frac{p(x_0;0)}{p'(x_0;0)} \frac{p'_{\text{tr}}(x_n;0)}{p_{\text{tr}}(x_n;0)} \right]. \tag{2.115}$$

Next, we consider the probability distribution to observe a certain entropy production during the forward process,

$$P(\sigma) \equiv \sum_{\mathbf{x}_n} \delta[\sigma - \sigma_{\text{fw}}(\mathbf{x}_n)] p(\mathbf{x}_n) = e^{\sigma/k_B} \sum_{\mathbf{x}_n} \delta[\sigma - \sigma_{\text{fw}}(\mathbf{x}_n)] p_{\text{tr}}(\mathbf{x}_n^\dagger), \tag{2.116}$$

where we used the definition of $\sigma_{\text{fw}}(\mathbf{x}_n)$ and replaced it by σ by virtue of the delta function. Switching the summation over $\mathbf{x}_n$ to $\mathbf{x}_n^\dagger$ and using that the operation $\dagger$ is an involution, we find the identity

$$P(\sigma) = e^{\sigma/k_B} \sum_{\mathbf{x}_n} \delta[\sigma - \sigma_{\text{fw}}(\mathbf{x}_n^\dagger)] p_{\text{tr}}(\mathbf{x}_n) \equiv e^{\sigma/k_B} Q(-\sigma). \tag{2.117}$$

Here, $Q(\sigma)$ denotes the probability distribution that, while performing the time–reversed process, the entropy production associated with the forward process takes on the value $\sigma = -\sigma_{\text{fw}}(\mathbf{x}_n^\dagger) = k_B \ln[p_{\text{tr}}(\mathbf{x}_n)/p(\mathbf{x}_n^\dagger)]$. Put differently, eqn (2.117) says that observing an entropy production $\sigma_{\text{fw}} > 0$ along a trajectory $\mathbf{x}_n$ during the forward process is exponentially more likely than associating an 'entropy production' $-\sigma_{\text{fw}} < 0$ to a corresponding trajectory in the time–reversed process. Furthermore, by multiplying eqn (2.117) with $e^{-\sigma/k_B}$ and integrating over σ, we can derive the integral fluctuation theorem (2.108).

Still, the form of eqn (2.117) does not yet appear very insightful as the interpretation of $Q(\sigma)$ is somewhat opaque. More insights can be gained by looking at the entropy production associated to the time–reversed process and by choosing $p_{\text{tr}}(x_0;0) = p'_{\text{tr}}(x_0;0)$. Then, eqn (2.115) implies that (note the exchanged role of $\mathbf{x}_n$ and $\mathbf{x}_n^\dagger$) $\sigma_{\text{fw}}(\mathbf{x}_n^\dagger) + \sigma_{\text{tr}}(\mathbf{x}_n) = k_B \ln[p(x_n;0)/p'(x_n;0)]$. This expression is in general non–zero because even if the time–reversed process starts with the final state of the forward process, the final state of the time–reversed process does not equal the original initial state of the forward process. However, if $\sigma_{\text{fw}}(\mathbf{x}_n^\dagger)+\sigma_{\text{tr}}(\mathbf{x}_n) = 0$, we immediately obtain from eqn (2.117) the following result:

Detailed fluctuation theorem for entropy production. *If for the choice (2.110) the final state of the backward process coincide with the initial state of the forward process, then*

$$\boxed{\frac{P(+\sigma)}{P_{\text{tr}}(-\sigma)} = e^{\sigma/k_B}.} \tag{2.118}$$

where $P_{\text{tr}}(\sigma)$ denotes the probability to observe a stochastic entropy production σ in the backward process for a classical Markov process obeying local detailed balance.

The interesting question is now: when is $\sigma_{\text{fw}}(\mathbf{x}_n^\dagger) + \sigma_{\text{tr}}(\mathbf{x}_n) = 0$ or, equivalently, $p(x_n; 0) = p'(x_n; 0)$? One class of situations that satisfies this criterion are systems with a periodic driving protocol $\lambda_t = \lambda_{t+T}$, which is time–symmetric in each period: $\lambda_{jT+t} = \lambda_{jT+T-t}$ for all $j \in \mathbb{N}$ and $t \in [0, T]$. Then, the system will eventually reach a periodic steady state $p(x_t; t) = p(x_{t+T}; t+T)$ and the same will also be true for the time–reversed probabilities: $p_{\text{tr}}(x_t; t) = p_{\text{tr}}(x_{t+T}; t+T)$. In fact, starting with a periodic steady state at time $t = 0$ and considering a trajectory of duration $t = jT$, one even finds that $p(\mathbf{x}_n) = p_{\text{tr}}(\mathbf{x}_n^\dagger)$. The detailed fluctuation theorem then simplifies to

$$P(-\sigma) = e^{-\sigma/k_B} P(+\sigma). \tag{2.119}$$

This result is particularly suited to explaining why observing violations of the second law of thermodynamics is so unlikely in our everyday macroworld. The probability of observing a negative entropy production $\sigma < 0$ is exponentially suppressed compared with the probability of observing the opposite value. In particular, notice that the exponential contains the entropy production measured in units of Boltzmann's constant, which explains why there was no experimental evidence for it until (roughly) the 2000s (we will treat experiments in greater detail in Section 2.9). Following this line of thought, the reader is asked to do the following exercise.

Exercise 2.17 Suppose the distribution $P(\sigma)$ is Gaussian with mean Σ and variance δ^2. Show that the integral fluctuation theorem fixes the variance to the value

$$\delta^2 = 2k_B\Sigma. \tag{2.120}$$

Show that this implies for the probability of observing a negative stochastic entropy production

$$\int_{-\infty}^{0} d\sigma P(\sigma) = \frac{1}{2}\text{erfc}\left(\frac{\sqrt{\Sigma/k_B}}{2}\right), \tag{2.121}$$

where erfc is the complementary error function. Plot this quantity to convince yourself that this probability is ridiculously small for $\Sigma \gg k_B$. To get a ballpark estimate, consider the textbook case of two ideal gases in a box at the same temperature and pressure initially separated by a dividing partition. After removing the partition, the average entropy production after the gases have finished mixing is $\Sigma = k_B N S_{\text{Sh}}[\{V_1/V, V_2/V\}]$ with V_1 (V_2) the volume initially occupied by gas 1 (2), $V = V_1 + V_2$ the total volume and N the number of particles. For a macroscopic system with $N \approx 10^{23}$ we see that Σ/k_B is ridiculously large.

We remark that a Gaussian distribution is generically expected to be a good description in the *linear response regime*, i.e. when the system is only slightly perturbed away from equilibrium. Equation (2.120) is then a manifestation of the famous **fluctuation–dissipation theorem**, which should not be confused with the fluctuation theorem. The fluctuation–dissipation theorem links fluctuations (δ^2) to dissipation (Σ) in the linear response regime, i.e. *close* to equilibrium. In contrast, the fluctuation theorem holds arbitrarily far from equilibrium.

Another class of interesting situations, where the detailed fluctuation theorem holds, are undriven systems coupled to multiple heat baths. If the system has reached a nonequilibrium steady state it is clear that we have $p(x_t;t) = p_{\text{tr}}(x_t,t)$, independent of t. Thus, the detailed fluctuation theorem holds again with $P_{\text{tr}}(\sigma) = P(\sigma)$. As in Section 2.4 we point out, however, that the treatment of multiple baths implicitly assumes that we know which bath ν has caused a jump, say, from state x_ℓ to $x_{\ell+1}$. This is an additional assumption, which is not easily met in an experiment. Nevertheless, fluctuation theorems involving multiple heat baths play an important role and we study them in detail in Chapter 4, whereas we focus in the remainder of this chapter on the single heat bath case.

Exercise 2.18 Convice yourself of the fact that you are able to repeat the steps above in presence of multiple heat baths.

Finally, an interesting consequence of our previous results is that we can express the average entropy production $\Sigma = \Delta S_S - Q/T$ alternatively as

$$\boxed{\Sigma = k_B \sum_{\mathbf{x}_n} p(\mathbf{x}_n) \ln \frac{p(\mathbf{x}_n)}{p_{\text{tr}}(\mathbf{x}_n^\dagger)} = k_B D[p(\mathbf{x}_n)|p_{\text{tr}}(\mathbf{x}_n^\dagger)],} \tag{2.122}$$

i.e. the entropy production equals (up to k_B) the relative entropy between the probability distributions of the forward and backward trajectories. Thus, for a classical Markov process obeying local detailed balance, the entropy production quantifies how easy it is to distinguish between the forward and the backward experiment with respect to the trajectory probabilities connected by eqn (2.110). This interpretation of entropy production turns out to be more generally valid also for quantum systems (see Chapter 4), but the trajectories one has to compare are then no longer the simple system trajectories $\mathbf{x}_n$. Moreover, while eqn (2.122) nicely demonstrates that entropy production quantifies time-reversal symmetry breaking at the macrolevel, eqn (2.122) does *not* prove the existence of an arrow of time. In fact, we started from a time-asymmetric situation (namely, a Markovian master equation), so it is clear that we get time-asymmetric results out. Moreover, as emphasized at the beginning, 'forwards' and 'backwards' are merely labels. We can also prove the increase in entropy for the backward dynamics using the same arguments, but calling the forward process backward process and vice versa. For further discussion see Appendix C.3.

Another word of caution is necessary. Despite the beauty of the result (2.102) and its consequences such as eqn (2.122), it is important to keep in mind that we derived them under the two assumptions of classicality and time-scale separation (which, in particular, implies Markovianity and local detailed balance). The relative entropy, which is by definition non-negative for *any* two trajectory probability distributions, is *not* necessarily linked to entropy production. In Section 2.8 we will even see explicitly that eqn (2.122) breaks down. Moreover, the next exercise shows that there is a certain *triviality* associated with the fluctuations theorem, which—when not taken carefully into account—can quickly lead to erroneous conclusions and 'derivations' of 'second laws' with little physical meaning.

Exercise 2.19 Integral and detailed fluctuation theorems can be trivially constructed if three mathematical requirements are met. First, we have two probability distributions $p(x)$ and $q(x)$ defined for $x \in X$, where X is some set. Second, let $p(x) = 0$ if and only if $q(x) = 0$. Third, assume we have an involution $\dagger : X \mapsto X$, which satisfies by definition $(x^\dagger)^\dagger = x$. Now, define the 'entropy production' $\sigma(x) \equiv \ln[p(x)/q(x^\dagger)]$. Show that the so-defined object always satisfies an integral and detailed fluctuation theorem of the form $\langle e^{-\sigma} \rangle = 1$ and $P(+\sigma)/Q(-\sigma) = e^\sigma$ with $P(\sigma) = \sum_x p(x)\delta[\sigma - \sigma(x)]$, $Q(\sigma) = \sum_x q(x)\delta[\sigma - \sigma_{\text{tr}}(x)]$ and $\sigma_{\text{tr}}(x) = \ln[q(x)/p(x^\dagger)]$. Thus, not every mathematical fluctuation theorem has a physical meaning and relates to the second law.

In the next section, we introduce another class of important fluctuation theorems.

2.7 Work Fluctuation Theorems

We now radically change the perspective. Instead of working with the reduced dynamics of an open system in terms of a classical Markov process obeying local detailed balance, we consider the system *and* the bath as one big isolated system described by classical Hamiltonian mechanics. Clearly, in the case where we can treat the bath using time-scale separation, the dynamics of the system matches our previous description in terms of a rate master equation. Time-scale separation is, however, a strong assumption and therefore the results reported in the following are considerably more general than our previous results. In this section we exclusively focus on the fluctuations of work, which we show to also satisfy fluctuation theorems. We return to the fluctuations of entropy production in the next section.

Work fluctuation in isolated systems

Before we turn to the system–bath paradigm, we start with an isolated system described by a Hamiltonian $H(\Gamma)$. Remember that in classical mechanics dynamics happens in a *phase space*. If the system has N particles with f degrees of freedom (typically $f = 3$), a point in phase space is denoted by $\Gamma = (q_1, \dots, q_{Nf}, p_1, \dots, p_{Nf}) \in \mathbb{R}^{2Nf}$, which requires one to microscopically specify the generalized coordinates q_i and conjugate momenta p_i of all particles. The time evolution of a phase-space point follows from Hamilton's equation of motion, the solution to which is written as $\Gamma^t = \phi^t(\Gamma^0)$ with the initial phase-space point Γ^0. Furthermore, a *trajectory* in phase-space defined by the set of points $\{\phi^t(\Gamma^0)|t \in [0, \tau]\}$ is denoted $\gamma^\tau = \gamma^\tau(\Gamma^0)$, where τ denotes some final time.

Remember that the dynamics of a system described by a time-independent Hamiltonian *conserves energy*, which follows from the fact that the Poisson bracket of the Hamiltonian with itself vanishes:

$$\frac{d}{dt}H(\Gamma^t) = \{H(\Gamma^t), H(\Gamma^t)\} \equiv \sum_{i=1}^{Nf}\left[\frac{\partial H(\Gamma^t)}{\partial q_i^t}\frac{\partial H(\Gamma^t)}{\partial p_i^t} - \frac{\partial H(\Gamma^t)}{\partial p_i^t}\frac{\partial H(\Gamma^t)}{\partial q_i^t}\right] = 0. \quad (2.123)$$

In order to change the energy of an isolated system, we have to make the Hamiltonian explicitly time dependent: $H(\Gamma) = H(\Gamma; \lambda_t)$. As before, we denote the external driving protocol by λ_t. Then, we find

$$\frac{d}{dt}H(\Gamma^t;\lambda_t) = \{H(\Gamma^t;\lambda_t), H(\Gamma^t;\lambda_t)\} + \left.\frac{\partial}{\partial t}\right|_{\Gamma^t} H(\Gamma^t;\lambda_t) = \left.\frac{\partial}{\partial t}\right|_{\Gamma^t} H(\Gamma^t;\lambda_t), \quad (2.124)$$

which is no longer zero. Here, $\partial_t|_{\Gamma^t}$ denotes a partial derivative with respect to time for a fixed Γ^t. Integrating the last expression and using $\partial_t|_{\Gamma^t}H(\Gamma^t;\lambda_t) = \dot{\lambda}_t\partial_{\lambda_t}H(\Gamma^t;\lambda_t)$ with $\dot{\lambda}_t \equiv \partial_t\lambda_t$, we obtain

$$\boxed{w(\gamma^\tau) \equiv H(\Gamma^\tau;\lambda_\tau) - H(\Gamma^0;\lambda_0) = \int_0^\tau dt\dot{\lambda}_t\frac{\partial H(\Gamma^t;\lambda_t)}{\partial\lambda_t}.} \quad (2.125)$$

This is the mechanical **work** done on the system along a trajectory γ^τ. This identification is justified if one recalls the phenomenological first law (2.1) and the fact that an isolated system does not exchange any heat ($Q = 0$). Thus, its change in energy must be work.

The work defined in Eq. (2.125) is a deterministic quantity, i.e. it is completely fixed by the initial point Γ^0 in phase space. In many situations, however, the precise initial microstate Γ^0 is not completely known. Then, one has to work with a probability distribution or 'ensemble' of states $\rho(\Gamma;0)$, which evolves in time according to

$$\rho(\Gamma;t) = \int d\Gamma^0\rho(\Gamma^0;0)\delta[\Gamma - \phi^t(\Gamma^0)]. \quad (2.126)$$

Consequently, the work $w(\gamma^\tau)$ done along a single trajectory becomes a random variable, characterized by its average

$$W(\tau) \equiv \langle w\rangle_{\gamma^\tau} \equiv \int d\Gamma^0 w[\gamma^\tau(\Gamma^0)]\rho(\Gamma^0;0) \quad (2.127)$$

and higher moments $\langle w^n\rangle_{\gamma^\tau}$ for $n > 1$.

Our next remarkable result tells us that work cannot fluctuate arbitrarily in a classical Hamiltonian system if the initial state is described by a canonical ensemble at inverse temperature β. The latter is defined in classical statistical mechanics by

$$\pi(\Gamma;\lambda) \equiv \frac{e^{-\beta H(\Gamma;\lambda)}}{h^{Nf}\mathcal{Z}(\lambda)}, \quad \mathcal{Z}(\lambda) \equiv \int\frac{d\Gamma}{h^{Nf}}e^{-\beta H(\Gamma;\lambda)}, \quad \mathcal{F}(\lambda) \equiv -k_BT\ln\mathcal{Z}(\lambda). \quad (2.128)$$

Here, h is Planck's constant and the additional factor h^{Nf} has been introduced in order to make the partition function $\mathcal{Z}(\lambda)$ dimensionless, such that we can take its logarithm to compute the **equilibrium free energy** $\mathcal{F}(\lambda)$ (also called the Helmholtz free energy). The factor h^{Nf} is typically interpreted as the phase-space volume of a microstate in classical statistical mechanics, otherwise the number of microstates becomes uncountable. However, below we are only interested in the *ratios* of partition functions or free energy *differences* such that this factor always cancels out. Moreover, we neglect the possibility of indistinguishable particles, which would introduce another factor $N!$. Again, this factor would also cancel out in all expressions below.

After these preliminary considerations, we can study the fluctuations of work for a system prepared in a canonical ensemble. The probability $p(w)$ of applying an amount of work w along a single realization of the process is

$$p(w) = \int d\Gamma^0 \delta[w - w(\gamma^\tau)]\pi(\Gamma^0; \lambda_0) \tag{2.129}$$

with $w(\gamma^\tau) = w[\gamma^\tau(\Gamma^0)]$. Note that we make no assumption about $H(\Gamma; \lambda_t)$ in our formulation: albeit starting in equilibrium, the system can be driven *far* from equilibrium during the evolution. Nevertheless, we obtain the following universal relation.

Integral work fluctuation theorem. *Work fluctuations of an isolated classical system prepared in a canonical ensemble at temperature T obey for any driving protocol*

$$\boxed{\left\langle e^{-\beta w}\right\rangle_\gamma \equiv \int dw e^{-\beta w} p(w) = e^{-\beta \Delta \mathcal{F}},} \tag{2.130}$$

where $\Delta\mathcal{F} \equiv \mathcal{F}(\lambda_\tau) - \mathcal{F}(\lambda_0)$ denotes the change in equilibrium free energy with respect to the initial temperature T.

The proof is straightforward. Using eqns (2.125) and (2.129), we find that

$$\int dw e^{-\beta w} p(w) = \frac{1}{\mathcal{Z}(\lambda_0)} \int d\Gamma^0 e^{-\beta H[\phi^\tau(\Gamma^0);\lambda_\tau]}. \tag{2.131}$$

Next, we use a central result in classical mechanics known as Liouville's theorem; see eqn (C.16). It allows us to change the variable Γ^0 in the integral to $\Gamma^\tau = \phi^\tau(\Gamma^0)$ without introducing any correction term since the Jacobian determinant associated with the transformation ϕ^τ equals one. Hence,

$$\int dw e^{-\beta w} p(w) = \frac{1}{\mathcal{Z}(\lambda_0)} \int d\Gamma^\tau e^{-\beta H(\Gamma^\tau;\lambda_\tau)} = \frac{\mathcal{Z}(\lambda_\tau)}{\mathcal{Z}(\lambda_0)}. \tag{2.132}$$

To complete the proof, we notice that $\mathcal{Z}(\lambda_\tau)/\mathcal{Z}(\lambda_0) = e^{-\beta\Delta\mathcal{F}}$.

Moreover, there is also a *detailed* work fluctuation theorem. To state it, we make, use of a *backward* or *time-reversed process*. Recall that, as detailed in Appendix C, an isolated Hamiltonian system obeys time-reversal symmetry. Furthermore, we now explicitly include the presence of an external magnetic field B in our description and denote the Hamiltonian by $H(\Gamma; \lambda, B)$. The corresponding canonical equilibrium state is denoted $\pi(\Gamma; \lambda, B)$. The forward process is defined as above and it is characterized by a work distribution (2.129). The backward or time-reversed process is defined by starting with the canonical ensemble $\pi(\Gamma; \lambda_\tau, -B)$ and by changing the driving protocol

$\lambda_t^\dagger \equiv \lambda_{\tau-t}$ in the opposite direction, starting from λ_τ and ending at λ_0. Furthermore, all this is done in the presence of an inverted magnetic field $-B$. The probability of applying a work w during the time-reversed process is

$$p_{\mathrm{tr}}(w) = \int d\Gamma^0 \delta[w - w_{\mathrm{tr}}(\gamma^\tau)]\pi(\Gamma^0; \lambda_\tau, -B), \tag{2.133}$$

where $w_{\mathrm{tr}}(\gamma^\tau) = H(\Gamma^\tau; \lambda_0, -B) - H(\Gamma^0; \lambda_\tau, -B)$ is the work done during the time-reversed process in unison with our definition (2.125). Note that in an actual experiment time also runs 'forwards' in the 'time-reversed' process. Therefore, we denote the initial microstate in the time-reversed process by Γ^0 and the final state by $\Gamma^\tau = \phi_{\mathrm{tr}}^\tau(\Gamma^0)$, as in the forward process, but with respect to a time-reversed flow ϕ_{tr}^t generated by $H(\Gamma; \lambda_t^\dagger, -B)$. Then, we obtain the following remarkable result.

Detailed work fluctuation theorem. *The distributions of work $p(w)$ and $p_{\mathrm{tr}}(w)$ for the forward and backward process obey the relation*

$$\boxed{\frac{p(w)}{p_{\mathrm{tr}}(-w)} = e^{\beta(w-\Delta\mathcal{F})}.} \tag{2.134}$$

The proof goes as follows. Using the definition (2.125) of work and the property of the Dirac delta function, we express $p(w)$ as

$$\begin{aligned} p(w) &= \int d\Gamma^0 \delta\big(w - [H(\Gamma^\tau; \lambda_\tau, B) - H(\Gamma^0; \lambda_0, B)]\big) \frac{e^{-\beta[H(\Gamma^\tau;\lambda_\tau,B)-w]}}{\mathcal{Z}(\lambda_0)} \\ &= e^{\beta(w-\Delta\mathcal{F})} \int d\Gamma^\tau \delta\big(w - [H(\Gamma^\tau; \lambda_\tau, B) - H(\Gamma^0; \lambda_0, B)]\big)\pi(\Gamma^\tau; \lambda_\tau, B). \end{aligned} \tag{2.135}$$

In the last step, we also used Liouville's theorem as above to switch the integration from Γ^0 to Γ^τ, interpreting $\Gamma^0 = (\phi^\tau)^{-1}(\Gamma^\tau)$ as a function of Γ^τ. Now, we use time-reversal symmetry to express the integral as the work probability distribution obtained in the time-reversed process. For this purpose, we need to recall some results from Appendix C. First, let $\Theta\Gamma$ denote the time reversal of a point Γ in phase space, which is obtained by flipping the sign of all momenta while keeping the positions fixed. In terms of that operation we can link the flows in the forward and backward dynamics via $\Theta(\phi^\tau)^{-1} = \phi_{\mathrm{tr}}^\tau\Theta$. Furthermore, we use that the Hamiltonian obeys the symmetry $H(\Gamma; \lambda, B) = H(\Theta\Gamma; \lambda, -B)$ for all λ. This allows us to write $p(w)$ as

$$\begin{aligned} p(w) &= e^{\beta(w-\Delta\mathcal{F})} \\ &\times \int d\Gamma^\tau \delta\big(w - [H(\Theta\Gamma^\tau; \lambda_\tau, -B) - H(\phi_{\mathrm{tr}}^\tau\Theta\Gamma^\tau; \lambda_0, -B)]\big)\pi(\Theta\Gamma^\tau; \lambda_\tau, -B). \end{aligned} \tag{2.136}$$

Next, we simply relabel $\Theta\Gamma^\tau$ by Γ^0 to make clear that this is the initial phase-space point in the backward dynamics. Thus,

$$\begin{aligned} p(w) &= e^{\beta(w-\Delta\mathcal{F})} \\ &\times \int d(\Theta\Gamma^0) \delta\big(w - [H(\Gamma^0; \lambda_\tau, -B) - H(\phi_{\mathrm{tr}}^\tau\Gamma^0; \lambda_0, -B)]\big)\pi(\Gamma^0; \lambda_\tau, -B). \end{aligned} \tag{2.137}$$

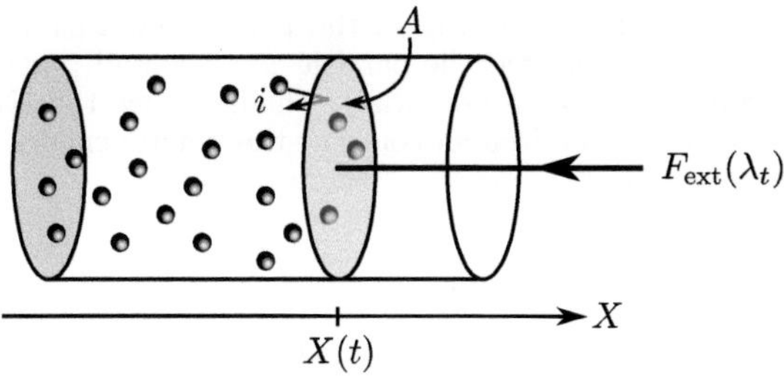

Fig. 2.8 Sketch of gas particles confined to a cylinder by a frictionless gliding piston with position $X(t)$ and surface area A. The bombardment of gas particles (here, the ith particle) against the surface area of the piston creates the pressure. This pressure can be compensated for by an external force $F_{\text{ext}}(\lambda_t)$ to control the piston's position X.

Here, we also used $\Theta^2 = 1$, which means that Θ is an *involution* because flipping the sign of the momentum twice does not change the state. This property also allows us to confirm that $\int d(\Theta\Gamma)f(\Gamma) = \int d\Gamma f(\Gamma)$ for any function f provided that the integral extends over the entire phase space. This allows us to conclude the proof since

$$\begin{aligned} p(w) &= e^{\beta(w-\Delta\mathcal{F})} \int d\Gamma^0 \delta(w - [H(\Gamma^0;\lambda_\tau,-B) - H(\phi_{\text{tr}}^\tau \Gamma^0;\lambda_0,-B)])\pi(\Gamma^0;\lambda_\tau,-B) \\ &= e^{\beta(w-\Delta\mathcal{F})} p_{\text{tr}}(-w). \end{aligned} \tag{2.138}$$

To summarize, the work fluctuation relations (2.130) and (2.134) are two remarkable identities linking a *non-equilibrium* concept (the work w applied to the system) to an *equilibrium* concept (the free energy difference $\Delta\mathcal{F}$). Their physical consequences are discussed below for the open system case and in later sections. Before we turn to open systems, however, the next exercise asks you to ponder the definition of work a bit further and to compare it with what you have learned at school.

Exercise 2.20 Consider a single particle with position vector $\boldsymbol{q}(t)$. If a force $\boldsymbol{F} = \boldsymbol{F}(\boldsymbol{q})$ acts on the particle, we learn in school that the work done on the particle is given by the scalar product of force multiplied by the displacement evaluated along the trajectory γ^τ of the particle: $w'(\gamma^\tau) \equiv \int_{\gamma^\tau} \boldsymbol{F}\cdot d\boldsymbol{q}$. Use Newton's law $\boldsymbol{F} = m\ddot{\boldsymbol{q}}(t)$ to deduce that $w' = \Delta T$, where $T = m\dot{\boldsymbol{q}}^2/2$ denotes the *kinetic energy* of the particle. Assuming that the total energy of the particle is $T + V$, where $V = V(\boldsymbol{q})$ is the potential energy, show that the difference between this and our definition of work is $w(\gamma^\tau) - w'(\gamma^\tau) = \Delta V$. Thus, in some sense, the schoolbook definition views the kinetic energy alone as the 'internal energy' of the system, whereas the potential in which the particle moves is something 'external'. As a consequence, 'work' is done on the particle even in a conservative and time-independent potential (e.g. an oscillating pendulum).

It is instructive to consider another textbook application: the compression of a gas in a cylinder by a piston (see Fig. 2.8 for a sketch). If P is the pressure of the gas occupying a volume V, thermodynamics tells us that the work done is $W = -\int PdV$. Now, let X denote the position of the piston with mass M, which obeys Newton's law $M\ddot{X} = \sum_i F_i + F_{\text{ext}}(\lambda_t)$. Here, F_i is the force exerted by the ith particle on the piston, which we assume to result from a very short-ranged but strong interaction (in practice, one could simulate it by reflecting the particles at the piston while keeping the piston-plus-particle momentum conserved). Show that the work done on the gas is $W = -\int PdV = -\sum_i \int F_i dX$. Derive that $W = \Delta E_{\text{tot}} -$

ΔT_{piston}, where T_{piston} is the kinetic energy of the piston and E_{tot} is the total energy of all particles and the piston (assume that the particle–particle and particle–piston forces are conservative, i.e. can be derived from a potential). Then, deduce that W is identical to our definition (2.125) if the system Hamiltonian contains the kinetic energy of the gas particles, the particle–particle potential and the particle–piston potential.

Work fluctuations in open systems

The idea of modelling the dynamics of an open classical system using classical Hamiltonian mechanics by incorporating the bath degrees of freedom is analogous to the quantum case, which was discussed in detail in Section 1.2. We therefore proceed with the mathematical description and write the classical system–bath Hamiltonian as

$$H_{SB}(\Gamma_{SB};\lambda_t) = H_S(\Gamma_S;\lambda_t) + H_B(\Gamma_B) + V_{SB}(\Gamma_{SB}). \tag{2.139}$$

Now, Γ_S (Γ_B) denotes a point in phase space of the system (bath) alone and $\Gamma_{SB} = \Gamma_S \oplus \Gamma_B$ is their direct sum. The system, bath and interaction energy are represented by the Hamiltonians $H_S(\Gamma_S;\lambda_t)$, $H_B(\Gamma_B)$ and $V_{SB}(\Gamma_{SB})$, respectively. Furthermore, we assume that the driving λ_t only affects the *system* Hamiltonian, which has important consequences in the following. Finally, the flow $\Gamma_{SB}^t = \phi^t \Gamma_{SB}^0$ in the global phase space also defines the system trajectory Γ_S^t by projection.

Before proceeding with work fluctuations, let us illustrate the abstract Hamiltonian (2.139) by the specific example of the Caldeira–Leggett model (see Exercise 1.2):

$$H_{SB}(\lambda_t) = \frac{p_s^2}{2} + V(q_s;\lambda_t) + \frac{1}{2}\sum_{k=1}^{n}\left[p_k^2 + \omega_k^2\left(q_k - \frac{c_k}{\omega_k^2}f(q_S)\right)^2\right]. \tag{2.140}$$

Here, the bath is modelled by a collection of n harmonic oscillators (in mass-weighted coordinates), such that we can identify $\Gamma_B = (q_1,\dots,q_n,p_1,\dots,p_n)$. The system is a single particle with phase-space coordinates $\Gamma_S = (q_s,p_s)$ moving in a time-dependent potential $V(q_s;\lambda_t)$. It is coupled to the positions of the bath oscillators via some function $f(q_s)$. Often, one simply chooses $f(q_s) = q_s$. Furthermore, the above Hamiltonian can be generalized to a three-dimensional world by attaching to every position and momentum an additional subscript $i \in \{1,2,3\}$ and by summing over i. The Hamiltonian (2.140) might appear simplistic with perhaps little use in practice. It turns out, however, that the opposite is true: eqn (2.140) is justified whenever the influence of the bath is Gaussian and it is successfully used to model the dynamics of a colloidal particle or more complicated molecules in water and other solvents (among many other applications). The resulting dynamics of the system is called *Brownian motion.* After this brief illustration, let us return to the abstract case again because, in fact, all our results below hold for any Hamiltonian of the form (2.139).

We proceed by noting that our general definition of work (2.125) reduces for a system–bath Hamiltonian, where only the system part is time dependent, to

$$\boxed{w(\gamma^\tau) = \int_0^\tau dt\dot{\lambda}_t \frac{\partial H_S(\Gamma_S^t;\lambda_t)}{\partial \lambda_t} = w(\gamma_S^\tau).} \tag{2.141}$$

We now recognize the importance of demanding that λ_t only affects the *system* Hamiltonian because we see that the work done on the system–bath composite can be inferred from *knowing only the system trajectories* $\gamma_S^\tau = \{\Gamma_S^t | t \in [0, \tau]\}$. This makes the experimental determination of work feasible and a specific example is studied in Section 2.9. We remark, however, that, in contrast to the isolated case, the system trajectory γ_S^τ now looks *noisy*: it is no longer deterministically determined by the initial condition Γ_S^0, albeit it is still determined by the initial point Γ_{SB}^0 in the *global* phase space.

For simplicity, we focus in this section on the *weak coupling regime*, in which the interaction Hamiltonian V_{SB} is assumed to be negligibly small compared with H_S and H_B. Then, the canonical ensemble (2.128) of the system–bath composite approximately factorizes as

$$\pi_{SB}(\Gamma_{SB}; \lambda) \approx \pi_S(\Gamma_S; \lambda)\pi_B(\Gamma_B). \tag{2.142}$$

In particular, the partition function splits into a product $\mathcal{Z}_{SB}(\lambda) = \mathcal{Z}_S(\lambda)\mathcal{Z}_B$, where only the system part depends on the control parameter λ.

Exercise 2.21 Derive eqn (2.142) and write down the explicit definitions for $\pi_S(\Gamma_S; \lambda)$ and $\pi_B(\Gamma_B)$.

Based on our previous considerations, it is then straightforward to show that the integral and detailed work fluctuation theorems, eqns (2.130) and (2.134), reduce for a system–bath Hamiltonian of the form (2.139) to the following form.

Work fluctuation theorems for open systems. *Consider a driven system weakly coupled to a bath prepared in a canonical ensemble (2.142) at temperature T. Let $p(w)$ ($p_{\mathrm{tr}}(w)$) be the probability distribution of the fluctuating work (2.141) in the forward (backward) process. Then, for any driving protocol λ_t, we have*

$$\boxed{\left\langle e^{-\beta w}\right\rangle_{\gamma_S} = e^{-\beta\Delta\mathcal{F}_S}, \quad \frac{p(w)}{p_{\mathrm{tr}}(-w)} = e^{\beta(w-\Delta\mathcal{F}_S)},} \tag{2.143}$$

where $\Delta\mathcal{F}_S \equiv \mathcal{F}_S(\lambda_\tau) - \mathcal{F}_S(\lambda_0)$ is the change in the system's equilibrium free energy with respect to temperature T.

Exercise 2.22 Derive eqn (2.143).

We devote the rest of this section to understanding the connection between the work fluctuation theorems derived here and the results of Section 2.6. For this purpose we return to the situaton of a classical Markov process obeying local detailed balance. The next exercise provides an independent derivation of eqn (2.143) for this case.

Exercise 2.23 Return to the setting from Section 2.6 and consider a forward and backward process by demanding that the initial state of the forward and backward process is described by a canonical ensemble, i.e. using our previous notation by $\pi_x(\lambda_0) = e^{-\beta E_x(\lambda_0)}/\mathcal{Z}_S(\lambda_0)$ and $\pi_x(\lambda_\tau) = e^{-\beta E_x(\lambda_\tau)}/\mathcal{Z}_S(\lambda_\tau)$, respectively. Use Crooks' lemma to derive

$$\ln \frac{p(\mathbf{x}_n)}{p_{\text{tr}}(\mathbf{x}_n^\dagger)} = \frac{w(\mathbf{x}_n) - \Delta\mathcal{F}_S}{k_B T}, \tag{2.144}$$

where $w(\mathbf{x}_n) = u(x_n, t_n) - u(x_0, 0) - q(\mathbf{x}_n)$ (identifying $t_n \equiv \tau$) is the stochastic work during the forward process. Use this result to derive eqn (2.143) after identifying $\boldsymbol{\gamma}_S$ with $\mathbf{x}_n$.

Next, we investigate the difference between the work fluctuation theorem and the entropy production fluctuation theorem. For this purpose, we return to the *average* second law. Remember that the integral fluctuation theorem for entropy production implies $\Sigma(t) = \Delta S_S(t) - Q(t)/T \geq 0$. We now rewrite this expression using the first law $\Delta U_S(t) = Q(t) + W(t)$

$$\Sigma(t) = \frac{W(t) - [\Delta U_S(t) - T\Delta S_S(t)]}{T} \geq 0. \tag{2.145}$$

This rewriting suggests introducing a **non-equilibrium free energy**, which plays an important role in the remainder of this book and is defined as

$$\boxed{F_S(t) \equiv U_S(t) - TS_S(t) = \sum_x E_x(\lambda_t)p_x(t) + k_B T \sum_x p_x(t) \ln p_x(t).} \tag{2.146}$$

In unison with the equilibrium free energy $\mathcal{F}_S(\lambda_t)$, it shares the property that it is defined with respect to some temperature T of a surrounding heat bath. Without reference to a heat bath at temperature T, eqn (2.146) becomes ill-defined. Apart from this temperature dependence and in contrast to the equilibrium free energy $\mathcal{F}_S(\lambda_t)$, which is a function of the energy landscape $E_x(\lambda_t)$ only, the non-equilibrium free energy $F_S(t)$ depends on the entire distribution $p_x(t)$, which can be out of equilibrium. Moreover, there is a remarkable connection between the two, which shows that the non-equilibrium free energy is always larger than the equilibrium free energy and their difference is quantified by the relative entropy:

$$\boxed{F_S(t) - \mathcal{F}_S(\lambda_t) = k_B T D[\mathbf{p}(t)|\boldsymbol{\pi}(\lambda_t)] \geq 0.} \tag{2.147}$$

Exercise 2.24 Derive eqn (2.147).

After these preliminary considerations, we return to our central question. First, with the help of definition (2.146) we can express the entropy production for a driven system coupled to a single heat bath as

$$\Sigma(t) = \frac{W(t) - \Delta F_S(t)}{T} \geq 0. \tag{2.148}$$

Next, assume that the system is initialized at equilibrium: $p_x(t_0) = \pi_x(\lambda_0)$. Then, using eqn (2.147), the second law can be bounded from above as

$$\Sigma(t) = \frac{W(t) - [F_S(t) - \mathcal{F}_S(\lambda_0)]}{T} \leq \frac{W(t) - \Delta\mathcal{F}_S(T, \lambda_t)}{T} \equiv \frac{W_{\text{diss}}(t)}{T}. \tag{2.149}$$

In this context W_{diss} is known as the **dissipated work**, which is always positive for a system starting in equilibrium. Remarkably, the inequality $W_{\text{diss}}(t) \geq 0$ is also implied by the work fluctuation theorem similar to the way $\Sigma(t) \geq 0$ is implied by the entropy production fluctuation theorem.

We can now answer the question of when does Σ coincide with W_{diss}/T, which happens in the following case: initialize the system at equilibrium, drive it out of equilibrium by changing λ_t and after time τ wait long enough by keeping λ_τ fixed such that at some time $\tau' \gg \tau$ the system has relaxed back to equilibrium, which implies $W_{\text{diss}}(\tau) = W_{\text{diss}}(\tau') = T\Sigma(\tau')$.

Thus, the work fluctuation theorem neglects in some sense the final non-equilibrium character of the system. However, the work fluctuation theorem holds even if we do not wait until time $\tau' \gg \tau$. In fact, eqn (2.143) continues to hold even if the system does not return to equilibrium at all. This point can cause confusion since it is the final *equilibrium* free energy $\mathcal{F}_S(\lambda_\tau)$ that appears in the work fluctuation theorem, albeit the final state of the system can be far from equilibrium. But this fact is perhaps also its most astonishing feature.

Equilibrium information from non-equilibrium dynamics. *Useful information about equilibrium properties is contained in thermodynamic fluctuations far from equilibrium.*

Finally, there is another difference from the *practical* point of view. The work fluctuation theorem requires only work along single trajectories to be measured, whereas the entropy production fluctuation theorem also requires knowledge about the non-equilibrium internal energy and system entropy at the beginning and at the end of the process. This can be challenging to obtain. For certain systems, such as those studied in Section 2.9, the work fluctuation theorem has been experimentally verified whereas the entropy production fluctuation theorem has not.

2.8 Strong Coupling Corrections

So far we have considered two situations in this chapter after the phenomenological description of Section 2.1. In Sections 2.2–2.6 we focused on systems described in terms of microstates or mesostates (collections of microstates) under the assumption that all other degrees of freedom are fast and time-scale separation can be applied. In the last section we dropped the assumption of time-scale separation and considered the non-Markovian dynamics of a system *weakly* coupled to a bath. Now, we overcome both assumptions and show that a stochastic thermodynamics framework can be established only based on the fact that the bath is initially thermal. No time-scale separation or weak coupling assumption is needed in the following. Despite this remarkable mathematical generality of the following results, there are some words of caution necessary at the end.

Work fluctuation theorems at strong coupling

The extension of eqn (2.143) to the strong coupling regime is straightforward if we recall the definition of the *Hamiltonian of mean force* from Section 1.3. Remember that strong coupling effects are responsible for deviations of the equilibrium state of an open system from a canonical Gibbs ensemble. These deviations are captured by the Hamiltonian of mean force. Using our classical notation, we define it as

$$\pi_S^*(\Gamma_S;\lambda) = \int d\Gamma_B \pi_{SB}(\Gamma_{SB};\lambda) \equiv \frac{e^{-\beta H_S^*(\Gamma_S;\lambda)}}{h^{N_S f}\mathcal{Z}_S^*(\lambda)}, \quad \mathcal{Z}_S^*(\lambda) \equiv \frac{\mathcal{Z}_{SB}(\lambda)}{\mathcal{Z}_B}. \tag{2.150}$$

Here, the factor $h^{N_S f}$ has been introduced for the same reasons as in eqn (2.128) and N_S denotes the number of system particles. Note that, since the Hamiltonians H_S, H_B and V_{SB} 'commute' in the classical case, the Hamiltonian of mean force can be further simplified compared with the quantum expression (1.22) and reads explicitly

$$H_S^*(\Gamma_S;\lambda) = H_S(\Gamma_S;\lambda) - \frac{1}{\beta}\ln\int d\Gamma_B e^{-\beta V_{SB}(\Gamma_{SB})}\pi_B(\Gamma_B). \tag{2.151}$$

Moreover, remember that the Hamiltonian of mean force depends on the bath temperature. This fact becomes important soon.

Next, let us return to our general work fluctuation theorems (2.130) and (2.134) derived for an arbitrary isolated system. Our goal is to apply them to the system–bath Hamiltonian (2.139) with an arbitary strong interaction V_{SB}. By definition of the partition function of the Hamiltonian of mean force we can express the global partition function as $\mathcal{Z}_{SB}(\lambda) = \mathcal{Z}_S^*(\lambda)\mathcal{Z}_B$ and we see that

$$\Delta\mathcal{F}_{SB} = -k_BT\ln\frac{\mathcal{Z}_{SB}(\lambda_t)}{\mathcal{Z}_{SB}(\lambda_0)} = -k_BT\ln\frac{\mathcal{Z}_S^*(\lambda_t)}{\mathcal{Z}_S^*(\lambda_0)} \equiv \Delta\mathcal{F}_S^*. \tag{2.152}$$

Here, $\mathcal{F}_S^*(\lambda_t) \equiv -k_BT\ln\mathcal{Z}_S^*(\lambda_t)$ denotes the **equilibrium free energy at strong coupling**. Note that, to arrive at the conclusion $\Delta\mathcal{F}_{SB} = \Delta\mathcal{F}_S^*$, we only need to assume that the bath Hamiltonian H_B is time independent. Then, eqns (2.130) and (2.134) immediately imply the **work fluctuation theorems at strong coupling**

$$\boxed{\left\langle e^{-\beta w}\right\rangle_\gamma = e^{-\beta\Delta\mathcal{F}_S^*}, \quad \frac{p(w)}{p_{\text{tr}}(-w)} = e^{\beta(w-\Delta\mathcal{F}_S^*)}.} \tag{2.153}$$

Non-Equilibrium thermodynamics at strong coupling

We now want to go one step further and study the entropy production for a system strongly coupled to a bath. As explained at the end of the last section, stochastic work defined in eqn (2.141) and equilibrium free energy $\mathcal{F}_S^*$ are insufficient to characterize entropy production in general. For a full thermodynamic understanding we also have to define heat, internal energy, system entropy and non-equilibrium free energy (not all of them are independent). Clearly, at strong coupling and without the assumption of time-scale separation, we are leaving the realm of well-established results. The following definitions have, however, appealing properties and we discuss at the end when it is safe to apply them.

Before we turn to the non-equilibrium thermo*dynamics*, it is useful to first reconsider some aspects of equilibrium thermo*statics* for a fixed value λ of the control protocol. For this purpose recall standard statistical mechanics using the Gibbs ensemble $\pi(\Gamma;\beta) = e^{-\beta H(\Gamma)}/h^{Nf}\mathcal{Z}(\beta)$. After introducing the free energy $\mathcal{F}(\beta) = -k_B T \ln \mathcal{Z}(\beta)$, a straightforward calculation reveals that the internal energy and system entropy become $\mathcal{U} = \int d\Gamma H(\Gamma)\pi(\Gamma) = \partial_\beta(\beta\mathcal{F})$ and $\mathcal{S} = -\int d\Gamma \pi(\Gamma)\ln[h^{Nf}\pi(\Gamma)] = k_B\beta^2\partial_\beta\mathcal{F}$, where we dropped the dependence on β for notational simplicity. Now, since we have already introduced an equilibrium free energy $\mathcal{F}_S^*(\lambda)$ at strong coupling, we *postulate* that the equilibrium internal energy and system entropy at strong coupling can be obtained by analogy,

$$\mathcal{U}_S^*(\lambda) \equiv \frac{\partial}{\partial\beta}[\beta\mathcal{F}_S^*(\lambda)], \quad \mathcal{S}_S^*(\lambda) \equiv k_B\beta^2\frac{\partial}{\partial\beta}\mathcal{F}_S^*(\lambda). \tag{2.154}$$

Explicit evaluation of the derivative reveals that

$$\mathcal{U}_S^*(\lambda) = \int d\Gamma_S \pi_S^*(\Gamma_S;\lambda)\left[H_S^*(\Gamma_S;\lambda) + \beta\frac{\partial}{\partial\beta}H_S^*(\Gamma_S;\lambda)\right], \tag{2.155}$$

$$\mathcal{S}_S^*(\lambda) = k_B\int d\Gamma_S \pi_S^*(\Gamma_S;\lambda)\left\{-\ln[h^{N_S f}\pi_S^*(\Gamma_S;\lambda)] + \beta^2\frac{\partial}{\partial\beta}H_S^*(\Gamma_S;\lambda)\right\}, \tag{2.156}$$

which clearly deviates from the standard canonical (weak coupling) case.

Exercise 2.25 Derive eqns (2.155) and (2.156). *Hint:* Remember that $H_S^*(\Gamma_S;\lambda)$ depends on the inverse temperature β.

An appealing property of the above definitions that directly follows from $\mathcal{Z}_S^*(\lambda) = \mathcal{Z}_{SB}(\lambda)/\mathcal{Z}_B$ is the relation

$$\mathcal{F}_S^*(\lambda) = \mathcal{F}_{SB}(\lambda) - \mathcal{F}_B. \tag{2.157}$$

To explore it, let us ignore the dependence on the parameter λ, which only affects the system, for a moment. Furthermore, recall that for a reversible process at constant temperature (i.e. an isothermal process) the change in free energy equals the work applied to the system. Now, suppose we start from a system *decoupled* from the bath with global free energy $\mathcal{F}_S + \mathcal{F}_B$. Then, we slowly switch on the coupling between the system and the bath up to some fixed value. At this point, according to Eq. (2.157), the global free energy can be decomposed as $\mathcal{F}_S^* + \mathcal{F}_B$. Taking their difference then reveals that $\mathcal{F}_S^* - \mathcal{F}_S$ is the work required to reversibly switch on the interaction from zero to some finite V_{SB}, i.e. it is the work required to put a system in contact with its environment, which is completely determined by the system partition functions $\mathcal{Z}_S^*$ and $\mathcal{Z}_S$.

Similar to eqn (2.157), the equilibrium internal energy and system entropy also obey the relations $\mathcal{U}_S^* = \mathcal{U}_{SB} - \mathcal{U}_B$ and $\mathcal{S}_S^* = \mathcal{S}_{SB} - \mathcal{S}_B$. This looks akin to the usual property of *additivity*—compare with eqn (2.2)—but additional care is required here, as the next exercise shows.

Exercise 2.26 To investigate whether additivity holds at strong coupling, imagine that the system S is split into two subsystems X and Y such that $S = XY$. Besides $\mathcal{Z}_S^*$ let us introduce the partition functions $\mathcal{Z}_X^* \equiv \mathcal{Z}_{XYB}/\mathcal{Z}_{YB}$ and $\mathcal{Z}_Y^* \equiv \mathcal{Z}_{XYB}/\mathcal{Z}_{XB}$. For both partition functions we could play the same game as above to arrive at the relations $\mathcal{F}_X^* = \mathcal{F}_{XYB} - \mathcal{F}_{YB}$ and $\mathcal{F}_Y^* = \mathcal{F}_{XYB} - \mathcal{F}_{XB}$. Moreover, eqn (2.157) still also holds. However, to show additivity, we also need to confirm that $\mathcal{F}_S^* = \mathcal{F}_{XY}^* = \mathcal{F}_X^* + \mathcal{F}_Y^*$. Convince yourself of the fact that this is in general only possible at weak coupling, where we can neglect the interactions V_{XY}, V_{XB} and V_{YB}. Thus, at strong coupling we can split the partition function *once* to arrive at eqn (2.157), but not multiple times.

Thus, additivity no longer holds at strong coupling and for this reason we do not associate equilibrium thermodynamic properties (such as temperature) with the system *alone*. Instead, equilibrium thermodynamic parameters here refer to the bath, which is assumed to be large. A more detailed discussion of the breakdown of additivity for small systems, which is linked to the violation of the zeroth law, is postponed to Section 3.6. We now proceed with non-equilibrium considerations.

For this purpose we extend the definitions of internal energy, system entropy and free energy to the non-equilibrium case by replacing the equilibrium system state $\pi_S^*(\Gamma_S)$ by an arbitrary system state $\rho_S(\Gamma;t)$ in eqns (2.155) and (2.156). Hence,

$$U_S^*(t) \equiv \int d\Gamma_S \rho_S(t) \left[H_S^*(\Gamma_S;\lambda_t) + \beta \frac{\partial}{\partial \beta} H_S^*(\Gamma_S;\lambda_t) \right], \tag{2.158}$$

$$S_S^*(t) \equiv k_B \int d\Gamma_S \rho_S(t) \left\{ -\ln[h^{N_S f} \rho_S(t)] + \beta^2 \frac{\partial}{\partial \beta} H_S^*(\Gamma_S;\lambda_t) \right\} \tag{2.159}$$

and the free energy follows from the standard definition $F_S^*(t) \equiv U_S^*(t) - TS_S^*(t)$. From these definitions we can directly read off definitions for the **stochastic internal energy** and **system entropy at strong coupling**,

$$u_S^*(\Gamma_S^t;t) \equiv H_S^*(\Gamma_S^t;\lambda_t) + \beta \frac{\partial}{\partial \beta} H_S^*(\Gamma_S^t;\lambda_t), \tag{2.160}$$

$$s_S^*(\Gamma_S^t;t) \equiv -k_B \ln \left[h^{N_S f} \rho_S(\Gamma_S^t;t) \right] + k_B \beta^2 \frac{\partial}{\partial \beta} H_S^*(\Gamma_S^t;\lambda_t), \tag{2.161}$$

and the **stochastic free energy** follows from $f_S^*(\Gamma_S^t;t) \equiv u_S^*(\Gamma_S^t;t) - Ts_S^*(\Gamma_S^t;t)$. Since we now have a definition for internal energy along a single system trajectory γ_S^t and also for work, which is still given by eqn (2.141), we invoke the first law to define the **stochastic heat at strong coupling**:

$$q_S^*(\gamma_S^t) = \Delta u_S^*(t) - w(\gamma_S^t). \tag{2.162}$$

Finally, we connect the notion of heat to the second law and define the stochastic entropy production

$$\sigma^*(\gamma_S^t) \equiv \Delta s_S^*(t) - \frac{q_S^*(\gamma_S^t)}{T}. \tag{2.163}$$

It remains to show that the average entropy production is positive: $\Sigma^*(t) \equiv \langle \sigma^*(\gamma_S^t) \rangle \geq 0$. Similar to the work fluctuation theorems, which require the initial state to be a global

Gibbs state, $\Sigma^*(t) \geq 0$ can also be derived only for a special class of initial states, which is, however, a little more general than global Gibbs states. For this purpose consider initial states of the form

$$\rho_{SB}(\Gamma_{SB};0) = \rho_S(\Gamma_S;0)\pi_B(\Gamma_B|\Gamma_S), \tag{2.164}$$

where the conditional state $\pi_B(\Gamma_B|\Gamma_S)$ of the bath is assumed to be equilibrated with respect to Γ_S:

$$\pi_B(\Gamma_B|\Gamma_S) \equiv \frac{\pi_{SB}(\Gamma_S,\Gamma_B;\lambda_0)}{\pi_S^*(\Gamma_S;\lambda_0)} = \frac{e^{-\beta[H_B(\Gamma_B)+V_{SB}(\Gamma_{SB})]}}{\int d\Gamma_B e^{-\beta[H_B(\Gamma_B)+V_{SB}(\Gamma_{SB})]}}. \tag{2.165}$$

The last identity can be derived from the expression for the Hamiltonian of mean force given in eqn (2.151). Note that $\pi_B(\Gamma_B|\Gamma_S)$ does not depend on λ_0. The experimental picture behind eqn (2.164) is to hold the system state fixed while waiting for the bath to equilibrate (in a macroscopic sense) with respect to the fixed system state. Afterwards, we drive the system according to the protocol λ_t. Then, we repeat the procedure many times to gather enough statistics with respect to the initial system state $\rho_S(\Gamma_S;0)$. As a result, the stochastic entropy production satisfies the following **integral fluctuation theorem at strong coupling**:

$$\boxed{\left\langle e^{-\sigma^*/k_B}\right\rangle_{\gamma_S^t} = 1,} \tag{2.166}$$

which implies $\Sigma^*(t) \geq 0$. Equipped with all the tools we have learned so far, the derivation of eqn (2.166) is left as an exercise.

Exercise 2.27 Use Liouville's theorem to derive eqn (2.166). *Hint:* An identity that could prove useful along the way is $\sigma^*(t)/k_B = \ln[\rho_S(\Gamma_S^0;0)\pi_B(\Gamma_B^0|\Gamma_S^0)/\rho_S(\Gamma_S^t;t)\pi_B(\Gamma_B^t|\Gamma_S^t)]$.

Another interesting identity shows that the entropy production *rate* at strong coupling can be written as

$$\dot{\Sigma}^*(t) = -k_B \left.\frac{\partial}{\partial t}\right|_{\lambda_t} D[\rho_S(\Gamma_S;t)|\pi_S^*(\Gamma_S;\lambda_t)]. \tag{2.167}$$

This expression makes a comparison with the weak coupling case particularly transparent, where the entropy production rate can be written in the same way with $\pi_S^*(\Gamma_S;\lambda_t)$ replaced by $\pi_S(\Gamma_S;\lambda_t)$; compare with eqn (2.61). Thus, at both weak and strong coupling, the entropy production rate characterizes the tendency to draw the system state closer to the instantaneous equilibrium distribution. Remarkably, the entropy production rate (2.167) does not always need to be positive, despite the fact that $\Sigma^*(t) = \int_0^t ds\dot{\Sigma}^*(s) \geq 0$. Indeed, whenever time-scale separation does *not* apply, we can have $\dot{\Sigma}^*(t) < 0$. If time-scale separation applies, the bath can be approximated as being in a conditional equilibrium state at each time t. For a fixed λ_t, the dynamics would then tend to draw the system state $\rho_S(\Gamma_S;t)$ closer to the respective equilibrium state $\pi_S^*(\Gamma_S;\lambda_t)$ within an infinitesimal time step dt.

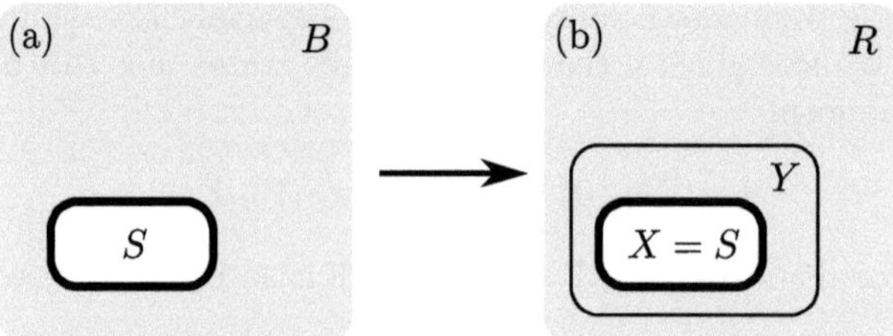

Fig. 2.9 (a) Sketch of a system S in strong contact with a bath B (indicated by the *thick* line separating S and B). (b) After identifying the parts Y of the bath B, which interact strongly with the system S (now labelled X), the composite 'supersystem' XY is in weak contact with the residual bath R (indicated by the *thin* line separating Y and R).

Exercise 2.28 Derive eqn (2.167). *Hint:* Use the relation $\partial_t H_S^*(\lambda_t) = \partial_t H_S(\lambda_t)$.

Strong coupling thermodynamics from coarse-graining

The above framework relies on a couple of postulates, such as eqn (2.154), that have to be handled with care, as Exercise 2.26 has already shown. Therefore, the question arises of whether one can further justify the definitions used above. The final part of this section shows that the thermodynamic framework above can be derived by starting from an unambiguous description at *weak coupling* and by using the argument of time-scale separation, which also clarifies the discussion below eqn (2.167).

For this purpose, we build on the coarse-graining framework developed in Section 2.5. To apply it, we imagine that the bath B is so large that not every degree of freedom of it can be strongly coupled to the system. Instead, we assume that only a small fraction $Y \subset B$ of the bath's degrees of freedom couples strongly to the system. The remaining degrees of freedom R of the bath, which we call the *residual bath*, are assumed to couple weakly to Y and the system S, which we relabel for the remainder of this exercise as $X \equiv S$. The original bath is therefore by construction split as $B = YR$. This is an *assumption*, but one can find situations where this holds approximately (think about, for example, a hydrophobic molecule X in aqueous solution R surrounded by surfactants Y). A sketch of the idea is provided in Fig. 2.9.

Therefore, we now switch the perspective and, instead of trying to directly describe X coupled to B, we start with a description of XY coupled to R. We also adapt the notation and assume that X and Y are described by a set of discrete states (this assumption is not necessary) and write the energy of XY as $E_{xy}(\lambda_t) = E_x(\lambda_t) + E_y + V_{xy}$. Since R couples weakly to XY, the equilibrium state of XY for a fixed λ_t is given by the Gibbs state $\pi_{xy}(\lambda_t)$. The reduced equilibrium state of X, denoted by $\pi_x^*(\lambda_t) = \sum_y \pi_{xy}(\lambda_t)$, can again be described by the Hamiltonian of mean force, which we denote in the present case by

$$E_x^*(\lambda_t) = E_x - k_B T \ln \langle e^{-\beta V_{xy}} \rangle_Y. \tag{2.168}$$

Here, $\langle \ldots \rangle_Y \equiv \sum_y \ldots \pi_y$ denotes an ensemble average with respect to the equilibrium state $\pi_y = e^{-\beta(E_y - \mathcal{F}_Y)}$ (with $\mathcal{F}_Y = -k_B T \ln \mathcal{Z}_Y$) of Y alone. This result echoes our findings in Exercise 2.11. Furthermore, the constrained equilibrium state of Y conditioned on a particular system state x reads in the present setting

$$\pi_{y|x} = \frac{\pi_{xy}(\lambda_t)}{\pi_x^*(\lambda_t)} = e^{-\beta[E_{xy}(\lambda_t)-\mathcal{F}_x(\lambda_t)]} \tag{2.169}$$

with $\mathcal{F}_x(\lambda_t) = -k_BT\ln\sum_y e^{-\beta E_{xy}(\lambda_t)}$. This follows from eqn (2.81) after identifying i_x (in the notation of Section 2.5) with xy, and it is the counterpart of eqn (2.165). The following neat relation links the concepts introduced above:

$$E_x^*(\lambda_t) = \mathcal{F}_x(\lambda_t) - \mathcal{F}_Y. \tag{2.170}$$

Exercise 2.29 Derive eqn (2.170).

Next, let us assume that Y evolves *fast*, such that we can apply time-scale separation. Then, the dynamics of X are Markovian but nevertheless strongly affected by Y. Theoretically, this assumption implies that we can approximate $p_{xy}(t) \approx \pi_{y|x}p_x(t)$ at all times t. Now, recall the definitions of stochastic internal energy and system entropy from Section 2.5, which we denote in the present context as $u_x(t) = E_x(\lambda_t) + \sum_y(V_{xy} + E_y)\pi_{y|x}(\lambda_t)$ and $s_x(t) = -k_B\ln p_x(t) - k_B\sum_y \pi_{y|x}(\lambda_t)\ln\pi_{y|x}(\lambda_t)$. It then follows from the assumption of time-scale separation that

$$\boxed{u_x(t) - u_x^*(t) = \mathcal{U}_Y, \quad s_x(t) - s_x^*(t) = \mathcal{S}_Y,} \tag{2.171}$$

where $u_x^*(t)$ and $s_x^*(t)$ follow from eqns (2.160) and (2.161) after replacing Γ_S, $\rho_S(\Gamma_S;t)$ and $H_S^*(\Gamma_S;\lambda_t)$ by x, $p_x(t)$ and $E_x^*(\lambda_t)$. Moreover, $\mathcal{U}_Y = \sum_y E_y\pi_y$ and $\mathcal{S}_Y = -k_B\sum_y \pi_y\ln\pi_y$ are the equilibrium internal energy and entropy of Y *alone*.

Exercise 2.30 Derive eqn (2.171). *Hint:* This is straightforward, but tedious. You are allowed to skip it if you understood the physical idea presented here.

Equation (2.171) allows us to draw the important conclusion that, apart from time-independent constants $\mathcal{U}_Y$ and $\mathcal{S}_Y$, which cancel out when taking differences, the thermodynamic description under coarse-graining and time-scale separation is *the same* as the one introduced above based on the Hamiltonian of mean force. We can therefore conclude that the entropy production rates also agree, $\dot{\Sigma}(t) = \dot{\Sigma}^*(t) \geq 0$, where $\dot{\Sigma}(t)$ was defined previously in eqn (2.94). Note that $\dot{\Sigma}^*(t)$ is always positive now; compare also with the discussion below eqn (2.167).

It is instructive to ask what can be said when time-scale separation does *not* apply. Then, the definitions based on the Hamiltonian of mean force for X alone no longer coincide with the thermodynamic description of XY. There is, however, a hierarchy of 'second laws', which bounds the dissipation at different levels of the description. For this purpose, let us denote by $\Sigma_{XY}(t) \equiv [W(t) - \Delta F_{XY}(t)]/T \geq 0$ with $F_{XY}(t) = \sum_{x,y} p_{xy}(t)[E_{xy}(\lambda_t) + k_BT\ln p_{xy}(t)]$ the entropy production that ensues from the joint description of X and Y in weak contact with the residual bath R. Now, assume that the initial state obeys $p_{xy}(0) = \pi_{y|x}p_x(0)$ in unison with our assumption (2.164). Then, we find

$$\Sigma^*(t) - \Sigma_{XY}(t) = k_B D[p_{xy}(t)|\pi_{y|x}p_x(t)] \geq 0. \tag{2.172}$$

Thus, the entropy production $\Sigma^*(t)$ computed within the Hamiltonian of mean force framework *overestimates* the entropy production $\Sigma_{XY}(t)$.

Exercise 2.31 Derive eqn (2.172).

On the other hand, we know from eqn (2.122) that the entropy production $\Sigma_{XY}(t)$ can be expressed as

$$\Sigma_{XY}(t) = k_B D[p(\mathbf{xy}_n)|p_{\text{tr}}(\mathbf{xy}_n^\dagger)]. \tag{2.173}$$

Here, $\mathbf{xy}_n$ denotes a trajectory of microstates in XY and $\mathbf{xy}_n^\dagger$ is its time reversal as defined in Section 2.6. Now, it is tempting to introduce an 'entropy production'

$$\Sigma_{\text{traj}}(t) \equiv k_B D[p(\mathbf{x}_n)|p_{\text{tr}}(\mathbf{x}_n^\dagger)] \tag{2.174}$$

obtained by comparing forward and backward trajectories of microstates in X alone. Obviously, $\Sigma_{\text{traj}}(t)$ is always positive and one could argue that it also measures some kind of dissipation. However, whereas $\Sigma_{XY}(t)$ and $\Sigma^*(t)$ can be linked to a clear notion of entropy and heat, this is no longer the case for $\Sigma_{\text{traj}}(t)$. Nevertheless, it is always true that $\Sigma_{\text{traj}}(t)$ *underestimates* $\Sigma_{XY}(t)$.

Exercise 2.32 Derive $\Sigma_{XY}(t) \geq \Sigma_{\text{traj}}(t)$. *Hint:* Show that you can construct a stochastic matrix T that maps the joint probability p_{xy} (seen as a vector) to its marginal $p_x = \sum_y p_{xy}$ (seen as a vector of lower dimension, which implies that T is not a square matrix). Then, use monotonicity of relative entropy (Theorem A.4) to deduce the desired result.

Thus, we obtain the hierarchy of inequalities

$$\boxed{\Sigma^*(t) \geq \Sigma_{XY}(t) \geq \Sigma_{\text{traj}}(t) \geq 0.} \tag{2.175}$$

If time-scale separation applies, then $\Sigma^*(t) = \Sigma_{XY}(t) = \Sigma_{\text{traj}}(t)$, but in general both $\Sigma^*(t)$ and $\Sigma_{\text{traj}}(t)$ deviate from the 'true' entropy production of XY. However, notice that $\Sigma^*(t)$ and $\Sigma_{\text{traj}}(t)$ can be constructed solely based on knowledge of the dynamics of X. In contrast, determination of $\Sigma_{XY}(t)$ requires knowledge of X *and* Y. In reality, one often has to deal with *hidden* degrees of freedom such as Y, which are experimentally inaccessible. Then, we can use $\Sigma^*(t)$ ($\Sigma_{\text{traj}}(t)$) to upper (lower) bound the entropy production $\Sigma_{XY}(t)$ by measuring only X.

Concluding discussion

We have demonstrated that, based solely on the assumption (2.164) about the initial state, a stochastic thermodynamics framework can be constructed without the need to assume time-scale separation, Markovianity or weak coupling. Definitions of internal energy, heat, work and system entropy were given along a single trajectory, satisfying a stochastic first law and a fluctuation theorem (and, hence, also a second law) and, at equilibrium, showing the same structure as their weak coupling counterpart. Furthermore, work fluctuation theorems hold for the same setting as well.

The mathematical generality of these results is remarkable, but care is required when identifying the microscopically defined quantities here with their phenomenological counterpart. We illustrated this by using the idea of coarse-graining, which shows that this identification is unambiguously justified only if time-scale separation applies. Nevertheless, even if the assumption of a fast and memoryless bath is not met, the present framework remains a promising and useful tool to study non-equilibrium thermodynamics at strong coupling as long as the bath is *large*.

In fact, the formal derivation of the fluctuation theorems (2.153) and (2.166) did not assume the bath to be large. However, from our knowledge of phenomenological non-equilibrium thermodynamics, we know that a second law of the form (2.6) is expected to emerge *only* in the limit of a weakly perturbed bath. This is no longer guaranteed if the bath is itself small and can be disturbed far away from its initial state. In this case, the correct second law should take on the form (2.4) or (2.5). This becomes transparent when we try to link the notion of heat to entropy changes in the bath, which we have not done here. We return to this problem in greater detail when we discuss the quantum case in Sections 3.7 and 3.8.

2.9 Measuring Free Energies of Complex Molecules

We conclude this chapter by discussing an important experimental application of the work fluctuation theorem. Remember that the work required to change the state of a system initially at temperature T between two configurations λ_0 and λ_τ obeys $W(\tau) \geq \mathcal{F}_S(\lambda_\tau) - \mathcal{F}_S(\lambda_0)$. In particular, in the presence of a heat bath, the equilibrium free energy determines whether a process can happen spontaneously (if $\Delta\mathcal{F}_S < 0$) or not (if $\Delta\mathcal{F}_S > 0$) and, therefore, knowledge of it is often crucial. Unfortunately, directly calculating $\mathcal{F}_S$ is challenging for complex systems (in terms of computational resources) and only in the reversible limit do we obtain $W(t) = \Delta\mathcal{F}_S$, allowing us to determine changes in equilibrium free energy by measuring work. However, experimentally implementing reversible state changes, in particular for small systems, is also challenging and typically involves additional interpolation procedures. In contrast, the work fluctuation theorems provide an exact equality for the change in equilibrium free energy, regardless of how fast or violent we drive the process. They therefore allow valuable equilibrium information to be extracted from non-equilibrium dynamics.

This insight was fruitfully used to reconstruct free energy landscapes of complex macromolecules, including DNA or RNA molecules. Theoretically calculating the free energy for these macromolecules, which are composed of hundreds of electrons and nuclei, is hard, but knowing these values is crucial to understanding processes in biophysics and biochemistry. In this section we theoretically discuss the basics needed to understand these remarkable 'single-molecule-pulling experiments' (see Fig. 2.10).

As in real cells in living organisms, the experiments are performed in aqueous solution at room temperature. Under these circumstances many properties of molecules are well described using *classical* statistical mechanics and we therefore describe the

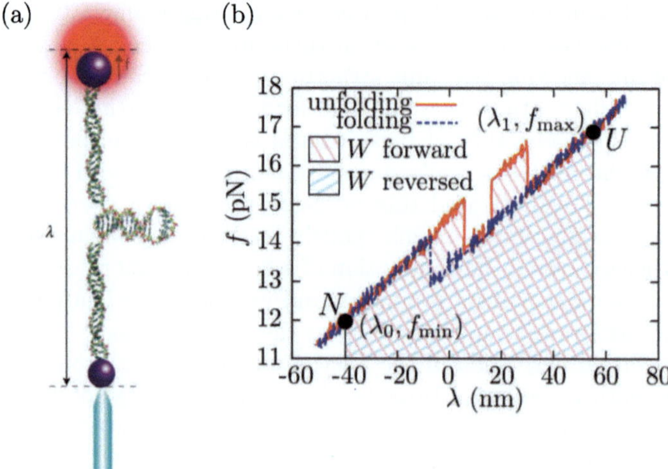

Fig. 2.10 (a) Sketch of a set-up for a single-molecule-pulling experiment taken from Alemany *et al.* (2012). A macromolecule (here a DNA hairpin) is glued to two beads at distance λ. The position of one bead is fixed, whereas the other feels a force f in the direction of the centre of the laser trap, which can be moved. (b) Force–distance curves for two exemplary repetitions of the experiment (here called 'forward' and 'reversed'). The work done on the molecule by going from one state to another (here labelled 'N' and 'U') is determined by the integral under the curve. Notice that the forces and displacements are in the pico-Newton and nanometer ranges, i.e. far away from the macroscopic and deterministic 19th century thermodynamics. (a) Reprinted by permission from Springer Nature: Springer Nature, Nature Physics (Experimental free-energy measurements of kinetic molecular states using fluctuation theorems, A. Alemany et al.), Copyright by Springer Nature (2012). (b) By courtesy of Felix Ritort.

macromolecule immersed in water with a classical Hamiltonian $H_S(\Gamma_S; \lambda_t)$, where Γ_S denotes the positions and momenta of all the atoms of the macromolecule. The parameter λ_t can be used to control the linear extension of the macromolecule by attaching a 'bead' to both ends of the molecule. One can imagine these beads as little balls glued to the molecule, which—in contrast to the molecule itself—are visible and controllable in the laboratory by using micropipettes and optical tweezers. We do not go into the physical details of how these beads function. The position of one bead, which is attached to the lower end of the molecule in Fig. 2.10, is *fixed* by using a micropipette, whereas the other bead, which is attached to the upper end of the molecule in Fig. 2.10, is trapped in a laser field (the 'optical tweezer'), whose position can be controlled and whose force on the bead can be measured. Therefore, the control parameter λ describes the distance between the fixed bead and the centre of the laser trap. The position x of the bead in the laser trap can fluctuate around its centre and it is determined by the microstate Γ_S of the molecule: $x = x(\Gamma_S)$. The general idea is now to vary the position $\lambda = \lambda_t$ of the optical tweezer, i.e. to *pull* the molecule, while recording at the same time the force exerted on the bead, which determines the work done on the system. This information at the end determines the free energy landscape of the molecule as a function of x, which allows one to, for example, measure the free energy difference between folded and unfolded states of the macromolecule.

In the following, we use the Hamiltonian of mean force $H_S^*(\Gamma_S;\lambda)$ in our calculation to include strong coupling effects; see eqn (2.151) for its definition. Moreover, to avoid notational overload, we drop all constants h^{Nf}, which are only there to ensure correct physical dimensions but which cancel out when taking differences. In fact, in real experiments measurement precision and data are *finite* such that all quantities below need to be discretized accordingly. Then, there is no longer any need for a factor h to define the size of a microstate.

We start our investigation by confirming the validity of the following identity:

$$\frac{e^{-\beta H_S^*(\Gamma_S;\lambda_\tau)}}{\mathcal{Z}_S^*(\lambda_0)} = \int d\Gamma_{SB}^0 \delta(\Gamma_S - \Gamma_S^\tau)\pi_{SB}(\Gamma_{SB}^0;\lambda_0)e^{-\beta w(\gamma_S^\tau)}, \tag{2.176}$$

where we used the notation of the previous two sections. To derive it, note that $w(\gamma_S^\tau) = H_{SB}(\Gamma_{SB}^\tau;\lambda_\tau) - H_{SB}(\Gamma_{SB}^0;\lambda_0)$ allows us to write

$$\frac{e^{-\beta H_S^*(\Gamma_S;\lambda_\tau)}}{\mathcal{Z}_S^*(\lambda_0)} = \int d\Gamma_{SB}^0 \delta(\Gamma_S - \Gamma_S^\tau)\frac{e^{-\beta H_{SB}(\Gamma_{SB}^\tau;\lambda_\tau)}}{\mathcal{Z}_{SB}(\lambda_0)}. \tag{2.177}$$

After using Liouville's theorem and evaluating the delta function, we confirm that

$$\frac{e^{-\beta H_S^*(\Gamma_S;\lambda_\tau)}}{\mathcal{Z}_S^*(\lambda_0)} = e^{-\beta H_S(\Gamma_S;\lambda_\tau)}\int d\Gamma_B^\tau \frac{e^{-\beta[V_{SB}(\Gamma_S,\Gamma_B^\tau)+H_B(\Gamma_B^\tau)]}}{\mathcal{Z}_{SB}(\lambda_0)}. \tag{2.178}$$

Now, eqn (2.176) follows straightforwardly after recalling the definition (2.151) of the Hamiltonian of mean force and the identity $\mathcal{Z}_{SB}(\lambda_0) = \mathcal{Z}_S^*(\lambda_0)\mathcal{Z}_B$.

Next, let us interpret the identity (2.176). Assuming the initial system–bath state to be in equilibrium, the right-hand side describes an ensemble average of the quantity $\delta(\Gamma_S - \Gamma_S^\tau)e^{-\beta w(\gamma_S^\tau)}$, which equals the exponentiated work (times $-\beta$) conditioned on having a certain system state Γ_S. The left-hand side, after multiplication with the trival factor $1 = \mathcal{Z}_S^*(\lambda_\tau)/\mathcal{Z}_S^*(\lambda_\tau)$, can be written as $e^{-\beta\Delta\mathcal{F}_S^*}\pi_S^*(\Gamma_S;\lambda_\tau)$, where $\pi_S^*(\Gamma_S;\lambda_\tau)$ was defined in eqn (2.150). Thus, we arrive at the identity

$$\boxed{e^{-\beta\Delta\mathcal{F}_S^*}\pi_S^*(\Gamma_S;\lambda_\tau) = \left\langle \delta(\Gamma_S - \Gamma_S^\tau)e^{-\beta w(\gamma_S^\tau)}\right\rangle_{\gamma_S^\tau}.} \tag{2.179}$$

Note that integrating this equation over Γ_S immediately implies the integral work fluctuation theorem at strong coupling; see eqn (2.153).

Let us now return to our single-molecule-pulling experiment. First, we notice that the assumption of an initial global equilibrium state is well justified for a macromolecule in aqueous solution if we have left it on its own for some time. Next, for definiteness we assume that the laser trap of the bead is characterized by a harmonic potential with stiffness k. The Hamiltonian of mean force is written as

$$H_S^*(\Gamma_S;\lambda_t) = \tilde{H}_S^*(\Gamma_S) + \frac{k}{2}(x-\lambda_t)^2, \tag{2.180}$$

where $\tilde{H}_S^*(\Gamma_S)$ denotes the Hamiltonian of mean force of the molecule in aqueous solution *without* any laser trap and $x = x(\Gamma_S)$ denotes the position of the bead in the optical tweezer, which depends on the microstate Γ_S of the macromolecule (i.e. the position of all the atoms forming the macromolecule).

Now, we multiply eqn (2.179) by $\delta[x' - x(\Gamma_S)]$ and integrate over Γ_S:

$$\begin{aligned}\int d\Gamma_S \delta[x' - x(\Gamma_S)] \frac{e^{-\beta\{\tilde{H}_S^*(\Gamma_S)+\frac{k}{2}[x(\Gamma_S)-\lambda_\tau]^2\}}}{\mathcal{Z}_S^*(\lambda_0)} \\ = \int d\Gamma_S \delta[x' - x(\Gamma_S)] \left\langle \delta[\Gamma_S - \Gamma_S^\tau] e^{-\beta w(\gamma_S)} \right\rangle_{\gamma_S^\tau}.\end{aligned} \tag{2.181}$$

The delta function allows one to pull out the factor $\exp[-\beta k(x' - \lambda_\tau)^2/2]$ on the left-hand side of eqn (2.181) and to shift it to the right-hand side:

$$\begin{aligned}\frac{\int d\Gamma_S \delta[x' - x(\Gamma_S)] e^{-\beta \tilde{H}_S^*(\Gamma_S)}}{\mathcal{Z}_S^*(\lambda_0)} &= \int d\Gamma_S \delta[x' - x(\Gamma_S)] \left\langle \delta(\Gamma_S - \Gamma_S^\tau) e^{-\beta \tilde{w}(\gamma_S^\tau)} \right\rangle_{\gamma_S^\tau} \\ &= \left\langle \delta[x' - x(\Gamma_S^\tau)] e^{-\beta \tilde{w}(\gamma_S^\tau)} \right\rangle_{\gamma_S^\tau},\end{aligned} \tag{2.182}$$

where we defined $\tilde{w}(\gamma_S^\tau) \equiv w(\gamma_S^\tau) - k(x - \lambda_\tau)^2/2$. This is the desired result after noting that $\tilde{\mathcal{Z}}_S^*(x') \equiv \int d\Gamma_S \delta[x' - x(\Gamma_S)] \exp[-\beta \tilde{H}_S^*(\Gamma_S)]$ yields the equilibrium free energy $\tilde{\mathcal{F}}_S^*(x') \equiv -k_B T \ln \tilde{\mathcal{Z}}_S^*(x')$ of the macromolecule *for a fixed extension* x'. Hence,

$$\boxed{\tilde{\mathcal{F}}_S^*(x') = -k_B T \ln \left\langle \delta[x' - x(\Gamma_S^\tau)] e^{-\beta \tilde{w}(\gamma_S^\tau)} \right\rangle_{\gamma_S^\tau} + \mathcal{F}_S^*(\lambda_0),} \tag{2.183}$$

which holds for any x' with respect to a fixed offset $\mathcal{F}_S^*(\lambda_0) = -k_B T \ln \mathcal{Z}_S^*(\lambda_0)$. Thus, if the experiment is initialized with the same λ_0, the term $\mathcal{F}_S^*(\lambda_0)$ cancels out when computing the free energy difference $\tilde{\mathcal{F}}_S^*(x_2) - \tilde{\mathcal{F}}_S^*(x_1)$ of the macromolecule for two different extensions x_2 and x_1. We remark that the extension x provides precise information about the *conformational* states of the macromolecule (e.g. whether it is folded or unfolded), which is relevant to determining its biochemical functionality. Experimentally, eqn (2.183) can be determined via the following procedure. Starting always with the same initial preparation and repeating the same protocol many times, one constructs histograms to approximate the joint probability distribution $p[\tilde{w}(\gamma_S^\tau), x_\tau]$, where x_τ denotes the final position of the bead in the laser trap. To determine the redefined work $\tilde{w}(\gamma_S^\tau) = w(\gamma_S^\tau) - k(x_\tau - \lambda_\tau)^2/2$, we need to measure two terms. The first term $w(\gamma_S^\tau) = -\int_0^\tau dt \dot{\lambda}_t k(x_t - \lambda_t)$ can be directly measured along a single trajectory by multiplying the force $f = -\dot{\lambda}_t k$ exerted on the bead in the laser trap by its displacement $(x_t - \lambda_t)$. Such force–distance curves are typically displayed in the literature on single-molecule-pulling experiments and two example trajectories are shown in Fig. 2.10. The second term $-k(x_\tau - \lambda_\tau)^2/2$ is directly determined by the final position of the bead. Finally, to determine the right-hand side of eqn (2.183), one considers the histogram $p[\tilde{w}(\gamma_S^\tau), x_\tau = x']$ for a fixed bead position x'. Multiplying this histogram by $e^{-\beta \tilde{w}(\gamma_S^\tau)}$, averaging over the work values $\tilde{w}(\gamma_S^\tau)$, taking the logarithm and multiplying by $-k_B T$ gives the desired equilibrium free energy $\tilde{\mathcal{F}}_S^*(x')$ from non-equilibrium work measurements (apart from an unimportant constant).

Further reading

2.1. Most of the material on phenomenological thermodynamics can be found in the groundbreaking work of Clausius (1865), who introduces the notion of entropy in science. The clarity of Clausius's writing is indeed astonishing given the age of the paper. A comprehensive modern treatment of phenomenological non-equilibrium thermodynamics, in particular that based on the local equilibrium assumption, is provided in the book by Kondepudi and Prigogine (2007). The briefly mentioned and much studied linear response regime is treated extensively, for example, by Pottier (2010).

2.2. We started with a discussion of equilibrium ensembles and readers interested in learning more about the equivalence of ensembles can look at Gallavotti (1999) and Touchette (2009). Next, we justified why Shannon entropy is a good candidate for thermodynamic entropy even out of equilibrium *provided* that certain assumptions are met. This question has a long history and Landauer (1961) was the first to argue for the validity of eqn (2.27). We remark that Landauer's principle is still subject to criticism, but it often seems semantic in nature. Clearly, not every two-level system, for which Landauer's principle applies, is suitable for storing one bit of *valuable information*, which needs to be reliably encoded, stored and read out. For many practical considerations one therefore finds a much higher work cost in erasing a bit of information (today's computers still operate far from the hypothetical Landauer bound). Nevertheless, this objection concerns the semantic question of whether the mathematical concept of Shannon entropy applied to a given physical set-up should be equated with the common-sense notion of information or not. It even seems that Landauer himself did not want to imply this statement *per se* and explicitly writes (Landauer, 1961): 'Note that our argument here does not necessarily depend upon connections, frequently made in other writings, between entropy and information'. There is now also a wealth of experimental evidence *in favour* of Landauer's principle (Bérut *et al.*, 2012; Orlov *et al.*, 2012; Jun *et al.*, 2014; Hong *et al.*, 2016; Peterson *et al.*, 2016; Gavrilov *et al.*, 2017). Among modern theoretical treatments of Landauer's principle, a transparent one was given by Deffner and Jarzynski (2013).

2.3. Here, we encountered for the first time the important concept of a master equation. According to van Kampen (2007), the origin of this terminology can be traced back to Nordsieck *et al.* (1940), where a probability balance equation of the form (2.31) was used to derive all other results. It then somehow got stuck in the community. But, indeed, if one is able to express the dynamics of an open system interacting with many other degrees of freedom in the simple closed form of eqn (2.31), this has something 'masterly'. It is not that simple to find clean derivations of the local detailed balance condition. For early work see van Kampen (1954) and Bergmann and Lebowitz (1955), modern and more rigorous treatments were given for classical systems by Maes and Netocny (2003) and for quantum systems by Strasberg *et al.* (2023). The requirements for a rate master equation to have a unique steady state (Exercise 2.4) are reviewed by van Kampen (2007). A detailed treatment of Fokker–Planck equations in connection with stochastic thermodynamics is given by Sekimoto (2010).

2.4. Sekimoto (1998) first formulated a first law at a trajectory level based on a Langevin equation, which describes the stochastic motion of a particle obeying on average a Fokker–Planck equation. The study of entropy along a single trajectory was mainly initiated by Seifert (2005). Furthermore, long before the study of *stochastic* thermodynamics, the average non-equilibrium thermodynamic description of master equations was established. An elegant way to do so is based on viewing the states x as a network and using graph theory (Schnakenberg, 1976). For a modern and extended account of this method see Polettini and Esposito (2019). The final comment in Exercise 2.9 hints at how this works.

2.5. Coarse-graining and time-scale separation are essential tools in non-equilibrium statistical mechanics since its inception. Treatments similar to ours were given, for example, by Seifert (2011) and Esposito (2012). The question of whether the dynamics is Markovian at the coarse-graining level (Exercise 2.10) was studied mathematically by Kemeny and Snell (1976) (note that they refer to 'coarse-graining' as 'lumping').

2.6. The quote 'the noise is the signal' comes from Landauer (1998). The terminology 'Crooks' lemma' is non-standard in the field of stochastic thermodynamics, yet I believe that it most aptly summarizes the importance of the paper by Crooks (1998). The study of entropy production fluctuation theorems for classical Markov processes obeying local detailed balance was pioneered by Crooks (1999) and Seifert (2005).

2.7. Also in Section 2.7 we used a terminology that is different from the standard one, and a few historical comments seem indispensable in this context. In fact, the integral work fluctuation theorem is often called *Jarzynski equality* owing to a seminal paper by Jarzynski (1997). On the other hand, eqn (2.130) could also be called the *Bochkov–Kuzovlev equality* because a very similar result was discovered much earlier by Bochkov and Kuzovlev (1977). There has been some debate about the question of who should be credited for deriving the work fluctuation theorem because the setting of Bochkov and Kuzovlev, strictly speaking, does not coincide with Jarzynski's setting (which agrees with our setting here). Their differences are discussed by Jarzynski (2007) and Bochkov and Kuzovlev (2013). Moreover, Jarzynski first derived the integral work fluctuation theorem for an *open* system. On the other hand, the two approaches are so closely related that it seems questionable to only credit Jarzynski for it, in particular given the circumstances that Bochkov and Kuzovlev were way *ahead of their time* when non-equilibrium fluctuations in small systems were not yet at the forefront of science. For this reason it seems better to use the less biased terminology of *work fluctuation theorems*. In addition, the detailed work fluctuation theorem is often called *Crooks' fluctuation theorem* owing to the work of Crooks (1999), but—apart from the same slight differences in the setting—it had also already appeared in the seminal work of Bochkov and Kuzovlev (1977). Early studies of the nonequilibrium free energy as used in eqn (2.146) were done by Donald (1987) and Gaveau and Schulman (1997). In *phenomenological* non-equilibrium thermodynamics, however, various notions of non-equilibrium free energies had already been introduced earlier.

2.8. The integral work fluctuation theorem at strong coupling was derived by Jarzynski (2004). The extension to a complete non-equilibrium thermodynamic framework, including the derivation of the integral fluctuation theorem (2.166), is due to Seifert (2016). A corresponding detailed fluctuation theorem can also be derived (Miller and Anders, 2017). Expression (2.167) appears in the work of Strasberg and Esposito (2019), where the sign of the entropy production rate is linked to the (non-)Markovianity of the dynamics. In the last part of the section we followed Strasberg and Esposito (2017), although the content of Exercise 2.32 was already noted by Gomez-Marin et al. (2008). We note that the idea to split the bath $B = YR$ into strongly coupled degrees of freedom Y and a residual Markovian bath R is *not* an assumption if the (classical) bath is described by a Caldeira–Leggett Hamiltonian. Then, such a transformation can always be achieved (Martinazzo *et al.*, 2011). For an introduction to such *Markovian embedding strategies* in the context of thermodynamic applications see Schaller and Nazir (2018).

2.9. The idea to use work fluctuation theorems to extract equilibrium information about macromolecules as well as relation (2.176) was first presented by Hummer and Szabo (2001). Shortly afterwards, the first experimental confirmation was reported (Liphardt *et al.*, 2002; Collin *et al.*, 2005). Since then single-molecule-pulling experiments have been successfully used to probe the equilibrium and non-equilibrium thermodynamics of small systems and this in turn contributed a lot to the wider acceptance of stochastic thermodynamics and fluctuation theorems in the scientific community. The sketch of the set-up shown in Fig. 2.10 is due to the more recent experiment by Alemany *et al.* (2012). A general overview of experimental progress in stochastic thermodynamics *beyond* single-molecule-pulling experiments was given by Ciliberto (2017).

3
Quantum Thermodynamics Without Measurements

Summary. We derive the basic laws of phenomenological non-equilibrium thermodynamics for small open systems, whose quantum nature can no longer be neglected. Emphasis is put from the beginning on deriving them from an underlying microscopic system–bath picture. Commonly considered approximation schemes (weak coupling master equations) are reviewed and their thermodynamics is studied. The role of the zeroth law at the nanoscale is discussed and exact identities for the entropy production, valid beyond weak coupling and the Markov approximation, are introduced. We also discuss the effect of finite baths and establish two frameworks to study non-equilibrium resources. We conclude with the study of particle transport and experimentally realized thermoelectric devices. This chapter focuses entirely on the isolated dynamics of a system coupled to a bath without any external interventions. Quantum measurements and fluctuations are discussed in later chapters.

3.1 Mechanical Work in the Quantum Regime

In contrast to heat, temperature and entropy, which are emergent concepts that appeared first in thermodynamics, internal energy and mechanical work are already well-known concepts from elementary courses in mechanics. It is therefore not too surprising that the definition of mechanical work in the quantum regime is unproblematic and parallels the classical case. This is at least true as long as we are not interested in fluctuations of work, whose discussion we postpone to later chapters.

We start our discussion as in Chapter 1 by considering an isolated system described by a state $\rho(t)$ and a Hamiltonian $H(\lambda_t)$. We note that the description of an isolated quantum system via a time-dependent Hamiltonian is a *semiclassical* approximation (we make this point more precise below). We also note that there is an unfortunate source of confusion here. In traditional thermodynamics a system is called 'isolated' if it can exchange neither energy nor particles with a bath; it is called 'closed' if it can exchange energy but not particles; and 'open' if it can exchange both energy and particles. We do not follow this terminology. Instead, we use the word 'isolated' for systems that evolve in a unitary way and all other systems are simply called 'open'.

Not very surprisingly, we identify the **internal energy** of the isolated system with the expectation value of its Hamiltonian:

$$\boxed{U(t) \equiv \mathrm{tr}\{H(\lambda_t)\rho(t)\}.} \tag{3.1}$$

Quantum Stochastic Thermodynamics. Philipp Strasberg, Oxford University Press.
 DOI: 10.1093/oso/9780192895585.003.0003

Note that we use this definition independent of the question of whether $\rho(t)$ describes an equilibrium state or not. Furthermore, since the system is isolated, it cannot exchange heat with the outside world ($Q = 0$) and the phenomenological first law then dictates that any change in energy must be due to work:

$$\Delta U(t) \equiv U(t) - U(0) = W(t). \tag{3.2}$$

Taking a closer look at this expression reveals that

$$\begin{aligned}\Delta U(t) &= \int_0^t ds \frac{d}{ds} \mathrm{tr}\{H(\lambda_s)\rho(s)\} \\ &= \int_0^t ds \left(\mathrm{tr}\left\{ \frac{\partial H(\lambda_s)}{\partial s}\rho(s) \right\} + \mathrm{tr}\left\{ H(\lambda_s)\frac{\partial \rho(s)}{\partial s} \right\} \right).\end{aligned} \tag{3.3}$$

Using the Liouville–von Neumann equation and the fact that the trace is cyclic, we see that the second term vanishes. Hence, the mechanical **work** reads

$$\boxed{W(t) = \int_{t_0}^t ds \mathrm{tr}\left\{ \frac{\partial H(\lambda_s)}{\partial s}\rho(s) \right\}.} \tag{3.4}$$

The term inside the integral $\dot{W}(s) = \mathrm{tr}\{[\partial_s H(\lambda_s)]\rho(s)\}$ is also called the **power** in the following. It describes the instantaneous change in the system's internal energy due to an external driving field. Notice the similarity of eqn (3.4) to the classical case, obtained after a phase-space average of eqn (2.125).

Next, we turn to the open system paradigm, where the global Hamiltonian of the isolated system–bath composite is written as $H_{SB}(\lambda_t) = H_S(\lambda_t) + H_B + V_{SB}$ in unison with Section 1.2. Now, however, we allow for a time-dependent system Hamiltonian similar to the classical case considered in Section 2.7. Since H_B and V_{SB} do *not* depend on time, it follows from eqn (3.4) without further approximations that the work (and power) done on the system is

$$\boxed{W(t) = \int_{t_0}^t ds \dot{W}(s) = \int_{t_0}^t ds \mathrm{tr}_S\left\{ \frac{\partial H_S(\lambda_s)}{\partial s}\rho_S(s) \right\}.} \tag{3.5}$$

Again, this definition is analogous to the classical case if we compare it with a phase-space average of eqn (2.141).

Notice that the expression on the right-hand side of eqn (3.5) can be computed solely based on knowledge of the open quantum system state $\rho_S(t)$. Therefore, it fits into our paradigm of Chapter 1, where we found out that the state $\rho_S(t)$ is experimentally accessible based on local quantum process tomography. For certain situations in thermodynamics, e.g. describing the act of coupling or decoupling a system and a bath, it is nevertheless necessary to consider a time-dependent interaction Hamiltonian $V_{SB} = V_{SB}(\lambda_t)$. The formula (3.5) for work then generalizes to

$$W(t) = \int_{t_0}^t ds \mathrm{tr}_S\left\{ \frac{\partial H_S(\lambda_s)}{\partial s}\rho_S(s) \right\} + \int_{t_0}^t ds \mathrm{tr}_{SB}\left\{ \frac{\partial V_{SB}(\lambda_s)}{\partial s}\rho_{SB}(s) \right\}. \tag{3.6}$$

Now, it is no longer possible to infer the work based solely on the local tomography of the open system. Instead, it typically requires additional theoretical modeling or assumptions to compute the term on the right-hand side of eqn (3.6).

The use of a time-dependent Hamiltonian $H(\lambda_t)$ provides an excellent approximation for many experiments involving, for example, manipulations of small systems by external laser fields. It is, however, instructive to see how a time-*dependent* Hamiltonian can emerge from a time-*independent* one, which explicitly includes the quantum mechanical description of the work reservoir. We now sketch such a derivation, followed by an exercise to make it more precise.

We consider a system labelled by S and a work reservoir labelled by W (the presence of a bath B can be included in the definition of S if desired). The global time-independent Hamiltonian is split as usual as $H_{SW} = H_S + H_W + V_{SW}$ and we assume an interaction Hamiltonian of the form $V_{SW} = A \otimes B$. The reduced dynamics of the system is given by tracing out W:

$$\frac{\partial}{\partial t}\rho_S(t) = -\frac{i}{\hbar}[H_S, \rho_S(t)] - \frac{i}{\hbar}[A, \mathrm{tr}_W\{B\rho_{SW}(t)\}]. \tag{3.7}$$

Now, assume that it is allowed to approximate $\rho_{SW}(t) \approx \rho_S(t) \otimes \rho_W(t)$ for all times t. Then, the previous equation reduces to

$$\frac{\partial}{\partial t}\rho_S(t) = -\frac{i}{\hbar}[H_S, \rho_S(t)] - \frac{i}{\hbar}\mathrm{tr}_W\{B\rho_W(t)\}[A, \rho_S(t)] \equiv -\frac{i}{\hbar}[H_S(\lambda_t), \rho_S(t)], \tag{3.8}$$

where we defined the time-dependent system Hamiltonian $H_S(\lambda_t) = H_S + \lambda_t A$ with the driving protocol $\lambda_t \equiv \mathrm{tr}_W\{B\rho_W(t)\}$. The question remains of when is the assumption $\rho_{SW}(t) \approx \rho_S(t) \otimes \rho_W(t)$ justified? Moreover, we would like to ensure that fluctuations in the operator B are negligible such that $\lambda_t = \mathrm{tr}_W\{B\rho_W(t)\}$ describes the same fixed driving protocol in every repetition of the experiment. It seems intuitive that these assumptions are satisfied if the work reservoir is a macroscopic device in a low entropy state, whose dynamics is dominated by a classical motion with negligible fluctuations. Therefore, we initially called the description using a time-dependent Hamiltonian $H(\lambda_t)$ 'semiclassical' since quantum (and also classical) fluctuations in the description of W are assumed negligible. However, precise and rigorous statements justifying $\rho_{SW}(t) \approx \rho_S(t) \otimes \rho_W(t)$ are hard to work out in general. Therefore, the following exercise makes our reasoning precise by considering a simple example.

Exercise 3.1 We consider an atom interacting with a single mode of the electromagnetic field, whose dynamics can be described in the simplest case by the **Jaynes–Cummings Hamiltonian**

$$H_{\mathrm{JC}} = \frac{\hbar\Omega}{2}\sigma_z + \hbar\omega a^\dagger a + \hbar g(\sigma_+ a + a^\dagger \sigma_-). \tag{3.9}$$

Here, the atom is described as a two-level system with energy gap $\hbar\Omega$ between the ground and excited states $|g\rangle$ and $|e\rangle$. The electromagnetic field is described as a harmonic oscillator with frequency ω and the operators $a^\dagger$ (a) create (annihilate) a single photon of the field. The atom–field interaction with strength g describes the emission (absorption) of a photon by the atom using the atomic lowering (raising) operator $\sigma_- = |g\rangle\langle e|$ $(\sigma_+ = |e\rangle\langle g|)$. The Jaynes–Cummings Hamiltonian presents a simplified but popular model in quantum optics because its dynamics can be solved exactly. To make our life simpler, we assume in the following that the atom is on resonance with the electromagnetic field: $\Omega = \omega$. Then, the Hamiltonian in the interaction picture with respect to $(\hbar\Omega/2)\sigma_z + \hbar\omega a^\dagger a$ becomes $\tilde{H}_{\mathrm{JC}} = \hbar g(\sigma_+ a + a^\dagger \sigma_-)$ and the unitary time evolution operator in the interaction picture reads

$$\begin{aligned}\tilde{U}(t) &= \cos\left(gt\sqrt{N+1}\right)|e\rangle\langle e| + \cos\left(gt\sqrt{N}\right)|g\rangle\langle g| \\ &\quad - i\frac{\sin\left(gt\sqrt{N+1}\right)}{\sqrt{N+1}}a|e\rangle\langle g| - ia^\dagger\frac{\sin\left(gt\sqrt{N+1}\right)}{\sqrt{N+1}}|g\rangle\langle e|.\end{aligned} \tag{3.10}$$

Here, $N = a^\dagger a$ denotes the photon number operator. Verify eqn (3.10), for instance, by checking $\partial_t\tilde{U}(t) = -\frac{i}{\hbar}\tilde{H}_{JC}\tilde{U}(t)$.

Now, let us regard the atom as the system S and the field as the work reservoir W. We consider an initially pure atom–field state of the form $|\psi(0)\rangle_S \otimes |\alpha\rangle_W$, where $|\psi(0)\rangle_S$ is an arbitrary state of the atom and $|\alpha\rangle_W \equiv e^{-|\alpha|^2/2}\sum_{n=0}^{\infty}\frac{\alpha^n}{\sqrt{n!}}|n\rangle$ denotes a *coherent* state of the field with amplitude $\alpha \in \mathbb{C}$. For simplicity, we assume $\alpha > 0$ below. The goal is to show that the electromagnetic field acts as a work reservoir for the atom if the amplitude of the coherent state is large enough: $\alpha \gg 1$. One way to do so is to look directly at the reduced atom state at time t:

$$\rho_A(t) = \mathrm{tr}_F\{\tilde{U}(t)|\psi(0)\rangle\langle\psi(0)|_S \otimes |\alpha\rangle\langle\alpha|_W\tilde{U}(t)^\dagger\}. \tag{3.11}$$

To evaluate the trace in the Fock basis, show first that (with $N|n\rangle = n|n\rangle$)

$$\begin{aligned}\langle n|\tilde{U}(t)|\alpha\rangle &= \left[\cos\left(gt\sqrt{n+1}\right)|e\rangle\langle e| + \cos\left(gt\sqrt{n}\right)|g\rangle\langle g| - i\alpha\frac{\sin\left(gt\sqrt{n+1}\right)}{\sqrt{n+1}}|e\rangle\langle g|\right]\langle n|\alpha\rangle \\ &\quad - i\frac{\sqrt{n}\sin\left(gt\sqrt{n+1}\right)}{\sqrt{n+1}}|g\rangle\langle e|\langle n-1|\alpha\rangle.\end{aligned} \tag{3.12}$$

Thus, the expression (3.11) for the reduced atom state contains terms proportional to $|\langle n|\alpha\rangle|^2$, $|\langle n-1|\alpha\rangle|^2$, $\langle n-1|\alpha\rangle\langle\alpha|n\rangle$ and its complex conjugate. Convince yourself numerically of the fact that all of these expressions are strongly peaked around $n \approx \alpha^2$ for $\alpha \gg 1$. You can also see this analytically by verifying for the probability distribution $p_n = |\langle n|\alpha\rangle|^2$ to find n photons in a coherent state that $\langle n\rangle = \sum_n np_n = \alpha^2$ and $\langle(n-\langle n\rangle)^2\rangle = \alpha^2$. Thus, the normalized standard deviation (also called the coefficient of variation) is $\sqrt{\langle(n-\langle n\rangle)^2\rangle}/\langle n\rangle = \alpha^{-1}$, which indicates that p_n is strongly peaked around its mean value for $\alpha \gg 1$. Then, use these insights together with the approximation $\sqrt{\alpha^2+1} \approx \alpha$, which is also justified for $\alpha \gg 1$, to confirm that $\rho_S(t) \approx \tilde{V}(t)|\psi(0)\rangle\langle\psi(0)|_S\tilde{V}(t)^\dagger$ with the unitary

$$\begin{aligned}\tilde{V}(t) &= \cos(gt\alpha)|e\rangle\langle e| + \cos(gt\alpha)|g\rangle\langle g| - i\sin(gt\alpha)|e\rangle\langle g| - i\sin(gt\alpha)|g\rangle\langle e| \\ &= \exp\left[-ig\alpha\left(\sigma_+ + \sigma_-\right)\right].\end{aligned} \tag{3.13}$$

Thus, the reduced state of the atom evolves approximately in a unitary way with respect to an interaction picture Hamiltonian $\hbar g\alpha(\sigma_+ + \sigma_-)$. Move out of the interaction picture with respect to $\frac{\hbar\Omega}{2}\sigma_z$ and show that the dynamics of the atom in the original frame is governed by the *time-dependent* Hamiltonian

$$H_S(\lambda_t) = \frac{\hbar\Omega}{2}\sigma_z + \hbar g\alpha(\sigma_+ e^{-i\omega t} + \sigma_- e^{i\omega t}), \tag{3.14}$$

where we used the resonance condition $\Omega = \omega$. Thus, we have shown that for $\alpha \gg 1$ the time-independent Hamiltonian (3.9) gives rise to a time-dependent Hamiltonian for the atom. Thus, the electromagnetic field acts as a work reservoir.

Finally, investigate how long the dynamics can be described by the approximated Hamiltonian (3.14). For this purpose assume that the atom is initialized in the ground state. Show that the probability of finding the field at time t still in the same coherent state $|\alpha\rangle$ is

$$\begin{aligned}\langle\alpha|\mathrm{tr}_A\{\tilde{U}(t)|g\rangle\langle g|\otimes|\alpha\rangle\langle\alpha|\tilde{U}(t)^\dagger\}|\alpha\rangle &= \cos^2(gt\alpha)+\frac{\alpha^2}{\alpha^2+1}\sin^2\left(gt\sqrt{\alpha^2+1}\right)\\ &= 1+\mathcal{O}\left(\frac{gt}{\alpha^2}\right)+\mathcal{O}\left(\frac{1}{\alpha^2}\right).\end{aligned} \tag{3.15}$$

Thus, as long as $t \ll \alpha^2/g$, it is justified to use the time-dependent Hamiltonian (3.14) to describe the dynamics of the atom.

Despite the usefulness of modelling a work reservoir with a time-dependent Hamiltonian, some words of caution are necessary before applying the above framework. In particular, care is required whenever one computes a *negative* work output $W(t) < 0$, which means that the system gives up energy to the work reservoir. This seems to imply that the system acts like an engine delivering useful work, but remember our semiclassical picture. If the quantum system is tiny, it is not easy to detect (and harness) its influence on a work reservoir, whose motion is assumed to be classical.

Because of this difficulty, we briefly mention two alternative approaches to work in the quantum regime. First, one possibility is to consider an inclusive approach, where the work output is treated quantum mechanically (as indicated by the exercise above). This requires us to explicitly model the storage medium of work, which is then called an *autonomous work reservoir* or a *quantum battery.* We will briefly consider this approach in Section 3.9. However, in most parts of this book we are interested in the fundamental thermodynamic principles of open quantum systems, for which it is important to understand the influence of time-dependent driving fields, which describe well many experiments. Instead, the design of particular storage models for work has larger relevance for concrete applications.

Second, thermodynamic work cannot only be of mechanical origin. As soon as different particle species with different chemical potentials are involved in the description, it becomes possible to extract *chemical work.* This form of work, which we will introduce in detail in Section 3.10, overcomes many problems because it is an autonomous form of work that does not rely on a time-dependent Hamiltonian and that can be produced by small quantum systems on a measurable scale.

3.2 Quantum Master Equations in the Weak Coupling Regime

In theory, to compute the work done on an open quantum system, eqn (3.5), or other relevant thermodynamic quantities, we need an efficient tool to compute the dynamics of $\rho_S(t)$ in the absence of any measurements or interventions. Such a tool is provided by a **(quantum) master equation**, by which we mean in general any differential equation describing the evolution of $\rho_S(t)$ in closed form. In this section, we derive quantum master equations in the *weak coupling regime*, which provide a basic tool in quantum thermodynamics and beyond, covering many interesting applications in non-equilibrium physics in general.

We consider a time-independent system–bath Hamiltonian $H_{SB} = H_S + H_B + V_{SB}$, keeping the generalization to the time-dependent case for Section 3.4. The goal is to find a closed description for the dynamics of $\rho_S(t) = \mathrm{tr}_B\{\rho_{SB}(t)\}$ under the assumption that V_{SB} is weak. To tackle this problem, we return to the Liouville–von Neumann equation (1.1), but switch to the interaction picture defined for an arbitrary system–bath operator O_{SB} as $\tilde{O}_{SB}(t) \equiv e^{i(H_S+H_B)t/\hbar} O_{SB} e^{-i(H_S+H_B)t/\hbar}$. The initial time, at which the Schrödinger, Heisenberg and interaction picture coincide, is $t_0 = 0$. The Liouville–von Neumann equation in the interaction picture reads

$$\frac{\partial}{\partial t}\tilde{\rho}_{SB}(t) = -\frac{i}{\hbar}[\tilde{V}_{SB}(t), \tilde{\rho}_{SB}(t)]. \tag{3.16}$$

A formal integration of it yields

$$\begin{aligned}\tilde{\rho}_{SB}(t) &= \tilde{\rho}_{SB}(0) - \frac{i}{\hbar}\int_0^t ds[\tilde{V}_{SB}(s), \tilde{\rho}_{SB}(s)] \\ &= \tilde{\rho}_{SB}(0) - \frac{i}{\hbar}\int_0^t ds[\tilde{V}_{SB}(s), \tilde{\rho}_{SB}(0)] \\ &\quad - \frac{1}{\hbar^2}\int_0^t ds \int_0^s ds'[\tilde{V}_{SB}(s), [\tilde{V}_{SB}(s'), \tilde{\rho}_{SB}(s')]],\end{aligned} \tag{3.17}$$

where the second equality arose from iteratively inserting the result of the first line. Taking the time derivative again and tracing out the bath degrees of freedom gives rise to the still formally exact equation

$$\frac{\partial}{\partial t}\tilde{\rho}_S(t) = -\frac{i}{\hbar}\mathrm{tr}_B\{[\tilde{V}_{SB}(t), \tilde{\rho}_{SB}(0)]\} - \frac{1}{\hbar^2}\int_0^t ds \mathrm{tr}_B\{[\tilde{V}_{SB}(t), [\tilde{V}_{SB}(s), \tilde{\rho}_{SB}(s)]]\}. \tag{3.18}$$

To proceed, it becomes necessary to make assumptions. Our first assumption, which is motivated by weak-coupling considerations, is that the initial state reads

$$\rho_{SB}(0) = \rho_S(0) \otimes \pi_B(\beta). \tag{3.19}$$

This state describes a system prepared in an arbitrary initial state $\rho_S(0)$ decoupled from the bath, which seems a reasonable first approximation at weak coupling. Furthermore, the bath is described by a canonical ensemble $\pi_B(\beta) = e^{-\beta H_B}/\mathcal{Z}_B$ at inverse temperature β. Note that we also used the initial state (3.19) for our numerical example in Section 1.2. Multiple heat baths at different temperatures are treated in Sections 3.4, 3.7 and 3.10 and more general initial states in Sections 3.7, 3.8 and 3.9. The next exercise shows that we can use the fact that the bath is prepared in a *fixed* state to get rid of the first term on the right-hand side of eqn (3.18).

Exercise 3.2 The claim is that we can set $\mathrm{tr}_B\{\tilde{V}_{SB}(t)\pi_B(\beta)\} = 0$ without loss of generality. Suppose that $\mathrm{tr}_B\{\tilde{V}_{SB}(t)\pi_B(\beta)\} \neq 0$. Show that you can introduce a redefined system Hamiltonian H'_S and interaction Hamiltonian V'_{SB} such that $\mathrm{tr}_B\{\tilde{V}'_{SB}(t)\pi_B(\beta)\} = 0$.

Furthermore, we assume that to lowest order in the interaction V_{SB} we can set

$$\tilde{\rho}_{SB}(t) = \tilde{\rho}_S(t) \otimes \pi_B(\beta) + \mathcal{O}(V_{SB}) \tag{3.20}$$

for all times t. This is called the **Born approximation** and it relies on the idea that the interaction is weak and the bath very large. Thus, we obtain from eqn (3.18) up to second order in V_{SB} the approximate equation

$$\frac{\partial}{\partial t}\tilde{\rho}_S(t) = -\frac{1}{\hbar^2}\int_0^t ds \mathrm{tr}_B\{[\tilde{V}_{SB}(t), [\tilde{V}_{SB}(s), \tilde{\rho}_S(s) \otimes \pi_B(\beta)]]\}. \tag{3.21}$$

This is a closed integro-differential equation for the system state $\rho_S(t)$. The fact that $\rho_S(t)$ appears in it in a time-convoluted form makes its application in practice difficult. However, notice that eqn (3.21) does not contain any zeroth-order term with respect to V_{SB}. To lowest order in V_{SB} we therefore have $\tilde{\rho}_S(t) = \tilde{\rho}_S(s) + \mathcal{O}(V_{SB})$ (remember that we are in the interaction picture). This insight turns eqn (3.21) into an ordinary, time-local differential equation of the form

$$\boxed{\frac{\partial}{\partial t}\tilde{\rho}_S(t) = -\frac{1}{\hbar^2}\int_0^t ds \mathrm{tr}_B\{[\tilde{V}_{SB}(t), [\tilde{V}_{SB}(s), \tilde{\rho}_S(t) \otimes \pi_B(\beta)]]\}.} \tag{3.22}$$

We call this equation the **weak coupling master equation**, a simple but powerful tool used for many computations in practice.

While being an efficient numerical tool, it is still hard to work with the weak coupling master equation analytically. To get further insights into the problem, we use additional assumptions. First of all, however, we make eqn (3.22) more explicit by decomposing the system–bath interaction as

$$V_{SB} = \sum_\alpha A_\alpha \otimes B_\alpha, \tag{3.23}$$

where A_α (B_α) are taken to be *Hermitian* system (bath) operators.

Exercise 3.3 We can always write $V_{SB} = \sum_\alpha A_\alpha \otimes B_\alpha$ for some system and bath operators A_α and B_α. The claim is that we can also assume A_α and B_α to be Hermitian without loss of generality. If A_α and B_α are not Hermitian, show that you can find Hermitian operators A'_α and B'_α such that $V_{SB} = \sum_\alpha A'_\alpha \otimes B'_\alpha$. *Hint:* Use that you can write any operator A as $A = A_1 + iA_2$ with A_1 and A_2 Hermitian.

An important role in the following plays the **bath correlation function**

$$C_{\alpha\alpha'}(t) \equiv \mathrm{tr}_B\{B_\alpha(t)B_{\alpha'}\pi_B(\beta)\}, \tag{3.24}$$

which determines the influence of the bath on the system to second order. To see this, we use $\mathrm{tr}_B\{B_\alpha(t)B_{\alpha'}(s)\pi_B(\beta)\} = \mathrm{tr}_B\{B_\alpha(t-s)B_{\alpha'}\pi_B(\beta)\}$ and that the trace is cyclic, which allows us to write the weak coupling master equation as

$$\frac{\partial}{\partial t}\tilde{\rho}_S(t) = \frac{1}{\hbar^2}\sum_{\alpha,\alpha'}\int_0^t ds \left\{C_{\alpha\alpha'}(t-s)\left[\tilde{A}_{\alpha'}(s)\tilde{\rho}_S(t)\tilde{A}_\alpha(t) - \tilde{A}_\alpha(t)\tilde{A}_{\alpha'}(s)\tilde{\rho}_S(t)\right]\right.$$
$$\left. + C_{\alpha'\alpha}(s-t)\left[\tilde{A}_\alpha(t)\tilde{\rho}_S(t)\tilde{A}_{\alpha'}(s) - \tilde{\rho}_S(t)\tilde{A}_{\alpha'}(s)\tilde{A}_\alpha(t)\right]\right\}. \tag{3.25}$$

Notice that $C^*_{\alpha\alpha'}(t) = C_{\alpha'\alpha}(-t)$, so the second line is the Hermitian conjugate of the first. Furthermore, after substituting $\tau \equiv t - s$ in the integral, we find

$$\frac{\partial}{\partial t}\tilde{\rho}_S(t) = \frac{1}{\hbar^2}\sum_{\alpha,\alpha'}\int_0^t d\tau C_{\alpha\alpha'}(\tau)\left[\tilde{A}_{\alpha'}(t-\tau)\tilde{\rho}_S(t)\tilde{A}_\alpha(t) - \tilde{A}_\alpha(t)\tilde{A}_{\alpha'}(t-\tau)\tilde{\rho}_S(t)\right]$$
$$+ \text{h.c.} \tag{3.26}$$

For an effectively infinitely large bath (i.e. a bath with a Poincaré recurrence time much larger than any time scale of interest; see the discussion in Section 1.3), one expects that $C_{\alpha\alpha'}(\tau)$ decays to zero for $\tau \to \infty$: after some time the bath can no longer remember some initial weak perturbation. The crucial idea of the following step is to assume that this decay is very *quick*, i.e. the bath is essentially *memoryless*. If this is the case, we can extend the upper limit of the integral in eqn (3.26) to infinity, which is known as the **Markov approximation**. Note the close analogy to the argument of time-scale separation, which we discussed extensively in Chapter 2 and which was also based on the assumption that the intrinsic time evolution of the bath is much faster than that of the system. Then, the weak coupling master equation reduces to the **Born–Markov equation** (also called **Redfield equation**):

$$\boxed{\begin{aligned}\frac{\partial}{\partial t}\tilde{\rho}_S(t) = \sum_{\alpha,\alpha'}\int_0^\infty \frac{d\tau}{\hbar^2} C_{\alpha\alpha'}(\tau)\left[\tilde{A}_{\alpha'}(t-\tau)\tilde{\rho}_S(t)\tilde{A}_\alpha(t) - \tilde{A}_\alpha(t)\tilde{A}_{\alpha'}(t-\tau)\tilde{\rho}_S(t)\right]\\ + \text{h.c.}\end{aligned}} \tag{3.27}$$

This seems like a small change compared with eqn (3.26), but the numerical and analytical treatment of eqn (3.27) is considerably simpler. Before proceeding, the next exercise checks when it is a good idea to use the Markov approximation.

Exercise 3.4 We consider the paradigmatic Caldeira–Leggett model discussed in Exercise 1.2 and Section 2.7 to describe a system coupled to a bath. The interaction Hamiltonian is of the form $V_{SB} = S \otimes B$ with $B = \sum_k c_k x_k$. Show that the bath correlation function reads explicitly

$$C(t) = \text{tr}_B\{B(t)B\pi_B\} = \sum_k \frac{\hbar c_k^2}{2\omega_k}\left[\cos(\omega_k t)\coth\left(\frac{\beta\hbar\omega_k}{2}\right) - i\sin(\omega_k t)\right]. \tag{3.28}$$

As discussed in Section 1.3, for any finite number of oscillators this function will never strictly decay to zero, but as long as the number of oscillators is very large any Poincaré recurrence in $C(t)$ can be neglected. Theoretically, this idea can be encoded in the assumption that the function $J(\omega) \equiv (\pi/2)\sum_k (c_k^2/\omega_k)\delta(\omega - \omega_k)$, which is known as the *spectral density* of the bath, is continuous. In terms of the spectral density the bath correlation function becomes

$$C(t) = \frac{\hbar}{\pi}\int_0^\infty d\omega\, J(\omega)\left[\cos(\omega t)\coth\left(\frac{\beta\hbar\omega}{2}\right) - i\sin(\omega t)\right]. \tag{3.29}$$

How $C(t)$ behaves in a specific case depends strongly on $J(\omega)$, which characterizes how the oscillators in the bath couple to the system. A particular important case is a spectral density of the form $J(\omega) = \gamma\omega\Theta(\omega_C - \omega)$, which is called *Ohmic*. Here, γ is an overall damping constant and the Heaviside function $\Theta(\omega_C - \omega)$ describes a cut-off that is responsible for a decay of $J(\omega)$ to zero for frequencies much larger than the cut-off frequency $\omega \gg \omega_C$. We note that, as long as ω_C is much larger than the system frequencies and as long as we are not interested in the ultrashort time behaviour of the system, it actually turns out that the specific choice of the cut-off function is unimportant.

The importance of the Ohmic spectral density comes from the fact that it often justifies the Markov approximation. For this purpose, the reader is asked to verify that for $J(\omega) \sim \omega$ one obtains $C(t) \sim \delta(t) + \mathcal{O}(\hbar)$ in the limit $\omega_C \to \infty$. *Hint:* It is useful to expand $\coth(x) = x^{-1} + \mathcal{O}(x)$, which yields

$$C(t) = \frac{1}{\pi}\int_0^\infty d\omega\, J(\omega)\cos(\omega t)\left[\frac{2}{\beta\omega} + \mathcal{O}(\beta\hbar^2)\right] - i\frac{\hbar}{\pi}\int_0^\infty d\omega\, J(\omega)\sin(\omega t). \tag{3.30}$$

This expansion shows that $C(t) \approx 2\int_0^\infty d\omega\, J(\omega)\cos(\omega t)/(\pi\beta\omega)$ is a good approximation in the classical regime for temperatures $T \gg \hbar\omega_C/k_B$.

Examples for the evolution of the bath correlation function for the Caldeira–Leggett model are shown in Fig. 3.1. As a rule of thumb, we keep in mind that the higher the temperature of the bath and the less structured the spectral density (compared with the Ohmic case), the more justified is the Markov approximation.

We remark that, even if the Markov approximation is not justified, we can use the Born–Markov master equation whenever we are only interested in the *long time limit* of the dynamics. Then, eqn (3.26) reduces automatically to eqn (3.27), which can be used to study transport properties in the steady-state regime at weak coupling (see, for example, Section 3.10). For other applications, e.g. in spectroscopy, the short time dynamics is instead important and the Markov approximation becomes questionable.

To proceed, we take a look at the system coupling operators A_α in the interaction picture. Using the eigenbasis of the system Hamiltonian $H_S = \sum_s \epsilon_s \Pi(\epsilon_s)$, we get

$$\tilde{A}_\alpha(t) = \sum_{s,s'} e^{-i(\epsilon_{s'}-\epsilon_s)t/\hbar}\Pi(\epsilon_s)A_\alpha\Pi(\epsilon_{s'}) \equiv \sum_\omega e^{-i\omega t}A_\alpha(\omega). \tag{3.31}$$

Here, $\omega = (\epsilon_{s'} - \epsilon_s)/\hbar$ is an index running over all *transition* frequencies of the system Hamiltonian and we defined $A_\alpha(\omega) \equiv \sum_{\epsilon_{s'}-\epsilon_s=\hbar\omega}\Pi(\epsilon_s)A_\alpha\Pi(\epsilon_{s'})$. Equation (3.31) can be regarded as a Fourier decomposition of the system coupling operator in the interaction picture. The Fourier components $A_\alpha(\omega)$ obey the following properties:

$$A_\alpha^\dagger(\omega) = A_\alpha(-\omega), \quad [A_\alpha(\omega), H_S] = \hbar\omega A_\alpha(\omega), \quad A_\alpha(\omega)\pi_S = e^{-\beta\hbar\omega}\pi_S A_\alpha(\omega). \tag{3.32}$$

In particular, the first property is used in the following without being mentioned explicitly.

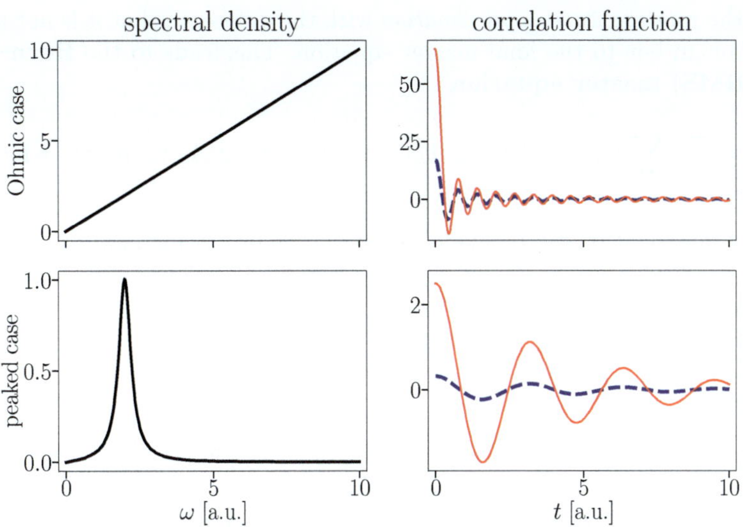

Fig. 3.1 Plot of the real part of the correlation function for the Caldeira–Leggett model for two different spectral densities displayed in the left column. The Ohmic case $J(\omega) = \gamma\omega\Theta(\omega_C - \omega)$ for $\gamma = 1$ and $\omega_C = 10$ and a peaked case $J(\omega) = d^2\gamma\omega/[(\omega^2 - \omega_0^2)^2 + \gamma^2\omega^2]$ for $d = 1$, $\gamma = 1/2$ and $\omega_0 = 2$. The correlation functions are displayed for a high-temperature case $T = 10$ (thin red lines) and a low-temperature case $T = 1$ (thick dashed blue lines). We set $k_B = 1$ and $\hbar = 1$.

Exercise 3.5 Derive the properties stated in eqn (3.32).

The Born–Markov equation can then be written as

$$\frac{\partial}{\partial t}\tilde{\rho}_S(t) = \sum_{\omega,\omega'}\sum_{\alpha,\alpha'} e^{i(\omega'-\omega)t}\Gamma_{\alpha\alpha'}(\omega)\left[A_{\alpha'}(\omega)\tilde{\rho}_S(t)A^\dagger_\alpha(\omega') - A^\dagger_\alpha(\omega')A_{\alpha'}(\omega)\tilde{\rho}_S(t)\right] + \text{h.c.}, \tag{3.33}$$

where we defined the *half-sided* Fourier transform of the bath correlation function

$$\Gamma_{\alpha\alpha'}(\omega) \equiv \frac{1}{\hbar^2}\int_0^\infty d\tau e^{i\omega\tau}C_{\alpha\alpha'}(\tau). \tag{3.34}$$

There is now one final approximation, which is often invoked and which is known as the **secular approximation**. It is based on the idea that the terms $e^{i(\omega-\omega')t}$ in eqn (3.33) oscillate quickly in time compared with the time evolution induced by the 'rates' $\Gamma_{\alpha\alpha'}(\omega)$. Note that $\Gamma_{\alpha\alpha'}$ has the dimension of a rate if we define the system coupling operators A_α to be dimensionless, which we assume to be done from here on; in general, however, $\Gamma_{\alpha\alpha'}$ is not real-valued. In any case, whenever $\omega - \omega' \gg |\Gamma_{\alpha\alpha'}(\omega)|$, the factor $e^{i(\omega-\omega')t}$ will average out during the evolution and we can set $e^{i(\omega-\omega')t} = \delta_{\omega,\omega'}$. Readers familiar with quantum optics might recognize this approximation as a variant of the *rotating wave approximation* with the difference that it is not applied to the Hamiltonian but to the final master equation. This leads to the **Born–Markov secular (BMS) master equation**

$$\frac{\partial}{\partial t}\tilde{\rho}_S(t) = \sum_\omega \sum_{\alpha,\alpha'} \Gamma_{\alpha\alpha'}(\omega)\left[A_{\alpha'}(\omega)\tilde{\rho}_S(t)A_\alpha^\dagger(\omega) - A_\alpha^\dagger(\omega)A_{\alpha'}(\omega)\tilde{\rho}_S(t)\right] + \text{h.c.} \quad (3.35)$$

The form of this equation can be made more intuitive by writing $\Gamma_{\alpha\alpha'}(\omega) = \frac{1}{2}\gamma_{\alpha\alpha'}(\omega) + \frac{i}{\hbar}\lambda_{\alpha\alpha'}(\omega)$, where $\gamma_{\alpha\alpha'}(\omega) = \gamma^*_{\alpha'\alpha}(\omega)$ and $\lambda_{\alpha\alpha'}(\omega) = \lambda^*_{\alpha'\alpha}(\omega)$ are elements of a Hermitian matrix. Their precise definition is

$$\gamma_{\alpha\alpha'}(\omega) \equiv \Gamma_{\alpha\alpha'}(\omega) + \Gamma^*_{\alpha'\alpha}(\omega) = \frac{1}{\hbar^2}\int_{-\infty}^{\infty} d\tau e^{i\omega\tau} C_{\alpha\alpha'}(\tau), \quad (3.36)$$

$$\lambda_{\alpha\alpha'}(\omega) \equiv \frac{\hbar}{2i}[\Gamma_{\alpha\alpha'}(\omega) - \Gamma^*_{\alpha'\alpha}(\omega)] = \frac{1}{2i\hbar}\int_{-\infty}^{\infty} d\tau \text{sgn}(\tau) e^{i\omega\tau} C_{\alpha\alpha'}(\tau) \quad (3.37)$$

with the sign function $\text{sgn}(\tau) = 2\Theta(\tau) - 1$. Thus, eqn (3.35) can also be expressed as

$$\boxed{\begin{aligned}\frac{\partial}{\partial t}\tilde{\rho}_S(t) =& -\frac{i}{\hbar}\sum_\omega\sum_{\alpha,\alpha'}\lambda_{\alpha\alpha'}(\omega)[A_\alpha^\dagger(\omega)A_{\alpha'}(\omega), \tilde{\rho}_S(t)] \\ &+ \sum_\omega\sum_{\alpha,\alpha'}\gamma_{\alpha\alpha'}(\omega)\left[A_{\alpha'}(\omega)\tilde{\rho}_S(t)A_\alpha^\dagger(\omega) - \frac{1}{2}\{A_\alpha^\dagger(\omega)A_{\alpha'}(\omega), \tilde{\rho}_S(t)\}\right].\end{aligned}} \quad (3.38)$$

This is the final result of this section. Before discussing in detail the properties of the BMS master equation in the next section, we finish with some general remarks.

First, by looking at eqn (3.38), we see that the influence of the bath on the system dynamics has two contributions. The first line describes environmentally induced modifications of the *unitary* part of the dynamics. The term

$$H_{\text{LS}} \equiv \sum_\omega\sum_{\alpha,\alpha'}\lambda_{\alpha\alpha'}(\omega)A_\alpha^\dagger(\omega)A_{\alpha'}(\omega) \quad (3.39)$$

is called the **Lamb shift Hamiltonian** and it renormalizes the eigenenergies of the system. In fact, one easily confirms that $[H_S, H_{\text{LS}}] = 0$. Thus, the Lamb shift Hamiltonian shifts the eigenenergies, but leaves the eigenbasis unchanged. In practice, computing the Lamb shift Hamiltonian can be cumbersome because of the involved integral (3.37) and it is therefore often neglected. There are two more reasons to neglect it. First, in practice, one often derives the system Hamiltonian H_S not from a first-principles (*ab initio*) calculation, but infers the transition frequencies ω experimentally. During this procedure, the system is usually in contact with the bath; hence, the so-inferred transition frequencies already contain the bath-induced renormalizations. Including H_{LS} in the final result would then count them twice. The second reason for neglecting H_{LS} relates to the fact that the perturbative hierarchy (beyond second order in V_{SB}) to compute H_{LS} can converge poorly. For instance, the fourth-order contribution can have a strong impact even at weak coupling and can tend to cancel out the second-order contribution. Indeed, benchmarks with exactly solvable models have shown that *excluding* H_{LS} can sometimes *improve* the accuracy of the master equation. Therefore, we take a pragmatic point of view on H_{LS} and drop it in all the following equations or, if its contribution is important, assume it to be included in the definition of H_S.

The second line of eqn (3.38) describes the—for us—more interesting *non-unitary* modifications of the system dynamics. It causes effects such as decoherence, dissipation and thermalization, which will become obvious soon. Therefore, the second line is often simply abbreviated as $\mathcal{D}\tilde{\rho}_S(t)$, where the superoperator $\mathcal{D}$ is known as the **dissipator**. Furthermore, the secular approximation guarantees that the dissipator commutes with the interaction picture transformation, i.e.

$$e^{iH_S t/\hbar}[\mathcal{D}\rho_S(t)]e^{-iH_S t/\hbar} = \mathcal{D}\tilde{\rho}_S(t). \tag{3.40}$$

Thus, going back to the Schrödinger picture, the BMS equation reads in short form

$$\frac{\partial}{\partial t}\rho_S(t) = -\frac{i}{\hbar}[H_S, \rho_S(t)] + \mathcal{D}\rho_S(t) \equiv \mathcal{L}\rho_S(t). \tag{3.41}$$

Here, we introduce the superoperator $\mathcal{L}$, which we call the **Liouvillian**, to have a compact notation for the entire master equation. We note that the Liouvillian $\mathcal{L}$ does not always refer to the BMS equation in the Schrödinger picture, instead $\partial_t\rho_S(t) = \mathcal{L}\rho_S(t)$ is often written as a short form for any kind of quantum master equation.

Of course, the question remains as to which of the various master equations derived in this section should be used in practice. This question has been around since the beginning of this field. In this book, we adopt the following philosophy.

First, for any particular *numerical* problem, the reader is advised to use the weak coupling master equation (3.22) or, if the short time dynamics is not of interest, the Born–Markov equation (3.27). The secular approximation should be used with care, in particular for complex open quantum systems involving multiple different transition frequencies. Its prediction compared with the exact dynamics can differ by *orders of magnitude* even at weak coupling. Furthermore, the secular approximation is uncontrolled in the sense that it can destroy physical symmetries present in the system–bath Hamiltonian.

Second, for *analytical* calculations, the reader is advised to first consider the BMS equation because it satisfies a number of important properties, which will be explored in various sections of this book. In contrast, these properties are not strictly satisfied by the Born–Markov equation, which makes it hard to obtain general analytical insights from it. The hope in practice is, of course, that the results derived for the BMS equation carry over to the weak coupling and Born–Markov equation, even if the secular approximation gives the wrong numerical predictions. Luckily, for a large class of systems and parameters, this hope has turned out to be true.

In this book, we will derive general results based on the BMS equation in Sections 3.3, 3.4, 4.4, 4.5 and 5.4. Furthermore, we will use the BMS equation for particular applications in Sections 3.9, 3.10 and 5.7. For all these applications, the BMS approximation is well justified experimentally.

3.3 Properties of the Born–Markov Secular Master Equation

This section recapitulates important properties of the BMS equation that are useful for various manipulations throughout this book. The derivation of most properties is simple and left as an exercise. Two properties that are not that simple to derive are stated without proof. More information is given in the list of references at the end.

First, we observe that the bath correlation functions obey a symmetry relation, which is independent of the question of whether we apply the secular approximation or not. This symmetry relation is known as the **Kubo–Martin–Schwinger condition**:

$$C_{\alpha\alpha'}(t) = C_{\alpha'\alpha}(-t - i\beta\hbar). \tag{3.42}$$

From this, we can deduce that the rates from eqn (3.36) obey **local detailed balance**:

$$\boxed{\gamma_{\alpha\alpha'}(\omega) = e^{\beta\hbar\omega}\gamma_{\alpha'\alpha}(-\omega).} \tag{3.43}$$

The reader should not be surprised by this result: we have already encountered a similar condition in Section 2.3, where we derived it in a mathematically different way by using time-scale separation (but no weak coupling).

Exercise 3.6 Derive eqns (3.42) and (3.43). Use the assumption that the bath correlation function is analytic in the complex τ plane for $\Im(\tau) \in [-i\beta\hbar, 0]$ and decays quickly to zero for $|\tau| \to \pm\infty$.

The previous result also implies that a possible steady state of the BMS equation is the Gibbs state $\pi_S = e^{-\beta H_S}/\mathcal{Z}_S$:

$$\mathcal{L}\pi_S = 0. \tag{3.44}$$

This is, of course, the result we expect based on arguments from equilibrium statistical mechanics. It is interesting to ask whether π_S is the only possible steady state. In principle, the existence of multiple steady states is possible and closely related to the question of whether there are additional conserved quantities, i.e. system observables O_S such that $\text{tr}_S\{O_S\partial_t\rho_S(t)\} = 0$ for any $\rho_S(t)$. However, for the explicit master equations appearing in this book there will be always a unique steady state.

Exercise 3.7 Show eqn (3.44). *Hint:* Recall eqn (3.32).

Next, we 'diagonalize' the dissipator $\mathcal{D}$ appearing in eqn (3.38). Recall that the $\gamma_{\alpha\alpha'}(\omega)$ are elements of a Hermitian matrix $\gamma(\omega)$. Hence, there exists a diagonal matrix $r(\omega)$ with eigenvalues $r_\alpha(\omega) \in \mathbb{R}$ such that $\gamma(\omega) = u^\dagger(\omega)r(\omega)u(\omega)$, where $u(\omega)$ is a unitary matrix. Furthermore, the local detailed balance condition (3.43) can be written as $\gamma(\omega) = e^{\beta\hbar\omega}\gamma^T(-\omega)$. Since $\gamma(-\omega) = u^\dagger(-\omega)r(-\omega)u(-\omega)$, we find

$$r_\alpha(\omega) = e^{\beta\hbar\omega}r_\alpha(-\omega), \quad u(-\omega) = u^*(\omega). \tag{3.45}$$

Now, we define system operators $S_\alpha(\omega) \equiv \sum_\alpha u_{\alpha\alpha'}(\omega)A_{\alpha'}(\omega)$, which obey the same relations (3.32) as $A_\alpha(\omega)$, in particular $S_\alpha^\dagger(\omega) = S_\alpha(-\omega)$. In terms of them, it is not too hard to show that the dissipator of the BMS equation takes on the simple form

$$\boxed{\mathcal{D}\rho_S = \sum_\omega \sum_\alpha r_\alpha(\omega) \left[S_\alpha(\omega)\rho_S S_\alpha^\dagger(\omega) - \frac{1}{2}\{S_\alpha^\dagger(\omega)S_\alpha(\omega), \rho_S\} \right].} \tag{3.46}$$

Finally, the eigenvalues $r_\alpha(\omega)$ of $\gamma(\omega)$ are not only real-valued, but also positive. This ultimatively justifies calling them *rates*. In the next exercise the reader is explicitly asked to verify this.

Exercise 3.8 Show that $\gamma(\omega)$ is a positive matrix, which implies $r_\alpha(\omega) \geq 0$, by showing that $\boldsymbol{v}^\dagger \cdot \gamma \cdot \boldsymbol{v} = \sum_{\alpha,\alpha'} v_\alpha^* \gamma_{\alpha\alpha'}(\omega) v_\alpha \geq 0$ for any vector $\boldsymbol{v}$. *Hint:* By evaluating the bath correlation function in the energy eigenbasis ($H_B|k\rangle = \epsilon_k|k\rangle$), you should obtain

$$\boldsymbol{v}^\dagger \cdot \gamma \cdot \boldsymbol{v} = 2\pi \sum_{k,\ell} \delta\left(\omega - \frac{\epsilon_\ell - \epsilon_k}{\hbar}\right) \frac{e^{-\beta\epsilon_k}}{\mathcal{Z}_B} \sum_\alpha v_\alpha^* B_{k\ell}^\alpha \sum_{\alpha'} (B_{k\ell}^{\alpha'})^* v_{\alpha'}. \tag{3.47}$$

Here, $B_{k\ell}^\alpha \equiv \langle k|B_\alpha|\ell\rangle = (B_{\ell k}^\alpha)^*$ (remember that $B_\alpha = B_\alpha^\dagger$) are the bath coupling operator elements in the energy eigenbasis. Deduce from eqn (3.47) the desired result.

The above results are important because they allow us to conclude that the dynamics generated by the BMS equation is *completely positive and trace-preserving* (CPTP) at all times. This is a consequence of an important mathematical theorem, which we state here without proof.

CPTP master equations. *Suppose the dynamics of a quantum system is described by a master equation of the form*

$$\begin{aligned} \frac{\partial}{\partial t}\rho(t) &= \mathcal{L}(t)\rho(t) \\ &\equiv -i[H(t), \rho(t)] + \sum_n \kappa_n(t) \left[J_n(t)\rho(t)J_n^\dagger(t) - \frac{1}{2}\{J_n^\dagger(t)J_n(t), \rho(t)\} \right]. \end{aligned} \tag{3.48}$$

Here, $H(t)$ and $J_n(t)$ are arbitrary time-dependent system operators with the constraint that $H(t)$ is Hermitian. Furthermore, $\kappa_n(t) \geq 0$ are arbitrary time-dependent but positive rates. Then, the dynamical map

$$\mathcal{E}(t_2, t_1) \equiv \exp_+ \left[\int_{t_1}^{t_2} \mathcal{L}(t)dt \right] \tag{3.49}$$

is CPTP for all $t_2 > t_1$.

The theorem on CPTP master equations should be compared with what we have discussed in Section 1.5, where we introduced the notion of CPTP maps, and Section 1.8, where we introduced the notion of dynamical maps. Recall that, provided the system is initially decorrelated from the environment, CPTP maps are the most general map to describe the change of a quantum system from one instance to another. The theorem above extends this statement, which was originally formulated for discrete time steps, to a differentiable dynamics. Interestingly, while every CPTP master equation implies an evolution in terms of CPTP maps, the converse is *not* true: not every CPTP map $\mathcal{E}(t_2, t_1)$ can be generated by a CPTP master equation as described in eqn (3.49).

It is also worth pointing out that the weak coupling and Born–Markov master equation in general do not have the form of a CPTP master equation. While it is straightforward to show that they preserve the trace, they can violate complete positivity (and even positivity) under exceptional circumstances. If that happens, one should interpret it as a signature that the entire weak coupling approximation is flawed.

The problem of breaking positivity is cured by the secular approximation, but the many useful properties of the BMS equation come at a cost. Besides its problems in accurately simulating complex open systems (as mentioned in the previous section), the next exercise also shows that not many quantum features are left in this approach.

Exercise 3.9 Consider sequential projective measurements of an open system in its energy eigenbasis $H_S = \sum_s \epsilon_s \Pi(\epsilon_s)$. Assume that in between the measurements the BMS approximation is justified and we can use the master equation $\partial_t \rho_S(t) = \mathcal{L}\rho_S(t)$, where $\mathcal{L}$ was introduced in eqn (3.41). Convince yourself that the non-normalized state conditioned on receiving the measurement outcomes $\epsilon_{s(n)}, \dots, \epsilon_{s(1)}$ at times $t_n > \dots > t_1 > 0$ reads

$$\tilde{\rho}_S(\epsilon_{s(n)}, \dots, \epsilon_{s(1)}) = \mathcal{P}(\epsilon_{s(n)}) e^{\mathcal{L}(t_n - t_{n-1})} \dots \mathcal{P}(\epsilon_{s(1)}) e^{\mathcal{L}t_1} \rho_S(0), \tag{3.50}$$

where $\mathcal{P}(\epsilon_{s(k)})$ is the projection superoperator associated with measurement result $\epsilon_{s(k)}$. Show that the measurement probabilities $p(\epsilon_{s(n)}, \dots, \epsilon_{s(1)}) = \text{tr}_S\{\tilde{\rho}_S(\epsilon_{s(n)}, \dots, \epsilon_{s(1)})\}$ satisfy the Kolmogorov consistency condition. Hence, they cannot be distinguished from a classical stochastic process owing to the inability of the BMS equation to create detectable coherence in the energy basis. *Hint:* Recall the content of Section 1.9, in particular what we called *classicality from incoherence.* Then, it is useful to show that the dephasing operation $\mathcal{D}_{H_S}$ in the energy basis (not to be mixed up with the dissipator $\mathcal{D}$) can be written as

$$\mathcal{D}_{H_S} \rho_S = \sum_s \Pi(\epsilon_s) \rho_S \Pi(\epsilon_s) = \lim_{T \to \infty} \frac{1}{T} \int_0^T dt e^{-iH_S t} \rho_S e^{iH_S t}. \tag{3.51}$$

Deduce and use that this implies that the dephasing operation commutes with the BMS dynamics, $[\mathcal{D}_{H_S}, \mathcal{L}] = 0$; compare this also with eqn (3.40).

Finally, we mention for completeness another property, which is known as **Davies' theorem**, whose derivation is very complicated. Colloquially stated, Davies' theorem ensures that, if all bath correlation functions $C_{\alpha\alpha'}(t)$ decay quickly enough for increasing t (at least as quick as t^{-x} with $x > 2$), then the BMS master equation describes the *exact* reduced dynamics of the system in the limit where the interaction strength V_{SB} becomes *infinitesimally small.* Davies' theorem is an important result in mathematical physics, but unfortunately its physical scope is limited. In experimental reality there is always a finite interaction V_{SB}, which might be small, but it cannot be scaled down to zero.

3.4 Thermodynamics in the Born–Markov Secular Approximation

In this section we derive the laws of non-equilibrium thermodynamics from the BMS equation. For this purpose we proceed in three steps. First, we consider the master equation as derived in Section 3.2. Afterwards, we discuss how far we can include a time-dependent Hamiltonian $H_S = H_S(\lambda_t)$ in the description. Finally, we extend our results to the case of multiple baths at different temperatures.

Single bath without driving

We consider the BMS equation (3.41) in the Schrödinger picture. Our thermodynamic investigation starts with the definition of the state functions **internal energy** and **system entropy**. Based on our experience with the weakly coupled classical case, it seems natural to use the definitions

$$\boxed{U_S(t) \equiv \text{tr}_S\{H_S\rho_S(t)\}, \quad S_S(t) \equiv k_B S_{\text{vN}}[\rho_S(t)] = -k_B\text{tr}_S\{\rho_S(t)\ln\rho_S(t)\},} \tag{3.52}$$

whose consistency and usefulness remains to be established in the following.

A straightforward calculation reveals for the change in internal energy that

$$\frac{d}{dt}U_S(t) = \text{tr}_S\left\{H_S\frac{\partial\rho_S(t)}{\partial t}\right\} = \text{tr}_S\{H_S\mathcal{D}\rho_S(t)\} \equiv \dot{Q}(t). \tag{3.53}$$

Since the Hamiltonian is time independent, there is only negligible work in unison with what we discussed in Section 3.1. Hence, the entire change in internal energy is identified with the heat flow *into* the system, which is—not very surprisingly—determined by the dissipator $\mathcal{D}$. Finally, we need to derive the second law of thermodynamics, which reads

$$\dot{\Sigma}(t) = \frac{d}{dt}S_S(t) - \frac{\dot{Q}(t)}{T} \geq 0. \tag{3.54}$$

Mathematically, the positivity of this expression follows from monotonicity of relative entropy (Theorem A.3), as explained in the next exercise.

Exercise 3.10 Show that you can write eqn (3.54) as $\dot{\Sigma}(t) = -k_B\partial_t D[\rho_S(t)|\pi_S]$. Next, remember the theorem about the CPTP master equation, eqn (3.49), which states that there exists a CPTP map $\mathcal{E}(dt)$ propagating the system state forwards in time: $\rho_S(t+dt) = \mathcal{E}(dt)\rho_S(t)$. Then, use $\mathcal{E}(dt)\pi_S = \pi_S$ (eqn (3.44)) to show that

$$\dot{\Sigma}(t) = k_B \lim_{dt\searrow 0}\frac{D[\rho_S(t)|\pi_S] - D[\mathcal{E}(dt)\rho_S(t)|\mathcal{E}(dt)\pi_S]}{dt}. \tag{3.55}$$

Conclude that this implies the positivity of $\dot{\Sigma}(t)$.

Single bath with driving

We now turn to the situation of a driven system Hamiltonian $H_S(\lambda_t)$. Writing $U_S(t,0) \equiv \exp_+[-i\int_0^t dsH_S(\lambda_s)ds/\hbar]$, the interaction picture for any system observable O_S is defined as $\tilde{O}_S(t,0) \equiv U_S^\dagger(t,0)O_SU_S(t,0)$. Consequently, all steps in Section 3.2 up to eqn (3.26) can be carried over to the driven case provided that the system–bath interaction is weak. Furthermore, provided that the bath correlation function decays quickly enough, eqn (3.27) also remains valid. However, the remaining steps to finally arrive at the BMS master equation can no longer be carried over in the same way. Nevertheless, one case in which we expect this to be still possible is the regime where $H_S(\lambda_t)$ changes *slowly* in time. We then expect that we can replace in eqn (3.38) ω by $\omega(\lambda_t) = \epsilon_{s'}(\lambda_t) - \epsilon_s(\lambda_t)$, which denotes the instantaneous transition frequencies of $H_S(\lambda_t)$. We denote the BMS equation (3.41) in this case by

$$\frac{\partial}{\partial t}\rho_S(t) = -\frac{i}{\hbar}[H_S(\lambda_t), \rho_S(t)] + \mathcal{D}(\lambda_t)\rho_S(t). \tag{3.56}$$

We remark that eqn (3.56) can be more rigorously derived and it does not necessarily imply that the system Hamiltonian changes *adiabatically slowly* in the sense that, if started in the initial Gibbs state $\rho_S(0) = \pi_S(\lambda_0)$, the system state at later times remains in equilibrium, $\rho_S(t) = \pi_S(\lambda_t)$. Nevertheless, the treatment of time-dependent fields using a master equation approach is a subtle topic. Since we will soon overcome the master equation picture in any case, we do not enter into further details here. Instead, we now proceed with the thermodynamic analysis of eqn (3.56).

First, our basic thermodynamic definitions (3.52) for internal energy and system entropy remain unchanged. Furthermore, we have already defined work for open systems in eqn (3.5). Combining these insights together with eqn (3.56) gives the first law for a driven open quantum system coupled to a single heat bath:

$$\frac{d}{dt}U_S(t) = \mathrm{tr}_S\left\{\frac{\partial H_S(\lambda_t)}{\partial t}\rho_S(t)\right\} + \mathrm{tr}_S\left\{H_S(\lambda_t)\frac{\partial \rho_S(t)}{\partial t}\right\} = \dot{W}(t) + \dot{Q}(t), \tag{3.57}$$

where the **heat flow** can be written as

$$\boxed{\dot{Q}(t) \equiv \mathrm{tr}_S\{H_S(\lambda_t)\mathcal{D}(\lambda_t)\rho_S(t)\}.} \tag{3.58}$$

Finally, we consider the second law, which is still of the same form as in eqn (3.54). The positivity of the entropy production rate in the driven case follows from similar steps as in Exercise 3.10 because

$$\dot{\Sigma}(t) = -k_B \left.\frac{\partial}{\partial t}\right|_{\lambda_t} D[\rho_S(t)|\pi_S(\lambda_t)] \geq 0, \tag{3.59}$$

where $\partial_t|_{\lambda_t}$ denotes a partial derivative with respect to time while keeping λ_t fixed. Note that we have already met the identity (3.59) in the classical case; see eqn (2.61).

We briefly discuss the possibility of also driving the coupling Hamiltonian $V_{SB}(\lambda_t)$. One example is an interaction Hamiltonian of the form $V'_{SB}(\lambda_t) = f(\lambda_t)V_{SB}$, where $f(\lambda_t) \in [-1, +1]$ is an arbitrary scalar function and V_{SB} is the interaction Hamiltonian used in the derivation of Section 3.2. Then, since the overall coupling $V'_{SB}(\lambda_t)$ remains weak, we can derive the BMS equation in the same way as before. The resulting master equation is obtained from eqn (3.41) by modifying the dissipator from $\mathcal{D}$ to $f(\lambda_t)^2\mathcal{D}$. This modification can have a drastic impact on the dynamics ; for example, $f(\lambda_t)$ could switch between zero and one and thereby model the effect of coupling (or decoupling) the system to (or from) the bath. However, As long as the dissipator remains at each time of the form (3.46), the basic definitions of internal energy, heat, work and entropy remain unchanged. In particular, in our ideal weak coupling regime, there is only negligible work associated with driving the coupling Hamiltonian $V'_{SB}(\lambda_t)$.

Multiple baths with driving

We turn to the most interesting case describing a driven system coupled to multiple heat baths as sketched in Fig. 3.2. This case is ideally suited to studying quantum heat engines, refrigerators, heat pumps and other devices. Furthermore, even in the absence of any driving, $\dot{\lambda}_t = 0$, the thermodynamic analysis is non-trivial as the system relaxes into a *non-equilibrium steady state* characterized by non-zero heat fluxes and a finite entropy production rate.

Before turning to the thermodynamic analysis, we briefly sketch the derivation of the master equation. The Hamiltonian for a system coupled to multiple baths labelled by an index ν is of the form

$$H_{SB}(\lambda_t) = H_S(\lambda_t) + \sum_\nu \left(V_{SB}^{(\nu)} + H_B^{(\nu)} \right). \tag{3.60}$$

Note that we assumed, as done throughout this book, that there is *no direct coupling* between two baths, i.e. no interaction Hamilonian $V_{B_\nu B_\mu}$. The different baths can only influence each other via the interaction with the system. We proceed by taking the initial state as an obvious generalization of our previous choice (3.19) to be

$$\rho_{SB}(0) = \rho_S(0) \bigotimes_\nu \pi_\nu(\beta_\nu), \tag{3.61}$$

i.e. each bath is described by a canonical state $\pi_\nu(\beta_\nu)$ at temperature β_ν. The claim is now that the resulting master equation, under the same assumptions that we have also used before, reads

$$\boxed{\frac{\partial}{\partial t}\rho_S(t) = -\frac{i}{\hbar}[H_S(\lambda_t), \rho_S(t)] + \sum_\nu \mathcal{D}_\nu(\lambda_t)\rho_S(t),} \tag{3.62}$$

where each dissipator $\mathcal{D}_\nu(\lambda_t)$ is of the same form as eqn (3.46), but with an additional index (ν) on all rates $r_\alpha(\omega)$ and system operators $S_\alpha(\omega)$. Thus, even if, say, bath ν and ν' have the same temperature, $\beta_\nu = \beta_{\nu'}$, it can be that $\mathcal{D}_\nu(\lambda_t) \neq \mathcal{D}_{\nu'}(\lambda_t)$ because the microscopic coupling Hamiltonians differ in general. Nevertheless, each dissipator always satisfies the relation

$$\mathcal{D}_\nu(\lambda_t)\pi_S(\lambda_t, \beta_\nu) = 0 \tag{3.63}$$

because the rates for each bath ν separately satisfy the local detailed balance condition $r_\alpha^{(\nu)}(\omega) = e^{\beta_\nu \hbar\omega} r_\alpha^{(\nu)}(-\omega)$. To summarize, eqn (3.62) describes the competition between multiple baths: each one tries to draw the system state closer to its respective equilibrium state. In the special case of only one heat reservoir, we recover the case discussed previously. Showing that each reservoir enters additively into the description is not complicated and left as an exercise.

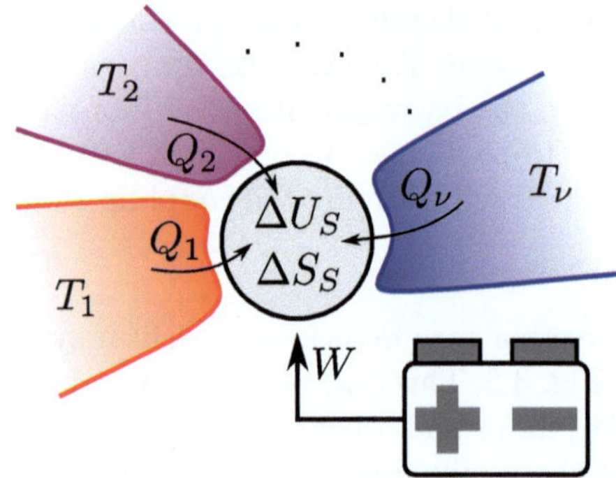

Fig. 3.2 Sketch of a system (circle in the middle) coupled to multiple heat baths at different temperatures T_ν and a work reservoir (sketched as a battery). Via the exchange of heat Q_ν and work W the internal energy and system entropy change as quantified by ΔU_S and ΔS_S.

Exercise 3.11 Show that eqn (3.62) follows from the derivation of Section 3.2 if for all $\nu \neq \mu$ the following cross-terms vanish: $\text{tr}_{\nu,\mu}\{\tilde{V}_{SB}^{(\nu)}(t)\tilde{V}_{SB}^{(\mu)}(s)\pi_{B_\nu}(\beta_\nu)\pi_{B_\mu}(\beta_\mu)\} = \text{tr}_\nu\{\tilde{V}_{SB}^{(\nu)}(t)\pi_{B_\nu}(\beta_\nu)\}\text{tr}_\mu\{\tilde{V}_{SB}^{(\mu)}(s)\pi_{B_\mu}(\beta_\mu)\} = 0$. Show that this follows from Exercise 3.2.

For the thermodynamic analysis we start with the investigation of the first law, which follows from eqn (3.62) and reads

$$\boxed{\frac{d}{dt}U_S(t) = \dot{W}(t) + \sum_\nu \dot{Q}_\nu(t).} \tag{3.64}$$

Here, the heat flux from bath ν is defined as $\dot{Q}_\nu(t) \equiv \text{tr}_S\{H_S(\lambda_t)\mathcal{D}_\nu(\lambda_t)\rho_S(t)\}$, which generalizes eqn (3.58) in the obvious way. Since each reservoir enters additively into the description, we also expect for the second law that

$$\boxed{\dot{\Sigma}(t) = \frac{d}{dt}S_S(t) - \sum_\nu \frac{\dot{Q}_\nu(t)}{T_\nu} \geq 0.} \tag{3.65}$$

This is shown as follows. First, note that

$$\frac{d}{dt}S_S(t) = -k_B\text{tr}_S\left\{\frac{\partial\rho_S(t)}{\partial t}\ln\rho_S(t)\right\} = -k_B\sum_\nu \text{tr}_S\{\mathcal{D}_\nu(\lambda_t)\rho_S(t)\ln\rho_S(t)\}. \tag{3.66}$$

The first equality is not entirely trivial, but it can be motivated by the classical identity $d_t\sum_x p_x(t)\ln p_x(t) = \sum_x[d_t p_x(t)]\ln p_x(t)$, which holds whenever the kernel of $p_x(t)$, i.e. the number of x with $p_x(t) = 0$, does not change (if it does, the Shannon entropy is not differentiable). The second equality follows from eqn (3.62) and the fact that the trace is cyclic. The next point to note is

$$-\frac{\dot{Q}_\nu(t)}{T_\nu} = k_B\text{tr}_S\{D_\nu(\lambda_t)\rho_S(t)\ln\pi_S(\lambda_t,\beta_\nu)\}, \tag{3.67}$$

which is easy to derive if one uses trace preservation: $\mathrm{tr}_S\{\mathcal{D}_\nu(\lambda_t)\rho_S\} = 0$ for any ρ_S. Putting the last two identities together, we recognize that

$$\dot{\Sigma}(t) = -k_B \sum_\nu \mathrm{tr}_S\{D_\nu(\lambda_t)\rho_S(t)[\ln \rho_S(t) - \ln \pi_S(\lambda_t, \beta_\nu)]\}. \tag{3.68}$$

Let us now define the CPTP maps $\mathcal{E}_\nu \equiv e^{D_\nu(\lambda_t)dt}$ for dt small enough such that the condition $\mathcal{E}_\nu \pi_S(\lambda_t, \beta_\nu) = \pi_S(\lambda_t, \beta_\nu)$ is satisfied. Then, it follows that

$$\dot{\Sigma}(t) = k_B \sum_\nu \lim_{dt \searrow 0} \frac{D[\rho_S(t)|\pi_S(\lambda_t, \beta_\nu)] - D[\mathcal{E}_\nu \rho_S(t)|\mathcal{E}_\nu \pi_S(\lambda_t, \beta_\nu)]}{dt}. \tag{3.69}$$

The entropy production rate is therefore expressed as a sum of positive terms, which follows from monotonicity of relative entropy (see also Exercise 3.10). Hence, $\dot{\Sigma}(t) \geq 0$, which concludes the derivation of the laws of thermodynamics in the presence of multiple baths. Note that the treatment of particle exchanges, where each bath is also characterized by a chemical potential μ_ν, is provided in Section 3.10.

To conclude, we emphasize the close similarity of this section to what we have discussed in the classical context in Section 2.4. In both cases the local detailed balance condition ensures that the Gibbs state is a steady state (with respect to bath ν), which can be used to derive the laws of thermodynamics. In fact, whenever the system Hamiltonian H_S contains no degeneracies, the description here is almost indistinguishable from its classical counterpart. Given the universality of the laws of thermodynamics, this is perhaps not too surprising. Still, the reader is asked to explicitly connect the quantum case to the classical description in the next exercise.

Exercise 3.12 Consider a non-degenerate system Hamiltonian $H_S = \sum_s \epsilon_s |s\rangle\langle s|$ with $\epsilon_s \neq \epsilon_{s'}$ for $s \neq s'$. Show that the populations $p_s(t) \equiv \langle s|\rho_S(t)|s\rangle$ of the energy eigenstates obey a closed equation of the form

$$\frac{d}{dt}p_s(t) - \sum_{s'} [R_{ss'}p_{s'}(t) - R_{s's}p_s(t)], \tag{3.70}$$

where we introduced the rates $R_{ss'} \equiv \sum_{\alpha,\alpha'} \gamma_{\alpha\alpha'}(\epsilon_{s'} - \epsilon_s)\langle s|A_{\alpha'}|s'\rangle\langle s'|A_\alpha|s\rangle$. Equation (3.70) is nothing but the classical rate master equation we met already in Section 2.3. Furthermore, use eqn (3.43) to confirm that the rates obey the local detailed balance condition (2.44) $\Gamma_{ss'}/\Gamma_{s's} = e^{-\beta(\epsilon_s - \epsilon_{s'})}$. Thus, the thermodynamics of the populations for a non-degenerate system can be analysed with the same tools as already developed in Section 2.4. Vice versa, we can view Section 3.2 as providing another alternative derivation of the local detailed balance condition. *Hint:* To derive eqn (3.70), remember that the 'jumps' caused by $A_\alpha^{(\dagger)}(\omega)$ in eqn (3.38) are strictly energy-perserving. Furthermore, remember that the coupling operators A_α are Hermitian.

What happens to the coherences $\rho_{ss'}(t) \equiv \langle s|\rho_S(t)|s'\rangle$, $s \neq s'$, of the density matrix? Show that, under the additional assumption that not only is the system Hamiltonian non-degenerate but also the set of transition frequencies, which means that $\epsilon_k - \epsilon_m = \epsilon_\ell - \epsilon_n$ implies $k = \ell$ and $m = n$, the coherences evolve according to the simple equation

$$\frac{d}{dt}\rho_{ss'}(t) = -\frac{i}{\hbar}(\epsilon_s - \epsilon_{s'})\rho_{ss'}(t) - \sum_{s''} \frac{\Gamma_{s''s} + \Gamma_{s''s'}}{2}\rho_{ss'}(t). \tag{3.71}$$

The solution of eqn (3.71) is an exponentially damped oscillation, thus causing all coherences to vanish in the long time limit. This effect is known as *decoherence*: an undriven quantum system weakly coupled to a thermal bath eventually becomes diagonal in the system's energy eigenbasis and loses all coherences.

3.5 Work Extraction from Small Systems

In this section we analyse simple thermodynamic processes for an open quantum system in contact with a *single* heat bath. We also discuss the maximum extractable work for a system in an arbitrary state ρ_S. Remember that, as stressed at the end of Section 3.1, care is required whenever one uses time-dependent external fields, which rely on a semiclassical description, to compute extractable work $W < 0$ from a small system. Nevertheless, it is pedagogically instructive to neglect these practical concerns for a moment. Then, the two main outcomes of this section are that we can define a non-equilibrium free energy, which characterizes the maximum extractable work for a quantum system in weak contact with a single bath, and that a quantum Carnot cycle operates with Carnot efficiency.

Before we start, we restate the laws of thermodynamics for a driven system in contact with a single heat bath in integrated form. We found that the definitions for internal energy and system entropy are $U_S = \text{tr}_S\{H_S(\lambda_t)\rho_S(t)\}$ and $S_S(t) = k_B S_{\text{vN}}[\rho_S(t)]$, and the total heat flow and work done on the system over a finite time become $Q(t) = \int_0^t ds\dot{Q}(s)$ and $W(t) = \int_0^t ds\dot{W}(s)$. The first law reads

$$\Delta U_S(t) = Q(t) + W(t). \tag{3.72}$$

To state the second law, it is useful to introduce the **non-equilibrium free energy**

$$\boxed{F_S(\rho_S, \lambda) \equiv U_S - TS_S = \text{tr}\{H_S(\lambda)\rho_S\} + k_B T \text{tr}_S\{\rho_S \ln \rho_S\}.} \tag{3.73}$$

This definition is analogous to the classical case, see eqn (2.146), and a straightforward calculation reveals the useful identity

$$F_S(\rho_S, \lambda) - \mathcal{F}_S(\lambda) = k_B T D[\rho_S|\pi_S(\lambda)] \geq 0, \tag{3.74}$$

where $\mathcal{F}_S(\lambda) = -k_B T \ln \mathcal{Z}_S(\lambda)$ is the equilibrium free energy of the system. Equation (3.74) is also analogous to the classical case, eqn (2.147). If the state $\rho_S(t)$ and the value of the control protocol λ_t are clear from context, we also write $F_S(t) = F_S[\rho_S(t), \lambda_t]$ in the following.

Now, using the first law (3.72) and definition (3.73), we can express the second law $\Sigma(t) = \Delta S_S(t) - Q(t)/T \geq 0$ in the alternative form

$$\boxed{\Sigma(t) = \frac{W(t) - \Delta F_S(t)}{T} \geq 0.} \tag{3.75}$$

Basic thermodynamic processes

We now discuss simple thermodynamic processes. The first is called an **isothermal process**. It is defined as follows. The system is initialized in a Gibbs state $\rho_S(0) =$

$\pi_S(\lambda_0)$ and, while being in contact with the heat bath, the system Hamiltonian $H_S(\lambda_t)$ is changed so slowly in time that the system state remains thermal at all times: $\rho_S(t) = \pi_S(\lambda_t)$ for all t. This process is *reversible*: the entropy production rate is zero at each time step, which becomes immediately transparent by recalling eqn (3.59). Thus, using the first law and $\Sigma(t) = 0$, we can characterize an isothermal process by

$$\text{isothermal process:} \quad \Delta S_S = \frac{Q}{T}, \quad W = \Delta \mathcal{F}_S = \mathcal{F}_S(\lambda_t) - \mathcal{F}_S(\lambda_0). \tag{3.76}$$

Next, we introduce **unitary processes**. They are characterized by the fact that the influence of the bath, i.e. the dissipator $\mathcal{D}$ in the master equation, can be neglected. This implies $Q = 0$. Since the system evolves unitarily, we also have $\Delta S_S = 0$. Thus, unitary processes are reversible. In principle, they can be realized in many ways.

One option is to assume that the time-dependent system Hamiltonian can be written as $H_S(\lambda_t) = H_0 + \hbar\delta(t-\tau)V$, where H_0 describes some baseline Hamiltonian, with respect to which we have derived the master equation, and $\hbar\delta(t-\tau)V$ denotes a sudden very short perturbation (note that V is dimensionless). Here, a perturbation qualifies as 'sudden' and 'very short' if we can neglect the intrinsic evolution time of the system (dictated by the transition frequencies of H_0) and the relaxation time of the system (dictated by the rates $r_\alpha(\omega)$ in eqn (3.46)) during the perturbation. Mathematically, this idea is modelled by a delta function, which allows us to focus only on the term $\hbar\delta(t-\tau)V$ at time $t = \tau$ and to neglect all other terms. Denoting by $\tau_\pm = \tau \pm \epsilon$ with $\epsilon \searrow 0$ a time shortly after or before τ, we find

$$\rho_S(\tau_+) = e^{-iV}\rho_S(\tau_-)e^{iV}. \tag{3.77}$$

Thus, the dissipative system dynamics is interrupted at time τ by a sudden 'kick', transforming the state $\rho_S(\tau_-)$ prior to the kick to $\rho_S(\tau_+)$ after the kick in a unitary way.

A second option to model a unitary process is to really decouple the system from the bath, i.e. to assume that $V_{SB} = 0$ for some time interval $[0, t]$. Then, only the system Hamiltonian $H_S(\lambda_t)$ influences the system dynamics. Consequently, its state changes according to the unitary transformation

$$\rho_S(t) = U_S(t,0)\rho_S(0)U_S^\dagger(t,0), \quad U_S(t,0) = \exp_+\left[-\frac{i}{\hbar}\int_0^t ds H_S(\lambda_s)\right]. \tag{3.78}$$

Finally, a third possibility to implement a unitary process is a **quench**. In this case one suddenly switches the system Hamiltonian from H_0 to H_1 at some time τ. Mathematically, this can be described as $H_S(\lambda_t) = \Theta(\tau - t)H_0 + \Theta(t-\tau)H_1$ and the master equation used to describe the dynamics before the quench for $t < \tau$ (after the quench for $t > \tau$) needs to be derived with respect to the Hamiltonian H_0 (H_1). Quenches are a somewhat special class of unitary processes because the system state does not change at time τ; instead, it simply evolves afterwards with respect to a different Hamiltonian and master equation.

In all three cases the thermodynamic interpretation of a unitary process is the *same*—it is only the way to compute the work that differs. We can summarize it as

$$\text{unitary process:} \quad Q = 0, \quad \Sigma = \Delta S_S = 0, \tag{3.79}$$

$$\text{sudden perturbation:} \quad \Delta U_S = W = \text{tr}\{H_S^{(0)}[\rho_S(\tau_+) - \rho_S(\tau_-)]\}, \tag{3.80}$$

$$\text{decoupled bath:} \quad \Delta U_S = W = \text{tr}\{H_S(\lambda_t)\rho_S(t) - H_S(\lambda_0)\rho_S(0)\}, \tag{3.81}$$

$$\text{quench:} \quad \Delta U_S = W = \text{tr}\{[H_S^{(1)} - H_S^{(0)}]\rho_S(\tau)\}. \tag{3.82}$$

Note that in the first and third case (sudden perturbation and quench) the internal energy changes by a *finite* amount although the process is modelled as taking *zero* time. Therefore, the power diverges at this point. This is, of course, a theoretical idealization, but it is convenient for many applications.

A subclass of unitary processes for a decoupled bath allows one to compute the work in a particularly simple way. They rely on the assumption that the system Hamiltonian $H_S(\lambda_t)$ is changed very slowly in time such that we can apply the *quantum adiabatic theorem*. In a nutshell, the quantum adiabatic theorem says the following: if the initial state of the system is $\rho_S(0) = \sum_s p_s|\epsilon_s(\lambda_0)\rangle\langle\epsilon_s(\lambda_0)|$, if no degeneracies are present at all times, i.e. $\epsilon_s(\lambda_t) \neq \epsilon_{s'}(\lambda_t)$ for $s \neq s'$, and if the system Hamiltonian is changed slowly enough, then the final state reads $\rho_S(t) = \sum_s p_s|\epsilon_s(\lambda_t)\rangle\langle\epsilon_s(\lambda_t)|$, i.e. it has not changed its populations p_s. The applied work during this process can be expressed as

$$\text{quantum adiabatic theorem:} \quad \Delta U_S = W = \sum_s [\epsilon_s(\lambda_t) - \epsilon_s(\lambda_0)]p_s. \tag{3.83}$$

Unfortunately, a source of confusion arises here because an **adiabatic process** in the thermodynamic sense is characterized only by demanding *no heat exchange*: $Q = 0$. Thus, an adiabatic process in the thermodynamic sense does *not* need to be slow. Moreover, for a macroscopic system with entropy S_S, an adiabatic process does *not* need to be reversible. Typically, $\Delta S_S > 0$ (consider, for example, the free expansion of a gas in a box or the mixing of ink and water). An adiabatic process that also satisfies $\Delta S_S = 0$ is called an **isentropic** or *reversible adiabatic* process in thermodynamics.

The fact that all adiabatic processes (in a thermodynamic sense) are reversible in our framework comes from our choice of system entropy, which equals (up to k_B) the von Neumann entropy. According to von Neumann, this entropy reflects the uncertainty 'computed from the perspective of an observer who can carry out all measurements that are possible in principle'. This choice of thermodynamic entropy is 'not applicable' for macrosopic systems (e.g. a box with 10^{23} gas particles or a pot containing ink and water). It is only a legitimate choice of thermodynamic entropy for *small systems* (e.g. a single spin), where it is meaningful to assume that one has full experimental control over it. We discuss this point further in Section 3.8.

Maximum extractable work from an open quantum system

According to the second law (3.75) it seems possible that $W(t) = \Delta F_S(t)$, i.e. the maximum amount of *extractable* work from a small open system equals *minus* the change in its non-equilibrium free energy. By combining what we have learned so far, we now demonstrate that this is indeed true and provide an explicit protocol for it.

To approach the problem, we first consider an initial thermal state $\rho_S(0) = \pi_S(\lambda_0)$. If we can transform the state to another thermal state at λ_t with a *lower* free energy, it is clear from eqn (3.76) that we can extract an amount of work $\mathcal{F}_S(\lambda_t)-\mathcal{F}_S(\lambda_0)<0$. However, it turns out that this procedure cannot be repeated in a cyclic way because to bring the final state back to its initial state we have to invest at least as much work as we have gained before. To see this, we require that the thermodynamic protocol starts and ends at the same value λ_0. Then, it follows directly from eqns (3.74) and (3.75) for an initial thermal state $\rho_S(0)=\pi_S(\lambda_0)$ that

$$W(t) \geq F_S[\rho_S(t),\lambda_0] - \mathcal{F}_S(\lambda_0) \geq 0. \tag{3.84}$$

Hence, we cannot extract any work from a thermal state in a cyclic way, which was to be expected if we recall the *Kelvin–Planck statement* of the second law of thermodynamics. This result is reassuring, but one might wonder whether this is a peculiarity of all the assumptions we have employed in order to derive the BMS master equation. The next exercise shows that this is not the case.

Exercise 3.13 The Gibbs state is characterized by the fact that it is impossible to extract work from it in a cyclic process even if we assume unlimited control, i.e. the ability to implement any unitary we like. This exercise gives insights into why this is the case.

For an isolated system with Hamiltonian H in the state ρ we define the maximum *extractable* work or **ergotropy** as

$$E(\rho) \equiv \max_U \mathrm{tr}\{H(\rho - U\rho U^\dagger)\} \geq 0, \tag{3.85}$$

where the maximization is over all possible unitaries U. Equation (3.85) describes a cyclic process, where the Hamiltonian H is the same at the beginning and at the end of the process. We now ask from which states ρ is it impossible to extract work, i.e. when is $E(\rho)=0$?

For simplicity, let us consider states diagonal in the energy eigenbasis. In fact, we show in the main text below that it is always possible to extract work from a state that is not diagonal in the energy eigenbasis. Our claim is now that $E(\rho)=0$ whenever ρ is *passive*. If we write $H=\sum_k \epsilon_k|k\rangle\langle k|$ with ordered energies $\epsilon_{k+1}\geq\epsilon_k$, then a passive state $\rho=\sum_k\lambda_k|k\rangle\langle k|$ satisfies by definition $\lambda_{k+1}\leq\lambda_k$ for all k, i.e. the populations in a passive state decrease with increasing energy. That it is impossible to extract work from such a passive state seems indeed intuitive. To prove it, we remark that $E(\rho)$ gets maximized if we minimize $\mathrm{tr}\{HU\rho U^\dagger\}$ for a fixed expectation value $\mathrm{tr}\{H\rho\}$. Then, verify that

$$\mathrm{tr}\{HU\rho U^\dagger\} = \sum_{k,\ell} \epsilon_k S_{k\ell}\lambda_\ell. \tag{3.86}$$

Here, the matrix S has components $S_{k\ell}=|U_{k\ell}|^2$ with $U_{k\ell}=\langle k|U|\ell\rangle$ and is *doubly stochastic*, which means that $S_{k\ell}\geq 0$ and $\sum_k S_{k\ell}=\sum_\ell S_{k\ell}=1$. Next, show that the set of doubly stochastic matrices is *convex*, which means that for any two doubly stochastic matrices S_1 and S_2 and any $\lambda\in[0,1]$ the matrix $\lambda S_1+(1-\lambda)S_2$ is also doubly stochastic. Then, convince yourself of the fact that the *extreme points* of this set are the permutation matrices. Here, an extreme point S of a convex set means that there exist no $S_1\neq S_2$ and $\lambda\in(0,1)$ such that $S=\lambda S_1+(1-\lambda)S_2$. Finally, you are allowed to use the following mathematical theorem: *the minimum of a linear function over a convex set always occurs at one of its extreme points.* Since eqn (3.86) is a linear function of S, our claim follows by recognizing that any permutation of populations of a passive state can never decrease its energy.

Surprisingly, it turns out that one can extract work from a passive state if one has sufficiently many *copies* of it available. Consider the n-fold tensor product $\rho^{\otimes n}=\rho\otimes\cdots\otimes\rho$ of passive states ρ. Suppose that the global Hamiltonian H_n is the sum of all local Hamiltonians H without any interaction terms. Then, for sufficiently large n it is possible that

$E(\rho^{\otimes n}) \equiv \max_U \text{tr}_n\{H_n(\rho^{\otimes n} - U\rho^{\otimes n}U^\dagger)\} > 0$, where the maximization over U involves all *global* unitaries and the trace is taken over the n-fold tensor product Hilbert space. There is only one state for which $E(\rho^{\otimes n}) = 0$ for all n and this is the Gibbs state $\rho = \pi(\beta) = e^{-\beta H}/\mathcal{Z}$ for any $\beta > 0$. The Gibbs state is therefore referred to as being **completely passive**. Readers interested in a proof will find a reference at the end of the chapter.

After having clarified that it is impossible to extract work from a Gibbs state in a cyclic process, we ask what happens if $\rho_S(0) \neq \pi_S(\lambda_0)$. Can we extract work by returning the nonequilibrium state $\rho_S(0)$ back to an equilibrium state $\pi_S(\lambda_0)$? Since $F_S[\rho_S(0), \lambda_0] - \mathcal{F}_S(\lambda_0) > 0$, the second law suggests that this is indeed possible and bounds the extractable work as

$$-W(t) \leq F_S[\rho_S(0), \lambda_0] - \mathcal{F}_S(\lambda_0). \tag{3.87}$$

Note that we demand that the control protocol is returned to its initial value λ_0, but the initial and final system state do not need to match. We now show that we can extract the maximum amount of work $-W(t) = F_S[\rho_S(0), \lambda_0] - \mathcal{F}_S(\lambda_0)$ in an idealized optimal process. For this purpose, we concatenate a unitary process and an isothermal process.

First, we perform a quench: if $\rho_S(0)$ is the state of the system, we switch the Hamiltonian from $H_S(\lambda_0)$ to $H_S(\lambda_0')$, where $H_S(\lambda_0')$ is constructed in such a way that $\rho_S(0) \sim e^{-\beta H_S(\lambda_0')}$, where β is fixed by the temperature of the bath. Thus, we set $H_S(\lambda_0') = -k_BT[\ln \rho_S(0) + \ln \mathcal{Z}_S(\lambda_0')]$, where we can choose the normalization constant $\mathcal{Z}_S(\lambda_0')$ arbitrarily. Since the process is unitary, we have $\Delta S_S = 0$ and with the specific choice of $H_S(\lambda_0')$ the work (3.82) becomes

$$W_{\text{unitary}} = \Delta U_S = \mathcal{F}_S(\lambda_0') - F_S[\rho_S(0), \lambda_0]. \tag{3.88}$$

Next, after we have transformed the non-equilibrium state to a thermal state, we use an isothermal process to transform the Hamiltonian $H_S(\lambda_0')$ back to $H_S(\lambda_0)$. In the case where this transformation is done very slowly, the work done on the system approaches the value of eqn (3.76) and reads $W_{\text{isothermal}} = \mathcal{F}_S(\lambda_0) - \mathcal{F}_S(\lambda_0')$. Adding this amount to eqn (3.88) yields the total amount of extractable work during this process:

$$-W_{\text{tot}} = -W_{\text{unitary}} - W_{\text{isothermal}} = F_S[\rho_S(0), \lambda_0] - \mathcal{F}_S(\lambda_0) > 0, \tag{3.89}$$

which equals the maximum possible value according to eqn (3.87). We remark that there exist alternative unitary processes, which yield the same final result.

Exercise 3.14 Consider an alternative unitary process with three steps: (i) transform the eigenbasis of $\rho_S(0)$ to the energy eigenbasis of $H_S(\lambda_0)$; (ii) reorder the populations of the energy states so that the new populations, which we denote by q_s, decrease monotonically with increasing energy; (iii) adjust the Hamiltonian, i.e. switch $\lambda_0 \mapsto \lambda_0'$, such that the separations between adjacent energy levels are set as $q_s = e^{-\beta\epsilon_s(\lambda_0')}/\mathcal{Z}_S(\lambda_0')$. The overall effect of these transformations is the same as before: the state after the three steps (i)–(iii) is a thermal equilibrium state with respect to $H_S(\lambda_0')$. The difference is now that $H_S(\lambda_0')$ has the same eigenbasis as $H_S(\lambda_0)$, which is in general not the case for our quench above. Verify that the amount of extracted work can be written in the same *form* as in eqn (3.88), but in general it has a different numerical value. Convince yourself that this difference cancels out when computing the total extractable work (3.89) after an isothermal transformation back to λ_0.

The above optimal process can also be reversed by transforming an equilibrium state $\pi_S(\lambda_0)$ to a non-equilibrium state $\rho_S(t)$ at a work cost of $W_{\text{tot}} = F_S[\rho_S(t), \lambda_0] - \mathcal{F}_S(\lambda_0) > 0$. Moreover, the initial and final values of the control protocol need not be the same. By concatenating the above protocols, we see that two states ρ_S and ρ'_S with Hamiltonian $H_S(\lambda)$ and $H_S(\lambda')$ can be transformed into each other at an optimal work cost $W_{\text{tot}} = F_S(\rho'_S, \lambda') - F_S(\rho_S, \lambda)$ by using only reversible processes. In retrospect, this also justifies the definition of the non-equillibrium free energy (3.73) because it is in full analogy with the equilibrium result (3.76) for an isothermal process. Thus, the non-equilibrium free energy characterizes the potential of a quantum state (with respect to a weakly coupled heat bath at temperature T) to deliver work under the assumption of *unlimited* control over the system Hamiltonian. The next exercise justifies a claim made in Exercise 3.13; namely, that it is always possible to extract work from a state with coherences in the energy eigenbasis.

Exercise 3.15 Recall definition (3.51) of the dephasing operation $\mathcal{D}_{H_S}$ and consider a state ρ_S with coherences in the energy eigenbasis, i.e. $\mathcal{D}_{H_S}\rho_S \neq \rho_S$. Show that $F_S(\mathcal{D}_{H_S}\rho_S) < F_S(\rho_S)$. Hence, it is possible to extract work when transforming ρ_S to $\mathcal{D}_{H_S}\rho_S$. *Hint:* If you have difficulties, recall Theorem A.2.

Thermodynamic cycles

The above analysis still does not allow us to extract work in a *repeatable* fashion. Starting with a non-equilibrium state $\rho_S(0)$, we can extract work only once. To extract work multiple times, we have to bring in at least one second heat bath at a different temperature. Here, one has two different options.

First, one can consider a system in permanent contact with multiple baths as treated at the end of the last section. Such set-ups are also called *continuous engines* as they are continuously coupled to the baths. We treat examples of this kind in Section 3.10, but here we want to remain within our single-bath framework.

Then, the only option left is to repeatedly couple and decouple the system to different baths in such a way that the system is in contact with at most one bath at any time or, in case of a unitary process, the system could also be decoupled from all baths. The time the system spends in contact with one bath (or no bath) defines a *stroke* and concatenating different strokes gives rise to a *quantum thermodynamic cycle.* Thermodynamic cycles, such as the Carnot, Otto or Stirling cycles, have played an important role in the historical development of thermodynamics because they allowed a complicated process to be broken up into strokes, which were easier to analyse. Also in quantum thermodynamics, cycles offer the advantage that they are easier to analyse. However, they also have the disadvantage that they need to repeatedly couple and decouple a system and a bath. While we said in Section 3.4 that this has in theory no thermodynamic cost, in any practical realization this still needs to be realized by macroscopic time-dependent fields, raising doubts about whether the work

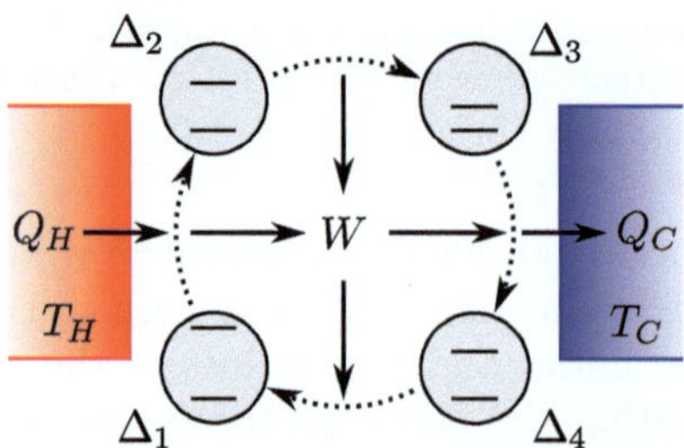

Fig. 3.3 Sketch of a quantum Carnot cycle for a two-level system with variable excited state energy Δ_t. The dotted arrows sketch the cycle. The solid arrows point towards the direction in which the flow of energy between the system and heat or work reservoir is *positive*.

output computed from the analysis of a quantum thermodynamic cycle equals the work output in the real world. Nevertheless, such cycles can present valuable toy models, in particular to compare the theory of quantum thermodynamics with macroscopic thermodynamics as demonstrated in the next exercise.

Exercise 3.16 The famous Carnot cycle consists of four steps. First, the working fluid (to be specific, let us imagine a gas in a box with variable volume V) in contact with a hot bath at temperature T_H undergoes an isothermal expansion, changing the volume from V_1 to $V_2 > V_1$. Afterwards, the gas is decoupled from the hot bath and experiences an isentropic (reversible adiabatic) expansion to a volume $V_3 > V_2$, thereby cooling down from T_H to T_C. In the third step the gas is put into contact with a cold bath at temperature T_C and compressed to a volume $V_4 < V_3$ in an isothermal process. Finally, the gas is decoupled from the cold bath and further compressed in an isentropic fashion to its initial volume $V_1 < V_4$, thereby heating up from T_C to T_H.

Your task is to construct an example of a **quantum Carnot cycle** as sketched in Fig. 3.3 by using a two-level system with Hamiltonian $H_S(\lambda_t) = \Delta_t|e\rangle\langle e|$ as a working medium. Here, $|e\rangle$ denotes the excited state with energy $\Delta_t \equiv \Delta(\lambda_t)$, which you can control at your convenience, and we set the energy of the ground state $|g\rangle$ to zero. Which kind of processes introduced in this section do you have to choose to realize a Carnot cycle? Show that, in order to have zero entropy production, the energy gap Δ_i at the four points $i \in \{1, 2, 3, 4\}$ needs to be chosen as follows:

$$\frac{\Delta_3}{\Delta_2} = \frac{\Delta_4}{\Delta_1} = \frac{T_C}{T_H}. \tag{3.90}$$

Show that in this reversible limit the efficiency of the quantum Carnot cycle equals the Carnot efficiency.

3.6 Correlations and the Zeroth Law of Thermodynamics

After having discussed the BMS equation and its consequences in detail, we now take a step back and consider a more general question: When does it make sense to associate an equilibrium temperature with a quantum system? As a by-product, we also shed some light on the questions: What is the effect of strong coupling and low temperature on the equilibrium states of open quantum systems? And: Under which circumstances is the Born approximation (3.20) justified?

To investigate these questions, we start with the phenomenological formulation of the **zeroth law of thermodynamics**, which is used to make the concept of an

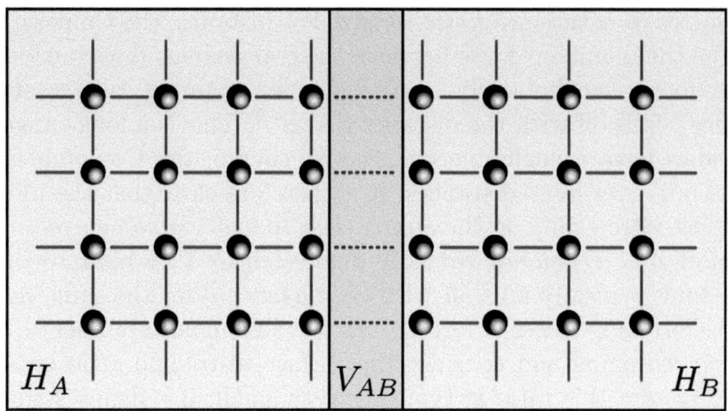

Fig. 3.4 Sketch of two interacting bodies A and B. Each body has its own Hamiltonian H_A or H_B describing interactions (solid lines) between its constituents (the balls). The contact region at a common surface is sketched as a dark-shaded region, where the interaction V_{AB} connects parts of the constituents of A with parts of the constituents of B (dashed lines).

equilibrium temperature well defined. It is usually formulated as follows. Imagine three thermodynamic systems labelled A, B and C and imagine that we can put them into contact such that they can exchange energy. Now, under the condition that there is no energy flow if A and B are put into contact and if B and C are put into contact, the zeroth law asserts that there is also no energy flow when A and C are put into contact. Under these conditions the systems A, B and C are said to be in mutual equilibrium, which motivates introducing a single scalar quantity to label all of them. This scalar quantity is called the equilibrium temperature, which is then the same not only for A, B and C alone but also for all composite systems AB, BC, AC and ABC. Note that the zeroth law does not fix the scale of temperature. Colloquially, Maxwell phrased the zeroth law as 'All heat is of the same kind'.

Thus, the zeroth law is related to the question of when the coupling between two systems is negligible at equilibrium. Coming back to statistical mechanics, this means in equations that we can compute all thermodynamically relevant quantities by approximating

$$\pi_{AB} \approx \pi_A \otimes \pi_B. \tag{3.91}$$

Here, $\pi_{AB} = e^{-\beta H_{AB}}/\mathcal{Z}_{AB}$ with $H_{AB} = H_A + H_B + V_{AB}$ denotes the joint canonical ensemble of AB and $\pi_A = e^{-\beta H_A}/\mathcal{Z}_A$ and $\pi_B = e^{-\beta H_B}/\mathcal{Z}_B$ denote the canonical ensembles of the isolated systems A and B, all parametrized by the same inverse temperature β. An immediately obvious case for which $\pi_A \otimes \pi_B = \pi_{AB}$ holds is given by $V_{AB} = 0$, but it is also obvious that this is a trivial case where A and B are not interacting (or were not yet 'put into contact'). In general, for $V_{AB} \neq 0$ one therefore expects eqn (3.91) to be only *approximately true*, but for macroscopic systems this approximation is often well justified. The reason for this is explained now.

Consider the two-dimensional sketch in Fig. 3.4 of two bodies A and B sharing a common surface, where they have been put into contact. For instance, we can imagine these bodies as crystals described by atoms sitting on a lattice. Now, recall that all fundamental forces in nature are *finite-ranged*. For instance, the Coulomb force, which will typically be the dominant force between the two systems (gravitational, weak and strong nuclear forces can be safely ignored for, for example, two crystals sharing a common surface), falls off with the distance d as d^{-2}. This is a long-ranged force, but, since most bodies have a negligible net electric charge, the Coulomb force tends to quickly 'cancel out' over large distances. It is therefore clear that the interaction V_{AB} has non-negligible effects only on the atoms *close to the common surface*. Atoms in the interior region of A or B remain virtually unaffected by V_{AB} because the correlation between two atoms typically falls off with the distance d for the same reason, namely the (effective) short-rangedness of forces. As a last step in our argumentation we invoke some geometric reasoning and consider the surface-to-volume ratio of A and B. For a macroscopic system this ratio is typically very small: if r denotes the diameter of the system, its surface scales as r^2 whereas its volume scales as r^3. Hence, their ratio $r^2/r^3 = r^{-1}$ is small in the macroscopic limit.

Thus, excluding exceptional cases such as systems with long-ranged correlations (which can appear, for instance, close to phase transitions), we conclude by purely geometric arguments that classic thermodynamics of macroscopic systems is a *weak coupling theory*. The approximation (3.91) is well justified and, together with the fact that the surface-to-volume ratio is small, we see that this implies that the internal energy and entropy of A and B are *additive*, which is related to the notion of *extensivity*; see eqn (2.2).

It is clear that this argumentation can no longer be applied if A or B (or both) become microscopically small. Putting aside the subtle question of how to precisely define the surface or volume of, for example, a single molecule, it is clear that their ratio is no longer small. One way out of this 'dilemma' was presented in the previous sections, where we assumed that the coupling between the system and the bath is small compared with the system and bath Hamiltonian itself and the bath is large. Thus, the bath itself was assumed to be an ideal thermodynamic object with a well-defined temperature obeying the zeroth law of thermodynamics. Then, this bath *imprints* its well-defined thermodynamic properties on the small system.

If both A and B are small, then the only way to save the zeroth law is to let their interaction V_{AB} become very small. Note that this assumption was *not* necessary for macroscopic systems: the coupling V_{AB} should be short-ranged, but it was allowed to strongly influence atoms at the surface. Thus, we have to confess that *there is no zeroth law for small systems in general*. The equilibrium thermodynamic properties of a small system depend sensitively on its 'volume' (no additivity in general) and the coupling to its environment. Therefore, whenever we talk in this book about equilibrium thermodynamic properties of a small system, we always mean them *relative* to some other large thermodynamic system such as a bath.

How large A and B have to be precisely and what kind of properties H_{AB} should have such that the zeroth law or eqn (3.91) are satisfied poses a challenging question. While progress has been achieved, the general question of how to define a 'good thermodynamic bath' is still open. We now present one simple argument to estimate the correlations in the state π_{AB}, which provides us with some interesting insights for the question when eqn (3.91) holds.

For this purpose consider the equilibrium free energy $\mathcal{F}_{AB}$ of π_{AB}. The non-equilibrium free energy of any state ρ_{AB} obeys $F_{AB}(\rho_{AB}) \geq \mathcal{F}_{AB}$ because of eqn (3.74). Let us choose $\rho_{AB} = \pi_A^* \otimes \pi_B^*$, where $\pi_A^* = \text{tr}_B\{\pi_{AB}\}$ and $\pi_B^* = \text{tr}_A\{\pi_{AB}\}$ denote the marginal states of A and B. Thus, ρ_{AB} is obtained from π_{AB} by neglecting all correlations. Using the identity $D(\pi_{AB}|\pi_A^* \otimes \pi_B^*) = I(\pi_{AB})$ with the mutual information I (see Appendix A), a short calculation reveals that $F_{AB}(\rho_{AB}) \geq \mathcal{F}_{AB}$ is equivalent to

$$I(\pi_{AB}) \leq \beta \text{tr}_{AB}\{V_{AB}(\pi_A^* \otimes \pi_B^* - \pi_{AB})\}. \tag{3.92}$$

This inequality bounds the correlations between A and B and it contains three relevant insights. Before discussing them, recall that the mutual information is always bounded as $0 \leq I(\rho_{AB}) \leq 2 \ln \min\{\dim \mathcal{H}_A, \dim \mathcal{H}_B\}$. The right-hand side of eqn (3.92) therefore always needs to be compared with the (logarithm of the) Hilbert space dimension of the smaller system. If this ratio is small, there is hope that the approximation (3.91) is valid. Of course, a small correlation $I(\pi_{AB})$ is not sufficient to justify eqn (3.91), which also requires $\pi_A^* \approx \pi_A$ and $\pi_B^* \approx \pi_B$. It is, however, much harder to obtain rigorous mathematical insights concerning the latter question.

The first insight we can gain from eqn (3.92) is that the correlations between A and B tend to be weak at weak coupling, $V_{AB} \approx 0$, which is perhaps not too surprising. The second insight, which is applicable to systems A and B with many constituents and a large Hilbert space dimension, is that correlations between A and B tend to be weak whenever V_{AB} is short-ranged, i.e. it affects only a very few constituents at the contact surface between A and B (as sketched in Fig. 3.4). The third insight revealed by eqn (3.92) is related to the prefactor β, which complements a discussion we started in Section 1.3 about the influence of low temperatures on deviations of the equilibrium state from the standard canonical form. By looking at eqn (3.92), it appears that correlations tend to be stronger the lower the temperature. While there is some truth in this statement, additional care is required here since the inverse temperature appears not only as a prefactor, but also in the terms π_A^*, π_B^* and π_{AB}. In particular, for small systems it is often justified to truncate their Hilbert space dimension at low temperatures because there are only a few excitations available. The correlations are then mainly of a quantum nature. In contrast, at high temperatures quantum correlations tend to diminish, but there is also more energy available such that more states can be occupied. This can give rise to increasing classical correlations, which can jeopardize our reasoning that correlations tend to be stronger at lower temperatures. A universally valid statement about the relations between temperature and correlations therefore does not exist.

We can now (partially) answer the questions posed at the beginning of this section. First, it only makes sense to associate an equilibrium thermodynamic temperature with *large* systems or systems in contact with a large bath. The exact meaning of 'large' is still not fully clear and depends in a complicated way on the system under consideration. A single spin is definitely too small, but a hundred spins interacting in a sufficiently complex way could already serve as a 'good' thermodynamic bath. The zeroth law therefore has to be applied with care and does not hold for very small systems, in particular not at strong coupling and for low temperatures. This distinguishes the theory of quantum thermodynamics from traditional thermodynamics.

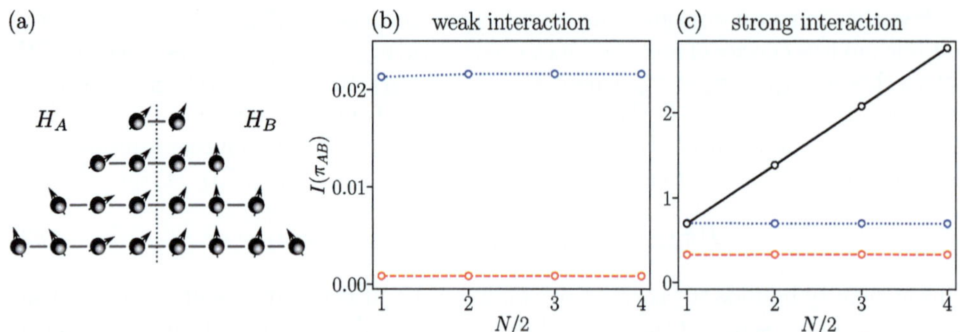

Fig. 3.5 (a) Sketch of the Ising chain with $N = 2, 4, 6$ and 8 spins, which we partition in the middle. (b) Plot of the mutual information between A and B as a function of N for weak interaction ($g = 0.2$) at low temperature $T = 0.5$ (blue dotted line) and at high temperature $T = 5$ (red dashed line). (c) The same at strong coupling ($g = 5$). In addition, and for comparison, we also plotted the maximum *classical* mutual information between A and B, which is $N \ln 2$ (black solid line). Other parameters: $k_B = 1, \hbar = 1, h = 1$.

Second, because of the increasing relevance of correlations at small scales, the Born approximation also becomes questionable in general.

The following exercise concludes this section and provides numerical evidence for our claims made above. It also exemplarily demonstrates what is known as an **area law** for quantum many-body systems.

Exercise 3.17 We consider the one-dimensional **Ising model**, which is a toy model for magnetic interactions in solid-state physics. It is described by the following Hamiltonian:

$$H_{\text{Ising}} = -h \sum_{i=1}^{N} \sigma_z^{(i)} - g \sum_{i=1}^{N-1} \sigma_x^{(i)} \sigma_x^{(i+1)}. \tag{3.93}$$

Here, N is the number of spins in the chain, $h \in \mathbb{R}$ denotes the strength of an external magnetic field and $g \in \mathbb{R}$ is the strength of the nearest-neighbour interactions between two spins. Remarkably, this model is exactly solvable and it exhibits a quantum phase transition for $N \to \infty$, where the ground state changes from a paramagnetic phase with all spins aligned along the external magnetic field in the z-direction (for $g < h$) to a ferromagnetic phase with all spins aligned in the x-direction (for $g > h$).

In this exercise we restrict ourselves, however, to studying the validity of eqn (3.91) numerically. Concretely, look at $N = 2, 4, 6$ and 8 spins and divide the chain in the middle into two equal halves labelled A and B (see Fig. 3.5a). Verify the following two insights numerically (compare also with the plots shown in Fig. 3.5).

First, in unison with what we said above, correlations are smaller for weaker interactions g and higher temperatures T (remember, however, that the latter result is related to the fact that the Hilbert space dimension is fixed and independent of temperature here).

Second, the correlations remain virtually constant as a function of N. Alternatively, if we 'normalize' the correlations by dividing $I(\pi_{AB})$ by the maximum possible amount of mutual information, these normalized correlations decrease as a function of N. This observation turns out to be true for a large class of many-body systems and it is known as an *area law*. The area law tells us that the amount of information that A shares with the outside world B is proportional to its *surface* (in our example the surface between A and B is just a point) and it does *not* grow proportional to its volume.

3.7 Exact Dissipation Inequalities

We continue our investigation of the laws of thermodynamics for open quantum systems beyond the weak coupling and Markovian assumption. The main outcomes of this section are two powerful inequalities, which are independent of the details of the system–bath Hamiltonian. The question of under which circumstances these inequalities can be identified with the second law of thermodynamics is, however, subtle and will be critically analysed in the next section.

Multiple baths and initial product state assumption

We start our consideration with the general case of a system coupled to multiple heat baths ν; see Fig. 3.2 for a sketch. The global Hamiltonian is taken to be the same as in eqn (3.60) with the difference that we also allow the coupling $V_{SB}^{(\nu)}(\lambda_t)$ to depend on time. Thus, $H_{SB}(\lambda_t) = H_S(\lambda_t) + \sum_\nu [V_{SB}^{(\nu)}(\lambda_t) + H_B^{(\nu)}]$. We ask: How do we have to define internal energy, heat and entropy such that we can derive the phenomenological laws of thermodynamics from Section 2.1? Furthermore, recall that the definition of work is fixed by eqn (3.6).

As it turns out, the condition ensuring a non-negative entropy production depends on the initial system–bath state $\rho_{SB}(0)$. This is related to the fact that the second law cannot be entirely explained dynamically, but requires a specific initial condition of the 'universe' (see also Appendix C.3). To illustrate this point, imagine a gas in a box that is initially homogeneously distributed, i.e. it has maximum entropy (Fig. 3.6). Now, assume that the initial velocities of all gas particles point to the left such that, at a later time, the distribution of gas particles is no longer homogeneous. Hence, the gas is in a state of lower entropy and the second law is 'violated'. Clearly, in reality we expect the initial condition shown in Fig. 3.6a to be a rare exception, which we cannot prepare in a reproducible way in a laboratory. However, if we want to investigate theoretically under which conditions the second law *always* holds, we have to restrict the class of initial states to exclude such 'hidden entropy sinks'.

In the present set-up of multiple baths, our choice for the initial state is $\rho_{SB}(0) = \rho_S(0) \bigotimes_\nu \pi_\nu(\beta_\nu)$, which we also used for the derivation of the weak coupling master equation; compare with eqn (3.61). This initial condition might not always be a good approximation (see also the discussion at the end of Section 3.6), but remember that the coupling Hamiltonian $V_{SB}^{(\nu)}(\lambda_t)$ can be time dependent. Thus, we can imagine, for instance, that at time $t = 0$ we switch on the coupling from zero to some finite value $V_{SB}^{(\nu)} \neq 0$ such that the global state up to time $t = 0$ is decorrelated. This is indeed justified for some experiments such as laser cooling of trapped ions, where the laser is responsible for coupling the motional and electronic degrees of freedom of the ions. Another question is whether it is also well justified to approximate the state of the bath by a Gibbs ensemble, which we assume to be granted here, but see Section 3.8 for extensions.

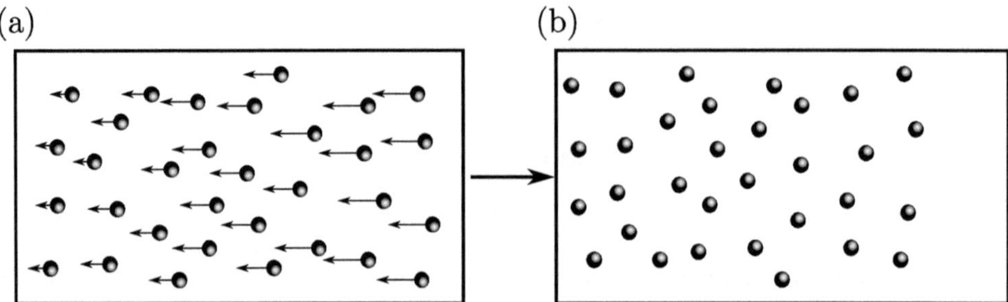

Fig. 3.6 (a) Homogeneously distributed particles of an ideal gas in a box with the arrows indicating the initial velocity. Although this state is highly improbable, it cannot be excluded *in principle.* (b) After some time all gas particles are closer to the left, thereby being in a state with lower entropy.

Turning to the first law, we need to split the global change in energy, which is identical to the work done, into a change in internal (system) energy and heat as

$$\begin{aligned} W(t) &= \mathrm{tr}_{SB}\{H_{SB}(\lambda_t)\rho_{SB}(t)\} - \mathrm{tr}_{SB}\{H_{SB}(\lambda_0)\rho_{SB}(0)\} \\ &= \Delta U_S(t) - \sum_\nu Q_\nu(t). \end{aligned} \tag{3.94}$$

In the weak coupling regime, we found that $U_S(t) = \mathrm{tr}_S\{H_S(\lambda_t)\rho_S(t)\}$. Since we could approximate in this case $V_{SB}^{(\nu)}(\lambda_t) \approx 0$, the first law implies that the heat flow into the system equals the negative change in bath energy, $Q_\nu(t) = -\mathrm{tr}_\nu\{H_B^{(\nu)}[\rho_\nu(t) - \rho_\nu(0)]\}$, where $\rho_\nu(t)$ is the state of bath ν at time t. But since the coupling energy is no longer negligible, we are left with the following puzzling question: Should we attribute $V_{SB}^{(\nu)}(\lambda_t)$ to the internal energy of the system or the bath or should we split it between the system and the bath and, if so, in which fraction? This non-trivial question has not yet been answered in full generality, but for the moment we identify the **strong coupling internal energy** with

$$U_S(t) \equiv \mathrm{tr}_{SB}\left\{\left[H_S(\lambda_t) + \sum_\nu V_{SB}^{(\nu)}(\lambda_t)\right]\rho_{SB}(t)\right\}, \tag{3.95}$$

which implies for the **heat**

$$Q_\nu(t) = -\mathrm{tr}_\nu\left\{H_B^{(\nu)}[\rho_\nu(t) - \rho_\nu(0)]\right\}. \tag{3.96}$$

This means that we attribute all of the coupling energy to the internal energy of the *system*, which is problematic from an operational perspective as eqn (3.95) can only be measured (and computed) if we have access to the bath degrees of freedom. On the other hand, the choice (3.95) is interesting in so far as it gives rise to the inequality

$$\Sigma(t) \equiv k_B \Delta S_{\mathrm{vN}}[\rho_S(t)] - \sum_\nu \frac{Q_\nu(t)}{T_\nu} \geq 0, \tag{3.97}$$

which we identify—under the conditions spelled out below—with the **entropy production** known from phenomenological non-equilibrium thermodynamics. The inequality (3.97) is remarkable as it follows solely from our initial state assumption and does not rely on any detail of the global Hamiltonian H_{SB}. Its positivity becomes evident by considering the following equivalent information-theoretic expressions:

$$\boxed{\begin{aligned}\Sigma(t) &= k_B D\left[\rho_{SB}(t)\,\middle|\,\rho_S(t)\bigotimes_\nu \pi_\nu(\beta_\nu)\right] \\ &= k_B\sum_\nu D[\rho_\nu(t)|\pi_\nu(\beta_\nu)] + k_B I_{\text{tot}}[\rho_{SB}(t)],\end{aligned}} \tag{3.98}$$

where $I_{\text{tot}}(\rho_{SB})$ is the *total information* (see Appendix A).

Exercise 3.18 Show that eqns (3.97) and (3.98) are equivalent by using that the global von Neumann entropy is conserved: $S_{\text{vN}}[\rho_{SB}(t)] = S_{\text{vN}}[\rho_{SB}(0)]$. Interpret the information-theoretic content of eqn (3.98) physically.

Let us now discuss how far it is justified to identify $\Sigma(t)$ with the entropy production known from phenomenological non-equilibrium thermodynamics. What is striking is that eqn (3.97) was derived without assuming the system to be small and the bath to be large. To take it to the extreme, the inequality $\Sigma(t) \geq 0$ continues to hold even if each 'bath' ν consists of a single spin coupled to a 'system' of 10^{23} spins. Of course, this completely reverses the system–bath logic. However, in order to identify $\Sigma(t)$ as the entropy production we not only have to show its positivity, but we also must be able to link each term of $\Sigma(t)$ to a *change in thermodynamic entropy.* Otherwise, eqn (3.97) is not the second law of thermodynamics, despite continuing to be a useful inequality for applications.

We start with the first term of eqn (3.97). Recall our observation that the von Neumann entropy (or its classical counterpart, the Shannon entropy) is a useful candidate for thermodynamic entropy if the system is *small*, where 'small' means that we assume the ability to accurately measure the system in any basis. This is clearly not a valid assumption for 10^{23} spins, but for a system with no more than 10 spins this could be justified. Next, we turn to the second term of eqn (3.97), which is more subtle. If we want to link $-Q_\nu/T_\nu$ to the change in entropy of the bath, we first of all need a definition for the thermodynamic entropy of the bath, which will be the content of Section 3.8. However, if we recall the phenomenological theory of thermodynamics (Section 2.1), we see that the second law (2.6) follows from the more general second law (2.5) only under the assumption that the bath is weakly perturbed by the system and stays *close* to equilibrium. Therefore, despite the fact that all mathematical identities given in this section remain valid for a system and bath of any size, the thermodynamic notions in this section are strictly valid only for a small system whose influence on the bath is negligible.

Single bath and initial correlations

Readers familiar with the content of Section 2.8, where we introduced a strong coupling framework for *classical* systems based on the *Hamiltonian of mean force*, might wonder about the different route taken in this section. Therefore, we now establish the quantum counterpart of the classical framework from Section 2.8.

In comparison with the framework above, its disadvantage is that we have to restrict the discussion to a *single* heat bath now. Thus, we consider the total Hamiltonian $H_{SB}(\lambda_t) = H_S(\lambda_t) + V_{SB}(\lambda_t) + H_B$ in the following. An advantage of the present framework is that it works for an initial system–bath state of the form

$$\rho_{SB}(0) = \pi_{SB}(\beta, \lambda_0). \tag{3.99}$$

This describes a *global* Gibbs state with non-zero system–bath correlations in general. For some experimental settings, this state is more realistic than a decorrelated state.

We now postulate the following definitions for the non-equilibrium **internal energy**, **system entropy** and **free energy**:

$$U_S^*(t) \equiv \mathrm{tr}_S \left\{ \rho_S(t) \left[H_S^*(\lambda_t) + \beta \frac{\partial}{\partial \beta} H_S^*(\lambda_t) \right] \right\}, \tag{3.100}$$

$$S_S^*(t) \equiv k_B \mathrm{tr}_S \left\{ \rho_S(t) \left[-\ln \rho_S(t) + \beta^2 \frac{\partial}{\partial \beta} H_S^*(\lambda_t) \right] \right\}, \tag{3.101}$$

$$F_S^*(t) \equiv U_S^*(t) - T S_S^*(t) = \mathrm{tr}_S \left\{ \rho_S(t) \left[H_S^*(\lambda_t) + k_B T \ln \rho_S(t) \right] \right\}. \tag{3.102}$$

These definitions agree with our classical findings with the difference that $H_S^*(\lambda_t)$ now denotes the quantum mechanical Hamiltonian of mean force known from eqn (1.22). This makes explicit computations harder, but formally one can show that the classical equilibrium relations (2.154) and (2.157) also hold in the quantum regime.

Exercise 3.19 Define $\mathcal{F}_S^*(\lambda) \equiv -k_B T \ln \mathcal{Z}_S^*(\lambda)$ and show that

$$\mathcal{U}_S^*(\lambda) \equiv \mathrm{tr}_S\{\pi_S^*[H_S^*(\lambda) + \beta \partial_\beta H_S^*(\lambda)]\} = \frac{\partial}{\partial \beta} \beta \mathcal{F}_S^*(\lambda) = \mathcal{U}_{SB}(\lambda) - \mathcal{U}_B, \tag{3.103}$$

$$\mathcal{S}_S^*(\lambda) \equiv k_B \mathrm{tr}_S\{\pi_S^*[-\ln \pi_S^* + \beta^2 \partial_\beta H_S^*(\lambda)]\} = k_B \beta^2 \frac{\partial}{\partial \beta} \mathcal{F}_S^*(\lambda) = \mathcal{S}_{SB}(\lambda) - \mathcal{S}_B. \tag{3.104}$$

Hint: Recall that $H_S^*(\lambda)$ depends on temperature. Furthermore, since $[H_S^*(\lambda), \partial H_S^*(\lambda)] \neq 0$ in general, this also implies that $\partial_\beta e^{-\beta H_S^*(\lambda)} \neq -e^{-\beta H_S^*(\lambda)}[H_S^*(\lambda) + \beta \partial_\beta H_S^*(\lambda)]$. Derive the results above by using that the trace is cyclic.

Next, we fix the definition of **heat** via the first law

$$Q^*(t) \equiv \Delta U_S^*(t) - W(t) \tag{3.105}$$

and it remains to be shown that the **entropy production** is non-negative:

$$\Sigma^*(t) = \Delta S_S^*(t) - \frac{Q^*(t)}{T} \geq 0. \tag{3.106}$$

Note that we use a star (*) on all quantities to distinguish them from the definitions introduced in the first part of this section. To show the positivity of eqn (3.106), we claim that eqn (3.106) is equivalent to the following information-theoretic expression:

$$\boxed{\Sigma^*(t) = k_B D[\rho_{SB}(t)|\pi_{SB}(\lambda_t)] - k_B D[\rho_S(t)|\pi_S^*(\lambda_t)].} \tag{3.107}$$

If this claim turns out to be true, then positivity of the entropy production (3.106) follows from monotonicity of relative entropy (Theorem A.3). Notice the formal similarity of eqns (3.98) and (3.107). Both ascribe the production of entropy to the question of how far (in an information-theoretic sense) the bath is pushed away from its initial equilibrium reference state.

Exercise 3.20 Show the equivalence of eqns (3.106) and (3.107).

Comparison of the two frameworks

To investigate the connection between the two different frameworks introduced here, we look at the difference in the entropy productions for the case of a single bath. Dropping the sum and the index ν in eqn (3.97), we find

$$\Sigma^*(t) - \Sigma(t) = \frac{\Delta\langle H_S(\lambda_t) - H_S^*(\lambda_t)\rangle(t)}{T} + \frac{\Delta\langle V_{SB}(\lambda_t)\rangle(t)}{T}, \tag{3.108}$$

where $\Delta\langle X(\lambda_t)\rangle(t) \equiv \text{tr}_{SB}\{X(\lambda_t)\rho_{SB}(t)\} - \text{tr}_{SB}\{X(\lambda_0)\rho_{SB}(0)\}$ for any operator X. We remark that eqn (3.108) holds regardless of the initial state $\rho_{SB}(0)$. Note that, for an arbitrary initial state, neither $\Sigma(t)$ nor $\Sigma^*(t)$ need be positive and their difference can also have either sign.

Now, taking into account that $H_S(\lambda_t) - H_S^*(\lambda_t)$ is only non-zero for non-zero coupling, we see that the differences in entropy production are completely determined by the *boundary effects* coming from the interaction term $V_{SB}(\lambda_t)$. While we said that $V_{SB}(\lambda_t)$ can be strong, it seems reasonable to assume it is finite-ranged. Thus, if the bath B is a large and extended object, we expect that $V_{SB}(\lambda_t)$ has a stronger influence on the system than on the bath seen as a whole.

Now, imagine that the small system is subjected to some periodic driving protocol λ_t with period τ for some time $n\tau$ with $n \gg 1$ large. In this case, it is likely that the system enters a (periodic) steady-state regime characterized by a constant dissipation of energy into the bath during each driving cycle. Thus, we expect that the bulk term $\text{tr}_B\{H_B[\rho_B(t) - \rho_B(0)]\}$ grows proportional with time t, whereas thermodynamic quantities describing what happens in and close to the system remain bounded. Then, it is reasonable to expect that $\Sigma^*(t) \approx \Sigma(t)$, i.e. both frameworks give similar predictions for a driven system for long times. Only if the driving protocol is short or constant, i.e. if $W(t)$ is small, do we expect eqn (3.108) to play a dominant role.

The next exercise sheds some additional light on this question and shows that $\Sigma^*(t) \approx \Sigma(t)$ holds if the zeroth law is satisfied.

Exercise 3.21 First, consider the situation where the coupling is initially zero, $V_{SB}(\lambda_0) = 0$, such that $H_S(\lambda_0) = H_S^*(\lambda_0)$. Show that this implies

$$\Sigma^*(t) - \Sigma(t) = k_B D[\rho_{SB}(t) \| \pi_{SB}(\lambda_t)] - k_B D[\rho_{SB}(t) \| \pi_S^*(\lambda_t) \otimes \pi_B]. \tag{3.109}$$

Thus, if eqn (3.91) holds, we have $\Sigma^*(t) = \Sigma(t)$. For a small system this is in general not the case, but for a large bath we still expect $\pi_B \approx \pi_B^*$. Then, eqn (3.109) effectively measures how different $\pi_{SB}(\lambda_t)$ is from $\pi_S^*(\lambda_t) \otimes \pi_B^*$, which is answered by inequality (3.92). Thus, the difference $\Sigma^*(t) - \Sigma(t)$ is related to a boundary term, which vanishes if the zeroth law holds.

Second, assume $V_{SB}(\lambda_0) \neq 0$ and an initial state of the form (3.99). Then, show that

$$\begin{aligned}\Sigma^*(t) - \Sigma(t) = {} & k_B D[\rho_{SB}(t) \| \pi_{SB}(\lambda_t)] - k_B D[\rho_{SB}(t) \| \pi_S^*(\lambda_t) \otimes \pi_B] \\ & + k_B I[\pi_{SB}(\lambda_0)] + k_B D[\pi_B^*(\lambda_0) \| \pi_B].\end{aligned} \tag{3.110}$$

Using the same reasoning as above, confirm that this term is related to a boundary term and that $\Sigma^*(t) = \Sigma(t)$ if the zeroth law holds.

Let us summarize this section. We found two inequalities, eqns (3.97) and (3.106), which hold for any system–bath Hamiltonian. For a small system only weakly influencing a large bath, both inequalities can be safely interpreted at the entropy production known from phenomenological thermodynamics. Their difference is related to the different choices of initial states, but their numerical difference is typically small if the system (coupled to a single bath) is driven for a long time, i.e. under strong non-equilibrium conditions. Both inequalities together with the associated definitions of internal energy (which fixes the heat) and system entropy (which fixes the non-equilibrium free energy) are used several times in the remainder of this book.

3.8 Nonequilibrium Entropy, Entropy Production and Finite Baths

In this section we continue with a question that we have already posed in Section 2.2, when we discussed two common candidates for entropy at equilibrium: the Boltzmann entropy and the Gibbs–Shannon–von Neumann entropy. Now, we ask how to define thermodynamic entropy out of equilibrium *in general*. The hope is that the answer to this question also provides the missing link between the heat flow and the entropy change of the bath that we identified in the previous section. The main outcome of this section is a derivation of the hierarchy of second laws that we know from phenomenological non-equilibrium thermodynamics. Specifically, we show that

$$\boxed{0 \le \Delta S_{SB}(\tau) \le \Delta S_S(\tau) + \Delta S_B(\tau) \le \Delta S_S(\tau) - \int_0^t \frac{đQ(t)}{T_t} \le \Delta S_S(\tau) - \frac{Q(\tau)}{T_0},} \tag{3.111}$$

where T_t denotes a microscopically defined time-dependent bath temperature. The difference between each of the inequalities is quantified by an information-theoretic-quantity, which becomes negligibly small in the case where one expects it to be in agreement with the phenomenological theory of thermodynamics.

Non-equilibrium entropy as observational entropy

Before we present our general definition for non-equilibrium entropy, we briefly repeat the motivation to overcome the notions of Boltzmann and Gibbs–Shannon–von Neumann entropy.

First, throughout this book we have collected evidence that the Gibbs–Shannon–von Neumann entropy provides a *legitimate* candidate for thermodynamic entropy if the system is *small*, where 'small' means that we must be able to control and measure all microscopic degrees of freedom that define the system. If the system is isolated and described by a density matrix $\rho(t)$, it then follows that its entropy is conserved, $S_{\text{vN}}[\rho(t)] = S_{\text{vN}}[\rho(0)]$, because $S_{\text{vN}}(U\rho U^\dagger) = S_{\text{vN}}(\rho)$ for any unitary U. This result makes sense for an isolated system for which we have complete control and knowledge at each time t. However, as soon as the system gets larger such that it is no longer possible to precisely control and measure all its microstates, the Gibbs–Shannon–von Neumann entropy concept faces problems. If we continue to identify $k_B S_{\text{vN}}[\rho(t)]$ with thermodynamic entropy, then we are inevitably led to the conclusion that there is no irreversibility in any isolated system. This is clearly in contradiction to the fact that, for instance, the free expansion of a gas, the mixing of two liquids or the evolution of the universe as a whole are not reversible.

One could object to the argument above that the evolution of an isolated system actually *is* reversible because the system returns to its initial state after the Poincaré recurrence time. But note that it is not possible to experimentally verify this statement because for a moderately sized system we would need to wait much longer than the age of the universe to see that happen. This attitude would also take away the predictive power of the second law, which indeed correctly explains a plethora of observed phenomena based on the *increase* in thermodynamic entropy.

Another stance on this problem is to use the Gibbs–Shannon–von Neumann entropy only for an appropriately chosen ensemble of states $\rho(t)$; for instance, one that maximizes $S_{\text{vN}}[\rho(t)]$ with respect to certain constraints. While that might work in principle, it does not explain how the laws of thermodynamics *emerge* from microscopic considerations, where the evolution of $\rho(t)$ is fixed given $\rho(0)$. It also requires us to introduce the notion of an 'ensemble' and to find a way to choose the right one.

In macroscopic systems, these problems are overcome by Boltzmann's entropy concept $S_B(X) = k_B \ln V(X)$, where $V(X)$ denotes the number of microstates compatible with the constraint X. In Boltzmann's view, there is no need to introduce an 'ensemble' because $S_B(X)$ is non-zero even for a pure state. Moreover, Boltzmann's entropy intuitively explains the emergence of the second law from microscopic considerations after realizing that the volume $V(X_{\text{eq}})$ corresponding to an equilibrium configuration X_{eq} is much larger than the volume $V(X_{\text{noneq}})$ corresponding to a non-equilibrium configuration X_{noneq}. Thus, it is very probable that the system evolves from X_{noneq} to X_{eq} and remains in X_{eq} for the overwhelming majority of times, thereby naturally explaining the second law.

However, Boltzmann's entropy is also not without drawbacks. Since the volume term for a single microstate equals one, its Boltzmann entropy is zero. Consequently, it does not explain why the Gibbs–Shannon–von Neumann entropy matches thermodynamic entropy for small systems. Thus, whereas Gibbs–Shannon–von Neumann

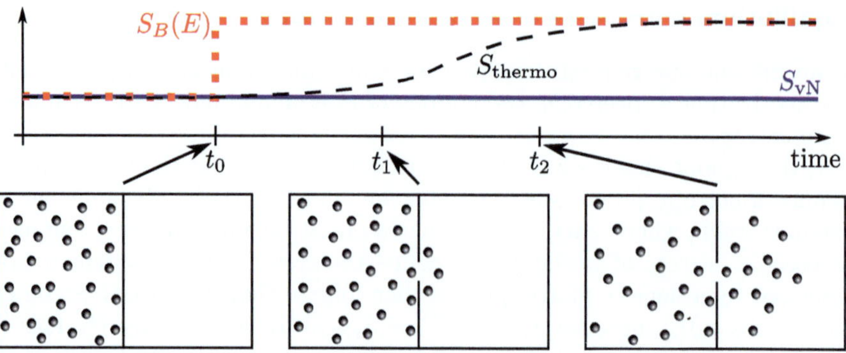

Fig. 3.7 Expansion of a gas from an initially constrained equilibrium state. Prior to time t_0, the gas is confined to the left half of the box and all notions of entropy coincide (up to small finite-size corrections). At time t_0, a small hole in the box is opened such that the gas can slowly expand into the right-hand volume. A naive application of the Boltzmann entropy concept predicts a sudden jump at time t_0 (dotted red line) since the number of available microstates increases discontinuously. In contrast, the Gibbs–Shannon–von Neumann entropy (solid blue line) stays constant since the gas evolves in a unitary way (we assume the box to be perfectly isolating). The real thermodynamic entropy should, however, smootly interpolate between the initial value at time t_0 and the Boltzmann entropy at late times since the particle number in the right-hand volume smoothly changes from zero to its equilibrium value $\approx N/2$, where N denotes the total number of gas particles.

entropy is *too fine-grained* in general, Boltzmann's entropy appears *too coarse-grained* to accurately reflect our increased technological abilities. Figure 3.7 summarizes the features and problems of these two different notions of entropy for the expansion of a gas.

As an intermediate solution between these two concepts, we look for a definition which explicitly incorporates the knowledge that an external agent can have about a thermodynamic system under reasonable experimental constraints. For this purpose, we introduce the notion of a *coarse-graining* X, characterized by a complete set of projectors $\{\Pi(x)\}$ obeying $\Pi(x)\Pi(x') = \delta_{x,x'}$ and $\sum_x \Pi(x) = 1$. Then, we define the following quantity, called **observational entropy**:

$$\boxed{S_{\text{obs}}^X(\rho) \equiv \sum_x p_x[-\ln p_x + \ln V(x)].} \tag{3.112}$$

Here, $p_x = \text{tr}\{\Pi(x)\rho\}$ is the probability of finding the system in state x and $V(x) \equiv \text{tr}\{\Pi(x)\}$ is a Boltzmann-like volume term. Thus, observational entropy combines uncertainty related to a hypothetical measurement of an observable with projectors $\{\Pi(x)\}$ (as quantified by its Shannon entropy) and an averaged Boltzmann entropy reflecting the remaining ignorance about the precise microstate even after knowing x.

Exercise 3.22 Which coarse-graining do you need to choose to ensure that $S_{\text{obs}}^X(\rho) = S_{\text{vN}}(\rho)$? When is $k_B S_{\text{obs}}^X(\rho) = S_B(X)$?

It is important to emphasize that eqn (3.112) is well defined for any state ρ and any coarse-graining X, but not every X is of thermodynamic relevance. Observational entropy does not equal thermodynamic entropy *per se*, but requires the correct choice of X, which—in the end—also has to be determined by the experimental conditions.

Before we turn to thermodynamic applications, we state a useful mathematical identity related to observational entropy. For that purpose, we introduce the following notation. First, we let $\rho(x) \equiv \Pi(x)\rho\Pi(x)/p(x)$ denote the post-measurement state given x. Second, we introduce the generalized microcanonical ensemble $\omega(x) \equiv \Pi(x)/V(x)$, which corresponds to the maximally uninformative state given x. Finally, we denote by $\mathcal{D}_X\rho \equiv \sum_x \Pi(x)\rho\Pi(x)$ the average post-measurement state or, alternatively, $\mathcal{D}_X$ can also be interpreted as a dephasing operation with respect to the coarse-graining X. Then, we have the following identity:

$$S_{\text{obs}}^X(\rho) - S_{\text{vN}}(\rho) = D(\rho|\mathcal{D}_X\rho) + \sum_x p(x)D[\rho(x)|\omega(x)]. \tag{3.113}$$

The term $\mathcal{D}(\rho|\mathcal{D}_X\rho)$ can be seen as a purely quantum contribution: it measures how much coherence is contained in the state ρ with respect to X. Instead, the second term is of classical origin: by evaluating it in the eigenbasis of $\rho(x)$, we see that $D[\rho(x)|\omega(x)]$ describes the distance of $\rho(x)$ (seen as a probability mixture) from the maximally mixed state $\omega(x)$. Note that eqn (3.113) implies $S_{\text{obs}}^X(\rho) \geq S_{\text{vN}}(\rho)$.

Exercise 3.23 Derive eqn (3.113).

Second law for a driven, homogeneous and isolated system

We now consider the first thermodynamic application and focus on an isolated system with Hamiltonian $H = \sum_\epsilon \epsilon|\epsilon\rangle\langle\epsilon|$. We consider the simple case where the relevant coarse-graining $X = E$ is defined by projectors

$$\Pi(E) \equiv \sum_{\epsilon\in[E,E+\delta)} |\epsilon\rangle\langle\epsilon|, \tag{3.114}$$

which correspond to a measurement of the energy with some uncertainty δ. One can clearly imagine more complicated scenarios, but the present one is sufficient for the open quantum system case studied below. Physically, the choice $X = E$ is justified if the isolated system is a large homogeneous chunk of matter, whose only relevant thermodynamic parameter is its total energy (or its temperature, respectively), assuming the volume to be known and fixed.

Next, recall that the relative entropy satisfies $D(\rho|\sigma) = 0$ if and only if $\rho = \sigma$. From our identity (3.113) we then infer that

$$S_{\text{obs}}^E(\rho) = S_{\text{vN}}(\rho) \quad \Leftrightarrow \quad \rho = \sum_E p_E\omega(E). \tag{3.115}$$

Here, p_E is an arbitrary probability distribution and $\omega(E)$ is the conventional microcanonical ensemble. Therefore, whenever p_E describes a microcanonical ensemble or approximates a canonical ensemble $\pi(\beta)$ with probabilities $\pi_E(\beta) \equiv \text{tr}\{\Pi(E)\pi(\beta)\} \approx V(E)e^{-\beta E}/\mathcal{Z}$ with $\mathcal{Z} \approx \sum_E V(E)e^{-\beta E}$, observational entropy coincides with the Boltzmann or von Neumann entropy up to small corrections. These statements hold only, of course, if we do not choose δ unreasonably large. In the following, we tacitly assume that δ is chosen small enough to be in agreement with equilibrium statistical mechanics such that $S_{\text{obs}}^E[\pi(\beta)] \approx S_{\text{vN}}[\pi(\beta)] \approx \mathcal{S}(\beta)/k_B$, where $\mathcal{S}(\beta)$ denotes the equilibrium thermodynamic entropy in the following.

Let us now consider a process where we change the Hamiltonian from $H(\lambda_0)$ to $H(\lambda_\tau)$ according to some prescribed driving protocol. Consequently, the coarse-graining $\{\Pi(E_t)\}$ also changes in time, where E_t is shorthand for $E(\lambda_t)$. Next, consider the class of initial states defined by the requirement (3.115): $\rho(0) = \sum_{E_0} p_{E_0}\omega(E_0)$. Since $S_{\text{vN}}[\rho(\tau)] = S_{\text{vN}}[\rho(0)]$, it follows immediately from eqn (3.113) that

$$k_B \Delta S_{\text{obs}}^{E_\tau}(\tau) \equiv k_B S_{\text{obs}}^{E_\tau}[\rho(\tau)] - k_B S_{\text{obs}}^{E_0}[\rho(0)] \geq 0. \tag{3.116}$$

This is the second law for an isolated and driven homogenous system.

Further interesting insights can be gained by introducing the concept of an effective **non-equilbrium temperature**, which will also play a decisive role to formulate Clausius' inequality for open quantum systems. For an arbitrary state $\rho(t)$ we define an inverse non-equilibrium temperature $\beta_t^* = (k_B T_t^*)^{-1}$ by demanding

$$\text{tr}\{H(\lambda_t)\rho(t)\} \equiv \text{tr}\{H(\lambda_t)\pi(\beta_t^*)\}, \tag{3.117}$$

i.e. we ask which inverse temperature a fictitious Gibbs state $\pi(\beta_t^*)$ needs to have such that its internal energy $\mathcal{U}(\beta_t^*)$ matches the true internal energy $U(t)$. In terms of our coarse-grained energies, definition (3.117) is expressed as (recall that we assume δ to be chosen sufficiently small)

$$\sum_{E_t} E_t p_{E_t}(t) \equiv \sum_{E_t} E_t \pi_{E_t}(\beta_t^*). \tag{3.118}$$

Operationally speaking, T_t^* can be defined in the following way. Suppose that we have an infinitely large 'superbath' at our disposal, whose equilibrium temperature T we can control and which we can put into weak contact with our system. Then, T_t^* matches the temperature T of the superbath if *no net heat flow* takes place when the system and the superbath are coupled to each other.

Ignoring any time–dependence λ_t for the moment, we find the familiar result

$$\frac{d\mathcal{U}(\beta^*)}{d\beta^*} = -\frac{\mathcal{C}(\beta^*)}{(\beta^*)^2}, \tag{3.119}$$

where $C(\beta^*) = (\beta^*)^2 [\text{tr}\{H^2\pi(\beta^*)\} - \text{tr}\{H\pi(\beta^*)\}^2]$ is the heat capacity. Since $C(\beta^*)/(\beta^*)^2 \geq 0$, this shows that $\beta^* = \beta^*(U)$ is a monotonically decreasing function of the internal energy U with $\beta^* = \infty$ if the system is in the ground state and $\beta^* = -\infty$ if the system is in the highest excited state (assuming the Hamiltonian H to be bounded from above, otherwise β^* remains positive). Thus, there is a one–to–one relationship between the internal energy and the nonequilbrium temperature.

Returning to the time–dependent case, definition (3.117) allows us to establish a remarkable connection between energetic changes of the isolated system and its associated 'equilibrium' entropy $\mathcal{S}(\beta_t^*)$, even if the system is out of equilibrium. To reveal this connection, we start by noting the identity

$$T_t^* d\mathcal{S}(\beta_t^*, \lambda_t) = dU(t) - \text{tr}\{[dH(\lambda_t)]\pi(\beta_t^*, \lambda_t)\}, \tag{3.120}$$

where $dU(t) = \text{tr}\{H(\lambda_{t+dt})\rho(t+dt)\} - \text{tr}\{H(\lambda_t)\rho(t)\}$ denotes the change in internal energy with respect to the nonequilibrium dynamics. If the Hamiltonian does not change in time, $dH(\lambda_t) = H(\lambda_{t+dt}) - H(\lambda_t) = 0$, we recover the familiar result $T_t^* d\mathcal{S}(\beta_t^*) = dU(t)$. Obviously, for an isolated system we then also have $dU(t) = 0$, but this will change for the open system case below.

Exercise 3.24 Derive eqn (3.120).

Now, recall that we identified in Section 3.1 $dU(t) = đW(t)$ as the mechanical work done on the system, but from a macroscopic point of view this identification might not seem entirely satisfactory. In fact, many macroscopic bodies *heat up* after they were subjected to an external driving protocol (e.g., think about your food put into a microwave or rub your hand about the surface of a table). We therefore introduce the concept of **remaining heat**, which quantifies the part of the mechanical work that is converted into 'heating up' the isolated system. It is defined as

$$đQ^{\text{rem}}(t) \equiv T_t^* d\mathcal{S}(\beta_t^*, \lambda_t). \tag{3.121}$$

It follows from this identification that $\mathcal{S}(\beta_\tau^*, \lambda_\tau) - \mathcal{S}(\beta_0^*, \lambda_0) = \int đQ^{\text{rem}}(t)/T_t^*$ and this result allows us to rewrite the second law (3.116) as

$$k_B S_{\text{obs}}^{E_\tau}[\rho(\tau)] - \mathcal{S}(\beta_\tau^*, \lambda_\tau) + \int \frac{đQ^{\text{rem}}(t)}{T_t^*} + \mathcal{S}(\beta_0^*, \lambda_0) - k_B S_{\text{obs}}^{E_0}[\rho(0)] \geq 0. \tag{3.122}$$

In particular, if the isolated system is prepared in a Gibbs state $\pi(\beta_0)$, the last two terms cancel. Furthermore, since the Gibbs state maximizes entropy with respect to a fixed energy, we can conclude $k_B S_{\text{obs}}^{E_\tau}[\rho(\tau)] \leq \mathcal{S}(\beta_\tau^*, \lambda_\tau)$. Consequently, we find

$$\int \frac{đQ^{\text{rem}}(t)}{T_t^*} \geq k_B \Delta S_{\text{obs}}^{E_\tau}(\tau) \geq 0, \tag{3.123}$$

which confirms the idea that $Q^{\text{rem}}(t)$ is *dissipated* during the process.

Finally, the complementary part of remaining heat is **recoverable work**. It is fixed via the first law and from eqn (3.120) we find explicitly that

$$đW^{\text{rec}}(t) = \text{tr}\{[dH(\lambda_t)]\pi(\beta_t^*, \lambda_t)\}. \tag{3.124}$$

The claim is now that $đW^{\text{rec}}(t)$ quantifies the part of the internal energy change $dU(t)$, which can be recovered from the system in a *macroscopic sense*. By this we mean the following: first, we assume to only know the average energy of the system; second, we are only allowed to change the protocol λ as specified in the Hamilonian (i.e., we are *not* allowed to implement arbitrary unitary operations on the system as assumed for a small open system in Section 3.5); and third, we are allowed to bring the system into weak contact with a superbath at an arbitrary temperature T. Clearly, if more (but not complete) knowledge and control is available, potentially more work can be extracted, but it is non-trivial to compute (or even estimate) the work quantitatively.

Exercise 3.25 Consider a process, where the system at time τ is put into contact with a superbath at temperature T_τ^*, and afterwards the temperature T_t^* and the driving protocol λ_t are *slowly* (i.e., reversibly) changed back to their initial values T_0^* and λ_0. Show that the extracted work during this process is $-\int_0^\tau đW^{\text{rec}}(t)$.

Hierarchy of second laws for an open system

We return to the system–bath paradigm. For simplicity, we consider only a single heat bath here, but the reader will encounter no problems when trying to fill in the gaps for multiple baths. Now, the spirit of the system–bath approach is rooted in the idea that the system is an object about which one has precise control, whereas the bath denotes all the irrelevant and inaccessible degrees of freedom. Following this approach, the most natural choice of coarse–graining to fix the observational entropy (3.112) is $X = S \otimes E_B$, where $S = \{|s\rangle\langle s|\}$ is an arbitrary set of rank-1 projectors acting on the system Hilbert space and the coarse-graining E_B of the bath is in analogy to eqn (3.114), describing a coarse-grained measurement of the bath energies with resolution δ. This reflects our limited information about the bath. We remark that, while $V(E_B) = \text{tr}_B\{\Pi(E_B)\}$ can be very large, we need to have at least some information about the bath—otherwise we could not even associate a temperature with it. Our definition (3.112) of observational entropy therefore reduces to

$$S_{\text{obs}}^{S,E_B}(\rho_{SB}) = \sum_{s,E_B} p_{s,E_B}[-\ln p_{s,E_B} + \ln V(E_B)] \tag{3.125}$$

with $p_{s,E_B} = \text{tr}_{SB}\{|s\rangle\langle s| \otimes \Pi(E_B)\rho_{SB}\}$. Equation (3.125) is identified with thermodynamic entropy (times k_B) in the following: $S_{SB} \equiv k_B S_{\text{obs}}^{S,E_B}(\rho_{SB})$.

Exercise 3.26 Consider a classical system coupled to a bath and relabel $s = E_x$, where E_x denotes the energy of a microstate x of the system. Furthermore, assume that the global energy E is fixed such that $E_B = E - E_x$ if the system is found in state x. Show that we have already identified in Chapter 2 the change in observational entropy (3.125) with the entropy production (without explicitly noticing it). *Hint:* Recall eqn (2.58).

Consider an initial system–bath state of the form

$$\rho_{SB}(0) = \sum_{s,E_B} p_{s,E_B}(0)|s\rangle\langle s| \otimes \omega(E_B). \tag{3.126}$$

Since $|s\rangle\langle s|$ is arbitrary, $\rho_{SB}(0)$ describes an arbitrary system state classically correlated with a bath that looks microcanonical in each energy shell E_B. Therefore, eqn (3.126) is more general than our previous choice for the initial state (see eqn (3.19)), which relied on a decorrelated state with a bath prepared in the canonical ensemble. It is simple to check that $S_{\text{obs}}^{S,E_B}[\rho_{SB}(0)] = S_{\text{vN}}[\rho_{SB}(0)]$. Then, from $S_{\text{vN}}[\rho_{SB}(\tau)] = S_{\text{vN}}[\rho_{SB}(0)]$ and our identity (3.113), we derive that

$$\boxed{\Delta S_{SB}(\tau) = k_B S_{\text{obs}}^{S_\tau,E_B}[\rho_{SB}(\tau)] - k_B S_{\text{obs}}^{S_0,E_B}[\rho_{SB}(0)] \geq 0.} \tag{3.127}$$

Note that in eqn (3.127) we allowed that the system coarse-graining $S_t = \{|s_t\rangle\langle s_t|\}$ can change in time without invalidating the second law. Equation (3.127) equals the *first* second law in our hierarchy (3.111) in unison with the phenomenological second law (2.3).

Next, we consider an initially decorrelated state $p_{s,e_B}(0) = p_s(0)p_{E_B}(0)$, as done many times before. This implies $S_{SB}(0) = S_{\text{obs}}^{S_0}[\rho_S(0)] + S_{\text{obs}}^{E_B}[\rho_B(0)] \equiv S_S(0) + S_B(0)$, where we introduced the thermodynamic entropy of S and B alone. At a later time τ we find instead $S_{SB}(\tau) = S_S(\tau) + S_B(\tau) - I_{S:E_B}(\tau)$, where $I_{S:E_B}(\tau)$ denotes the mutual information of the final probability distribution $p_\tau(s, E_B)$. Since $I_{S:E_B}(\tau) \geq 0$, we obtain

$$\boxed{\Delta S_S(\tau) + \Delta S_B(\tau) \geq 0,} \tag{3.128}$$

which coincides with the phenomenological second law (2.4). The difference from the more general second law (3.127) for an initially decorrelated state is

$$\Delta S_S(\tau) + \Delta S_B(\tau) - \Delta S_{SB}(\tau) = I_{S:E_B}(\tau) \geq 0. \tag{3.129}$$

This confirms the second equality in the hierarchy (3.111). As expected, whenever system–bath correlations are small, entropy becomes additive, such that $\Delta S_{SB}(\tau) \approx \Delta S_S(\tau) + \Delta S_B(\tau)$.

We proceed by linking the change $\Delta S_B(\tau)$ in bath entropy to the heat flux. We do so by applying the definition of non-equilibrium temperature from eqn (3.117) to the *bath*. Thus, let T_t^* be the temperature obtained from eqn (3.117) by adding a subscript B to the Hamiltonian and the state. It follows from eqn (3.121) that

$$\mathcal{S}_B(\beta_\tau^*) - \mathcal{S}_B(\beta_0^*) = \int \frac{đQ_B^{\text{rem}}(t)}{T_t^*}, \tag{3.130}$$

where $Q_B^{\text{rem}}(t)$ is the remaining heat of the bath. Notice that $đQ_B^{\text{rem}}(t) = dU_B(t) = \text{tr}\{H_B[\rho_B(t+dt) - \rho_B(t)]\}$ if the bath Hamiltonian is time–independent as assumed in all previous sections. In unison with our sign convention to count heat positively if it increases the energy of the open system, we set $đQ(t) \equiv -đQ^{\text{rem}}(t)$ in the following. Moreover, as also done in previous sections, we now assume that the bath is initially prepared in a Gibbs state at inverse temperature β_0, $\rho_B(0) = \pi_B(\beta_0)$. The change in bath entropy is then given by $\Delta S_B(\tau) = S_B(\tau) - \mathcal{S}_B(\beta_\tau^*) - \int đQ(t)/T_t^* \leq -\int đQ(t)/T_t^*$. Thus, we confirm **Clausius's inequality**

$$\boxed{\Delta S_S(\tau) - \int \frac{đQ(t)}{T_t^*} \geq 0,} \tag{3.131}$$

which coincides with the phenomenological second law (2.5). The difference from the more general second law (3.128) for a bath prepared in a Gibbs state is

$$\Delta S_S(\tau) - \int \frac{đQ(t)}{T_t^*} - [\Delta S_S(\tau) + \Delta S_B(\tau)] = \mathcal{S}_B(\beta_\tau^*) - S_B(\tau) \geq 0. \tag{3.132}$$

By using the defining equation (3.118) of β_τ^*, we can write this term in a more transparent form as

$$\mathcal{S}_B(\beta_\tau^*) - S_B(\tau) = k_B D[p_{E_B}(\tau)|\pi_{E_B}(\beta_\tau^*)] \geq 0, \tag{3.133}$$

where $D[\cdot|\cdot]$ here denotes a classical relative entropy. Thus, if the bath is well described by a Gibbs state with a time-dependent temperature, we find $\Delta S_{SB}(\tau) \approx \Delta S_S(\tau) - \int đQ(t)/T_t^*$. Equation (3.132) confirms the third inequality in the hierarchy (3.111).

Finally, based on eqn (3.131) it is not hard to see that our previously derived second law (3.97) follows in the case where the bath temperature $T_t^* \approx T_0$ remains approximately constant. Then, we obtain the familiar result

$$\Delta S_S(\tau) - \frac{Q(\tau)}{T_0} \geq 0, \tag{3.134}$$

which coincides with the phenomenological second law (2.6). Remarkably, the difference from the more general second law (3.131) is given by

$$\boxed{\Delta S_S(\tau) - \frac{Q(\tau)}{T_0} - \left[\Delta S_S(\tau) - \int \frac{đQ(t)}{T_t^*}\right] = k_B D\left[\pi_B(\beta_\tau^*)|\,\pi_B(\beta_0)\right] \geq 0,} \tag{3.135}$$

which confirms the fourth and final inequality in the hierarchy (3.111). From this result it is evident that eqn (3.134) coincides with the change in thermodynamic entropy, and thus with the second law, only if the bath remains close to the initial equilibrium state, i.e. if $\rho_B(\tau) \approx \pi_B(\beta_0)$, then $\Delta S_{SB}(t) \approx \Delta S_S(\tau) - Q(\tau)/T_0$. We have already anticipated this result in Section 3.7.

Exercise 3.27 Derive eqn (3.135) by using eqn (3.130).

This section finishes our endeavour to derive the laws of phenomenological non-equilibrium thermodynamics, as introduced in Section 2.1. We found consistent microscopic definitions for internal energy, heat, work, entropy and temperature, which allow us to study the laws of thermodynamics for small quantum systems coupled to (semiclassical) work reservoirs and multiple heat baths, even if the baths show finite size effects and are characterized by time-dependent temperatures. In particular, since mutual information can be also expressed as a relative entropy, observe that eqns (3.113), (3.129), (3.133) and (3.135) demonstrate that each second law in the hierarchy (3.111) can be expressed as a (sum of) relative entropies, quantifying the amount of disregarded information at each step. Note, however, that this rewriting is only possible for a particular class of initial states. For arbitrary (perhaps pure) initial states the entropy production involves differences of relative entropies and can have either sign. However, entropy production is always well quantified by looking at the total change of entropy ΔS_{SB}, which could become negative for specific initial states such as the one shown in Fig. 3.6.

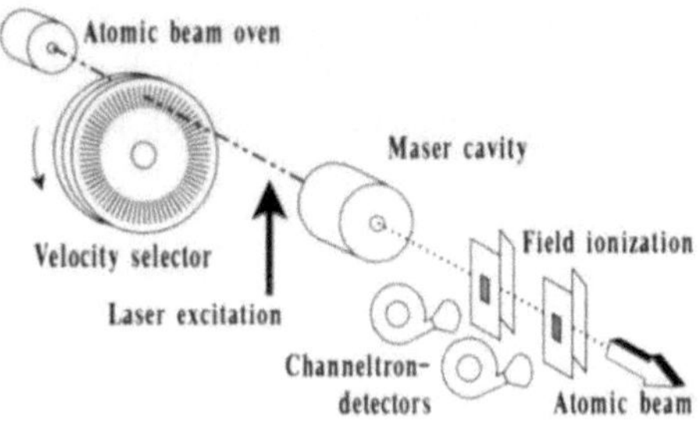

Fig. 3.8 Sketch of a micromaser taken from one of the early experiments. Reprinted figure with permission from G. Rempe, F. Schmidt-Kaler, and H. Walther, Phys. Rev. Lett. 64, 2783 (1990). Copyright (2021) by the American Physical Society.

In the rest of this chapter, we introduce a versatile framework to deal with non-equilibrium resources different from driving fields or temperature gradients and we show how to incorporate particle transport in our description above. Furthermore, phenomenological thermodynamics is a theory about averages or mean values. How to include fluctuations in the description, as in classical stochastic thermodynamics, is the content of Chapters 4 and 5.

3.9 Non-Equilibrium Resources and Repeated Interactions

The traditional notion of a system coupled to idealized heat baths and work reservoirs is sufficient to explain thermodynamic engines from the 19th century, but it is no longer adequate to describe the plethora of quantum experiments carried out today. Nowadays, experimentalists often use much more sophisticated *resources* to probe or control the dynamics of small systems. These resources are themselves out of equilibrium and cannot be described by a single temperature or mechanical degree of freedom. In Section 3.8, we provided a recipe to deal with certain non-equilibrium features in the bath, which decrease the amount of entropy produced. However, in spirit the bath was still treated as a relatively large and structureless object. This section introduces a framework to incorporate microscopically small non-equilibrium resources in the laws of thermodynamics in a controlled way and it plays an important role in Chapter 5. Nevertheless, the general treatment of non-equilibrium resources remains a big open challenge in quantum thermodynamics.

Motivation and idea

Experimentally, the present framework can be motivated by considering the **micromaser** in Fig. 3.8. The open quantum system under consideration is a *cavity*. For our theoretical purposes a cavity is simply a box of mirrors, which supports a discrete set of electromagnetic modes (standing waves) with particular resonance frequencies. Since no mirror is perfect, the photons of each mode can be exchanged with the outside world, which makes the cavity an *open* quantum system.

In the simplest case, on which we exclusively focus here, the cavity contains a single mode in one dimension with resonance frequency ω_c (this can be realized with two convex mirrors facing each other). The isolated cavity Hamiltonian is $H_S = \hbar\omega_c a^\dagger a$

with the photon creation (annihilation) operators $a^\dagger$ (a) obeying the commutation relation $[a, a^\dagger] = 1$. If the cavity is of sufficiently high quality, the coupling to the outside electromagnetic field can be regarded as a weak perturbation. In this case, the cavity master equation reduces to

$$\begin{aligned}\frac{\partial}{\partial t}\rho_S(t) = &- i[\omega_c, \rho_S(t)] \\ &+ \frac{1+N_{\text{th}}}{2\tau_c}\left[a\rho_S(t)a^\dagger - \frac{1}{2}\{a^\dagger a, \rho_S(t)\}\right] + \frac{N_{\text{th}}}{2\tau_c}\left[a^\dagger\rho_S(t)a - \frac{1}{2}\{aa^\dagger, \rho_S(t)\}\right].\end{aligned} \tag{3.136}$$

Here, $N_{\text{th}} = (e^{\beta\hbar\omega_c} - 1)^{-1}$ is the mean thermal photon number of the outside field at frequency ω_c, which equals the Bose–Einstein distribution, and τ_c is known as the cavity lifetime, which determines the rate of dissipation. Readers unfamiliar with this theoretical description are asked to microscopically derive the cavity master equation.

Exercise 3.28 We model the photon exchanges between the system with Hamiltonian $H_S = \hbar\omega_c a^\dagger a$ and its environment with the Hamiltonian

$$H_{SB} = H_S + \hbar\sum_k g_k(a + a^\dagger)(b_k + b_k^\dagger) + \hbar\sum_k \omega_k b_k^\dagger b_k. \tag{3.137}$$

Here, the index k labels all the information about the field modes of the environment (i.e. their frequency $\omega_k > 0$, polarization, direction, etc.). The operators $b_k^\dagger$ and b_k create and annihilate a photon in mode k and they obey $[b_k, b_{k'}^\dagger] = \delta_{kk'}$ and $[a^{(\dagger)}, b_k^{(\dagger)}] = 0$. The bilinear coupling to the system is described by the real-valued parameter g_k. First, compare eqn (3.137) with the Caldeira–Leggett Hamiltonian (1.25) for $S = a + a^\dagger$. Convince yourself of the fact that both Hamiltonians have the same form apart from a correction term, which is negligible for small g_k. Second, derive eqn (3.136) using the BMS approximation as detailed in Section 3.2 (neglect any Lamb shift terms). Show that the cavity lifetime τ_c is microscopically determined by $\tau_c^{-1} = 4\pi\sum_k g_k^2\delta(\omega_c - \omega_k)$. In view of the largeness of the environment, which contains a *continuum* of modes, this expression can also be written as $\tau_c^{-1} = 4\pi\int_0^\infty d\omega\rho(\omega)g(\omega)^2\delta(\omega_c - \omega)$, where $\rho(\omega)$ is the density of field modes as a function of the frequency. Investigate which processes are described by the second line of eqn (3.136). Do the rates obey local detailed balance?

We remark that eqn (3.136) describes the experimentally observed dynamics very well for a large range of parameters provided that g_k is sufficiently small. Thus, the Markov and secular approximation are well justified for a high-quality cavity.

The master equation (3.136) describes a cavity coupled to an ideal heat bath: in the long run the cavity will equilibrate, $\rho_S(t) \to \pi_S$, and does not show any interesting non-equilibrium features. To introduce them, the dynamics of the cavity is modified by shooting atoms through it; see again Fig. 3.8. These atoms are prepared in an oven, selected according to their velocity and excited by laser light before they enter the cavity. Ideally, the atoms can be described as two-level systems and the cavity–atom dynamics is generated by the Jaynes–Cummings Hamiltonian (3.9). As a consequence, if the atoms are prepared in an excited state and interact with the cavity for the right amount of time (which can be tuned via their velocity), the atoms can emit one photon into the cavity. To detect the atom state, one applies an external field in such a way that the ground and excited states are ionized at different times. Measuring when the atom is ionized then reveals whether it was in the ground or excited state after the interaction with the cavity. This, in turn, reveals information about the state of the cavity field.

Interesting non-equilibrium quantum dynamics can be observed whenever the cavity lifetime τ_c is larger than the mean waiting time between two atoms. Then, it is possible to build up a field inside the cavity and to create interesting quantum states. Remarkably, it is possible to build up a cavity field with a photon number above its thermal expectation value with *less than one atom* in the cavity on average. In contrast, in conventional lasers the cavity is pumped using an external macroscopic field. Historically, the study of the micromaser has experimentally answered many questions in quantum optics, quantum measurement theory and quantum foundations, but from a thermodynamic point of view the micromaser is an unconventional device. Since the atoms generate a non-equilibrium state in the cavity, we know from Section 3.5 that we can extract work from the cavity in a repeatable way, although the cavity is coupled only to a single heat bath. Does this violate the second law of thermodynamics? Clearly not since we also have to take into account the atoms in the energetic and entropic balance. But the atoms seem to act neither like a heat bath (they are not even prepared in a thermal state) nor like a work reservoir (there is no macroscopic limit for a single atom, which would allow a time-dependent field to be generated, as explored in Exercise 3.1). We therefore need to treat the reservoir of atoms in a different and novel way. Inspired by the micromaser, we now introduce an abstract framework which is able to do so. We return to the specific example of the micromaser at the end of this section again. Another related application will be treated in detail in Section 5.7.

The stream of atoms flying through the cavity forms a new type of environment, to which we also refer as a non-equilibrium reservoir in the following. It is characterized by four distinctive features.

(i) While the total amount of atoms can be large, the cavity ideally interacts at each time with at most one of them.

(ii) The initial state of the atoms can be different from a thermal state.

(iii) The atom–cavity interaction time or strength can be manipulated.

(iv) After the interaction with an atom, the atom never returns and interacts with the cavity again.

It is clear that the first two properties are very different from the case of a heat bath. The third property might also be satisifed for a heat bath, but for the micromaser experiment the possibility to control the interaction between the system and different parts of the reservoir is *essential*. Finally, the last property makes the description simpler and amounts to performing a Markov approximation: since the atoms—once interacted with the cavity—never come back, the cavity–atom correlations built up during the interaction do not influence the future dynamics of the cavity. Taken together, the properties (i)–(iv) constitute the framework of **repeated interactions**, which is also called a **collisional model**.

Thermodynamic description at strong system–bath coupling

We describe the entire situation with the following Hamiltonian:

$$H_{SAB}(\lambda_t) = H_S(\lambda_t) + V_{SB} + H_B + V_{SA}(\lambda_t) + H_A. \tag{3.138}$$

Here, A is a subscript associated with the physical degrees of freedom of our new reservoir (e.g. the atoms in the case of the micromaser). The bare Hamiltonian of the reservoir is written as $H_A = \sum_k H_{A(k)}$, where $k \in \{0, 1, \ldots, n\}$ is some index, which—according to properties (i) and (iv)—labels the parts of the reservoir interacting one by one with the system (e.g. the kth atom in the case of the micromaser). Consequently, $[H_{A(k)}, H_{A(\ell)}] = 0$ for all k, ℓ, and we have $V_{SA}(\lambda_t) = \sum_k V_{SA(k)}(\lambda_t)$. To complete the set-up, the time dependence of $V_{SA}(\lambda_t)$ has to obey property (iv), which can be ensured by writing

$$V_{SA}(\lambda_t) = \sum_k \Theta(t - t_k)\Theta(t'_k - t)v_{SA(k)}. \tag{3.139}$$

Here, the t_k and t'_k are times ordered according to $0 = t_0 < t'_0 < t_1 < t'_1 < \cdots < t_n < t'_n$. Thus, $v_{SA(k)}$ describes the interaction between the system and the reservoir during the kth interaction interval $[t_k, t'_k)$. Note that none of the interaction intervals is overlapping and that the system sometimes can only be coupled to the bath B but not to the reservoir A. Furthermore, one often considers regular and equidistant time intervals such that $t_k = k\tau$ and $t'_k = t_k + \tau'$ with $\tau' < \tau$, but this is not assumed in the following. Finally, we fix the initial state of the set-up, which we take to be

$$\rho_{SAB} = \pi_{SB}(\lambda_0) \bigotimes_k \rho_{A(k)}(0). \tag{3.140}$$

Here, the $\rho_{A(k)}(0)$ are arbitrary states of the reservoir part $A(k)$ that are only assumed to be decorrelated from $A(\ell)$ $(\ell \neq k)$. This allows us to include a large class of initial non-equilibrium states into the thermodynamic description in the following.

If the reader jumps back and looks at Fig. 1.6, a certain similarity between the theoretical description given here and the one in Chapter 1 becomes evident. In fact, in Chapter 1 we used a stream of *ancillas* to implement quantum measurements or more general control operations on the open system. These interventions were assumed to happen instantaneously, which can be considered as a limiting case of the more general interaction (3.139). Furthermore, the ancillas were subjected to a quantum measurement after the interaction with the system. For simplicity, this case is excluded here, but we return to it in Chapter 5. Here, we only focus on arbitrary initially decorrelated ancillas interacting with the system in an arbitrary way without any explicit modelling of the final measurement. Nevertheless, the resulting (thermo)dynamic framework is very rich. Since we only used the micromaser as a motivating example and now consider the framework of repeated interactions at an abstract level, we refer to the $A(i)$ as ancillas instead of atoms in the following. Luckily, the words ancilla and atom start with the same letter and do not require any change of notation.

We now turn to the thermodynamic description of the repeated interactions framework. As usual, we start with the definition of mechanical work. We have two sources of time dependence in our set-up such that it is convenient to split the total amount of work as $W(t) = W_S(t) + W^{\text{ctrl}}(t)$. The first term is the conventional one and comes from the time dependence of the system Hamiltonian: $W_S(t) = \int_0^t ds \mathrm{tr}_S\{[\partial_s H_S(\lambda_s)]\rho_S(s)\}$. The second contribution follows from the time dependence in eqn (3.139). Taking into account that $\partial_t \Theta(t) = \delta(t)$, we find

$$\boxed{W^{\mathrm{ctrl}}(t) = \sum_k \mathrm{tr}_{SA(k)}\{v_{SA(k)}[\rho_S(t_k) \otimes \rho_{A(k)}(t_k) - \rho_{SA(k)}(t'_k)]\}.} \tag{3.141}$$

This describes the amount of work required to switch on and off the interaction between the system and the ancilla. The superscript 'ctrl' reminds us that this work can be used to implement a control operation in view of our framework from Chapter 1. Note that, at the point where we switch on the interaction, the system–ancilla state is decorrelated, $\rho_{SA(k)}(t_k) = \rho_S(t_k) \otimes \rho_{A(k)}(t_k)$, because of eqn (3.140) and properties (i) and (iv) above. We remark that the Heaviside step function is, of course, a theoretical idealization. In reality, $V_{SA(k}(\lambda_t)$ also depends smoothly on time. Often, however, the time scales are such that eqn (3.139) is a good approximation, which simplifies the theoretical treatment.

To proceed, we now use our strong coupling framework developed in the second part of Section 3.7. Recall that all identities stated there hold for an *arbitrary* system. The crucial idea is therefore to apply the strong coupling framework to the (super)system $S' \equiv SA$, which consists of the actual system of interest as well as all ancillas. Since the system S is in contact with at most one ancilla at any given time, the Hamiltonian of mean force for S' simplifies to

$$H^*_{SA}(\lambda_t) = \begin{cases} H^*_{SA(k)}(\lambda_t) + \sum_{\ell \neq k} H_{A(\ell)} & \text{for} \quad t_k \leq t < t'_k, \\ H^*_S(\lambda_t) + H_A & \text{for} \quad t'_k \leq t < t_{k+1}. \end{cases} \tag{3.142}$$

Based on this expression, we can generalize the strong coupling non-equilibrium free energy from eqn (3.102) to the supersystem:

$$F^*_{SA}(t) = \mathrm{tr}_{SA}\left\{\rho_{SA}(t)\left[H^*_{SA}(\lambda_t) + k_B T \ln \rho_{SA}(t)\right]\right\}. \tag{3.143}$$

Likewise, we can also extend the definitions (3.100) and (3.101) for the internal energy and system entropy to the supersystem. The first law can then be expressed as

$$\Delta U^*_{SA}(t) = W_S(t) + W^{\mathrm{ctrl}}(t) + Q^*(t). \tag{3.144}$$

Furthermore, the second law also holds. This can be expressed in general as

$$\Sigma^*_{SA}(t) = \Delta S^*_S(t) - \frac{Q^*(t)}{T} = \frac{W(t) - \Delta F^*_{SA}(t)}{T} \geq 0. \tag{3.145}$$

Similar to our findings in Section 3.7, its positivity follows from confirming that

$$\Sigma^*_{SA}(t) = D[\rho_{SAB}(t)|\pi_{SAB}(\lambda_t)] - D[\rho_{SA}(t)|\pi^*_{SA}(\lambda_t)]. \tag{3.146}$$

Exercise 3.29 Confirm eqn (3.146).

The second law above is useful if we regard the system and all ancillas as one big supersystem. However, to compute the supersystem's entropy we must be able to keep track of all correlations between the system and all ancillas. In practice, this is often cumbersome. What is typically accessible is the initial and the final state of each *single* ancilla; see the experimental set-up of the micromaser again. The goal in the following is to make the role of a single ancilla more transparent in the description. For this purpose, we focus from now on on a time t such that $t'_n \leq t < t_{n+1}$, i.e. the system has already interacted with $n+1$ ancillas (recall that the first ancilla is labelled by $n=0$), but it is not yet in contact with the next ancilla $A(n+1)$.

The initial state (3.140) implies $S_{\text{vN}}[\rho_{SA}(0)] = S_{\text{vN}}[\rho_S(0)] + \sum_k S_{\text{vN}}[\rho_{A(k)}(0)]$. At time t, the von Neumann entropy of the supersystem becomes

$$S_{\text{vN}}[\rho_{SA}(t)] = S_{\text{vN}}[\rho_S(t)] + \sum_k S_{\text{vN}}[\rho_{A(k)}(t)] - I_{\text{tot}}[\rho_{SA}(t)], \tag{3.147}$$

where $I_{\text{tot}}[\rho_{SA}(t)] \geq 0$ is the total information. From eqn (3.142) we deduce that the change in the non-equilibrium free energy (3.143) becomes

$$\Delta F^*_{SA}(t) = \Delta F^*_S(t) + \sum_{k=0}^{n} \Delta F_{A(k)}(t) + I_{\text{tot}}[\rho_{SA}(t)]. \tag{3.148}$$

Here, $\Delta F^*_S(t)$ denotes the change in strong coupling non-equilibrium free energy (3.102) of the system alone, whereas $\Delta F_{A(k)}(t)$ denotes the change in the *weak coupling* non-equilibrium free energy of ancilla $A(k)$. It appears here because, by construction, the ancillas are decoupled from the system (and, hence, also from the bath) at the time t that we consider. Equation (3.148) allows us to reformulate the second law (3.145) as

$$\boxed{\Sigma^*_S(t) \equiv \frac{W(t) - \Delta F^*_S(t) - \sum_{k=0}^{n} \Delta F_{A(k)}(t)}{T} \geq 0} \tag{3.149}$$

and its positivity follows by noting $\Sigma^*_S(t) = \Sigma^*_{SA}(t) + I_{\text{tot}}[\rho_{SA}(t)] \geq 0$. In contrast to Σ^*_{SA}, Σ^*_S quantifies the entropy production from a *system-intrinsic* point of view: it *neglects* experimentally inaccessible correlations in the system–ancilla state, which yields a larger but experimentally easier accessible entropy production $\Sigma^*_S(t) \geq \Sigma^*_{SA}(t)$. In the context of the micromaser, eqn (3.149) can be determined by quantum state tomography of the final atoms and by estimating the state of the cavity. The latter can be done by using the master equation (3.136) together with a microscopic model for the atom–cavity interaction such as the Jaynes–Cummings Hamiltonian from eqn (3.9).

Coming back to the general interpretation of the repeated interactions framework, we can infer from eqn (3.149) that our new kind of reservoir acts as a *resource of non-equilibrium free energy*, allowing us to extract work indefinitely as long as the ancillas provide enough free energy. To better focus on the new thermodynamic features of this framework, we switch to the weak coupling description in the remainder.

Thermodynamic description at weak coupling

The weak coupling case follows from the previous treatment by dropping the superscript '$*$' from all quantities. Then, an alternative version of eqn (3.149) reads

$$\Sigma_S(t) = \Delta S_S(t) + \sum_k \Delta S_{A(k)}(t) - \frac{Q(t)}{T} \geq 0. \tag{3.150}$$

This form is instructive to compare the repeated interactions framework with how we treated a heat bath in Section 3.7. To make a comparison meaningful, assume that the ancillas are prepared in a thermal state at temperature T: $\rho_{A(k)}(0) = \pi_{A(k)}$. From eqn (3.74) we infer that $\Delta S_{A(k)}(t) = \Delta U_{A(k)}(t)/T - k_B D[\rho_{A(k)}(t)|\pi_{A(k)}]$ and, hence,

$$\Sigma_S(t) = \Delta S_S(t) + \frac{\sum_k \Delta U_{A(k)}(t) - Q(t)}{T} - \sum_k k_B D[\rho_{A(k)}(t)|\pi_{A(k)}] \geq 0. \tag{3.151}$$

Furthermore, in Section 3.7 we identified the change in energy of the environment, which in the present scenario includes the bath B *and* all ancillas, as minus the heat. Thus, let us label for the moment $Q_{\text{tot}}(t) \equiv Q(t) - \sum_k \Delta U_{A(k)}(t)$. Since the relative entropy is positive, we find that

$$\Delta S_S(t) - \frac{Q_{\text{tot}}(t)}{T} \geq \Sigma_S(t) \geq 0. \tag{3.152}$$

Thus, the identification of heat in Section 3.7 yields an inequality that *overestimates* the entropy production $\Sigma_S(t)$ identified in this section. This resonates well with the discussion we had already in Section 3.8. The identification of heat (divided by T) as the *only* contribution to the change in bath entropy requires that the state of the bath can be well approximated by an equilibrium state with constant temperature, which is typically only the case for a large bath. Moreover, we showed in eqn (3.132) that the final non-equilibrium distribution of the bath decreases the amount of entropy produced, similar to eqn (3.152). Since the ancillas were assumed to be small microscopic systems, we have a much higher degree of control and knowledge about them than for a conventional heat bath. This is reflected in eqn (3.150) by using the (fine-grained) von Neumann entropy for the ancillas.

Finally, let us consider the steady-state regime assuming that the system–ancilla interaction time $t'_k - t_k \equiv \tau'$ and the total time interval $t_{k+1} - t_k \equiv \tau > \tau'$ are independent of k, and also assuming that $H_S(\lambda_t) = H_S(\lambda_{t+\tau})$ is periodic. Then, after many system–ancilla interaction cycles, say, after time $t_{n-1} = (n-1)\tau$, the system will have reached a steady state characterized by $U_S(t_{n-1}) = U_S(t_n)$ and $S_S(t_{n-1}) = S_S(t_n)$. The first and second laws of the repeated interactions framework then simplify to

$$\boxed{0 = W^{(n)} + Q^{(n)}(t) + \Delta U^{(n)}_{A(n)}, \quad \Sigma^{(n)}_S = \frac{W^{(n)} - \Delta F^{(n)}_{A(n)}}{T} \geq 0.} \tag{3.153}$$

Here, $W^{(n)} = W(t_n) - W(t_{n-1})$ and $Q^{(n)} = Q(t_n) - Q(t_{n-1})$ are the work applied and the heat flow during the nth interval. Furthermore, $\Delta U^{(n)}_{A(n)}$ and $\Delta F^{(n)}_{A(n)}$ are the change in internal energy and non-equilibrium free energy of ancilla $A(n)$ during the nth interval. Equation (3.153) allows for *repeated* work extraction bounded by $-W^{(n)} \leq -\Delta F^{(n)}_{A(n)}$. This should be contrasted with the conventional second law obtained without ancillas (see eqn (3.75)), where work extraction from a non-equilibrium system state is possible only *once*. This justifies our claim above that the reservoir in the repeated interactions framework is a resource of non-equilibrium free energy.

In particular, the change in non-equilibrium free energy $\Delta F_A = \Delta U_A - T\Delta S_A$ of the ancillas can be dominated by the entropic contribution, i.e. $|T\Delta S_A| \gg |\Delta U_A|$, or by the energetic contribution, $|\Delta U_A| \gg |T\Delta S_A|$. In the former case, one talks about an *information reservoir* or *ideal memory*, to which we return in Chapter 5. In the latter case, one calls the ancillas a *work reservoir*. The final (slightly longer) exercise examines the thermodynamics of the micromaser and shows that it contains aspects of both an information and a work reservoir. To conclude, the repeated interaction picture provides a clean framework able to intuitively explain the thermodynamics of modern experiments, which make use of resources beyond the traditionally considered heat baths and work reservoirs.

Exercise 3.30 First, we need to model the cavity dynamics under the influence of the atoms. In general, this is non-trivial, but for the micromaser it turns out that the atom–cavity interaction time τ' is very short compared with the cavity lifetime τ_c. Compared with the time scale of the dissipative dynamics (3.136), the atom–cavity interaction represents a sudden and short perturbation similar to the delta kick studied around eqn (3.77). Thus, we treat the atom–cavity dynamics separately from the master equation and combine them later.

The cavity dynamics resulting from the atom–cavity interaction can be modelled by the CPTP map $\rho_S^+ = \mathrm{tr}_A\{U_{\mathrm{JC}}\rho_S^- \otimes |e\rangle\langle e|_A U_{\mathrm{JC}}^\dagger\}$, where $\rho_S^\pm$ denotes the cavity state before or after the interaction, respectively, $|e\rangle$ is the initially excited state of the atom and $U_{\mathrm{JC}} = e^{-iH_{\mathrm{JC}}\tau'/\hbar}$ is the unitary time evolution operator resulting from the Jaynes–Cummings Hamiltonian (3.9). Assume the atom and cavity to be at exact resonance, write $\rho_S^- = \sum_{m,n}\rho_{m,n}^-|m\rangle\langle n|$ and use the interaction picture time evolution operator (3.10) to derive that

$$\begin{aligned}\rho_S^+ = \sum_{m,n} e^{-iH_S\tau'/\hbar}\, &\{\sin\left(gt\sqrt{m+1}\right)\sin\left(gt\sqrt{n+1}\right)|m+1\rangle\langle n+1| \\ &+ \cos\left(gt\sqrt{m+1}\right)\cos\left(gt\sqrt{n+1}\right)|m\rangle\langle n|\}\, e^{iH_S\tau'/\hbar}\rho_{m,n}^-.\end{aligned} \tag{3.154}$$

For simplicity, we focus from now on on the case where the initial cavity state $\rho_S(0)$ contains no coherence in the Fock basis, i.e. $\rho_{m,n}(0) = \delta_{m,n}P_n(0)$. Recall Exercise 3.12, which shows that the master equation (3.136) cannot generate any coherence in the Fock basis, and show that the CPTP map (3.154) also does not generate any coherence. Thus, the state of the cavity is completely described by the probabilities $P_n(t)$ of finding n photons at time t, which simplifies the algebra. In particular, show that eqn (3.136) reduces to the rate master equation $\partial_t P_m(t) = \sum_n R_{m,n}P_n(t)$ with rate matrix

$$R_{m,n} = \frac{1+N_{\mathrm{th}}}{\tau_c}[(m+1)\delta_{m+1,n} - m\delta_{m,n}] + \frac{N_{\mathrm{th}}}{\tau_c}[m\delta_{m-1,n} - (m+1)\delta_{m,n}], \tag{3.155}$$

and eqn (3.154) can be written as $P_m^+ = \sum_n T_{m,n}P_n^-$ with transition matrix

$$T_{m,n} = \delta_{m,n}\cos^2\left(g\tau'\sqrt{n+1}\right) + \delta_{m,n+1}\sin^2\left(g\tau'\sqrt{n+1}\right). \tag{3.156}$$

Therefore, if we assume a constant waiting time $\tau \gg \tau'$ between two subsequent atoms, the probability vector $\boldsymbol{P}$ evolves according to

$$\boldsymbol{P}(k\tau) = \underbrace{e^{R\tau}T\dots e^{R\tau}T}_{k\text{ times}}\boldsymbol{P}(0). \tag{3.157}$$

Note that the dynamics starts with an atom–cavity interaction and we neglect the very small interaction time τ' in the notation. Equation (3.157) describes a stroboscopic time evolution where the master equation dynamics is interrupted by sudden interactions with the atoms.

Finally, to investigate the thermodynamics of the micromaser, we also need access to the state of the atoms after the interaction. Use that the cavity state contains no coherences in the Fock basis to show that this state reads

$$\begin{aligned}\rho_A^+ &= \mathrm{tr}_S\{U_{\mathrm{JC}}\rho_S^- \otimes |e\rangle\langle e|_A U_{\mathrm{JC}}^\dagger\} \\ &= \sum_n \left[\cos^2\left(g\tau'\sqrt{n+1}\right)|e\rangle\langle e| + \sin^2\left(g\tau'\sqrt{n+1}\right)|g\rangle\langle g|\right] P_n^-. \end{aligned} \tag{3.158}$$

In the last part of this exercise you are asked to numerically study the dynamics and thermodynamics of the micromaser for the following realistic experimental values. The cavity is characterized by a frequency of $\omega_c/2\pi = 51.1$ GHz (a high-frequency microwave) and the environment has a temperature of $T = 0.8$ K. The mean occupation number of the cavity is therefore at equilibrium $N_{\mathrm{th}} \approx 0.05$. The cavity has a lifetime of $\tau_c = 65$ ms and the atom–cavity interaction time is taken to be $\tau' = 9.55$ μs, which implies $\tau' \ll \tau_c$ and ensures the validity of the stroboscopic dynamics (3.157). The atom–cavity interaction strength in eqn (3.9), which equals half the *vacuum Rabi frequency* at exact resonance, equals $g/\pi = 47.9$ kHz. The waiting time between two atoms is taken to be $\tau = 16.4$ ms. As the initial state, choose the thermal state: $P_n(0) = \pi_n$.

For these parameters a few selected results are shown in Fig. 3.9. Figure. 3.9a reveals that after a few interactions the cavity reaches a steady state with a mean photon number of approximately $\langle n\rangle \approx 2.3$, which is much larger than N_{th}. Figure 3.9b shows the energetic changes in atom i due to the interaction with the cavity. Since the atoms are prepared in the excited state, we always have $\Delta U_{A(i)} \leq 0$. At steady state, roughly half of the initially excited atoms emit a photon in the cavity. The other half remains in the excited state. The state of the outgoing atoms is therefore well approximated by $\rho_A' \approx I_A/2$ and corresponds to the maximum von Neumann entropy (Fig. 3.9c). Note that this state is still far away from its corresponding thermal state at temperature T, where one finds the atom in the excited state with probability 0.002. Comparing the energetic and entropic balance of the atom, we find $|\Delta U_{A(i)}|/|T\Delta S_{A(i)}| \approx 2.3$ at steady state. Thus, the thermodynamics of the micromaser is more strongly influenced by the energetics, but the atomic reservoir is still far away from the ideal limit of a work reservoir.

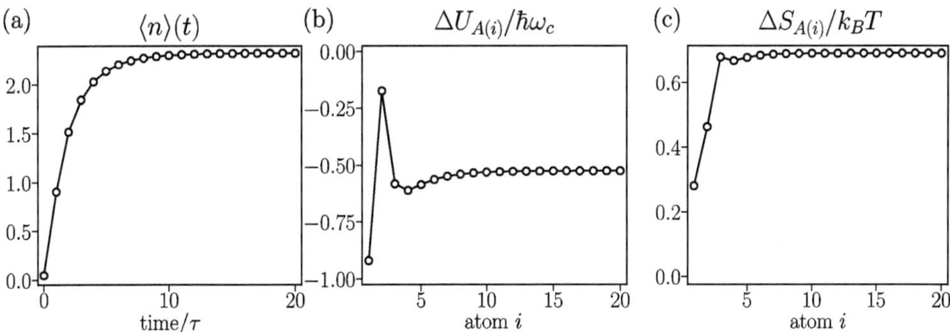

Fig. 3.9 (a) Plot of the average photon number $\langle n\rangle(t) = \sum_n nP_n(t)$ in the cavity as a function of time in units of τ. (b) Plot of the dimensionless energetic change $\Delta U_{A(i)}/\hbar\omega_c$ as a function of atom i. (c) Plot of the change in von Neumann entropy $\Delta S_{A(i)}/k_BT$ as a function of atom i.

3.10 Particle Transport and Thermoelectric Devices

Open quantum systems not only exchange energy with their environment but also *matter* or *particles*. To describe the exchange of particles, one uses the concept of **chemical potentials**, which determine the change in internal energy per added particle of that species. Different particle species with different associated chemical potentials are relevant for chemical reactions, but here we focus only on a single species, namely *electrons*. Exchange (or transport) of electrons is a ubiquituous phenomenon in nature and they are clearly the most relevant particle species (with non-zero chemical potential) for many experiments in quantum nanoscience. This final section of the chapter gives a concise introduction to some of the many interesting applications that open up when one considers transport of electrons. Luckily, the general formalism developed so far only needs slight modifications to account for that.

General thermodynamic considerations

We start with equilibrium considerations. If a system with Hamiltonian H and particle number operator $\hat{N}$ is in contact with an environment at inverse temperature β and chemical potential μ, and if the system can exchange energy and particles with it, its equilibrium state is given by the **grand canonical ensemble**:

$$\Xi(\beta,\mu) = \frac{e^{-\beta(H-\mu\hat{N})}}{\mathcal{Z}(\beta,\mu)}. \tag{3.159}$$

The (grand) partition function is defined as $\mathcal{Z}(\beta,\mu) = \mathrm{tr}\{e^{-\beta(H-\mu\hat{N})}\}$, and we remark that we neglected any strong coupling corrections here. Furthermore, we use a 'hat' on the particle number operator $\hat{N}$ to distinguish it from its expectation value $N \equiv \mathrm{tr}\{\hat{N}\rho\}$. Next, consider that the grand canonical state is infinitesimally perturbed away from its state $\Xi(\beta,\mu)$ to another state ρ such that $\Delta \equiv \rho - \Xi(\beta,\mu) = \mathcal{O}(\epsilon)$ with ϵ a small expansion parameter. Since we are close to equilibrium, we equate the von Neumann entropy (times k_B) with thermodynamic entropy. Its change to first order in ϵ reveals the following relation, which we use later on:

$$dS = -k_B\mathrm{tr}\{\Delta \ln \Xi(\beta,\mu)\} = \frac{1}{T}\mathrm{tr}\{H\Delta\} - \frac{\mu}{T}\mathrm{tr}\{\hat{N}\Delta\} = \frac{1}{T}dU - \frac{\mu}{T}dN. \tag{3.160}$$

To set up a general thermodynamic framework that is valid out of equilibrium, we consider a system coupled to multiple baths as at the end of Section 3.4 or at the beginning of Section 3.7. The Hamiltonian is $H_{SB}(\lambda_t) = H_S(\lambda_t) + \sum_\nu [V_{SB}^{(\nu)}(\lambda_t) + H_B^{(\nu)}]$, but now we also allow for the exchange of particles and associate particle number operators $\hat{N}_S$ and $\hat{N}_\nu$ to the system and bath ν. The initial state is assumed to be

$$\rho_{SB}(0) = \rho_S(0) \bigotimes_\nu \Xi_\nu(\beta_\nu,\mu_\nu), \tag{3.161}$$

which generalizes our previous choice (3.61) to include particle exchanges. Thus, eqn (3.161) describes a system in an arbitrary initial state $\rho_S(0)$ decorrelated from multiple baths ν, which are described by grand canonical ensembles (3.159) at inverse temperatures β_ν and chemical potenials μ_ν.

We start with the first law and keep the definition of the system internal energy $U_S(t)$ as in eqn (3.95). Furthermore, the power is defined as usual as

$$\dot{W}_{\text{mech}} = \text{tr}_{SB}\left\{\rho_{SB}(t)\frac{\partial}{\partial t}H_{SB}(\lambda_t)\right\}, \tag{3.162}$$

but we added a subscript 'mech' to emphasize the *mechanical* origin of this term, coming from a time dependence in the Hamiltonian. The change in internal energy can then be expressed as

$$\frac{d}{dt}U_S(t) = \dot{W}_{\text{mech}}(t) - \sum_\nu \frac{d}{dt}U_\nu(t), \tag{3.163}$$

where $U_\nu(t) \equiv \text{tr}_\nu\{H_\nu \rho_\nu(t)\}$ denotes the energy of bath ν. Based on our previous experience, it is tempting to equate $d_t U_\nu(t)$ with minus the heat flow into the system, but remember that the defining property of heat is that it is linked to changes in entropy. We now return to eqn (3.160), which suggests that the heat flow *from* bath ν *into* the system should be defined as

$$\boxed{\dot{Q}_\nu(t) \equiv -\left(\frac{d}{dt}U_\nu(t) - \mu_\nu \frac{d}{dt}N_\nu(t)\right)} \tag{3.164}$$

such that $đQ_\nu = TdS_\nu$. Accepting this definition of heat, the energy balance (3.163) turns into

$$\boxed{\frac{d}{dt}U_S(t) = \dot{W}_{\text{mech}}(t) + \sum_\nu \dot{Q}_\nu(t) + \dot{W}_{\text{chem}}(t),} \tag{3.165}$$

where we defined the rate of **chemical work**

$$\boxed{\dot{W}_{\text{chem}}(t) \equiv -\sum_\nu \mu_\nu \frac{d}{dt}N_\nu(t).} \tag{3.166}$$

Chemical work is a novel form of work. It is associated with the possibility that electrons can charge a condensator or a battery and the potential energy stored therein can be used, for example, to perform useful mechanical work. Explicit examples are treated below, but for the moment we only remark that, similar to W_{mech}, also W_{chem} is defined to be positive if we perform work on the system.

Now, if our new definitions introduced above are correct, then we should be able to derive the second law in the conventional form:

$$\boxed{\Sigma(t) = \Delta S_S(t) - \sum_\nu \frac{Q_\nu(t)}{T_\nu} \geq 0.} \tag{3.167}$$

Here, the system entropy $S_S(t) = k_B S_{\text{vN}}[\rho_S(t)]$ is still the same as in Section 3.7, but the heat flux $Q_\nu(t)$ is now defined by eqn (3.164). Of course, eqn (3.167) equals the entropy production known from phenomenological non-equilibrium thermodynamics only if the baths are large enough that they remain close to their initial equilibrium state; compare with the discussion in Section 3.8. This also becomes transparent by recalling the way we derived eqn (3.160). The positivity of eqn (3.167) follows by noting an identity closely related to eqn (3.98):

$$\Sigma(t) = D\left[\rho_{SB}(t)\middle|\rho_S(t)\bigotimes_\nu \Xi_\nu(\beta_\nu,\mu_\nu)\right]. \tag{3.168}$$

Exercise 3.31 Derive eqn (3.168).

If we can establish the laws of thermodynamics similar to Section 3.7 without assuming weak coupling or Markovianity, then it is reasonable to expect that we find results similar to Section 3.4 in the case where we can apply the BMS approximation. In fact, we could repeat here the derivation of the BMS master equation as in Section 3.2; we only need to replace the canonical by the grand canonical ensemble when doing the Born approximation. However, establishing the laws of thermodynamics in full generality as in Section 3.4 is more involved now because the bath correlation functions no longer obey the simple Kubo–Martin–Schwinger condition (3.42). Therefore, we prefer to focus on examples here. This is worthwhile since we have gathered so far little experience with the master equation framework *in practice.* Moreover, these models have direct experimental relevance and they allow us to study the effect of *thermoelectricity.* This effect is only made possible by the transport of particles and it allows us to extract work without any time-dependent driving. Therefore, all Hamiltonians are time independent in the following. Hence, $W_{\text{mech}} = 0$ in the remainder.

General theoretical modelling

The theoretical description of electron transport is often based on a system–bath Hamiltonian H_{SB}, which is known as a *tunnel Hamiltonian.* For a coupling to a single bath, the tunnel Hamiltonian reads

$$H_{SB} = H_S + \sum_k \epsilon_k c_k^\dagger c_k + \hbar \sum_{s,k} (t_{s,k} d_s^\dagger c_k + t_{s,k}^* c_k^\dagger d_s), \tag{3.169}$$

which deserves some comments. First, the bath Hamiltonian $\sum_k \epsilon_k c_k^\dagger c_k$ models a sum of non-interacting electrons. The index k summarizes the information necessary to specify the electron mode (energy ϵ_k, wave vector, spin, etc.). Since electrons are fermions, the creation and annihilation operators $c_k^\dagger$ and c_k obey the *anti-commutation* relations $\{c_k, c_\ell^\dagger\} = \delta_{k,\ell}$ and $\{c_k, c_\ell\} = \{c_k^\dagger, c_\ell^\dagger\} = 0$. Furthermore, we left the explicit form of the system Hamiltonian unspecified in eqn (3.169) (examples follow below), but we assumed that the $d_s^\dagger$ and d_s are fermionic creation and annihilation operators associated with the system. Clearly, they must satisfy the same anti-commutation relations: $\{d_s, d_{s'}^\dagger\} = \delta_{s,s'}$, $\{d_s, d_{s'}\} = \{d_s^\dagger, d_{s'}^\dagger\} = 0$ and also $\{c_k^{(\dagger)}, d_s^{(\dagger)}\} = 0$. The coupling between the system and the bath is described by complex coefficients $t_{s,k}$, which describe the process of an electron k leaving the bath and jumping into the system, thereby creating a system electron s (the opposite process is described by $t^*_{s,k}$). In the context of electron transport, this process is typically referred to as tunnelling. Furthermore, the baths are sometimes called *leads* or *terminals* in the literature, but we stick to our previous terminology.

One might wonder why the bath is modelled as a set of *non-interacting* electrons since electrons are always interacting via Coulomb repulsion. However, the bath in this context is often metallic in nature. For low excitation energies close to the Fermi energy (which is another word for the chemical potential at zero temperature), the bath can be well described as a set of non-interacting (quasi)electrons.

Furthermore, for the tunnel Hamiltonian (3.169) the particle number operator for the bath reads $\hat{N}_B = \sum_k c_k^\dagger c_k$. By adding a subscript ν to all quantities related to the bath, it is also clear how to extend (3.169) to multiple baths.

Finally, since eqn (3.169) describes a quantum many-body Hamiltonian in 'second quantization', it is not defined on a tensor product structure $\mathcal{H}_S \otimes \mathcal{H}_B$, but on its anti-symmetrized part (for a fixed particle number). One might therefore wonder whether the tools developed so far, which relied on a tensor product structure as in eqn (3.161), can be used to deal with eqn (3.169). As the next exercise shows, the answer to this question is yes!

Exercise 3.32 The **Jordan–Wigner transformation** maps a set of N fermions with annihilation operators f_i to a set of Pauli matrices $\tilde{f}_i$ acting on different spins with a tensor product structure. Show that the operators

$$\tilde{f}_i = \underbrace{\sigma_z \otimes \cdots \otimes \sigma_z}_{i-1} \otimes \sigma_- \otimes \underbrace{I \otimes \cdots \otimes I}_{N-i} \tag{3.170}$$

satisfy the anti-commutation relations $\{\tilde{f}_i, \tilde{f}_j^\dagger\} = \delta_{i,j}$ and $\{\tilde{f}_i, \tilde{f}_j\} = \{\tilde{f}_i^\dagger, \tilde{f}_j^\dagger\} = 0$. Use the identity $\sum_{k=0}^N \binom{N}{k} = 2^N$ to show that the Hilbert space dimension of N spins equals the Fock space dimension of N fermions. Finally, confirm that the system–bath Hamiltonian (3.169) has the desired tensor product structure after Jordan–Wigner transformation since

$$H_{SB} = H_S \otimes I_B + I_S \otimes \sum_k \epsilon_k \tilde{c}_k^\dagger \tilde{c}_k - \hbar \sum_{s,k} (t_{s,k} \tilde{d}_s^\dagger \otimes \tilde{c}_k + t^*_{s,k} \tilde{d}_s \otimes \tilde{c}_k^\dagger). \tag{3.171}$$

Here, $\tilde{d}_s$ and $\tilde{c}_k$ act only on $\mathcal{H}_S$ or $\mathcal{H}_B$, respectively.

Example 1: The single-electron transistor

We start with the simplest non-equilibrium transport set-up one can imagine within the current framework. It is specified by the system–bath Hamiltonian

$$H_{SB} = \epsilon_0 d^\dagger d + \sum_{\nu \in \{L,R\}} \sum_k \left[\epsilon_{\nu,k} c_{\nu,k}^\dagger c_{\nu,k} + \hbar (t_{\nu,k} d^\dagger c_{\nu,k} + t^*_{\nu,k} c_{\nu,k}^\dagger d) \right]. \tag{3.172}$$

Here, the system Hamiltonian describes a single non-interacting fermionic level with energy ϵ_0. The system is tunnel-coupled with a Hamiltonian of the form (3.169) to two baths, which we label L and R for 'left' and 'right'. We call the model specified by eqn (3.172) a **single-electron transistor**, which we justify below. Another common name for it is the *single-resonant-level model.*

Can this simple model actually describe any real physical device? Remarkably, it can—at least approximately—describe quite a variety of situations, ranging from impurities in solid-state environments to artifically engineered quantum nanotechnologies. Here, we are mainly interested in the latter application, so we briefly try to motivate eqn (3.172) from that perspective, but leave its detailed justification for the specialized literature. In this context, the system S is also called a **quantum dot**, which is a general terminology for any artificially engineered nanostructure with discrete energy levels (an 'artificial atom'). Such an artificial nanostructure can be created by spatially confining the system in all three dimensions to a small *box.* This is similar to the photon cavity studied in the Section 3.9, but electrons in a box can show a discrete energy spectrum for two reasons. One reason is quantum mechanical in nature: confining a particle to a box potential (even if its shape is not perfect) gives rise to a discrete set of eigenenergies, where the discreteness becomes the more pronounced the smaller the box is. This effect also happens to the electromagnetic modes in a cavity. The second reason is 'classical': physical systems free from any constraints have a natural tendency to be *electrically neutral.* Adding a charged particle (such as an electron) to an electrically neutral system increases its electrostatic energy. The discreteness of energy is here due to the fact that electric charge is *quantized.* This also helps to explain why it can be sufficient to consider only a single-electron level for the system in eqn (3.172): with the addition of each electron to the system, its electrostatic energy grows *quadratically* and, if temperatures are low and external fields weak enough, it can be sufficient to restrict the actual many-body Hamiltonian to a few effective levels only. This is known as **Coulomb blockade** and in the extreme case, where one restricts the system Hamiltonian to a single-electron level, one also speaks of *ultrastrong* Coulomb blockade. With today's nanotechnological possibilities, one can create boxes, where all these effects become important. Finally, one might wonder where are the spin degrees of freedom in eqn (3.172). Indeed, we here restrict the discussion to spinless electrons and exclude the rich physics emerging from magnetic interactions, which goes beyond the scope of this book. Most experimental features for the models studied here are, however, well described by models of spinless electrons. Furthermore, formally including the spin degrees of freedom within a master equation approach does not require any new techniques, but actual computations (owing to the growth of the system Hilbert space dimension) are harder.

The rest of the Hamiltonian (3.172) then takes into account that electrons can tunnel out of the box into the environment, as already described below eqn (3.169). An atomic force microscope image of a set-up, which is well described by the Hamiltonian of a single-electron transistor, is shown in Fig. 3.10. Remarkably, experimental progress over the past 20 years has made it possible to monitor the occupation of a quantum dot in real time and with a resolution at the single-electron level. This is achieved by using a **quantum point contact**. A quantum point contact is an external electric circuit where electrons have to flow through a very narrow confinement. This confinement is so narrow that a slight perturbation of it has an observable consequence for the electron current flowing through it. Now, one places this confinement in proximity to

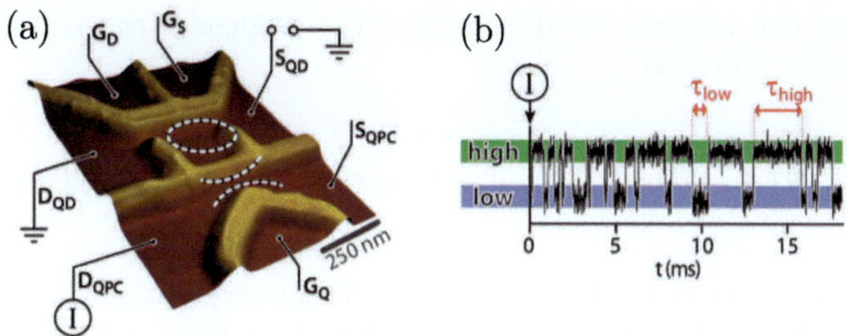

Fig. 3.10 (a) Atomic force microscope image of a quantum dot (dashed white circle) coupled to two electron baths, here labelled as 'source' S_{QD} and 'drain' D_{QD}. The two curved dashed white lines indicate the narrow confinement of the quantum point contact, which is coupled to its own electron source S_{QPC} and drain D_{QPC}. The so-called 'gates' G_{D}, G_{S} and G_{Q} allow the energy landscape such as the tunnel barriers or the energy of the quantum dot to be changed. The electrons in this set-up are confined to the reddish-brown regions (a two-dimensional electron gas). To understand the flow of electrons in this set-up, the reader can imagine filling this three-dimensional landscape with a tiny amount of water with the convention that the water in a source is kept at a higher level than in a drain. (b) Exemplary trace of the current I flowing through the quantum point contact. The difference between a high current (corresponding to an empty quantum dot) and a low current (filled dot) is clearly visible and serves as the signal to detect single electrons in real time. Both figures taken from Flindt *et al.* (2009).

the quantum dot. If the quantum dot is *empty*, i.e. if it does not contain any (excess) electrons and is electrically neutral, then it does not influence the electrons flowing through the confinement. In contrast, if the quantum dot is *filled*, i.e. if it is charged with one excess electron, then the confinement becomes even more narrow owing to Coulomb repulsion and this lowers the current of electrons flowing through it, which can be detected in a laboratory.

After having motivated the model (3.172) experimentally, we now turn to its theoretical description. We assume weak coupling, i.e. the tunnel amplitudes $t_{\nu,k}$ are very small, and we also assume the Markov approximation to be justified. In the experimental set-up discussed above, it is indeed possible to control $t_{\nu,k}$ very well with external gates such that these assumptions are often justified. The claim is now that under these circumstances the master equation in the interaction picture becomes

$$\begin{aligned}\frac{\partial}{\partial t}\tilde{\rho}_S(t) =& \sum_\nu \Gamma_\nu(\epsilon_0)[1-f_\nu(\epsilon_0)]\left[d\tilde{\rho}_S(t)d^\dagger - \frac{1}{2}\{d^\dagger d, \tilde{\rho}_S(t)\}\right] \\ &+ \sum_\nu \Gamma_\nu(\epsilon_0) f_\nu(\epsilon_0)\left[d^\dagger\tilde{\rho}_S(t)d - \frac{1}{2}\{dd^\dagger, \tilde{\rho}_S(t)\}\right],\end{aligned} \tag{3.173}$$

where we introduced the bare tunnelling rates $\Gamma_\nu(\epsilon_0) \equiv 2\pi\hbar\sum_k |t_{\nu,k}|^2\delta(\epsilon_0 - \epsilon_{\nu,k})$ and the Fermi function $f_\nu(\epsilon_0) \equiv (e^{\beta_\nu(\epsilon_0-\mu_\nu)}+1)^{-1}$. Denoting by $|F\rangle$ ($|E\rangle$) the state of the quantum dot when it is filled with one electron (empty with zero electrons) and introducing the corresponding probabilities $p_F(t) = \langle F|\rho_S(t)|F\rangle$ and $p_E(t) =$

$\langle E|\rho_S(t)|E\rangle$, we can capture the dynamics of the single-electron transistor with the simple rate master equation

$$\frac{d}{dt}\begin{pmatrix} p_F(t) \\ p_E(t) \end{pmatrix} = \sum_\nu \Gamma_\nu(\epsilon_0) \begin{pmatrix} -[1-f_\nu(\epsilon_0)] & f_\nu(\epsilon_0) \\ 1-f_\nu(\epsilon_0) & -f_\nu(\epsilon_0) \end{pmatrix} \begin{pmatrix} p_F(t) \\ p_E(t) \end{pmatrix}. \tag{3.174}$$

Exercise 3.33 Convince yourself of the validity of the master equation for the single-electron transistor. If it does not look intuitive, derive it explicitly. You can do so by using the results from Section 3.2 and Exerice 3.32 or, which is probably a less cumbersome way, directly work with the fermionic Hamiltonian (3.172) and perform the weak coupling and Markov approximation. The latter road also reveals that the rate master equation (3.174) is valid *without* performing the secular approximation.

The rate master equation (3.174) describes two processes in the system: either an electron can jump into the system from bath ν, which is proportional to the probability $f_\nu(\epsilon_0)$ of finding an electron at energy ϵ_0 in the bath, or an electron can jump out of the system into bath ν, which is proportional to the probablity $1 - f_\nu(\epsilon_0)$ of having no electron in the bath at energy ϵ_0. The ratio of these two processes is

$$\frac{f_\nu(\epsilon_0)}{1-f_\nu(\epsilon_0)} = e^{-\beta_\nu(\epsilon_0-\mu_\nu)}. \tag{3.175}$$

This result should be familiar as it can be interpreted as the condition of **local detailed balance for fermions**. The next exercise explains in detail why.

Exercise 3.34 Consider a rate master equation $d_t p_x = \sum_{x'} R_{x,x'} p_{x'}$ as also studied in Chapter 2. In contrast to Chapter 2, however, we now allow that this rate master equation also describes exchanges of particles and not only of energy. Assume $R_{x,x'}$ was derived from a system in contact with an ideal single heat bath with temperature T and chemical potential μ (multiple heat baths follow by additivity of $R_{x,x'} = \sum_\nu R^{(\nu)}_{x,x'}$). The claim is that

$$\frac{R_{x,x'}}{R_{x'x}} = e^{-\beta[\epsilon_x - \epsilon_{x'} - \mu(n_x - n'_x)]}, \tag{3.176}$$

where ϵ_x and n_x ($\epsilon_{x'}$ and $n_{x'}$) describe the energy and particle number of state x (x'). Note that eqn (3.176) reduces to eqn (3.175) for the single-electron transistor. Show that eqn (3.176) is identical to

$$\frac{R_{x,x'}}{R_{x'x}} = \exp\left[\frac{S_B(E-\epsilon_x, N-n_x) - S_B(E-\epsilon_{x'}, N-n_{x'})}{k_B}\right], \tag{3.177}$$

where E and N are the macroscopic total energy and particle number and $\epsilon_{x/x'}$ and $n_{x/x'}$ can be regarded as small perturbations. This relation is therefore analogous to eqn (2.43), linking the ratio of transition rates with a corresponding change in entropy, which defines the local detailed balance condition. *Hint:* You can make use of the standard thermodynamic relations $T^{-1} = \partial_E S_B(E, N)$ (at constant volume V and particle number N) and $-\mu/T = \partial_N S_B(E, N)$ (at constant V and E).

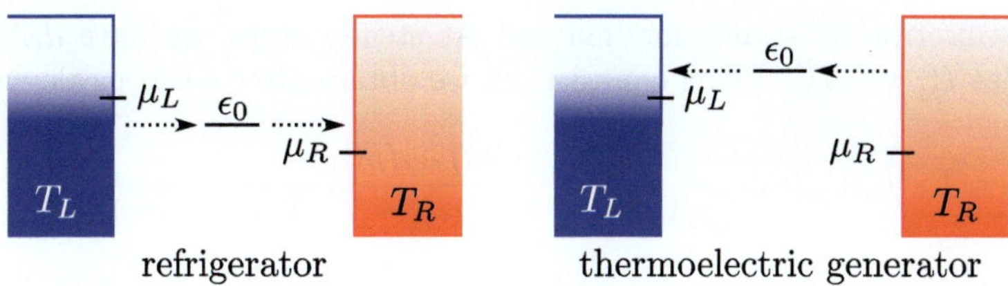

Fig. 3.11 Two sketches of the single-electron transistor corresponding to two different modes of operation. We indicate the relative position of the dot energy ϵ_0 with respect to the chemical potentials μ_L and μ_R in the sketch. Together with the temperatures T_L and T_R, this determines the preferred transport direction (dotted arrows). The shading in the reservoirs sketches the Fermi function $f_\nu(\epsilon)$ as a function of energy ϵ (dark shading implies $f_\nu(\epsilon) \approx 1$; no shading implies $f_\nu(\epsilon) \approx 0$).

We now start to see how the physics and thermodynamics emerge from eqn (3.174); see also the sketch in Fig. 3.11. If $\mu_L = \mu_R \equiv \mu$ but $T_L \neq T_R$, electrons have a tendency to move in the direction of lower (higher) temperature if $\epsilon_0 > \mu$ ($\epsilon_0 < \mu$). In contrast, if $\mu_L \neq \mu_R$ but $T_L = T_R \equiv T$, electrons have a tendency to move in the direction of lower chemical potential. For illustration, consider the simple case of $T = 0$ (strictly speaking, a rate master equation description is no longer adequate for zero temperature). Then, the Fermi functions become Heaviside step functions $f_\nu(\epsilon_0) = \Theta(\mu_\nu - \epsilon_0)$. Transport of electrons can then only happen if the quantum dot energy lies in the *transport window* $\mu_L > \epsilon_0 > \mu_R$. This explains the terminology single-electron transistor. Remember that a transistor is a device whose conductance can be controlled by an external gate. Since the energy ϵ_0 can be experimentally well controlled by an external gate voltage (compare with Fig. 3.10), we can change the transport properties of our single-electron device as in a transistor. Now, in the general case when $\mu_L \neq \mu_R$ *and* $T_L \neq T_R$, the competition of the above-mentioned tendencies can lead to interesting effects and, for definiteness, we assume $\mu_L \geq \mu_R$ and $T_L \leq T_R$ in the following.

For simplicity, we focus on the steady state-regime, which is determined by setting eqn (3.174) to zero and which yields the steady-state probabilities

$$\bar{p}_E = \frac{\sum_\nu \Gamma_\nu(\epsilon_0)[1 - f_\nu(\epsilon_0)]}{\sum_\nu \Gamma_\nu(\epsilon_0)[1 - f_\nu(\epsilon_0)] + \sum_\nu \Gamma_\nu(\epsilon_0) f_\nu(\epsilon_0)}, \quad \bar{p}_F = 1 - \bar{p}_E. \tag{3.178}$$

Unless $T_L = T_R$ and $\mu_L = \mu_R$, in which case the steady state reduces to the grand canonical ensemble from eqn (3.159), this describes a *non-equilibrium* steady state characterized by non-zero currents and a positive entropy production rate. In principle, there are four currents: a particle current from the left and right bath and an energy current from the left and right bath. The particle currents can be identified by looking at the change in particle number of the system:

$$\begin{aligned} &\frac{d}{dt} N_S(t) = \frac{d}{dt} p_F(t) = \sum_\nu I_M^{(\nu)}(t), \\ &\text{with} \quad I_M^{(\nu)}(t) = \Gamma_\nu(\epsilon_0) f_\nu(\epsilon_0) p_E(t) - \Gamma_\nu(\epsilon_0)[1 - f_\nu(\epsilon_0)] p_F(t), \end{aligned} \tag{3.179}$$

where the subscript M stands for 'matter'. At steady state, we have $d_t N_S(t) = 0$, which implies $I_M^L = -I_M^R$. Using eqn (3.178), we obtain after a short calculation

$$I_M^L = \Gamma_L(\epsilon_0)\Gamma_R(\epsilon_0)\frac{f_L(\epsilon_0) - f_R(\epsilon_0)}{\sum_\nu \Gamma_\nu(\epsilon_0)[1 - f_\nu(\epsilon_0)] + \sum_\nu \Gamma_\nu(\epsilon_0) f_\nu(\epsilon_0)}. \tag{3.180}$$

The sign of I_M^L is therefore determined by the sign of $f_L(\epsilon_0) - f_R(\epsilon_0)$.

Energy currents can be identified by looking at the change in internal energy of the system or, equivalently, by using the formulas derived in Section 3.4 and the master equation (3.173). The result is that the energy current from bath ν is *proportional* to the matter current from bath ν:

$$I_E^{(\nu)} = \epsilon_0 I_M^{(\nu)}. \tag{3.181}$$

The reason for this is that each transferred particle transfers the same amount of energy, which is ϵ_0. Property (3.181) is called **tight coupling** and it greatly simplifies the analysis. At steady state, all currents are proportional: $I_E^L = -I_E^R = \epsilon_0 I_M^L = -\epsilon_0 I_M^R$. From eqn (3.164) the heat flows follow as $\dot{Q}_\nu = I_E^{(\nu)} - \mu_\nu I_M^{(\nu)}$ and the entropy production *rate* becomes

$$\dot{\Sigma} = -\sum_\nu \frac{\dot{Q}_\nu}{T_\nu} = \left(\frac{\epsilon_0}{T_R} - \frac{\epsilon_0}{T_L} + \frac{\mu_L}{T_L} - \frac{\mu_R}{T_R}\right) I_M^L \geq 0. \tag{3.182}$$

The non-negativity of the entropy production rate is derived in the next exercise.

Exercise 3.35 For the case of the single-electron transistor, show explicitly that $\dot{\Sigma} \geq 0$ by showing that the sign of I_M^L is determined by the sign of $\frac{\epsilon_0}{T_R} - \frac{\epsilon_0}{T_L} + \frac{\mu_L}{T_L} - \frac{\mu_R}{T_R}$.

In the general case, assume that you have a quantum master equation of the form $\partial_t \rho_S(t) = -i[H_S, \rho_S(t)]/\hbar + \sum_\nu \mathcal{D}_\nu \rho_S(t)$ with $\mathcal{D}_\nu \Xi_S(\beta_\nu, \mu_\nu) = 0$. Can you show that the entropy production rate

$$\dot{\Sigma}(t) = k_B \frac{d}{dt} S_{\text{vN}}[\rho_S(t)] - \sum_\nu \frac{\dot{Q}_\nu(t)}{T_\nu} \geq 0 \tag{3.183}$$

is always non-negative? *Hint:* Recall how we derived eqn (3.65).

If we operate the single-electron transistor in a configuration where $\mu_L > \mu_R$ and $T_L < T_R$, then there are only two modes of operation possible. Either $I_M^L > 0$, in which case our device acts like a **refrigerator** and transports energy against the temperature gradient (Fig. 3.11a), or $I_M^L < 0$, in which case our device acts like a **thermoelectric** device transporting particles against the chemical potential gradient (Fig. 3.11b). In the first case, the chemical power $\dot{W}_{\text{chem}} = (\mu_L - \mu_R) I_M^L$ is positive (work is done on the system to cool the cold reservoir). In the second case, the chemical power is negative (work is extracted by using an energy flow from hot to cold). In principle, there is also a third case in which $I_M^L = 0$, although $\mu_L > \mu_R$ and $T_L < T_R$, which is determined by the condition $f_L(\epsilon_0) = f_R(\epsilon_0)$ as follows from eqn (3.180). This characterizes the turning point where the device changes from being a refrigerator to being a thermoelectric generator. Finally, it is instructive to rewrite the entropy production rate in terms of the chemical power,

$$T_L\dot{\Sigma} = \dot{W}_{\text{chem}} + \eta_C\dot{Q}_R \geq 0, \tag{3.184}$$

where $\eta_C = 1 - T_L/T_R$ denotes the Carnot efficiency. The latter equation reveals that the efficiency of the single-electron transistor when operated as a thermoelectric device with $\dot{W}_{\text{chem}} < 0$ obeys

$$\eta = \frac{-\dot{W}_{\text{chem}}}{\dot{Q}_R} = \eta_C - \frac{T_L\dot{\Sigma}}{\dot{Q}_R} \leq \eta_C. \tag{3.185}$$

Thus, in agreement with phenomenological thermodynamics, our single-electron device can also not operate with an efficiency higher than Carnot.

The single-electron transistor is an exceptionally simple device. A particular problem for its design in the real world is related to the temperature gradient: on such tiny length scales it is hard to keep up a large enough temperature imbalance $T_L < T_R$ without causing other parasitic heat flows, which are typically dominated by phonons (lattice vibrations), through the device. In the final part of this section, we briefly introduce a device that is a little more complicated than the single-electron transistor. As a consequence, however, it also shows more interesting effects and overcomes at least some of the problem just mentioned.

Example 2: Coulomb-coupled quantum dots

The device consists of two parts. First, a single-electron transistor as before with the difference that the baths are kept at the same temperature. Second, another quantum dot, which is coupled to its own bath with a different temperature and which interacts with the quantum dot of the single-electron transistor via Coulomb repulsion. The device is sketched is Fig. 3.12 and the system Hamiltonian reads

$$H_S = \epsilon_S d_S^\dagger d_S + \epsilon_D d_D^\dagger d_D + U d_S^\dagger d_S d_D^\dagger d_D. \tag{3.186}$$

The subscript S labels the quantum dot of the single-electron transistor with on-site energy ϵ_S and the subscript D labels the quantum dot with on-site energy ϵ_D, which we interprete below as a 'detector' in a suitable parameter regime. The Coulomb repulsion is modelled by the parameter U. Note that the Hamiltonian forbids electrons to jump between the quantum dots S and D. The coupling to the three reservoirs is modelled by the standard tunnel Hamiltonian (3.169): for the quantum dot S this equals eqn (3.172), with d replaced by d_S, and for the quantum dot D we have

$$V_{SB}^D + H_B^D = \hbar\sum_k (t_{D,k}d_D^\dagger c_{D,k} + t_{D,k}^* c_{D,k}^\dagger d_s) + \sum_k \epsilon_k c_{D,k}^\dagger c_{D,k}. \tag{3.187}$$

The system Hilbert space can be described with the basis vectors $|\delta\sigma\rangle$, where $\delta \in \{0,1\}$ and $\sigma \in \{E,F\}$ denote the state of the dot D and S, respectively, which can be either empty (0 or E) or filled (1 or F). The system Hamiltonian is diagonal in that basis: $H_S = \epsilon_S|0F\rangle\langle 0F| + \epsilon_D|1E\rangle\langle 1E| + (\epsilon_S + \epsilon_D + U)|1F\rangle\langle 1F|$. Within the weak coupling regime the system dynamics is again well described by a rate master

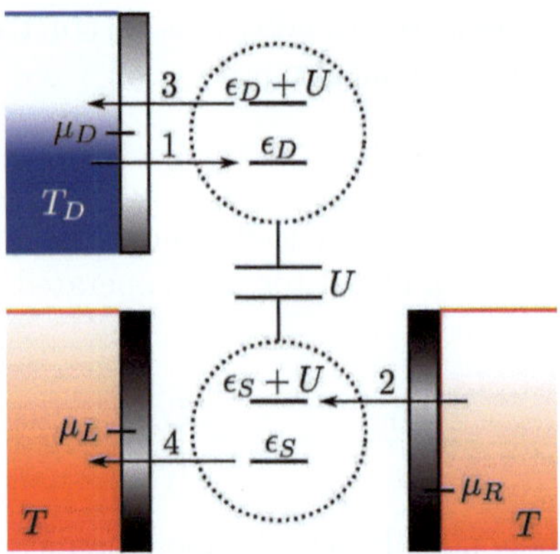

Fig. 3.12 Sketch of a single-electron transistor, which interacts via Coulomb repulsion (sketched as a capacitor with strength U) with a second quantum dot above it, which is in contact with a single bath only. We indicate the temperatures and chemical potentials as well as the two energies $\epsilon_{S/D}$ and $\epsilon_{S/D} + U$ in the dots (dotted circles). As in Fig. 3.11, the shading in the baths indicates the value of the Fermi function. Moreover, we also indicate the strength of the tunnel amplitudes $t_{\nu,k}$ as a function of energy $\epsilon_{\nu,k}$ in the shaded grey boxes. White means high transparency, i.e. large $|t_{\nu,k}|$, and black indicates low transparency, i.e. small $|t_{\nu,k}|$. The numbered arrows sketch an exemplary trajectory, which becomes likely in the limit considered below, where the device acts as a thermoelectric generator. (1) An electron enters dot D (no electron in dot S). (2) Owing to the higher transparency of the right barrier at energy $\epsilon_S + U$, an electron enters dot S from the right bath. The electron in dot D now has energy $\epsilon_D + U$ owing to Coulomb repulsion. (3) The electron in dot D jumps out and the energy of the electron in dot S is lowered. (4) Because of the higher transparency of the left barrier at energy ϵ_S, the electron leaves dot S and moves into the left bath. The net effect is that one electron is transferred from right to left and an energy U is transferred from the single-electron transistor to the cold bath.

equation of the form $d_t \boldsymbol{p} = R\boldsymbol{p}$ with an additive rate matrix $R = \sum_\nu R^{(\nu)}$. Using the ordered basis $\boldsymbol{p} = (p_{0E}, p_{1E}, p_{0F}, p_{1F})^T$, the rate matrices can be written as

$$R^{L/R} = \begin{pmatrix} -\Gamma_{L/R} f_{L/R} & 0 & \Gamma_{L/R}(1 - f_{L/R}) & 0 \\ 0 & -\Gamma^U_{L/R} f^U_{L/R} & 0 & \Gamma^U_{L/R}(1 - f^U_{L/R}) \\ \Gamma_{L/R} f_{L/R} & 0 & -\Gamma_{L/R}(1 - f_{L/R}) & 0 \\ 0 & \Gamma^U_{L/R} f^U_{L/R} & 0 & -\Gamma^U_{L/R}(1 - f^U_{L/R}) \end{pmatrix},$$
$$R^D = \begin{pmatrix} -\Gamma_D f_D & \Gamma_D(1 - f_D) & 0 & 0 \\ \Gamma_D f_D & -\Gamma_D(1 - f_D) & 0 & 0 \\ 0 & 0 & -\Gamma^U_D f^U_D & \Gamma^U_D(1 - f^U_D) \\ 0 & 0 & \Gamma^U_D f^U_D & -\Gamma^U_D(1 - f^U_D) \end{pmatrix}. \tag{3.188}$$

Here, we introduced the following abbreviations. The dynamics of dot S is influenced by the bath $\nu \in \{L, R\}$ with bare tunnelling rates $\Gamma_\nu = 2\pi\hbar \sum_k |t_{\nu,k}|^2 \delta(\epsilon_S - \epsilon_{\nu,k})$ and $\Gamma^U_\nu = 2\pi\hbar \sum_k |t_{\nu,k}|^2 \delta(\epsilon_S + U - \epsilon_{\nu,k})$ and Fermi functions $f_\nu = (e^{\beta_\nu(\epsilon_S - \mu_\nu)} + 1)^{-1}$ and $f^U_\nu = (e^{\beta_\nu(\epsilon_S + U - \mu_\nu)} + 1)^{-1}$. Here, the superscript U emphasizes that the rates and Fermi functions change if dot D is occupied. Furthermore, for the dynamics of dot D we simply need to replace the subscripts ν and S in the previous equations by D. The next exercise makes the master equation above intuitive.

Exercise 3.36 Make a sketch where each state $|\delta\sigma\rangle$ is a point and draw arrows between two states whenever the baths allow a transition between them. Associate the arrows with processes described in the rate matrix (3.188) and check whether local detailed balance is satisfied. If you still have doubts about the validity of eqn (3.188), derive it explicitly.

As for the single-electron transistor, we now focus only on the steady-state thermodynamic properties of the rate master equation. In this case there are three energy currents $I_E^{(\nu)}$ and three matter currents $I_M^{(\nu)}$, $\nu \in \{L, R, D\}$. Luckily, not all of them are independent. Instead, they obey the constraints

$$I_E^L + I_E^R + I_E^D = 0, \quad I_M^L + I_M^R = 0, \quad I_M^D = 0. \tag{3.189}$$

The first equality comes from energy conservation; the second and third are a consequence of matter conservation and the fact that electrons cannot be exchanged between the dots, i.e. there is no flow of electrons from bath D to bath L or R or vice versa. One way to show $I_M^D = 0$ directly is to add the first and third component of the vector $R\boldsymbol{p}$ and to use the steady-state condition.

From now on, we focus on the case $\beta_L = \beta_R \equiv \beta$ and we assume $\beta_D > \beta$ (i.e. $T_D < T$) and $\mu_L > \mu_R$ as sketched in Fig. 3.12. Then, the entropy production rate reads

$$\dot{\Sigma} = -\beta(\dot{Q}_1 + \dot{Q}_2) - \beta_D\dot{Q}_D = (\beta - \beta_D)I_E^D + \beta(\mu_L - \mu_R)I_M^L \geq 0, \tag{3.190}$$

which describes the competition of two terms. On the one hand, it is possible to use a flow of energy into the cold bath, $(\beta - \beta_D)I_E^D > 0$, to transport electrons against the bias, $\beta(\mu_L - \mu_R)I_M^L < 0$. On the other hand, we could also use the flow of electrons with the bias, $\beta(\mu_L - \mu_R)I_M^L > 0$, in order to cool the cold reservoir, $(\beta - \beta_D)I_E^D < 0$. As for the single-electron transistor above, the first mode of operation characterizes a thermoelectric device and the second a refrigerator. For the present device it is also possible that both terms are positive: $(\beta - \beta_D)I_E^D > 0$ *and* $\beta(\mu_L - \mu_R)I_M^L > 0$. In this case, the device is not doing anything useful from a thermodynamic perspective.

While the second law allows for these three possibilities, it does not tell us which of them is actually realized in a given configuration. In general, the exact values of the energy and matter currents depend in a complicated way on all the parameters entering the rate matrix R above. For the present device, it is still possible to obtain explicit expressions in a straightforward way since the rate matrix is only a 4×4 matrix. Nevertheless, these expressions are already quite lengthy. Therefore, we follow another strategy and consider a special case instead. Although this special case involves some unrealistic parameter choices, it offers more physical intuition. It also gives general insights into the working mechanism of these devices.

For definiteness, we focus on the operation as a thermoelectric device. Furthermore, to reduce the amount of parameters we assume that we can tune the chemical potentials as $\mu_L = \epsilon_S + eV/2$ (where e is the electron charge, which we will set to

$e \equiv 1$ in the following), $\mu_R = \epsilon_S - V/2$ and $\mu_D = \epsilon_D + U/2$. From these definitions, it follows that $\mu_L - \mu_R = V$. Consequently, we interpret V as the *voltage bias* across the single-electron transistor.

To obtain a simple description of the dynamics, we assume that the dynamics associated with dot D is much faster than the dynamics associated with dot S. In equations, we assume $\Gamma_D^{(U)} \gg \Gamma_L^{(U)}, \Gamma_R^{(U)}$. This assumption should be familiar: recalling the content of Section 2.5, we see that this is the assumption of time-scale separation, which allows us to coarse-grain the dynamics of dot D in order to obtain an effective master equation for S alone. This works as follows. From our rate master equation we obtain the time evolution of the probability $p_E(t) = p_{0E}(t) + p_{1E}(t)$ to find the dot S empty,

$$\begin{aligned}\frac{d}{dt}p_E(t) = &- \sum_{\nu\in\{L,R\}} \left[\Gamma_\nu f_\nu p_{0E}(t) + \Gamma_\nu^U f_\nu^U p_{1E}(t)\right] \\ &+ \sum_{\nu\in\{L,R\}} \left[\Gamma_\nu(1-f_\nu)p_{0F}(t) + \Gamma_\nu^U(1-f_\nu^U)p_{1F}(t)\right],\end{aligned} \tag{3.191}$$

and the probability of finding it filled follows from $p_F(t) = 1 - p_E(t)$. We now write $p_{\delta\sigma}(t) = p_{\delta|\sigma}(t)p_\sigma(t)$ and use the assumption of time-scale separation to replace the time-dependent conditional probabilities $p_{\delta|\sigma}(t)$ for dot D with their steady-state values $\bar{p}_{\delta|\sigma}$, which read explicitly

$$\bar{p}_{0|0} = 1 - f_D, \quad \bar{p}_{1|0} = f_D, \quad \bar{p}_{0|1} = 1 - f_D^U, \quad \bar{p}_{1|1} = f_D^U. \tag{3.192}$$

For simplicity, let us further consider the limit where the bath of dot D is very cold, i.e. we let $T_D \to 0$. Together with the choice $\mu_D = \epsilon_D + U/2$, this implies $(\bar{p}_{0|0}, \bar{p}_{1|0}, \bar{p}_{0|1}, \bar{p}_{1|1}) = (0,1,1,0)$, i.e. the dot S and dot D are perfectly *anti-correlated*: the dot D is empty whenever dot S is filled and vice versa. This also justifies calling the dot D a *detector*: in the limit considered here, we can infer the state of dot S solely by looking at the state of dot D. In this parameter regime, the so far exact coarse-grained master equation (3.191) reduces to the approximate but much simplified equation

$$\frac{d}{dt}p_E(t) = -\sum_{\nu\in\{L,R\}} \Gamma_\nu^U f_\nu^U p_E(t) + \sum_{\nu\in\{L,R\}} \Gamma_\nu(1-f_\nu)p_F(t). \tag{3.193}$$

This equation is now the starting point for our thermodynamic analysis.

The first striking feature of eqn (3.193) is that local detailed balance is *broken*. The ratio of the rates of jumping in and out of the dot no longer obeys eqn (3.176). This is because we coarse-grained a degree of freedom in a *non-equilibrium* set-up. This broken detailed balance condition is also responsible for the fact that we can have an electron current against the bias: $I_M^L < 0$. To see this, consider the steady-state probabilities $\bar{p}_E$ and $\bar{p}_F$ of eqn (3.193) and the steady-state current

$$I_M^L = \Gamma_L^U f_L^U \bar{p}_E - \Gamma_L(1-f_L)\bar{p}_F = \frac{\Gamma_L^U \Gamma_R f_L^U(1-f_R) - \Gamma_L \Gamma_R^U (1-f_L) f_R^U}{\sum_\nu [\Gamma_\nu^U f_\nu^U + \Gamma_\nu(1-f_\nu)]}, \tag{3.194}$$

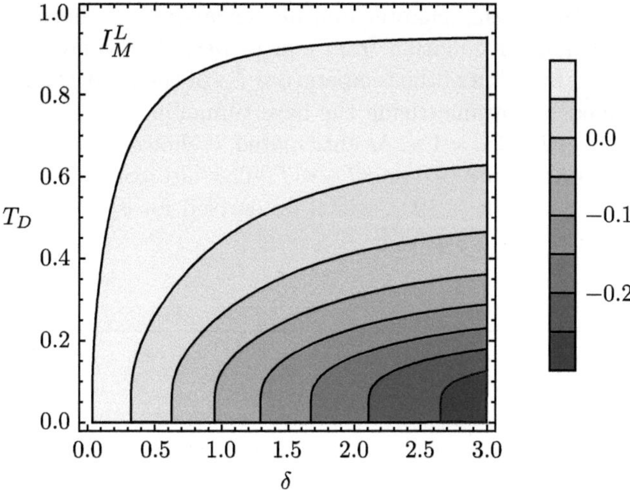

Fig. 3.13 Contour plot of the electric current from bath L for $T_L = T_R = 1$, $V = 1$, $\Gamma_L = \Gamma_R = 1$, $U = 1$ and we also set $k_B = 1$. Furthermore, we chose $\Gamma_D = \Gamma_D^U = 100 \gg \Gamma_{L/R}$ to ensure that time-scale separation applies.

which does not simplify to our previous result (3.180). Owing to the presence of the second dot D, the current is modified and can become negative. This is the case exactly when

$$\Gamma_L^U \Gamma_R f_L^U (1 - f_R) < \Gamma_L \Gamma_R^U (1 - f_L) f_R^U \quad \Leftrightarrow \quad 1 < \frac{\Gamma_L}{\Gamma_L^U} \frac{\Gamma_R^U}{\Gamma_R} \frac{1 - f_L}{f_L^U} \frac{f_R^U}{1 - f_R}. \tag{3.195}$$

One can check that the factor $(1 - f_L) f_R^U / [f_L^U (1 - f_R)]$ is always smaller than one. Hence, in order to have $I_M^L < 0$, this must be compensated by the factor $\Gamma_L \Gamma_R^U / (\Gamma_L^U \Gamma_R)$.

We now investigate this final condition by returning to the microscopic definitions of the rates. We found that, for example,

$$\Gamma_L = 2\pi\hbar \sum_k |t_{L,k}|^2 \delta(\epsilon_S - \epsilon_{L,k}), \quad \Gamma_L^U = 2\pi\hbar \sum_k |t_{L,k}|^2 \delta(\epsilon_S + U - \epsilon_{L,k}) \tag{3.196}$$

and similarly for Γ_R and Γ_R^U. Thus, in order to have $\Gamma_L > \Gamma_L^U$, the microscopic coupling $t_{L,k}$ must be larger around the energy ϵ_S and smaller around the energy $\epsilon_S + U$. Put differently, the coupling of the left bath must act like an *energy filter*, which preferably allows electrons to tunnel at a lower energy. Vice versa, for the right bath we need the opposite case ($\Gamma_R^U > \Gamma_R$): it should preferably allow electrons to tunnel at the higher energy $\epsilon_S + U$ instead of ϵ_S. We have also sketched this in Fig. 3.12. This is an important prerequisite for a thermoelectric generator: in order to transport electrons from right to left ($I_M^L < 0$), the *left–right symmetry needs to be broken.* In the single-electron transistor studied previously, this was accomplished by the temperature bias. Since we have no left–right temperature bias here, we need to engineer the tunnelling rates such that the symmetry is broken.

Figure 3.13 concludes this chapter and our excursion on thermoelectric devices. By solving the full master equation from eqn (3.188), it displays a contour plot of $I_M^L = I_M^L(T_D, \delta)$ as a function of the temperature T_D of the bath D and the *asymmetry parameter* δ, defined by parametrizing the bare tunnelling rates as $\Gamma_L^U \equiv e^{-\delta}\Gamma_L$ and $\Gamma_R^U \equiv e^{\delta}\Gamma_R$ and by setting $\Gamma_L = \Gamma_R$. As anticipated, it shows that transport against the voltage becomes possible for $\delta > 0$ and $T_D < T$. This also confirms that our simplified argumentation based on eqn (3.193), which we derived for $T_D = 0$, continues to hold for finite temperatures of the bath D.

Further reading

3.1. Our definition of work based on a time-dependent Hamiltonian is widely used in non-equilbrium statistical mechanics. In Exercise 3.1 we made use of some elementary results from quantum optics, which is a large field better covered elsewhere (Scully and Zubairy, 1997). Readers interested in quantum batteries can take a look at the introductory article by Campaioli *et al.* (2018).

3.2. The derivation of the BMS equation is standard and we closely followed the notation of Breuer and Petruccione (2002). Readers interested in the accuracy of perturbative master equations will easily find plenty of literature. Some recent studies focusing in particular on the failure of the secular approximation were done by Purkayastha *et al.* (2016) and Hartmann and Strunz (2020). That the secular approximation can violate important symmetries of the problem has been known for a long time (Suárez *et al.*, 1992; Kohen *et al.*, 1997).

3.3. Many of the properties of the BMS equation that we reviewed have been known for a long time and are covered in textbooks on open quantum systems (Breuer and Petruccione, 2002; Rivas and Huelga, 2012). The fact that every CPTP master equation gives rise to a CPTP map, but that the converse is not true, was investigated by Wolf *et al.* (2008*a*). Davies' theorem briefly mentioned at the end of the section is very challenging to derive (Davies 1974, 1976). A mathematically rigorous extension of it to non-vanishing coupling strengths was only recently reported by Merkli (2020) and co-workers (see references therein).

3.4. That the BMS equation can be used to derive a consistent framework of non-equilibrium thermodynamics was realized a long time ago (McAdory and Schieve, 1977; Spohn, 1978; Alicki, 1979; Spohn and Lebowitz, 1979; Lindblad, 1983). Today, the BMS equation has become the workhorse for many studies in quantum thermodynamics. Readers interested in microscopic derivations of weak coupling master equations for time-dependent Hamiltonians can find a wealth of literature and I recommend taking a first look at the work of Yamaguchi *et al.* (2017). Furthermore, albeit looking like an innocent assumption, modelling the effect of multiple baths by simply *adding* the dissipators of each bath to the master equation (as we did) can be problematic even at weak coupling (Mitchison and Plenio, 2018).

3.5. The idealized processes introduced in Section 3.5 play an important role in reconciling quantum thermodynamics with the traditional theory of phenomenological thermodynamics and they are widely used in the field. We quoted von Neumann (1929) (for an English translation see von Neumann (2010)) about the applicability of his entropy concept to problems in statistical mechanics. An optimal thermodynamic protocol to extract the maximum amount of work from a non-equilibrium state was introduced by Jacobs (2009). The definition of ergotropy goes back to Allahverdyan *et al.* (2004). Readers interested in an accessible proof of the fact that the Gibbs state is completely passive (Exercise 3.13) should consult Skrzypczyk *et al.* (2015); for the orginal work see Pusz and Woronowicz (1978) and Lenard (1978).

3.6. Our discussion of the zeroth law of thermodynamics was supported by inequality (3.92), which was derived by Wolf *et al.* (2008*b*). Mathematically rigorous insights into the question of at which length scales it makes sense to talk about a local equilibrium temperature are discussed by Hartmann (2006) and Kliesch *et al.* (2014). Readers interested in area laws are referred to a review article (Eisert *et al.*, 2010). Maxwell's quote at the beginning is taken from his book (Maxwell, 1871).
3.7. Deriving the laws of thermodynamics in a Hamiltonian way has a long history dating back to Boltzmann. Within the nowadays conventionally considered open quantum system paradigm, an early microscopic derivation of the laws of thermodynamics can be found in a (seemingly forgotten) paper by Bassett (1978). The first part of the section builds on a paper by Esposito *et al.* (2010), but identical manipulations for a single heat bath had already been carried out by Lindblad (1983). The second law (3.106) within the Hamiltonian of mean force approach was derived by Strasberg and Esposito (2019).
3.8. We largely followed the exposition of Strasberg and Winter (2021), where the hierarchy of second laws (3.111) was derived and discussed. In a quantum mechanical context, the definition (3.112) of observational entropy appears first in the work of von Neumann (1929), but it has been only recently revived as a general candidate for thermodynamic entropy in isolated quantum many–body systems (Šafránek *et al.*, 2019a; Šafránek *et al.*, 2019b). Our definition of nonequilibrium temperature was first discussed as a phenomenological concept by Muschik (1977), who called it a "nonequilibrium contact temperature." Readers interested in the problem of work extraction under control restrictions will find more information in the article by Wilming *et al.* (2016). For a derivation of eqn (3.135) and a discussion of its consequences see Strasberg *et al.* (2020).
3.9. We motivated the framework of repeated interactions by the example of the micromaser, which is treated in many textbooks on quantum optics (Scully and Zubairy, 1997). The sketch of the experimental set-up in Fig. 3.8 is taken from one of the early experiments in the field (Rempe *et al.*, 1990). In quantum thermodynamics, the framework of repeated interactions gained wider attention with the work of Barra (2015), who used it to provide a consistent thermodynamic description for master equations, which were previously believed not to have such a consistent thermodynamic description. The potential of the repeated interactions framework to provide a consistent thermodynamic description of non-equilibrium resources, which explain a wide variety of phenomena, was realized by Strasberg *et al.* (2017) for weak system–bath coupling. In our exposition we started, however, with the strong coupling case (Strasberg, 2019*b*). Readers interested in a detailed analysis of the micromaser from a thermodynamic perspective can consult the work of Cresser (2019).
3.10. In the final section about electron transport and thermoelectricity in nanostructures, we could only scratch the surface of this fascinating field, which bears much potential for future applications. A thorough recent review covering many different theoretical aspects was written by Benenti *et al.* (2017). For an introductory article about thermoelectricity in quantum dot nanostructures see the work of Sothmann *et al.* (2015). The book by Schaller (2014) focuses in particular on the master equation approach and treats many examples in detail, whereas the book by Nazarov and Blanter (2009) presents the field from a broader perspective. Readers interested in a general derivation of local detailed balance for grand canonical reservoirs can look at the work of Bulnes Cuetara *et al.* (2016). Our first example, the single-electron transistor, has become the simplest workhorse model in the field. Figure 3.10 is taken from the experiment by Flindt *et al.* (2009) and recent experimental results showed that the single-electron transistor can work with high efficiency (Josefsson *et al.*, 2018). Furthermore, it is interesting to remark that our rate master equation (3.174) becomes exact in the *infinite bias regime*, i.e. when $\mu_L - \mu_R \to \infty$, even at low temperatures and with finite system–bath coupling strength (Gurvitz and Prager, 1996). This result indicates that quantum master equations work par-

ticularly well under *strong* non-equilibrium conditions. Our second example, the two Coulomb-coupled quantum dots, was first proposed as a thermoelectric generator by Sánchez and Büttiker (2011). It has been realized in a variety of experiments (Hartmann *et al.*, 2015; Roche *et al.*, 2015; Thierschmann *et al.*, 2015). Our theoretical treatment followed closely the work of Strasberg *et al.* (2013), who used the model as a minimal description of an autonomous Maxwell demon (we treat Maxwell's demon in detail in Section 5.5). A closely related model was also experimentally realized (Koski *et al.*, 2015).

4
Quantum Fluctuation Theorems

Summary. We introduce a theoretical technique known as the two-point measurement scheme, which allows us to derive a variety of quantum fluctuation theorems with the same form as in classical stochastic thermodynamics. Alternatively, we show how to study fluctuations based on a quantum master equation approach combined with counting field or quantum jump trajectory methods. Although quantum fluctuation theorems have the same form as classical fluctuation theorems, they are more limited in their scope and practical applicability. Nevertheless, we study one important theoretical consequence of them, known as the thermodynamic uncertainty relation. This relation is used to discuss the impossibility of Carnot efficiency at finite power. The latter point also naturally connects to the third law of thermodynamics.

4.1 Two-Point Measurement Scheme

From this chapter onwards, we focus on fluctuations of thermodynamic quantities in quantum systems. In principle, measuring the fluctuations of some observable X seems straightforward in quantum mechanics. One performs projective measurements of X on many copies of identically prepared systems. Afterwards, the outcomes allow one to construct the probability distribution $p(x)$, where x is an eigenvalue of X, from which one can compute all moments $\langle X^n \rangle$ for $n \geq 1$ and there is nothing more we can know about it. Unfortunately, this simple picture only holds for single-time measurements, where $\langle X^n \rangle = \text{tr}\{X^n \rho\}$ is evaluated always with the same ρ. Thus, this procedure reveals only the *static* fluctuations associated with the state ρ. In thermodynamics, we are, however, interested in dynamically evolving quantities such as *changes* in energy or entropy. Accessing dynamical fluctuations turns out to be much harder than accessing static fluctuations. Many different methods have been developed and in general they *differ* in their predictions. The reason for their discrepancy is always the same: quantum measurements are *disturbing*. Depending on how the measurement is performed and how the fluctuations are defined, the results can be different.

In this section we introduce the two-point measurement scheme from an abstract perspective; thermodynamic applications of it will be discussed in subsequent sections. The two-point measurement scheme is a theoretically powerful framework with the benefit that the fluctuations defined in this scheme obey the same fluctuation theorems that we know already from the classical context. It has, however, conceptual shortcomings, which we will overcome in Chapter 5.

Throughout this section, we consider an isolated system with Hamiltonian $H(\lambda_t)$. This isolated system can, of course, describe a system–bath composite, but explicit

Quantum Stochastic Thermodynamics. Philipp Strasberg, Oxford University Press.
 DOI: 10.1093/oso/9780192895585.003.0004

connections to the system–bath paradigm will be done only in later sections. Furthermore, let us denote by $X = \sum_x x\Pi(x)$ an arbitrary observable with eigenvalues x and projectors $\Pi(x)$ with rank $V(x) = \text{tr}\{\Pi(x)\} \geq 1$. The projectors are orthogonal $\Pi(x)\Pi(x') = \delta_{x,x'}\Pi(x)$ and obey the completeness relation $\sum_x \Pi(x) = 1$. The observable is also allowed to depend explicitly on time through the driving protocol λ_t: $X = X(\lambda_t)$. In fact, in Section 4.2 we will actually choose $X = H(\lambda_t)$, but for the moment we leave X unspecified. To keep the notation concise, we write x_t instead of $x(\lambda_t)$, such that $X(\lambda_t) = \sum_{x_t} x_t\Pi(x_t)$.

As the name suggests, the two-point measurement scheme asks for the statistics of the observable $X(\lambda_t)$ obtained at *two* times, which we label τ and 0 where $\tau > 0$ is taken to be the later time. Furthermore, the observable is measured projectively. Consequently, the probability of obtaining outcome x_0 at time 0 and x_τ at time τ is

$$p(x_\tau, x_0) = \text{tr}\{\Pi(x_\tau)U(\tau,0)\Pi(x_0)\rho(0)\Pi(x_0)U^\dagger(\tau,0)\}, \tag{4.1}$$

where $U(\tau,0)$ is the unitary time evolution operator and $\rho(0)$ is the initial state. The two-time probability $p(x_\tau, x_0)$ completely characterizes the statistics, which can be obtained from the two projective measurements. If the projectors are of rank 1, i.e. $\Pi(x_t) = |x_t\rangle\langle x_t|$, the two-time probability can be written as

$$p(x_\tau, x_0) = |\langle x_\tau|U(\tau,0)|x_0\rangle|^2\langle x_0|\rho(0)|x_0\rangle. \tag{4.2}$$

Note that by convention the time evolution operator acts to the right in a scalar product, i.e. a cleaner notation would be $\langle x_\tau|[U(\tau,0)|x_0\rangle]$ instead of $\langle x_\tau|U(\tau,0)|x_0\rangle$.

In all cases considered below, it turns out that the initial state is of the form

$$\rho(0) = \sum_{x_0} \mu(x_0)\Pi(x_0), \tag{4.3}$$

where the $\mu(x_0)$ are non-negative numbers, which we call *weights*. These weights do not sum up to one unless all projectors have rank 1. In general, the probability of obtaining outcome x_0 is $p(x_0) = \mu(x_0)V(x_0)$, such that $\sum_{x_0} \mu(x_0)V(x_0) = 1$.

Based on what we said above, we can already derive our first formal fluctuation theorem. Consider an initial state of the form (4.3) and an arbitrary set of weights $\mu(x_\tau)$, such that $\sum_{x_\tau} \mu(x_\tau)V(x_\tau) = 1$. It then follows that

$$\boxed{\left\langle e^{-\ln[\mu(x_0)/\mu(x_\tau)]}\right\rangle_{x_\tau,x_0} \equiv \sum_{x_\tau,x_0} e^{-\ln[\mu(x_0)/\mu(x_\tau)]}p(x_\tau,x_0) = 1,} \tag{4.4}$$

which we call the *abstract integral fluctuation theorem*. The derivation of eqn (4.4) simply relies on cancelling terms and we leave it therefore as an exercise.

Exercise 4.1 Prove eqn (4.4).

The validity of the abstract integral fluctuation theorem solely relies on the choice of initial state, eqn (4.3). The weights $\mu(x_\tau)$ are arbitrary as long as $\sum_{x_\tau} \mu(x_\tau)V(x_\tau)$

$= 1$, the final observable $X(\lambda_\tau)$ is also arbitrary and even the unitary time evolution $U(\tau, 0)$ appearing in eqn (4.1) is arbitrary. In fact, eqn (4.4) remains valid even if we replace the actual time evolution operator $U(\tau, 0)$ by *any other* unitary operator. Put differently, eqn (4.4) is almost tautologous. We made a similar observation in Exercise 2.19. The fact that we can later extract many physical insights from the abstract integral fluctuation theorems hinges crucially on two points. First, we need to choose the observable $X(\lambda_0)$ appropriately such that our assumption (4.3) for the initial state is justified. Often, the initial state turns out to be a Gibbs ensemble. Second, we will see below that the exponent $\ln[\mu(x_0)/\mu(x_\tau)]$ in eqn (4.4) can be linked to interesting physical quantities such as the dissipated work or the entropy production during the process. The first point ensures that eqn (4.4) is valid for a large class of situations and the second point gives eqn (4.4) physical substance.

Before we study applications of eqn (4.4), we derive an abstract detailed fluctuation theorem. In Chapter 2, we saw that the detailed fluctuation theorem is related to the notion of a time-reversed process. This also turns out to be the case for quantum systems. Therefore, we make extensive use of the properties of the time-reversal operator Θ in the following. All of these properties are derived in Appendix C. Note that, in reality, also the 'time-reversed' (or backward) process is implemented in a laboratory in the forward direction of increasing time. We label, however, states or measurement outcomes in the backward process in the reversed order, which simplifies the overall notation. Thus, for instance, the initial state in the time-reversed process is denoted $\rho_{\mathrm{tr}}(\tau)$. At the risk of redundancy, we emphasize again that these are merely labels and have nothing to do with the time an actual clock shows in a laboratory!

Also in the time-reversed process we perform two measurements. Now, however, we start with the measurement of the observable $X_\Theta(\lambda_\tau) = \Theta X(\lambda_\tau)\Theta^{-1}$, which is the time-reversal of $X(\lambda_\tau)$. Importantly, $X_\Theta(\lambda_\tau)$ and $X(\lambda_\tau)$ have the same spectrum, i.e. the same eigenvalues, but perhaps different eigenvectors (Exercise C.5). Hence, we write $X_\Theta(\lambda_\tau) = \sum_{x_\tau} x_\tau \Pi_\Theta(x_\tau)$ and, even though $\Pi_\Theta(x_\tau) \neq \Pi(x_\tau)$ in general, it always holds that $\mathrm{tr}\{\Pi_\Theta(x_\tau)\} = \mathrm{tr}\{\Pi(x_\tau)\} = V(x_\tau)$. Finally, the time-reversed process ends with a measurement of $X_\Theta(\lambda_0) = \Theta X(\lambda_0)\Theta^{-1} = \sum_{x_0} x_0 \Pi_\Theta(x_0)$. In between the first and the second measurement, the system evolves in time according to the time-reversed Hamiltonian $H_\Theta(\lambda_t^\dagger) = \Theta H(\lambda_t^\dagger)\Theta^{-1}$ with a time-reversed driving protocol $\lambda_t^\dagger \equiv \lambda_{\tau-t}$. If $t \in [0, \tau]$ denotes the actual increasing time during the implementation of the time-reversed process, this means that the value of the control protocol at time t is $\lambda_{\tau-t}$. Thus, the time-reversed experiment starts with the Hamiltonian $H_\Theta(\lambda_\tau)$ and ends with $H_\Theta(\lambda_0)$. In analogy with the initial state (4.3) in the forward process, the initial state of the time-reversed process is assumed to be

$$\rho_{\mathrm{tr}}(\tau) = \sum_{x_\tau} \mu(x_\tau)\Pi_\Theta(x_\tau). \tag{4.5}$$

Note that in general $\rho_{\mathrm{tr}}(\tau) \neq \Theta\rho(\tau)\Theta^{-1}$ with $\rho(\tau)$ the final state of the forward process. The arbitrary weights $\mu(x_\tau)$ obey again $\sum_{x_\tau} \mu(x_\tau)V(x_\tau) = 1$ and the probability of first measuring outcome x_τ and then x_0 in the time-reversed process becomes

$$p_{\mathrm{tr}}(x_0, x_\tau) = \mathrm{tr}\{\Pi_\Theta(x_0)U_\Theta(\tau, 0)\Pi_\Theta(x_\tau)\rho_{\mathrm{tr}}(\tau)\Pi_\Theta(x_\tau)U_\Theta^\dagger(\tau, 0)\}. \tag{4.6}$$

Here, the unitary time evolution operator is $U_\Theta(\tau,0) = \exp_+[-i\int_0^\tau ds H_\Theta(\lambda_{\tau-s})/\hbar]$.

It turns out that $p(x_\tau, x_0)$ and $p_{\rm tr}(x_0, x_\tau)$ are connected via the identity

$$p_{\rm tr}(x_0, x_\tau) = \frac{\mu(x_\tau)}{\mu(x_0)} p(x_\tau, x_0), \tag{4.7}$$

which is crucial in the following. Its derivation is simple if one recalls eqn (C.33):

$$\mathrm{tr}\{\Pi_\Theta(x_0)U_\Theta(\tau,0)\Pi_\Theta(x_\tau)U_\Theta^\dagger(\tau,0)\} = \mathrm{tr}\{\Pi(x_\tau)U(\tau,0)\Pi(x_0)U^\dagger(\tau,0)\}. \tag{4.8}$$

Exercise 4.2 Derive eqn (4.7).

Finally, for many applications below, it turns out that we are not interested in directly comparing $p(x_0, x_\tau)$ and $p_{\rm tr}(x_0, x_\tau)$, but we are rather interested in how a function $f = f(x)$ of the measurement outcomes varies in time: $\Delta f = f(x_\tau) - f(x_0)$. Whatever f is, the probability of observing a change Δf in the forward or time-reversed process can be computed from $p(x_0, x_\tau)$ and $p_{\rm tr}(x_0, x_\tau)$ as

$$P(\Delta f) = \sum_{x_\tau, x_0} \delta(\Delta f - [f(x_\tau) - f(x_0)]) p(x_\tau, x_0), \tag{4.9}$$

$$P_{\rm tr}(\Delta f) = \sum_{x_0, x_\tau} \delta(\Delta f - [f(x_0) - f(x_\tau)]) p_{\rm tr}(x_0, x_\tau). \tag{4.10}$$

Now, consider the function $f(x) = -\ln\mu(x)$, where $\mu(x)$ are the weights specified above. Hence, $e^{f(x_\tau)-f(x_0)} = \mu(x_0)/\mu(x_\tau)$, and it follows from eqn (4.7) that

$$\begin{aligned} P(\Delta f) &= \sum_{x_\tau, x_0} \delta(\Delta f - [f(x_\tau) - f(x_0)]) \frac{\mu(x_0)}{\mu(x_\tau)} p_{\rm tr}(x_0, x_\tau) \\ &= e^{\Delta f} \sum_{x_\tau, x_0} \delta(\Delta f - [f(x_\tau) - f(x_0)]) p_{\rm tr}(x_0, x_\tau), \end{aligned} \tag{4.11}$$

where the delta function allowed us to pull the factor $e^{\Delta f}$ out of the summation. Then, by definition (4.10), we obtain $P(\Delta f) = e^{\Delta f} P_{\rm tr}(-\Delta f)$. This is the *abstract detailed fluctuation theorem*, which is typically written in the form

$$\boxed{\frac{P(\Delta f)}{P_{\rm tr}(-\Delta f)} = e^{\Delta f}.} \tag{4.12}$$

Note that eqn (4.12) implies the abstract integral fluctuation theorem (4.4) since

$$\begin{aligned} \left\langle e^{-\ln[\mu(x_0)/\mu(x_\tau)]} \right\rangle_{x_\tau, x_0} &= \sum_{x_\tau, x_0} e^{-[f(x_\tau)-f(x_0)]} p(x_\tau, x_0) \\ &= \sum_{\Delta f} e^{-\Delta f} \sum_{x_\tau, x_0} \delta(\Delta f - [f(x_\tau) - f(x_0)]) p(x_\tau, x_0) \\ &= \sum_{\Delta f} e^{-\Delta f} P(\Delta f) = \sum_{\Delta f} P_{\rm tr}(-\Delta f) = 1. \end{aligned} \tag{4.13}$$

Furthermore, for a relevant class of applications, it turns out that the probability distributions to observe a change Δf in the forward and time-reversed process are equal: $P(\Delta f) = P_{\text{tr}}(\Delta f)$. Under these circumstances, the abstract detailed fluctuation theorem simplifies to

$$\boxed{\frac{P(\Delta f)}{P(-\Delta f)} = e^{\Delta f}.} \tag{4.14}$$

The next exercise explores when this is the case.

Exercise 4.3 Derive that $P(\Delta f) = P_{\text{tr}}(\Delta f)$ if (a) $\rho_{\text{tr}}(\tau) = \rho(0)$, (b) the measured observables are invariant under time-reversal, i.e. $X_\Theta(\lambda_0) = X(\lambda_0)$ and $X_\Theta(\lambda_\tau) = X(\lambda_\tau)$, (c) the Hamiltonian is invariant under time-reversal, and (d) the driving protocol is time-symmetric, which means that $\lambda_t = \lambda_{\tau-t}$ for all $t \in [0, \tau]$.

4.2 Work Fluctuation Theorems for Quantum Systems

We turn to the first application of our abstract framework developed above and consider measurements of the energy $X(\lambda_t) \equiv H(\lambda_t) = \sum_{\epsilon_t} \epsilon_t \Pi(\epsilon_t)$. We start by working out the mathematical consequences of this choice before we discuss its physical interpretation. The initial state is assumed to be a Gibbs state,

$$\rho(0) = \pi(\lambda_0) = \frac{e^{-\beta H(\lambda_0)}}{\mathcal{Z}(\lambda_0)} = \sum_{\epsilon_0} \Pi(\epsilon_0) e^{-\beta[\epsilon_0 - \mathcal{F}(\lambda_0)]}, \tag{4.15}$$

where $\mathcal{F}(\lambda_0) = -k_B T \ln \mathcal{Z}(\lambda_0)$ denotes the equilibrium free energy. Hence, the probability of obtaining outcome ϵ_0 is $p(\epsilon_0) = \text{tr}\{\Pi(\epsilon_0)\pi(\lambda_0)\} = V(\epsilon_0) e^{-\beta[\epsilon_0 - \mathcal{F}(\lambda_0)]}$, such that $\mu(\epsilon_0) = e^{-\beta[\epsilon_0 - \mathcal{F}(\lambda_0)]}$, where $\mu(\epsilon_0)$ denotes the weights appearing in eqn (4.3).

According to our abstract integral fluctuation theorem (4.4), we are free to choose any set of final weights $\mu(\epsilon_\tau)$ obeying $\sum_{\epsilon_\tau} \mu(\epsilon_\tau) V(\epsilon_\tau) = 1$ and we choose $\mu(\epsilon_\tau) = e^{-\beta[\epsilon_\tau - \mathcal{F}(\lambda_\tau)]}$. These weights correspond to a Gibbs state $\pi(\lambda_\tau)$, but notice that this does *not* mean that the state $\rho(\tau) = U(\tau, 0)\rho(0)U^\dagger(\tau, 0)$ must coincide with $\pi(\lambda_\tau)$. In fact, in general $\rho(\tau)$ can be far away from $\pi(\lambda_\tau)$. Nevertheless, our purely formal result (4.4) continues to hold and implies

$$\left\langle e^{-\beta(\epsilon_\tau - \epsilon_0)} \right\rangle_{\epsilon_\tau, \epsilon_0} = e^{-\beta \Delta \mathcal{F}}, \tag{4.16}$$

where $\Delta\mathcal{F} = \mathcal{F}(\lambda_\tau) - \mathcal{F}(\lambda_0)$ is a constant, which can be taken out of the average $\langle \dots \rangle_{\epsilon_\tau, \epsilon_0}$. Furthermore, notice that $\mu(\epsilon_\tau)$ also equals the weights appearing in the Gibbs state $\pi_\Theta(\lambda_\tau)$ of the time-reversed Hamiltonian $H_\Theta(\lambda_\tau)$ since it has the same spectrum as $H(\lambda_\tau)$. Thus, our abstract detailed fluctuation theorem (4.12) implies

$$\frac{P(\Delta\epsilon)}{P_{\text{tr}}(-\Delta\epsilon)} = e^{\beta(\Delta\epsilon - \Delta\mathcal{F})}. \tag{4.17}$$

In view of our previous notation, we have $\Delta f = \beta(\Delta\epsilon - \Delta\mathcal{F})$, but we set $P(\Delta f) \equiv P(\Delta\epsilon)$ for brevity since $\Delta\epsilon$ is the only fluctuating quantity—β and $\Delta\mathcal{F}$ are constants instead.

Before we interpret eqns (4.16) and (4.17) from a physical perspective, we note that we could have also chosen different final weights $\mu(\epsilon_\tau)$. This can indeed lead to interesting consequences, as the next exercise shows.

Exercise 4.4 During the driving protocol the internal energy changes *on average* from the initial equilibrium value $\mathcal{U}(0) = \text{tr}\{H(\lambda_0)\pi(\lambda_0)\}$ to some final non-equilibrium value $U(\tau) = \text{tr}\{H(\lambda_\tau)\rho(\tau)\} = W(\tau) + \mathcal{U}(0)$, where $W(\tau) = \int_0^\tau dt\text{tr}\{[\partial_t H(\lambda_t)]\rho(t)\}$ is the work done on the system. Let us define an inverse temperature β_τ^* in such a way that a fictitious Gibbs ensemble $\pi(\beta_\tau^*, \lambda_\tau) = e^{-\beta_\tau^* H(\lambda_\tau)}/\mathcal{Z}(\beta_\tau^*, \lambda_\tau)$ has the same internal energy as the final non-equilibrium state $\rho(\tau)$, i.e. $U(\tau) = \mathcal{U}(\beta_\tau^*, \tau) \equiv \text{tr}\{H(\lambda_\tau)\pi(\beta_\tau^*, \lambda_\tau)\}$ (observe that we have used this definition already in eqn (3.117)). Now, choose the weights $\mu(\epsilon_\tau) = e^{-\beta_\tau^*[\epsilon_\tau - \mathcal{F}(\beta_\tau^*, \lambda_\tau)]}$ and show that

$$\left\langle e^{-\beta_\tau^*[\epsilon_\tau - \mathcal{F}(\beta_\tau^*, \lambda_\tau)] + \beta[\epsilon_0 - \mathcal{F}(\beta, \lambda_0)]} \right\rangle_{\epsilon_\tau, \epsilon_0} = 1. \tag{4.18}$$

Use $e^x \geq 1 + x$ to deduce from here that $\beta_\tau^*[U(\tau) - \mathcal{F}(\beta_\tau^*, \lambda_\tau)] \geq \beta[\mathcal{U}(0) - \mathcal{F}(\beta, \lambda_0)]$. Show that this inequality implies $\mathcal{S}(\beta_\tau^*, \lambda_\tau) \geq \mathcal{S}(\beta, \lambda_0)$, where $\mathcal{S}(\beta_\tau^*, \lambda_\tau)$ is the equilibrium thermodynamic entropy of the fictitious Gibbs state $\pi(\beta_\tau^*, \lambda_\tau)$. What is the interpretation of this result?

To interpret the above result physically, we start by spelling out in detail the steps that would be necessary to confirm eqn (4.16) or (4.17) experimentally. First, we must ensure that we have a sufficiently isolated system, whose initial state is well approximated by a canonical Gibbs ensemble. There are many ways to justify a Gibbs ensemble and the most appealing one is perhaps that a system, which is weakly coupled to a bath, is described by a Gibbs state if the system *and* bath are in equilibrium, at least from a macroscopic point of view.[1] However, in the derivation above, we have considered an *isolated* system, so the initial Gibbs ensemble is typically justified by assuming that the system was in the past for $t < 0$ in contact with a bath for a sufficiently long time. The coupling with the bath is then assumed to be either controllable (and switched off) or very weak, such that its influence is negligible during the period $[0, \tau]$. The second experimental requirement that we then need to add is the ability to perform projective measurements of the Hamiltonian $H(\lambda_t)$ at the initial and final time of the protocol. This gives rise to the stochastic outcomes ϵ_0 and ϵ_τ, which can be used to confirm eqn (4.16). If one also wants to confirm eqn (4.17) experimentally, one needs to ensure that the same assumptions are also justified for the time-reversed experiment. Whether these assumptions are realistic or not strongly depends on the system under consideration; we explore this question further below.

Readers familiar with the classical work fluctuation theorem (Section 2.7) will notice that eqns (4.16) and (4.17) look identical to the classical results $\langle e^{-\beta w} \rangle = e^{-\beta \Delta \mathcal{F}}$ and $p(w) = e^{\beta(w - \Delta \mathcal{F})} p_{\text{tr}}(-w)$ if we equate $\Delta\epsilon = w$, i.e. if we interpret the change in energy $\epsilon_\tau - \epsilon_0$ obtained from the two projective measurements as the stochastic work done on the system. We now investigate in detail whether this is justified by splitting the total change in internal energy during the process, starting from the initial state $\pi(\lambda_0)$ and ending with the final result ϵ_τ, into three parts:

[1] When we write 'from a macroscopic point of view,' we mean that all realistic (i.e., coarse–grained) measurements give rise to the same well–defined temperature (or other thermodynamic variable) because the measurements are incapable of sensing differences in the microscopic system state.

$$\Delta u = \Delta u_{\text{meas 1}} + \Delta u_{\text{drive}} + \Delta u_{\text{meas 2}}. \tag{4.19}$$

The first part captures the influence of the first measurement:

$$\Delta u_{\text{meas 1}} = \epsilon_0 - \text{tr}\{H(\lambda_0)\pi(\lambda_0)\}. \tag{4.20}$$

This change in energy is due to an update of our state of knowledge: after receiving the measurement outcome ϵ_0, the pre-measurement state $\pi(\lambda_0)$ with internal energy $\text{tr}\{H(\lambda_0)\pi(\lambda_0)\}$ changes to the post-measurement state $\omega(\epsilon_0) \equiv \Pi(\epsilon_0)/V(\epsilon_0)$ with internal energy ϵ_0. Note that this change in internal energy, which comes from picking an initial member of the canonical ensemble, is also not counted as work in the classical case and therefore disregarded in the following (we return to it in Chapter 5).

Next, we have the term

$$\Delta u_{\text{drive}} = \text{tr}\left\{H(\lambda_\tau)U(\tau,0)\omega(\epsilon_0)U^\dagger(\tau,0)\right\} - \epsilon_0. \tag{4.21}$$

This captures the change in internal energy due to the driving protocol λ_t starting from the post-measurement state $\omega(\epsilon_0)$. Equation (4.21) exactly *coincides* with our definition of work from the previous chapter (see eqn (3.4)), which we derived for any initial state and any unitary evolution generated by a time-dependent Hamiltonian $H(\lambda_t)$. Therefore, eqn (4.21) can be unambiguously identified as work.

The third and most subtle part captures the energetic change due to the second measurement,

$$\Delta u_{\text{meas 2}} = \epsilon_\tau - \text{tr}\left\{H(\lambda_\tau)U(\tau,0)\omega(\epsilon_0)U^\dagger(\tau,0)\right\}, \tag{4.22}$$

which arises because individual 'collapses of the wavefunction' need not preserve energy. Note that $\Delta u_{\text{meas 2}}$ vanishes in the classical case, i.e. when the pre-measurement state $U(\tau,0)\omega(\epsilon_0)U^\dagger(\tau,0)$ contains no coherences in the eigenbasis of $H(\lambda_\tau)$. Furthermore, we also cannot rely on our reasoning from Chapter 3 to identify $\Delta u_{\text{meas 2}}$ as work because the change in internal energy is not due to a unitary time evolution. Therefore, the least biased decision would probably be to keep $\Delta u_{\text{meas 2}}$ explicit in all the following equations. However, there are at least two pragmatic reasons why it is a good idea to count eqn (4.22) as work (further discussion about this topic can be found in Chapter 5). The first reason is simply to avoid notational overload. Second, if we only identify $w = \Delta u_{\text{drive}}$ as the stochastic work, we find in general that $\langle e^{-\beta w}\rangle \neq e^{-\beta\Delta\mathcal{F}}$, in contrast to the classical case.

Thus, to summarize, if we interpret $w = \Delta u_{\text{drive}} + \Delta u_{\text{meas 2}}$ as the stochastic work associated with the two-point measurement scheme, then the following **quantum work fluctuation theorems** hold:

$$\boxed{\left\langle e^{-\beta w}\right\rangle = e^{-\beta\Delta\mathcal{F}}, \quad \frac{P(w)}{P_{\text{tr}}(-w)} = e^{\beta(w-\Delta\mathcal{F})}.} \tag{4.23}$$

In fact, whenever the terminology 'two-point measurement scheme' appears in the literature, it is typically used not only to refer to the abstract process introduced in Section 4.1, but also to imply that work fluctuations are identified via the prescription $w = \Delta u_{\text{drive}} + \Delta u_{\text{meas 2}}$.

We now turn to the case of open quantum systems by assuming that the Hamiltonian of the isolated system can be written as $H_{SB}(\lambda_t) = H_S(\lambda_t) + V_{SB} + H_B$. Importantly, as in previous chapters we assume the bath Hamiltonian H_B to be time independent. A time-dependent interaction Hamiltonian V_{SB} is treated in a moment, but we first consider the situation of a constant but very weak interaction V_{SB}. To a good approximation, the thermal system–bath state at time $t = 0$ can then be written as

$$\rho_{SB}(0) = \pi_{SB}(\lambda_0) \approx \pi_S(\lambda_0) \otimes \pi_B. \tag{4.24}$$

We now perform a first projective measurement of the system and bath energy $H_S(\lambda_0)+H_B$ and denote the outcome by $\epsilon_S^0 + \epsilon_B^0$. Afterwards, we execute an arbitrary driving protocol λ_t followed at time τ by a second projective measurement of $H_S(\lambda_\tau) + H_B$. The outcome of this measurement is denoted by $\epsilon_S^\tau + \epsilon_B^\tau$ and we define the stochastic work as $w = (\epsilon_S^\tau + \epsilon_B^\tau) - (\epsilon_S^0 + \epsilon_B^0)$. We then obtain the work fluctuation theorems

$$\boxed{\left\langle e^{-\beta w}\right\rangle = e^{-\beta\Delta\mathcal{F}_S}, \qquad \frac{P(w)}{P_{\mathrm{tr}}(-w)} = e^{\beta(w-\Delta\mathcal{F}_S)}.} \tag{4.25}$$

Equation (4.25) is a direct consequence of eqn (4.23) for a system–bath Hamiltonian in the weak coupling limit and the fact that $\Delta\mathcal{F}_{SB} = \Delta\mathcal{F}_S$, which follows from the time independence of H_B.

The work fluctuation theorems (4.25) can also be derived if the system is allowed to interact strongly with the bath for a finite time. This is modelled by a time-dependent interaction Hamiltonian $V_{SB}(\lambda_t)$, which satisfies $V_{SB}(\lambda_0) = V_{SB}(\lambda_\tau) = 0$, i.e. the system and bath are at the initial and final time decoupled. If the interaction Hamiltonian V_{SB} is non-negligible also at the initial and final times, we can still derive a strong coupling work fluctuation theorem by using the Hamiltonian of mean force, as shown in the next exercise.

Exercise 4.5 Recall the concept of the Hamiltonian of mean force $H_S^*(\lambda_t)$ introduced in Section 1.3. Consider an arbitrary system–bath Hamiltonian of the form $H_{SB}(\lambda_t) = H_S(\lambda_t)+V_{SB}(\lambda_t) + H_B$. Use relation (1.21) to confirm that $\mathcal{F}_{SB}(\lambda_t) = \mathcal{F}_S^*(\lambda_t) + \mathcal{F}_B$, where $\mathcal{F}_S^*(\lambda_t) \equiv -k_B T \ln \mathcal{Z}_S^*(\lambda_t)$ is the strong coupling equilibrium free energy. Then, show that the following strong coupling quantum work fluctuation theorems

$$\left\langle e^{-\beta w}\right\rangle = e^{-\beta\Delta\mathcal{F}_S^*}, \qquad \frac{P(w)}{P_{\mathrm{tr}}(-w)} = e^{\beta(w-\Delta\mathcal{F}_S^*)}, \tag{4.26}$$

with $\Delta\mathcal{F}_S^* = \mathcal{F}_S^*(\lambda_t) - \mathcal{F}_S^*(\lambda_0)$, are a consequence of eqn (4.23).

Relations (4.23), (4.25) and (4.26) are the quantum counterparts of the classical work fluctuation theorems (2.130), (2.134), (2.143) and (2.153). Note that they have *exactly the same form*. Putting aside the subtle question about the interpretation of eqn (4.22), this demonstrates that statistical mechanics is a universal theory, the basic concepts of which apply to classical and quantum systems alike. There is, however, also one *big conceptual difference* between quantum and classical work fluctuation theorems.

Recall that the ensemble average $\langle \ldots \rangle$ as well as the probability distribution $P(w)$ in eqns (4.23), (4.25) or (4.26) are constructed by performing projective measurements of the Hamiltonian $H(\lambda_t)$ of an isolated system, of the decoupled system–bath Hamiltonian $H_S(\lambda_t) + H_B$ or of the full system–bath Hamiltonian $H_{SB}(\lambda_t)$. Now, consider one of the specific examples we have treated so far in this book, a macromolecule in aqueous solution (Section 2.9), a cavity coupled to the outside electromagnetic modes (Section 3.9) or a quantum dot coupled to a reservoir of electrons (Section 3.10). In all of these examples, the number of particles or modes of the environment is enormous, say, about 10^{20}. This implies that the dimension of the corresponding Hilbert space is of the order of $\exp(10^{20})$ and that a projective measurement of the corresponding Hamiltonian has of the order of $\exp(10^{20})$ different outcomes. It is virtually *impossible* to implement such a fine-grained projective measurement in a laboratory.

In contrast, classical work fluctuation theorems have been experimentally confirmed for a variety of system–bath set-ups with a Hamiltonian of the form $H_S(\lambda_t) + V_{SB} + H_B$. How can that be? The reason for this is that, classically, the work probability distribution $P(w)$ can be inferred by observing *only the system.* That is to say, the work fluctuations are accessible locally, which is immediately transparent by looking at eqn (2.141). Therefore, one never needs to measure the energy of the bath in the classical context.

This result might come as a surprise if one recalls that the expected work $W(t)$ can be computed based only on knowledge of the reduced system state $\rho_S(t)$; see eqn (3.5). Therefore, one might wonder whether there are other *local* measurement strategies to extract the probability distribution $P(w)$ of stochastic work w, which satisfies the conventional fluctuation theorem. As it turns out, despite 20 years of research, nobody has found such a local measurement strategy. Of course, this does *not* imply that quantum stochastic thermodynamics makes no experimentally testable predictions. Quantum fluctuation theorems imply a variety of important consequences (see, for example, Sections 4.6 and 4.7) and in Chapter 5 we will go beyond quantum fluctuation theorems and set up a framework to study fluctuating thermodynamic quantities based on local measurements only.

4.3 Exchange and Further Fluctuation Theorems

We continue the derivation of quantum fluctuation theorems and focus on a system exchanging energy and particles with multiple baths (Fig. 4.1). In this case, the work fluctuation theorem is not applicable. Instead, we derive fluctuation theorems for the entropy production and the exchanges of energy and particles.

We consider the conventional system–bath Hamiltonian $H_{SB}(\lambda_t) = H_S(\lambda_t) + \sum_\nu [V_{SB}^{(\nu)}(\lambda_t) + H_B^{(\nu)}]$ and assume the initial state to be

$$\rho_{SB}(0) = \rho_S(0) \bigotimes_\nu \Xi_\nu. \tag{4.27}$$

Here, $\rho_S(0)$ is an arbitrary system state and $\Xi_\nu = \exp[-\beta_\nu(H_B^{(\nu)} - \mu_\nu \hat{N}_\nu)]/\mathcal{Z}_\nu$ the grand canonical ensemble of bath ν with the (grand) partition function $\mathcal{Z}_\nu = \mathrm{tr}_\nu\{\exp[-\beta_\nu(H_B^{(\nu)} - \mu_\nu \hat{N}_\nu)]\}$. We assume $[H_B^{(\nu)}, \hat{N}_\nu] = 0$ below. Readers unfamiliar with particle transport can consult Section 3.10 before continuing reading.

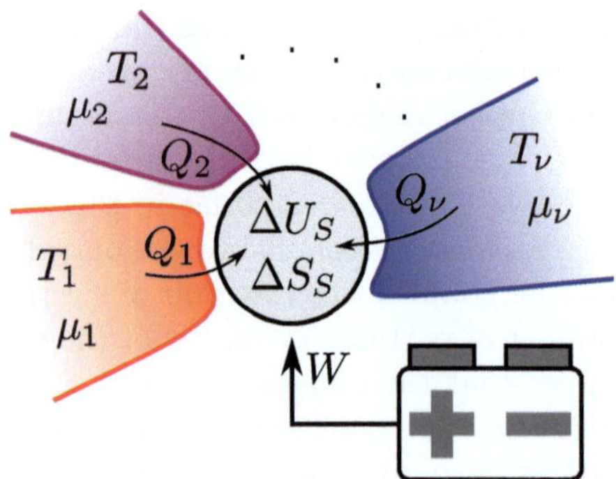

Fig. 4.1 Sketch of a system (circle in the middle) coupled to multiple heat baths at different temperatures T_ν and chemical potentials μ_ν and a work reservoir (sketched as a battery). Via the exchange of heat Q_ν and work W the internal energy and system entropy change, as quantified by ΔU_S and ΔS_S.

We now write the initial state (4.27) in its eigenbasis. The system state is $\rho_S(0) = \sum_{s_0} p(s_0)|s_0\rangle\langle s_0|$ and we introduce projectors $\Pi(\epsilon_\nu, n_\nu)$ corresponding to eigenenergy ϵ_ν and particle number n_ν of bath ν. Thus, $\Xi_\nu = \sum_{\epsilon_\nu, n_\nu} e^{-\beta(\epsilon_\nu - \mu_\nu n_\nu)}\Pi(\epsilon_\nu, n_\nu)/\mathcal{Z}_\nu$. After these preparations, it is now easy to recast the initial state (4.27) into the form of eqn (4.3). Namely, by defining the list $\mathbf{x}_0 = (s_0; \epsilon_1^0, n_1^0; \dots; \epsilon_\nu^0, n_\nu^0; \dots)$, we can write $\rho_{SB}(0) = \sum_{\mathbf{x}_0} \mu(\mathbf{x}_0)\Pi(\mathbf{x}_0)$ with

$$\mu(\mathbf{x}_0) = p(s_0)\prod_\nu \frac{e^{-\beta(\epsilon_\nu^0 - \mu_\nu n_\nu^0)}}{\mathcal{Z}_\nu}, \quad \Pi(\mathbf{x}_0) = |s_0\rangle\langle s_0| \bigotimes_\nu \Pi(\epsilon_\nu^0, n_\nu^0). \tag{4.28}$$

It is essential that the weights $\mu(\mathbf{x}_0)$ introduced in Section 4.1 are not confused with the chemical potentials μ_ν.

In the spirit of the two-point measurement scheme, we assume that an initial projective measurement of $\mathbf{x}_0$ is performed, i.e. of the system state in its eigenbasis and of the energy and particle number of each bath. Since all of these observables commute, we can perform this measurement simultaneously. After the first projective measurement, we let the system evolve in time according to a prescribed driving protocol λ_t. Finally, we perform a second projective measurement. Again, we measure the energy and particle number of each bath and we measure the system in the eigenbasis of the time-evolved reduced density matrix $\rho_S(\tau) = \sum_{s_\tau} p(s_\tau)|s_\tau\rangle\langle s_\tau| = \mathrm{tr}_B\{U(\tau,0)\rho_{SB}(0)U^\dagger(\tau,0)\}$. Note that $\rho_S(\tau)$ denotes the *unconditional* time-evolved state, i.e. the time-evolved state obtained from the initial condition $\rho_{SB}(0)$ *without* performing any measurement. Alternatively, since the first measurement is done in the eigenbasis of $\rho_{SB}(0)$, we can also say that $\rho_S(\tau)$ denotes the final state of the system *averaged* over all possible initial measurement outcomes. The result of the second measurement is denoted by $\mathbf{x}_\tau = (s_\tau; \epsilon_1^\tau, n_1^\tau; \dots; \epsilon_\nu^\tau, n_\nu^\tau; \dots)$ and the weights associated with it are chosen as

$$\mu(\mathbf{x}_\tau) = p(s_\tau)\prod_\nu \frac{e^{-\beta(\epsilon_\nu^\tau - \mu_\nu n_\nu^\tau)}}{\mathcal{Z}_\nu}. \tag{4.29}$$

Next, we compute

$$\ln \frac{\mu(\mathbf{x}_0)}{\mu(\mathbf{x}_\tau)} = -\ln p(s_\tau) + \ln p(s_0) + \sum_\nu \beta_\nu[\epsilon_\nu^\tau - \epsilon_\nu^0 - \mu_\nu(n_\nu^\tau - n_\nu^0)]. \tag{4.30}$$

The first two terms are related to the change in stochastic entropy of the system and we equate $-\ln p(s_\tau) + \ln p(s_0) \equiv \Delta s_S/k_B$. In fact, in the absence of coherences, this definition is identical to the definition of stochastic entropy in the classical case; compare with eqn (2.56). Furthermore, $\epsilon_\nu^\tau - \epsilon_\nu^0$ and $n_\nu^\tau - n_\nu^0$ denote the change in energy and particle number of bath ν. Recalling the definition of heat in the presence of a chemical potential, eqn (3.164), we equate $\epsilon_\nu^\tau - \epsilon_\nu^0 - \mu_\nu(n_\nu^\tau - n_\nu^0) = -q_\nu$ with minus the stochastic heat flow from bath ν. Therefore, we obtain

$$\ln \frac{\mu(\mathbf{x}_0)}{\mu(\mathbf{x}_\tau)} = \frac{\Delta s_S}{k_B} - \sum_\nu \beta_\nu q_\nu = \frac{\sigma}{k_B}, \tag{4.31}$$

where we introduced the stochastic entropy production $\sigma \equiv \Delta s_S - \sum_\nu q_\nu/T_\nu$.

From our abstract fluctuation theorems in Section 4.1 it now follows immediately that the following **integral entropy production fluctuation theorem** holds:

$$\boxed{\left\langle e^{-\sigma/k_B}\right\rangle_{\mathbf{x}_\tau,\mathbf{x}_0} = 1.} \tag{4.32}$$

This fluctuation theorem presents a significant refinement of results we found in Chapter 3. As discussed there, for a small system coupled to large thermal baths, the identification of stochastic entropy s_S, stochastic heat q_ν and stochastic entropy production σ as done above is also justified.

Exercise 4.6 Show that eqn (4.32) implies the second laws (3.97) and (3.167).

Similar to Section 2.6 the identification of a detailed entropy production fluctuation theorem is subtle. Recall that the change in stochastic system entropy $\Delta s_S/k_B = -\ln p(s_\tau) + \ln p(s_0)$ in eqn (4.31) is defined with respect to the *forward* dynamics, i.e., the $p(s_\tau)$ are the eigenvalues of $\rho_S(\tau)$ provided the initial system state has eigenvalues $p(s_0)$. If we start the backwards dynamics with the weights appearing in eqn (4.29), there is no guarantee that the final state in the backward dynamics has the same probabilities $p(s_0)$ as the initial state for the forward dynamics. To be clear about this, we write as in Section 2.6

$$\frac{P(\sigma)}{Q(-\sigma)} = e^{\sigma/k_B}, \tag{4.33}$$

where $Q(\sigma)$ is the probability to obtain the 'entropy production' (4.31), which is defined with respect to the forward dynamics, during the backward dynamics. There exists, however, a large class of detailed fluctuation theorems where the cumbersome boundary term $-\ln p(s_\tau) + \ln p(s_0)$ does not appear and where the entropy production (4.31) reduces to $\sigma = -\sum_\nu q_\nu/T_\nu$.

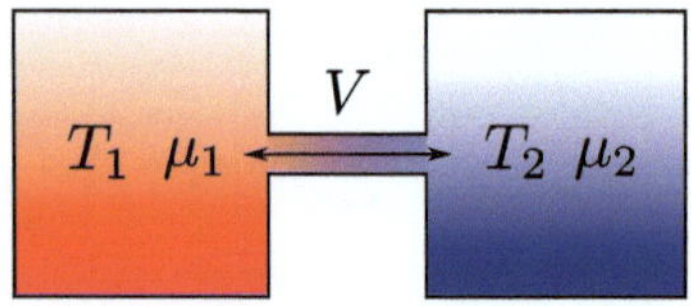

Fig. 4.2 Two baths prepared at different temperatures and chemical potentials are put into contact via some interaction V for a finite time τ. Note that the red and blue colour only indicate the initial temperatures. After the interaction, the temperatures of bath 1, 2 or both need not be well defined.

To reveal them, we take a step back and reconsider the above situation in the *absence* of any system. This means we assume a Hamiltonian of the form $\sum_\nu H_B^{(\nu)} + V(\lambda_t)$ describing different baths ν, which are directly put into contact via some time-dependent interaction $V(\lambda_t)$. The initial state is assumed to be a tensor product of grandcanonical ensembles Ξ_ν. We further introduce the shorthand notation $\boldsymbol{\Delta\epsilon} = (\Delta\epsilon_1, \dots, \Delta\epsilon_\nu, \dots)$ and $\boldsymbol{\Delta n} = (\Delta n_1, \dots, \Delta n_\nu, \dots)$ to denote the changes in energy and particle number of all baths. Then, it follows from our findings above that the following fluctuation theorem holds:

$$\frac{P(\boldsymbol{\Delta\epsilon}, \boldsymbol{\Delta n})}{P_{\text{tr}}(-\boldsymbol{\Delta\epsilon}, -\boldsymbol{\Delta n})} = \exp\left[\sum_\nu \beta_\nu(\Delta\epsilon_\nu - \mu_\nu \Delta n_\nu)\right]. \tag{4.34}$$

In addition, we now consider the time-reversal symmetric case for which $\Theta H_B^{(\nu)} \Theta^{-1} = H_B^{(\nu)}$ and $\Theta V(\lambda_t)\Theta^{-1} = V(\lambda_{\tau-t})$. It then follows that $P_{\text{tr}}(\boldsymbol{\Delta\epsilon}, \boldsymbol{\Delta n}) = P(\boldsymbol{\Delta\epsilon}, \boldsymbol{\Delta n})$ (see Exercise 4.3) and the above fluctuation theorem simplifies to

$$\boxed{\frac{P(\boldsymbol{\Delta\epsilon}, \boldsymbol{\Delta n})}{P(-\boldsymbol{\Delta\epsilon}, -\boldsymbol{\Delta n})} = \exp\left[\sum_\nu \beta_\nu(\Delta\epsilon_\nu - \mu_\nu \Delta n_\nu)\right].} \tag{4.35}$$

This relation is known as the **exchange fluctuation theorem**.

To understand the significance of the exchange fluctuation theorem, we consider the special case of two baths, $\nu \in \{1, 2\}$; see Fig. 4.2. We also use the natural assumption that the total particle number operator $\hat{N}_{\text{tot}} \equiv \hat{N}_1 + \hat{N}_2$ commutes with the global Hamiltonian $H(\lambda_t) = H_B^{(1)} + H_B^{(2)} + V(\lambda_t)$ at all times. This implies

$$[\hat{N}_{\text{tot}}, U(\tau, 0)] = 0, \tag{4.36}$$

which follows immediately from the explicit form of $U(\tau, 0) = \exp_+[-i\int_0^\tau dt H(\lambda_t)/\hbar]$. Equation (4.36) implies that

$$P(\Delta\epsilon_1, \Delta\epsilon_2, \Delta n_1, \Delta n_2) = P(\Delta\epsilon_1, \Delta\epsilon_2, \Delta n_1, -\Delta n_1) \tag{4.37}$$

or in short: $\Delta n_2 = -\Delta n_1$. In words, since the total number of particles is constant, the number of particles lost (gained) in bath 1 must equal the number of particles gained (lost) in bath 2.

Exercise 4.7 Derive eqn (4.37).

Let us now try to apply the same argumentation for the distribution of energy outcomes. Clearly, if it were true that $[H_B^{(1)}+H_B^{(2)},U(\tau,0)]=0$, we would obtain $\Delta\epsilon_1 = -\Delta\epsilon_2$. In general, however, $[H_B^{(1)}+H_B^{(2)},V(\lambda_t)]\neq 0$ implies $[H_B^{(1)}+H_B^{(2)},U(\tau,0)]\neq 0$, even if we switch off the interaction at the beginning and at the end of the protocol, $V(\lambda_0)=V(\lambda_\tau)=0$. Therefore, $\Delta\epsilon_1\neq-\Delta\epsilon_2$ in general.

At this point it is instructive to consider the weak coupling limit again. Then, the amount of energy stored in the interaction is negligible and the work invested in switching on a weak interaction is also negligible. Thus, to a good approximation, we can conclude that $\Delta\epsilon_1\approx-\Delta\epsilon_2$. Consequently, the probability distribution for the energy and particle exchange statistics depends effectively only on two variables, instead of four, because of conservation of energy and particles. We write

$$P(\Delta\epsilon_1,\Delta n_1)\approx P(\Delta\epsilon_1,-\Delta\epsilon_1,\Delta n_1,-\Delta n_1) \tag{4.38}$$

and the exchange fluctuation theorem (4.35) becomes within a good approximation

$$\frac{P(\Delta\epsilon_1,\Delta n_1)}{P(-\Delta\epsilon_1,-\Delta n_1)}=e^{(\beta_1-\beta_2)\Delta\epsilon_1-(\beta_1\mu_1-\beta_2\mu_2)\Delta n_1}. \tag{4.39}$$

Let us consider two special cases of eqn (4.39) for illustration. First, the case $\beta_1=\beta_2\equiv\beta$. Then, the exchange statistics are completely determined by the change in particle number of bath 1 and we obtain

$$\frac{P(\Delta n_1)}{P(-\Delta n_1)}=e^{-\beta(\mu_1-\mu_2)\Delta n_1}. \tag{4.40}$$

Note that this relation is exact as it does not rely on our assumption that $\Delta\epsilon_1\approx-\Delta\epsilon_2$. Now, suppose that the chemical potential gradient is positive: $\mu_1-\mu_2>0$. Then, the fluctuation theorem (4.40) tells us that the tendency of particles to flow *against* the gradient ($\Delta n_1>0$) is exponentially suppressed compared with the tendency to flow *with* the gradient ($\Delta n_1<0$). This explains the emergence of the second law on average in particle transport set-ups. Note, however, that eqn (4.40) also tells us that there must exist processes for which $\Delta n_1>0$, i.e. in some run of the experiment we have to observe particles flowing against the bias.

A similar interpretation is also reached for the second special case, for which we assume $\mu_1=\mu_2$. The exchange fluctuation theorem then reduces to

$$\frac{P(\Delta\epsilon_1)}{P(-\Delta\epsilon_1)}=e^{(\beta_1-\beta_2)\Delta\epsilon_1}, \tag{4.41}$$

which, as emphasized above, holds only in the weak coupling limit or for a precisely tuned interaction such that $[H_B^{(1)}+H_B^{(2)},U(\tau,0)]=0$.

It is worth briefly commenting on the probability distributions $P(\Delta n_1)$ and $P(\Delta\epsilon_1)$ appearing in eqns (4.40) and (4.41). They can be obtained by taking the marginal of the joint probability distribution $P(\Delta\epsilon_1, \Delta n_1)$:

$$P(\Delta n_1) = \sum_{\Delta\epsilon_1} P(\Delta\epsilon_1, \Delta n_1), \quad P(\Delta\epsilon_1) = \sum_{\Delta n_1} P(\Delta\epsilon_1, \Delta n_1). \tag{4.42}$$

It is, however, not necessary to go this way and to measure both energy and particle exchanges first, and to marginalize afterwards. Instead, one can obtain, for instance, $P(\Delta n_1)$ directly from particle measurements *without* the need to measure energy exchanges, i.e.

$$P(\Delta n_1) = \sum_{n_1^\tau, n_1^0} \delta[\Delta n - (n_1^\tau - n_1^0)] p(n_1^\tau, n_1^0), \tag{4.43}$$

where $p(n_1^\tau, n_1^0) = \text{tr}\{\Pi(n_1^\tau)U(\tau,0)\Pi(n_1^0)\Xi_1 \otimes \Xi_2 \Pi(n_1^0)U^\dagger(\tau,0)\}$ is the two-point probability. This is possible because the measurements are *non-disturbing*: all observables commute with each other and with the initial state.

Finally, we consider the open system scenario again (Fig. 4.1), but we assume the system Hamiltonian H_S to be time independent. There are different ways to derive the following statement, but perhaps the simplest case is to assume that the system is also initialized in a grandcanonical ensemble Ξ_S with inverse temperature β_S and chemical potential μ_S. If we measure the energy and particle number of the system and all baths, our results above imply the fluctuation theorem (assuming time-reversal symmetry of all Hamiltonians and the driving protocol)

$$\frac{P(\Delta\epsilon_S, \Delta\boldsymbol{\epsilon}, \Delta n_S, \Delta\mathbf{n})}{P(-\Delta\epsilon_S, -\Delta\boldsymbol{\epsilon}, -\Delta n_S, -\Delta\mathbf{n})} = e^{\beta_S(\Delta\epsilon_S - \mu_S \Delta n_S) + \sum_\nu \beta_\nu(\Delta\epsilon_\nu - \mu_\nu \Delta n_\nu)}. \tag{4.44}$$

The fluctuating energy and particle changes in the system and the baths obey the relation

$$\Delta\epsilon_S + \sum_\nu \Delta\epsilon_\nu \approx 0, \quad \Delta n_S + \sum_\nu \Delta n_\nu = 0, \tag{4.45}$$

where the first relation is only approximately true at weak coupling, whereas the second equality is exact.

Now, consider the traditional open quantum system scenario, where the system is very small (e.g. a two-level system), but the baths are very large (e.g. 10^{23} two-level systems). Then, after we have switched on the interaction between the system and the baths, the system will reach a *non-equilibrium steady state* after a short transient time t_{relax}. In this regime, the statistical properties of the system no longer change, whereas the baths are still exchanging energy and particles through the system (until they reach a global equilibrium state after very long time, which we assume to be much longer than the observation time τ). In this case, the fluctuations in energy and particle number of the system, $\Delta\epsilon_S$ and Δn_S, are bounded whereas the bath statistics $\Delta\epsilon_\nu$ and Δn_ν scale *extensively* with τ (meaning that their averages obey $\langle\Delta\epsilon_\nu\rangle \sim \tau$

and $\langle \Delta n_\nu \rangle \sim \tau$). To a good approximation, we can therefore neglect the fluctuations in $\Delta\epsilon_S$ and Δn_S in eqn (4.45) and arrive at the conclusion

$$\boxed{\text{for } \tau \gg t_{\text{relax}}: \quad \frac{P(\Delta\boldsymbol{\epsilon}, \Delta\mathbf{n})}{P(-\Delta\boldsymbol{\epsilon}, -\Delta\mathbf{n})} = \exp\left[\sum_\nu \beta_\nu(\Delta\epsilon_\nu - \mu_\nu \Delta n_\nu)\right].} \tag{4.46}$$

This relation is called the **steady-state (exchange) fluctuation theorem**. In contrast to previous fluctuation theorems, which were exact statements about *finite-time* processes, the steady-state fluctuation theorem focuses on the long time limit. Its advantage is that it is independent of the system fluctuations: only the statistics of the bath enter it. In fact, eqn (4.46) has been confirmed experimentally in electronic transport set-ups through quantum dot structures, which we reviewed in Section 3.10. While these set-ups are well described by a classical rate master equation, the experimental confirmation of the steady-state fluctuation theorems also has theoretical impact. Indeed, to arrive at eqn (4.46) we relied on a number of sketchy arguments (weak coupling, long time limit) and it is hard to make them theoretically rigorous in general. In that respect, Section 4.4, where we will derive the steady-state fluctuation theorem using a quantum master equation perspective, is also reassuring. Such master equations, when applied to quantum dot set-ups in electronic transport, indeed explain the experiments.

4.4 Counting Field Methods for Quantum Master Equations

In this and the following section we return to the master equation paradigm studied in detail in Chapter 3. The goal is to introduce general tools to access the statistics of energy exchanges *beyond* its mean value (for ease of notation, we exclude particle exchanges here, which can be treated with the same tools). Moreover, at the end of this section, we derive the steady-state fluctuation theorem (4.46).

Full counting statistics

For convenience, we start by writing down the BMS equation again for a single heat bath in the interaction picture (compare with eqn (3.46)):

$$\frac{\partial}{\partial t}\tilde{\rho}_S(t) = \sum_{\omega,\alpha} r_\alpha(\omega)\left[S_\alpha(\omega)\tilde{\rho}_S(t)S_\alpha^\dagger(\omega) - \frac{1}{2}\{S_\alpha^\dagger(\omega)S_\alpha(\omega), \tilde{\rho}_S(t)\}\right] = \mathcal{D}\tilde{\rho}_S(t). \tag{4.47}$$

We recall that the rates $r_\alpha(\omega) = e^{\beta\hbar\omega}r_\alpha(-\omega)$ satisfy local detailed balance, where ω denotes a transition frequency in the system. The system operators $S_\alpha(\omega)$ obey the symmetry $S_\alpha(\omega) = S_\alpha^\dagger(-\omega)$ and describe transitions from one energy eigenvalue $\epsilon_{s'}$ to another energy eigenvalue $\epsilon_s = \epsilon_{s'} - \hbar\omega$. If $\hbar\omega > 0$, the system *loses* energy; otherwise, it *gains* energy.

Note that we assume the system Hamiltonian H_S to be time independent in this section; otherwise, the system does not reach a steady state in general. Hence, it also cannot exhibit a *steady-state* fluctuation theorem. Exceptions are driving protocols, which are adiabatically slow (and easy to treat) or periodic drivings (which are complicated from a technical point of view). Nevertheless, many of the formal methods developed here still prove useful in situations with driving.

To derive fluctuation theorems, we need to be able to extract information about dynamical fluctuations from the master equation (4.47). *A priori*, it is not immediately clear how this can be done: since the master equation only describes the evolution of $\rho_S(t)$, we cannot, for instance, make use of our previous formalism.

Our starting point is a detailed look at the master equation (4.47). We can see that only the first term proportional to $S_\alpha(\omega)\tilde{\rho}_S(t)S_\alpha^\dagger(\omega)$ describes a change in system energy. Namely, if the system starts with energy $\epsilon_{s'}$, its energy afterwards is $\epsilon_s = \epsilon_{s'} - \hbar\omega$. We therefore define $\mathcal{J}(\omega)\rho_S \equiv \sum_\alpha r_\alpha(\omega)S_\alpha(\omega)\rho_S S_\alpha^\dagger(\omega)$ and call $\mathcal{J}(\omega)$ a **jump (super)operator**. The master equation is then rewritten as

$$\frac{\partial}{\partial t}\tilde{\rho}_S(t) = \mathcal{L}_0\tilde{\rho}_S(t) + \sum_\omega \mathcal{J}(\omega)\tilde{\rho}_S(t), \tag{4.48}$$

where the superoperator $\mathcal{L}_0$ summarizes all the anti-commutator terms. It does not contain any energy-changing terms.

Next, we write down a formal solution of eqn (4.48) in the form of a *Dyson series*:

$$\begin{aligned}\tilde{\rho}_S(t) = &\sum_{n=0}^{\infty}\sum_{\omega_n,\dots,\omega_2,\omega_1}\int_0^t dt_n \cdots \int_0^{t_2} dt_1 \\ &\times e^{\mathcal{L}_0(t-t_n)}\mathcal{J}(\omega_n)e^{\mathcal{L}_0(t_n-t_{n-1})}\dots\mathcal{J}(\omega_2)e^{\mathcal{L}_0(t_2-t_1)}\mathcal{J}(\omega_1)e^{\mathcal{L}_0 t_1}\rho_S(0).\end{aligned} \tag{4.49}$$

Here, $\rho_S(0)$ is the initial system state and the term in the summation for $n = 0$ is defined as $e^{\mathcal{L}_0 t}\rho_S(0)$.

Exercise 4.8 Prove that eqn (4.49) solves eqn (4.48).

The Dyson series offers a nice interpretation of the dynamics of a quantum master equation. By looking at the term in the second line of eqn (4.49) for a fixed n, we see that it describes the state of the system at time t conditioned on n jumps, which happened at times $t_n > \cdots > t_2 > t_1$ and released an energy ω_i at time t_i into the bath. We denote this conditional state by $\rho(t|\omega_n,\dots,\omega_2,\omega_1)$ and write

$$\tilde{\rho}_S(t) = \sum_{n=0}^{\infty}\sum_{\omega_n,\dots,\omega_2,\omega_1}\int_0^t dt_n \cdots \int_0^{t_2} dt_1 \tilde{\rho}(t|\omega_n,\dots,\omega_2,\omega_1). \tag{4.50}$$

Again, the term for $n = 0$ is interpreted as the state of the system conditioned on *no jump* having happened.

For most applications, the conditional state $\tilde{\rho}(t|\omega_n,\dots,\omega_2,\omega_1)$ contains too much information. At the end, we are often only interested in the *net* heat transfer, which we define to be positive if it comes *from* the bath. Therefore, we introduce the state

$$\tilde{\rho}_S(t|q) = \sum_{n=0}^{\infty}\sum_{\omega_n,\dots,\omega_1}\delta(q + \hbar\omega_n + \cdots + \hbar\omega_1)\int_0^t dt_n \cdots \int_0^{t_2} dt_1 \tilde{\rho}(t|\omega_n,\dots,\omega_1), \tag{4.51}$$

conditioned on having a net heat flow $q = -\hbar\omega_n - \cdots - \hbar\omega_1$ into the system. Taking the time derivative of $\rho_S(t|q)$, we verify that it evolves in time according to the *generalized heat-resolved* master equation

$$\frac{\partial}{\partial t}\tilde{\rho}_S(t|q) = \mathcal{L}_0\tilde{\rho}_S(t|q) + \sum_\omega \mathcal{J}(\omega)\tilde{\rho}_S(t|q+\hbar\omega). \tag{4.52}$$

Note that $\tilde{\rho}_S(t|q)$ is no longer normalized; instead, $p_t(q) = \mathrm{tr}_S\{\tilde{\rho}_S(t|q)\}$ is the probability that the system absorbed a net amount of heat q until time t. The master equation (4.52) therefore contains much more information than the master equation (4.47) or (4.48) with which we started. The price for this is that eqn (4.52) acts on a much larger space: it acts on the tensor product of the system state space and the space of probability distributions for the random variable q. In general, q can be unbounded and, therefore, eqn (4.52) acts on an infinite space.

To overcome the problem of infinity, we Fourier transform eqn (4.52). We define

$$\tilde{\rho}_S(\chi, t) \equiv \int_{-\infty}^{\infty} đq e^{iq\chi}\tilde{\rho}_S(t|q), \tag{4.53}$$

where χ is called the **counting field**. Consequently, $\rho_S(t,\chi)$ evolves in time according to the counting field master equation

$$\boxed{\frac{\partial}{\partial t}\tilde{\rho}_S(\chi, t) = \mathcal{L}_0\tilde{\rho}_S(\chi, t) + \sum_\omega \mathcal{J}(\omega)e^{-i\hbar\omega\chi}\tilde{\rho}_S(\chi, t) \equiv \mathcal{L}(\chi)\tilde{\rho}_S(\chi, t).} \tag{4.54}$$

We refer to $\mathcal{L}(\chi)$ as the (counting field) Liouvillian in the following. We remark that $\mathcal{L}(\chi)$ preserves neither the trace nor the complete positivity of the dynamics. Furthermore, observe that for $\chi = 0$ eqn (4.54) reduces to our starting point—the BMS master equation (4.47). Since $p_t(q) = \mathrm{tr}_S\{\tilde{\rho}_S(t|q)\}$ is the probability of a net heat flow q up to time t,

$$M(\chi, t) \equiv \mathrm{tr}_S\{\tilde{\rho}(\chi, t)\} = \int_{-\infty}^{\infty} đq e^{iq\chi}p_t(q) \tag{4.55}$$

is the *moment generating function*. It obeys the symmetry

$$M(\chi, t) = M^*(-\chi^*, t). \tag{4.56}$$

Furthermore, we also define the *cumulant generating function* $C(\chi, t) \equiv \ln \mathrm{tr}_S\{\tilde{\rho}(\chi, t)\}$. Readers unfamiliar with these notions are advised to do the next exercise.

Exercise 4.9 Consider a probability distribution $p(q)$ and its Fourier transform $M(\chi) = \int đq e^{iq\chi}p(q)$. Show that the moments $\langle q^n\rangle = \int đq q^n p(q)$ for any $n \in \mathbb{N}$ can be obtained from $M(\chi)$ via differentiation:

$$\langle q^n\rangle = (-i)^n \left.\frac{\partial^n}{\partial\chi^n}M(\chi)\right|_{\chi=0}. \tag{4.57}$$

Hence, the name 'moment generating function'. It is worth remarking that mathematicians call $M(\chi)$ the characteristic function and $M(-i\chi)$ the moment generating function, which is, however, non-standard in the physics community. Similarly, $C(\chi) = \ln M(\chi)$ generates all cumulants via differentiation. Show explicitly that the first two cumulants (the mean value and variance) follow as

$$\kappa_1 \equiv \langle q \rangle = -i \frac{\partial}{\partial \chi} C(\chi)\bigg|_{\chi=0}, \quad \kappa_2 \equiv \langle q^2 \rangle - \langle q \rangle^2 = (-i)^2 \frac{\partial^2}{\partial \chi^2} C(\chi)\bigg|_{\chi=0}. \tag{4.58}$$

Higher-order cumulants κ_n ($n \geq 3$) characterize deviations from a Gaussian probability distribution. It is useful to keep in mind the mathematical fact that either $\kappa_n = 0$ for all $n \geq 3$ (which defines a Gaussian) or $\kappa_n \neq 0$ for all $n \geq 3$.

Thus, the solution of eqn (4.54), which can be written as $\tilde{\rho}_S(\chi, t) = e^{\mathcal{L}(\chi)t}\rho_S(0)$, gives access to the open system dynamics since $\tilde{\rho}_S(t) = \tilde{\rho}_S(\chi = 0, t)$ *and* it reveals the entire statistics of energy exchanges with the bath since $M(\chi, t) = \mathrm{tr}_S\{e^{\mathcal{L}(\chi)t}\rho_S(0)\}$. This is called **full counting statistics**.

As a cross-check, we consider the heat current $\dot{Q}(t)$, which is the time derivative of the first moment $\langle q \rangle(t)$. The foregoing exercise implies the formula

$$\dot{Q}(t) = -i \frac{\partial}{\partial t} \frac{\partial}{\partial \chi} M(\chi, t)\bigg|_{\chi=0} = -\hbar \sum_{\omega,\alpha} \omega r_\alpha(\omega) \mathrm{tr}_S\{S_\alpha^\dagger(\omega) S_\alpha(\omega) \tilde{\rho}_S(t)\}. \tag{4.59}$$

At first sight, this formula looks quite different from the one we have conventionally used: $\dot{Q}(t) = \mathrm{tr}_S\{H_S \mathcal{D}\rho_S(t)\}$; see, for example, eqn (3.53). Both are, however, identical, as the next exercise is asking you to demonstrate.

Exercise 4.10 Show that eqn (4.59) equals the conventional definition. *Hint:* Recall eqn (3.32), which also holds for $S_\alpha(\omega)$.

The above construction generalizes to the case of multiple bath $\nu \in \{1, \ldots, n\}$ by simply adding the contribution of each bath to the final Liouvillian,

$$\frac{\partial}{\partial t} \tilde{\rho}_S(\boldsymbol{\chi}, t) = \mathcal{L}(\boldsymbol{\chi}) \tilde{\rho}_S(\boldsymbol{\chi}, t) = \sum_{\nu=1}^{n} \mathcal{L}_\nu(\chi_\nu) \tilde{\rho}_S(\boldsymbol{\chi}, t), \tag{4.60}$$

where $\boldsymbol{\chi} = (\chi_1, \ldots, \chi_n)$ denotes a vector of counting fields. Here, each Liouvillian $\mathcal{L}_\nu(\chi_\nu)$ describes the coupling to one bath and reads explicitly

$$\mathcal{L}_\nu(\chi_\nu)\rho_S = \sum_{\omega,\alpha} r_{\alpha,\nu}(\omega) \left[e^{-i\hbar\omega\chi_\nu} S_{\alpha,\nu}(\omega) \rho_S S_{\alpha,\nu}^\dagger(\omega) - \frac{1}{2}\{S_{\alpha,\nu}^\dagger(\omega) S_{\alpha,\nu}(\omega), \rho_S\} \right]. \tag{4.61}$$

Note that we have already met this additive structure for a master equation with respect to different baths in Section 3.4. There, we argued that this structure is justified if the global system–bath state remains approximately of product form with the baths staying close to their initial equilibrium states. That is to say, the initial state (4.27) also has to remain a good approximation at later times.

Derivation of the steady-state fluctuation theorem

Let us now investigate the steady-state fluctuation theorem based on eqn (4.60). The moment generating function is generalized to $M(\boldsymbol{\chi},t) = \text{tr}_S\{\tilde{\rho}(\boldsymbol{\chi},t)\}$, which is linked to the probability distribution $p_t(\mathbf{q}) = p_t(q_1,\ldots,q_n)$ via n-dimensional Fourier transformation. It then turns out that there is a remarkable relation between the fluctuation theorem and symmetries in the moment generating function:

$$\boxed{\frac{p_t(\mathbf{q})}{p_t(-\mathbf{q})} = \exp\left(\sum_{\nu=1}^{n} a_\nu q_\nu\right) \quad \Leftrightarrow \quad M(\boldsymbol{\chi},t) = M(i\mathbf{a}-\boldsymbol{\chi},t) \;\; (\text{for } \boldsymbol{\chi}\in\mathbb{R}^n).} \tag{4.62}$$

Here, $\mathbf{a} = (a_1,\ldots,a_n)$ is some (not yet further specified) vector. The proof of this relation is left as an exercise.

Exercise 4.11 Show eqn (4.62) by starting with

$$p_t(\mathbf{q}) = \frac{1}{(2\pi)^n}\int_{-\infty}^{\infty} d\chi_1 \cdots \int_{-\infty}^{\infty} d\chi_n e^{-iq_1\chi_1}\ldots e^{-iq_n\chi_n} M(\boldsymbol{\chi},t). \tag{4.63}$$

Hint: Use Cauchy's integral theorem and assume that the moment generating function is sufficiently well behaved (analytic and decaying to zero for $|\chi_\nu| \to \infty$).

For any finite t we do not expect the symmetry (4.62) to be exactly valid, but instead it should emerge in the long time limit $t \to \infty$. Quite problematically, all moments and cumulants become proportional to t in the long time limit and therefore tend to diverge. On physical grounds, this is particularly easy to see for the first moment or cumulant. Suppose the system reaches after some time a non-equilibrium steady state ρ_{ness} characterized by non-vanishing steady-state heat fluxes $\dot{Q}_\nu^{\text{ness}} \neq 0$. These fluxes must match the total heat flux divided by t in the long time limit, i.e.

$$\dot{Q}_\nu^{\text{ness}} = \lim_{t\to\infty} \frac{Q_\nu(t)}{t} = \lim_{t\to\infty} \frac{\langle q\rangle(t)}{t}. \tag{4.64}$$

Hence, the first cumulant must scale linearly with time: $\langle q\rangle(t) \sim t$. As a consequence, it is more meaningful to work with the scaled cumulant generating function

$$S(\boldsymbol{\chi}) \equiv \lim_{t\to\infty} \frac{C(\boldsymbol{\chi},t)}{t} \tag{4.65}$$

to access the long time limit and the steady-state fluctuations. In fact, if this limit exists (which we assume from now on), the moment generating function becomes in the long time limit $M(\boldsymbol{\chi},t) \to e^{S(\boldsymbol{\chi})t}$ and the steady-state probabilities $\bar{p}(\mathbf{q})$ for the heat exchange statistics become

$$\bar{p}(\mathbf{q}) = \frac{1}{(2\pi)^n}\int_{-\infty}^{\infty} d\chi_1 \cdots \int_{-\infty}^{\infty} d\chi_n e^{-iq_1\chi_1}\ldots e^{-iq_n\chi_n} e^{S(\boldsymbol{\chi})t}. \tag{4.66}$$

A symmetry for $S(\boldsymbol{\chi})$ of the form (4.62) therefore implies a fluctuation theorem valid at steady state and we now show that

$$\boxed{\frac{\bar{p}(\mathbf{q})}{\bar{p}(-\mathbf{q})} = \exp\left(-\sum_{\nu=1}^{n} \beta_\nu q_\nu\right) \quad \Leftrightarrow \quad S(\boldsymbol{\chi}) = S(-i\boldsymbol{\beta} - \boldsymbol{\chi}) \ \ (\text{for } \boldsymbol{\chi} \in \mathbb{R}^n).} \tag{4.67}$$

Here, $\boldsymbol{\beta} = (\beta_1, \ldots, \beta_n)$ is the vector of inverse temperatures of the baths. Equation (4.67) corresponds to the steady-state exchange fluctuation theorem (4.46) in the absence of particle transport if we remember that $q_\nu = -\Delta\epsilon_\nu$ (the heat from bath ν is positive if it decreases the bath energy).

To show eqn (4.67), we diagonalize the counting field Liouvillian $\mathcal{L}(\boldsymbol{\chi})$ in superoperator space. Using the mapping from Appendix B.1, we can write the Liouvillian as an ordinary matrix as

$$\begin{aligned}\hat{\mathcal{L}}(\boldsymbol{\chi}) &= \sum_{\nu,\omega,\alpha} r_{\alpha,\nu}(\omega) e^{-i\hbar\omega\chi_\nu} S_{\alpha,\nu}(\omega) \otimes S^*_{\alpha,\nu}(\omega) \\ &\quad - \frac{1}{2} \sum_{\nu,\omega,\alpha} r_{\alpha,\nu}(\omega) \left\{ S^\dagger_{\alpha,\nu}(\omega) S_{\alpha,\nu}(\omega) \otimes I + I \otimes S^T_{\alpha,\nu}(\omega) S^*_{\alpha,\nu}(\omega) \right\}.\end{aligned} \tag{4.68}$$

This tedious and long expression does not look very encouraging, but luckily we only need to use a simple property of it later on. Furthermore, this expression is ready for numerical application. Unfortunately, $\hat{\mathcal{L}}(\boldsymbol{\chi})$ is not Hermitian and therefore there is no guarantee that it can be diagonalized in the conventional sense. However, it can always be put into Jordan normal form: $\hat{\mathcal{L}}(\boldsymbol{\chi}) = Q(\boldsymbol{\chi}) J(\boldsymbol{\chi}) Q(\boldsymbol{\chi})^{-1}$. Here, $Q(\boldsymbol{\chi})$ is some invertible transformation matrix and $J(\boldsymbol{\chi})$ is a block diagonal matrix of the form

$$J(\boldsymbol{\chi}) = \begin{pmatrix} J_0(\boldsymbol{\chi}) & & \\ & \ddots & \\ & & J_K(\boldsymbol{\chi}) \end{pmatrix}, \quad J_i(\boldsymbol{\chi}) = \begin{pmatrix} \lambda_i(\boldsymbol{\chi}) & 1 & & \\ & \lambda_i(\boldsymbol{\chi}) & \ddots & \\ & & \ddots & 1 \\ & & & \lambda_i(\boldsymbol{\chi}) \end{pmatrix}. \tag{4.69}$$

Each $J_i(\boldsymbol{\chi})$ is called a *Jordan block*, which contains the eigenvalues $\lambda_i(\boldsymbol{\chi})$ on the diagonal and every entry on the diagonal above the diagonal is a one (all other elements are zero). The eigenvalues are determined as usual by the roots of the characteristic polynomial $\det[\hat{\mathcal{L}}(\boldsymbol{\chi}) - \lambda I]$, where I is the identity matrix in superoperator space.

Next, we return to the moment generating function, which becomes

$$M(\boldsymbol{\chi}, t) = \mathrm{tr}_S\{e^{\mathcal{L}(\boldsymbol{\chi})t} \rho_S(0)\} = \langle\langle 1 | Q(\boldsymbol{\chi}) e^{J(\boldsymbol{\chi})t} Q(\boldsymbol{\chi})^{-1} | \rho_S(0) \rangle\rangle, \tag{4.70}$$

where $\rho_S(0)$ is an arbitrary initial state and $|\rho_S(0)\rangle\rangle$ denotes its vectorized version in superoperator space. For simplicity, we assume that the eigenvalue with the largest real part, denoted $\lambda_0(\boldsymbol{\chi})$ and obeying $\Re[\lambda_0(\boldsymbol{\chi})] > \Re[\lambda_k(\boldsymbol{\chi})]$ for all $k \neq 0$, is non-degenerate, i.e. the corresponding Jordan block $J_0(\boldsymbol{\chi})$ is a 1×1 matrix. Physically speaking, this is equivalent to assuming that there is one *unique* steady state of the dynamics. Consequently, the moment generating function can be written as

$$M(\boldsymbol{\chi}, t) = e^{\lambda_0(\boldsymbol{\chi})t} \langle\langle 1 | Q(\boldsymbol{\chi}) e^{[J(\boldsymbol{\chi}) - \lambda_0(\boldsymbol{\chi}) I]t} Q(\boldsymbol{\chi})^{-1} | \rho_S(0) \rangle\rangle, \tag{4.71}$$

which implies for the scaled cumulant generating function

$$\frac{C(\chi,t)}{t} = \lambda_0(\chi) + \frac{1}{t}\ln\langle\langle 1|Q(\chi)e^{[J(\chi)-\lambda_0(\chi)I]t}Q(\chi)^{-1}|\rho_S(0)\rangle\rangle. \tag{4.72}$$

We continue with a close look at the second term. First, notice that the matrix exponential of a Jordan block is easy to compute because $J_k(\chi) = \lambda_k(\chi)I + N$, where N is a *nilpotent* matrix, which means that $N^k = 0$ for some integer k. Thus,

$$e^{[J_k(\chi)-\lambda_0(\chi)I]t} = e^{[\lambda_k(\chi)-\lambda_0(\chi)]t}\left(I + Nt + \cdots + \frac{1}{(k-1)!}(Nt)^{k-1}\right) \tag{4.73}$$

and the long time limit of this term is zero since $\Re[\lambda_k(\chi) - \lambda_0(\chi)] < 0$. Therefore, in the long time limit $e^{[J(\chi)-\lambda_0(\chi)I]t}$ becomes time independent and only the 1×1 block containing the eigenvalue $\lambda_0(\chi)$ contributes. This implies for the scaled cumulant generating function

$$\boxed{S(\chi) = \lim_{t\to\infty}\frac{C(\chi,t)}{t} = \lambda_0(\chi).} \tag{4.74}$$

To conclude the derivation of the steady-state fluctuation theorem, we look at the eigenvalues of the Liouvillian, which are determined by the roots of the characteristic polynomial $\det[\hat{\mathcal{L}}(\chi) - \lambda I]$. Now, it turns out that this characteristic polynomial obeys the following symmetry relation:

$$\det[\hat{\mathcal{L}}(\chi) - \lambda I] = \det[\hat{\mathcal{L}}(\chi - i\boldsymbol{\beta}) - \lambda^* I]^*, \tag{4.75}$$

which follows from $\hat{\mathcal{L}}(\chi - i\boldsymbol{\beta}) = \hat{\mathcal{L}}(\chi)^\dagger$.

Exercise 4.12 Show that $\hat{\mathcal{L}}(\chi - i\boldsymbol{\beta}) = \hat{\mathcal{L}}(\chi)^\dagger$.

Consequently, this symmetry is also shared by all eigenvalues, $\lambda_k(\chi) = \lambda_k^*(\chi - i\boldsymbol{\beta})$, and hence also by the scaled cumulant generating function: $S(\chi) = S(\chi - i\boldsymbol{\beta})^*$. This symmetry is not yet of the desired form (4.67), but now we recall eqn (4.56), which finally implies $S(\chi) = S(\chi - i\boldsymbol{\beta})^* = S(-\chi - i\boldsymbol{\beta})$ for $\chi \in \mathbb{R}^n$. Thus, the steady-state fluctuation theorem follows.

In summary, the counting field method allows one to access the energy exchange statistics with the baths. It yields fluctuation theorems of the expected form based on our previous experience. Therefore, the following question arises: Is the energy exchange statistics obtained from the counting field method *identical* to the statistics obtained from the two-point measurement scheme? In short, if the BMS equation describes the dynamics accurately, the answer is: yes! Readers interested in proving this will find a helpful reference at the end of this chapter.

Finally, we remark that counting field methods can be applied to studying not only energy exchange statistics, but also particle exchange statistics. The last exercise of this section explores this in detail for the example of the single-electron transistor.

Exercise 4.13 Consider the single-electron transistor from Section 3.10 for $\beta_L = \beta_R \equiv \beta$, the dynamics of which is described by the rate master equation (see eqn (3.174))

$$\frac{d}{dt}\begin{pmatrix} p_F(t) \\ p_E(t) \end{pmatrix} = \sum_\nu \Gamma_\nu(\epsilon_0) \begin{pmatrix} -[1-f_\nu(\epsilon_0)] & f_\nu(\epsilon_0) \\ 1-f_\nu(\epsilon_0) & -f_\nu(\epsilon_0) \end{pmatrix} \begin{pmatrix} p_F(t) \\ p_E(t) \end{pmatrix}. \tag{4.76}$$

In the following, we set for simplicity $\Gamma_L(\epsilon_0) = \Gamma_R(\epsilon_0) \equiv \Gamma$, $\mu_L = \epsilon_0 + V/2$ and $\mu_R = \epsilon_0 - V/2$, where V denotes the voltage bias. We also set the electron charge to one: $e \equiv 1$.

We are now interested in counting the number of electron jumps n_ν from bath $\nu \in \{L, R\}$ into the system. The state of the system conditioned on $\boldsymbol{n} = (n_L, n_R)$ jumps is denoted by $p_\sigma(t|\boldsymbol{n})$, where $\sigma \in \{E, F\}$ denotes the state of the quantum dot (empty or filled). Since the number of electron jumps is discrete, $n_\nu \in \mathbb{Z}$, it is convenient to define the counting field by

$$\rho_\sigma(\boldsymbol{\chi}, t) = \sum_{\boldsymbol{n}} e^{i\boldsymbol{n}\cdot\boldsymbol{\chi}} p_\sigma(t|\boldsymbol{n}) \quad \Leftrightarrow \quad p_\sigma(t|\boldsymbol{n}) = \int_{-\pi}^{\pi} \frac{d\chi_L}{2\pi} \int_{-\pi}^{\pi} \frac{d\chi_R}{2\pi} e^{-i\boldsymbol{n}\boldsymbol{\chi}} \rho_\sigma(\boldsymbol{\chi}, t) \tag{4.77}$$

with $\boldsymbol{\chi} = (\chi_L, \chi_R)$. Deduce that the master equation with counting fields reads

$$\frac{d}{dt}\begin{pmatrix} \rho_F(\boldsymbol{\chi}, t) \\ \rho_E(\boldsymbol{\chi}, t) \end{pmatrix} = \Gamma \sum_\nu \begin{pmatrix} -[1-f_\nu(\epsilon_0)] & e^{i\chi_\nu} f_\nu(\epsilon_0) \\ e^{-i\chi_\nu}[1-f_\nu(\epsilon_0)] & -f_\nu(\epsilon_0) \end{pmatrix} \begin{pmatrix} \rho_F(\boldsymbol{\chi}, t) \\ \rho_E(\boldsymbol{\chi}, t) \end{pmatrix}. \tag{4.78}$$

Next, compute the two eigenvalues $\lambda_\pm(\boldsymbol{\chi})$ of the rate matrix with counting fields. Confirm that both obey the symmetries

$$\lambda_\pm(\chi_L, \chi_R) = \lambda_\pm(\chi_L - \chi_R, 0), \quad \lambda_\pm(\chi_L, \chi_R) = \lambda_\pm\left(-\chi_L + i\frac{\beta V}{2}, -\chi_R - i\frac{\beta V}{2}\right). \tag{4.79}$$

Show that the first one implies at steady state the conservation law $n_L + n_R = 0$. Furthermore, show that the second one implies the exchange fluctuation theorem from eqn (4.40).

4.5 Stochastic Quantum Jump Trajectories

Section 4.4 showed how to study energy and particle exchange statistics beyond their mean value using a quantum master equation. However, in contrast to the classical case, the notion of a 'single stochastic trajectory' is not yet transparent. Furthermore, solving either eqn (4.52) or eqn (4.54) can be cumbersome in practice. The goal of this section is to develop a method that overcomes these difficulties.

Dynamical description

For simplicity, we start with the quantum master equation

$$\frac{\partial}{\partial t}\rho_S(t) = J\rho_S(t)J^\dagger - \frac{1}{2}\{J^\dagger J, \rho_S(t)\}, \tag{4.80}$$

where J is an arbitrary system operator. Hence, there is only one jump superoperator $\mathcal{J}\rho_S = J\rho_S J^\dagger$. Counting the number n of jumps, we find with the method used in Section 4.4 the jump-resolved master equation

$$\frac{\partial}{\partial t}\rho_S(t|n) = J\rho_S(t|n-1)J^\dagger - \frac{1}{2}\{J^\dagger J, \rho_S(t|n)\}. \tag{4.81}$$

Despite its simplicity, there are relevant physical scenarios which can be described with a single-jump operator. The next exercise provides an intuitive example, which is helpful to keep in mind to motivate our subsequent manipulations. Of course, at the end of this section, we generalize the theory properly.

Exercise 4.14 In Exercise 3.28 we derived the master equation of a damped cavity. Show that eqn (3.136) is of the form (4.80) in the parameter limit $k_B T/\hbar\omega_c \to 0$, which is often a good approximation. What is the physical meaning of J and n in the case of a damped cavity?

The probability of a quantum jump (or, in view of the previous exercise, a photon leaking out of the cavity) at time t in the next infinitesimal time step dt is

$$p_{\text{jump}}(t) = \text{tr}_S\{\mathcal{J}\rho_S(t)\}dt = \text{tr}_S\{J^\dagger J\rho_S(t)\}dt. \tag{4.82}$$

In view of the generalized quantum measurement theory developed in Section 1.4, $\mathcal{M}_1 \equiv \mathcal{J}dt$, being a CP map, can be seen as a measurement superoperator associated with the detection event of, for example, a photon. However, in order to describe a quantum measurement, there must be at least a second superoperator $\mathcal{M}_0$ such that $\mathcal{M}_0 + \mathcal{M}_1$ is not only a CP but a CPTP map. Thus, we choose

$$\mathcal{M}_0\rho_S \equiv \left(1 - \frac{J^\dagger J}{2}dt\right)\rho_S\left(1 - \frac{J^\dagger J}{2}dt\right). \tag{4.83}$$

Then, the map $\mathcal{M}_0 + \mathcal{M}_1$ is CP and also trace-preserving to first order in dt. Since dt was assumed to be infinitesimal, the terms of $\mathcal{O}(dt^2)$ are negligible. Physically speaking, if $\mathcal{M}_1$ corresponds to the detection of a jump, $\mathcal{M}_0$ is the complementary superoperator corresponding to the detection of 'no jump'. The probability of no jump is

$$p_{\text{no jump}}(t) = \text{tr}_S\{\mathcal{M}_0\rho_S(t)\} = 1 - p_{\text{jump}}(t) + \mathcal{O}(dt^2). \tag{4.84}$$

Remember that J is completely arbitrary in our argumentation. The next exercise shows that a quantum master equation can be seen as being equivalent to a (so far fictitious) quantum measurement process, as just introduced.

Exercise 4.15 Consider the average post-measurement state $\rho_S(t+dt) = (\mathcal{M}_0+\mathcal{M}_1)\rho_S(t)$. Show that for infinitesimal dt the time evolution $\rho_S(t+dt) = (\mathcal{M}_0 + \mathcal{M}_1)\rho_S(t)$ is identical to the prediction of the master equation (4.80).

The picture that emerges from these considerations is therefore in unison with the one emerging from our previous result (4.49), where we wrote $\rho_S(t)$ in the form of a Dyson series: the system evolution is interrupted at random times by sudden jumps $\mathcal{J}$. However, this random character does not yet become completely transparent by looking at the jump-resolved master equation (4.81), which is a deterministic differential equation on a large space. Therefore, our goal is now to find an evolution equation for the system where this randomness becomes transparent.

For this purpose, let $n(t)$ denote the number of jumps up to time t, which we have seen in a single run of a (fictitious) experiment. Furthermore, let us assume that the state at time t is pure and denoted by $|\psi(t)\rangle$ (for simplicity, we drop the subscript S on the wavefunction $|\psi(t)\rangle$ in the following). We now ask: How can $n(t)$ change in the next infinitesimal time step, i.e. what is $dn(t) \equiv n(t+dt) - n(t)$? Clearly, there can be either one jump or no jump and hence $dn(t) \in \{0, 1\}$. This argument rests on the fact that dt is infinitesimal, otherwise multiple jumps can happen in a finite time interval. Furthermore, we expect a jump with probability $p_{\text{jump}}(t)$. These two rules can be summarized as

$$\boxed{dn(t)^2 = dn(t), \quad \mathbb{E}_{\psi(t)}[dn(t)] = p_{\text{jump}}(t) = \langle J^\dagger J\rangle_{\psi(t)} dt.} \tag{4.85}$$

Here, $\mathbb{E}_{\psi(t)}[\dots]$ denotes the expectation value with respect to the random jumps for a *fixed* state $|\psi(t)\rangle$ and $\langle J^\dagger J\rangle_{\psi(t)} \equiv \langle\psi(t)|J^\dagger J|\psi(t)\rangle$ denotes the quantum mechanical expectation value. A process described by a stochastic increment $dn(t)$ obeying the conditions (4.85) is called a *point process*.

Based on the previous considerations, we can also compute how $|\psi(t)\rangle$ changes in the next time step. For that purpose, we write the measurement superoperators as $\mathcal{M}_0\rho_S = M_0\rho_S M_0^\dagger$ and $\mathcal{M}_1\rho_S = M_1\rho_S M_1^\dagger$ with the measurement operators

$$M_0 = 1 - \frac{J^\dagger J}{2}dt, \quad M_1 = \sqrt{dt}J. \tag{4.86}$$

Thus, if a jump happens, $|\psi(t)\rangle$ changes to

$$|\psi(t+dt)\rangle = \frac{M_1|\psi(t)\rangle}{\sqrt{\langle M_1^\dagger M_1\rangle_{\psi(t)}}} = \frac{J|\psi(t)\rangle}{\sqrt{\langle J^\dagger J\rangle_{\psi(t)}}}. \tag{4.87}$$

Here, we made sure that $|\psi(t+dt)\rangle$ is normalized, which explains the denominator. Likewise, if no jump happens, $|\psi(t)\rangle$ changes to

$$|\psi(t+dt)\rangle = \frac{M_0|\psi(t)\rangle}{\sqrt{\langle M_0^\dagger M_0\rangle_{\psi(t)}}} = \left[1 + \frac{dt}{2}\left(\langle J^\dagger J\rangle_{\psi(t)} - J^\dagger J\right) + \mathcal{O}(dt^2)\right]|\psi(t)\rangle, \tag{4.88}$$

where we used a Taylor expansion up to first order in dt for the second equality.

Now, we can combine eqns (4.87) and (4.88) into a single equation by using that $dn(t)$ is one (zero) in case of a jump (no jump). Hence,

$$|\psi(t+dt)\rangle = dn(t)\frac{J|\psi(t)\rangle}{\sqrt{\langle J^\dagger J\rangle_{\psi(t)}}} + [1 - dn(t)]\left[1 + \frac{dt}{2}\left(\langle J^\dagger J\rangle_{\psi(t)} - J^\dagger J\right)\right]|\psi(t)\rangle, \tag{4.89}$$

where we neglected terms of order dt^2. In fact, this also allows us to neglect terms proportional to $dn(t)dt$ because the expected number of jumps in an infinitesimal time interval dt is itself of order dt. Furthermore, subtracting $|\psi(t)\rangle$ from the previous equation and denoting $d|\psi(t)\rangle \equiv |\psi(t+dt)\rangle - |\psi(t)\rangle$, we arrive at

$$\boxed{d|\psi(t)\rangle = \left[dn(t)\left(\frac{J}{\sqrt{\langle J^\dagger J\rangle_{\psi(t)}}} - 1\right) + \frac{dt}{2}\left(\langle J^\dagger J\rangle_{\psi(t)} - J^\dagger J\right)\right]|\psi(t)\rangle.} \tag{4.90}$$

This is known as a **stochastic Schrödinger equation**. It is an insightful and useful computational tool to describe the dynamics of open quantum systems in terms of a stochastic trajectory of quantum jumps. Hence, a particular solution of the stochastic Schrödinger equation is called a **quantum jump trajectory**. The process of decomposing a master equation such as eqn (4.80) into a stochastic evolution in the form of eqn (4.90) is called **unravelling**. Finally, notice that eqn (4.90) is *nonlinear* with respect to $|\psi(t)\rangle$ in contrast to the conventional Schrödinger equation. The reason one calls it a 'Schrödinger equation' comes from the fact that the state of the system remains *pure* along a quantum jump trajectory, which you are asked to prove later on.

A simple way to apply eqn (4.90) in practice works as follows. First, one discretizes time and replaces the infinitesimal dt by a finite but small δt. Next, one starts with some initial pure state $|\psi(0)\rangle$, which might be drawn at random from an ensemble of initial states $\rho_S(0) = \sum_k \lambda_k(0)|\psi_k(0)\rangle\langle\psi_k(0)|$. Then, at each time step, a random number $r(t)$ is chosen uniformly from the unit interval $[0, 1]$. If

$$r(t) < p_{\text{jump}}(t) = \langle J^\dagger J\rangle_{\psi(t)}\delta t, \tag{4.91}$$

one sets $dn(t) = 1$ and computes $|\psi(t+\delta t)\rangle$ according to eqn (4.90). Otherwise, if $r(t) \geq p_{\text{jump}}(t)$, one sets $dn(t) = 0$ and computes $|\psi(t+\delta t)\rangle$ using the same equation.

Repeating this procedure many times, we obtain an *ensemble* of quantum jump trajectories and the average over this ensemble is denoted by $\mathbb{E}[\ldots]$. If we set $t = N\delta t$ for some $N \in \mathbb{N}$, we can write the ensemble average more explicitly as

$$\mathbb{E}[\ldots] = \mathbb{E}_{\lambda_k(0)}[\mathbb{E}_{\psi_k(0)}[\mathbb{E}_{\psi_k(\delta t)}[\ldots\mathbb{E}_{\psi_k(N\delta t)}[\ldots]\ldots]]]. \tag{4.92}$$

In the case that the initial system state is mixed, $\rho_S(0) = \sum_k \lambda_k(0)|\psi_k(0)\rangle\langle\psi_k(0)|$, the first ensemble average $\mathbb{E}_{\lambda_k(0)}[\ldots]$ describes the average of randomly sampling an initial pure state $|\psi_k(0)\rangle$ according to the distribution $\{\lambda_k(0)\}$. This average is then followed by averages over the point process with respect to the time-evolved state.

We now consider the ensemble average of the stochastic density matrix $\sigma(t) \equiv |\psi(t)\rangle\langle\psi(t)|$ at time t, where $|\psi(t)\rangle$ is obtained from the stochastic Schrödinger equation. The claim is that

$$\rho_S(t) = \mathbb{E}[\sigma(t)], \tag{4.93}$$

where $\rho_S(t)$ is the solution of the master equation (4.80) we started with. Clearly, given the way in which we constructed the stochastic Schrödinger equation, this should be the case, but it is intructive to explicitly verify it.

To this end, we consider the stochastic evolution equation for $\sigma(t)$. Denoting $d\sigma(t) = \sigma(t+dt) - \sigma(t)$, one finds to first order in dt

$$d\sigma(t) = dn(t)\left[\frac{J\sigma(t)J^\dagger}{\langle J^\dagger J\rangle_{\sigma(t)}} - \sigma(t)\right] + dt\left[\langle J^\dagger J\rangle_{\sigma(t)}\sigma(t) - \frac{1}{2}\{J^\dagger J, \sigma(t)\}\right], \tag{4.94}$$

where $\langle J^\dagger J\rangle_{\sigma(t)} = \mathrm{tr}_S\{J^\dagger J\sigma(t)\} = \langle J^\dagger J\rangle_{\psi(t)}$.

Exercise 4.16 Derive eqn (4.94). *Hint:* In stochastic calculus, the product rule of differentiation needs to be changed to

$$d\sigma(t) = |d\psi(t)\rangle\langle\psi(t)| + |\psi(t)\rangle\langle d\psi(t)| + |d\psi(t)\rangle\langle d\psi(t)|. \tag{4.95}$$

In contrast to standard calculus, the last term is not necessarily negligible, i.e. it is not necessarily of higher order in dt. This becomes transparent by recalling that the stochastic increment obeys $dn(t)^2 = dn(t)$. Use eqn (4.95) to derive eqn (4.94).

Furthermore, eqn (4.94) preserves the purity of the initial state, i.e. if $\sigma(0)$ is initially pure, then it remains pure for all times. In hindsight, this observation justifies working with a stochastic Schrödinger equation (4.90) instead of a stochastic master equation such as eqn (4.94). This also simplifies the numerical implementation as $|\psi(t)\rangle$ has only $\dim \mathcal{H}_S$ many components instead of $\sigma(t)$, which has $(\dim \mathcal{H}_S)^2$ many.

Exercise 4.17 Show that eqn (4.94) preserves the purity of a state. To this end, suppose that $\sigma(t) = |\psi(t)\rangle\langle\psi(t)|$ is pure. Then, show that this implies $\sigma^2(t+dt) = \sigma(t+dt)$, i.e. $\sigma(t+dt) = |\psi(t+dt)\rangle\langle\psi(t+dt)|$ is also pure.

Now, to show eqn (4.93), we take the ensemble average of eqn (4.94), which yields

$$d\varrho(t) = \mathbb{E}\left[dn(t)\frac{J\sigma(t)J^\dagger}{\langle J^\dagger J\rangle_{\sigma(t)}} - dn(t)\sigma(t) + dt\langle J^\dagger J\rangle_{\sigma(t)}\sigma(t)\right] - \frac{dt}{2}\{J^\dagger J, \varrho(t)\}, \quad (4.96)$$

where we denoted $\varrho(t) \equiv \mathbb{E}[\sigma(t)]$. It remains to evaluate the ensemble average of the three terms in the square bracket. To do so, we recall that $dn(t)$ is a stochastic increment of a point process, which depends only on the actual state of the system $\sigma(t)$, but not on any previous states. Since the last ensemble average in eqn (4.92) averages over the point process for a *fixed* state $\sigma(t)$, we can replace $dn(t)$ under the ensemble average $\mathbb{E}$ by its expectation value $\langle J^\dagger J\rangle_{\sigma(t)}$. Thus, $\mathbb{E}[dn(t)\sigma(t)] = \mathbb{E}[\langle J^\dagger J\rangle_{\sigma(t)}\sigma(t)]dt$, which implies that the second and third terms in the square of eqn (4.96) cancel. The remaining one becomes, according to the same philosophy,

$$\mathbb{E}\left[dn(t)\frac{J\sigma(t)J^\dagger}{\langle J^\dagger J\rangle_{\sigma(t)}}\right] = \mathbb{E}[J\sigma(t)J^\dagger]dt = J\varrho(t)J^\dagger dt. \quad (4.97)$$

Taken together, we obtain

$$d\varrho(t) = dt J\varrho(t)J^\dagger - \frac{dt}{2}\{J^\dagger J, \varrho(t)\}. \quad (4.98)$$

After dividing by dt, we see that $\varrho(t)$ obeys the same master equation as $\rho_S(t)$, eqn (4.80). Thus, if the initial conditions coincide, eqn (4.93) follows.

In the next exercise, readers are asked to explicity confirm our reasoning above by applying the formalism to the case of the cavity master equation.

Exercise 4.18 Exercise 4.14 has shown that the master equation of an open cavity in the low-temperature and high-frequency regime obeys eqn (4.80) with $J = \sqrt{\gamma}a$ for some rate γ. Show that the average number of photons in the cavity decays exponentially to zero: $\langle a^\dagger a\rangle(t) = e^{-\gamma t}\langle a^\dagger a\rangle(0)$. Now, reproduce this behaviour using the stochastic Schrödinger equation. An example is shown in Fig. 4.3.

The above analysis can be generalized to a quantum master equation of the form

$$\frac{\partial}{\partial t}\rho_S(t) = -\frac{i}{\hbar}[H, \rho_S(t)] + \sum_k J_k\rho_S(t)J_k^\dagger - \frac{1}{2}\{J_k^\dagger J_k, \rho_S(t)\}, \quad (4.99)$$

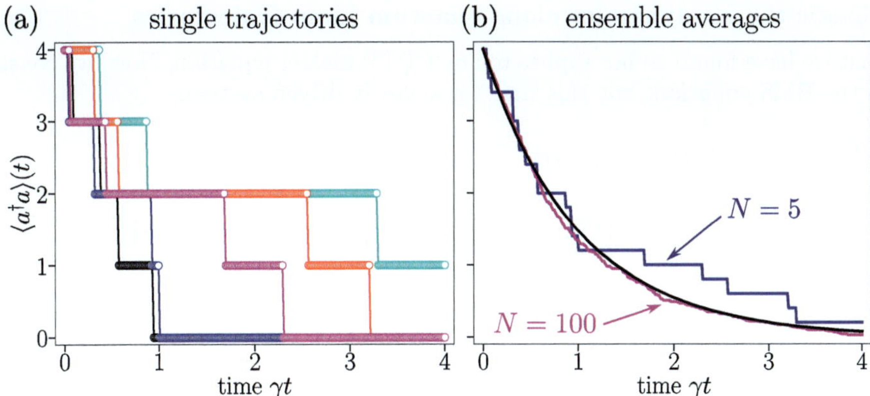

Fig. 4.3 Quantum trajectory simulation of a damped cavity initialized in a Fock state with four photons: $a^\dagger a|\psi(0)\rangle = 4|\psi(0)\rangle$. The plots show the time evolution of the photon number expectation value $\langle a^\dagger a\rangle(t)$ over (dimensionless) time γt. (a) $\langle a^\dagger a\rangle(t)$ evaluated along five examples of quantum jump trajectories, each obtained for a single realization of the process. (b) Ensemble average of $\langle a^\dagger a\rangle(t)$ for the five trajectories shown in (a) (noisy blue curve) and for 100 trajectories (noisy pink curve). Furthermore, we also plot the solution of the master equation (smooth black curve).

which contains a coherent part with $H = H^\dagger$ and multiple jump superoperators $\mathcal{J}_k\rho_S = J_k\rho_S J_k^\dagger$. The stochastic Schrödinger equation, which ensues from unravelling eqn (4.99), reads

$$\boxed{\begin{aligned} d|\psi(t)\rangle = \sum_k &\left[dn_k(t)\left(\frac{J_k}{\langle J_k^\dagger J_k\rangle_{\psi(t)}^{1/2}} - 1\right) + \frac{dt}{2}\left(\langle J_k^\dagger J_k\rangle_{\psi(t)} - J_k^\dagger J_k\right)\right]|\psi(t)\rangle \\ &- dt\frac{i}{\hbar}H|\psi(t)\rangle. \end{aligned}} \tag{4.100}$$

The point process is now defined by the relations

$$\boxed{dn_k(t)dn_\ell(t) = \delta_{k,\ell}dn_k(t), \quad \mathbb{E}_{\psi(t)}[dn_k(t)] = \langle J_k^\dagger J_k\rangle_{\psi(t)}dt.} \tag{4.101}$$

Furthermore, when simulating eqn (4.100), one uses one random number $r(t)$ to decide whether any jump at all happens according to (see eqn (4.91))

$$r(t) < p_{\text{jump}}(t) = \sum_k \langle J_k^\dagger J_k\rangle_{\psi(t)}\delta t, \tag{4.102}$$

and a second random number $r'(t)$ drawn uniformly from the unit interval to decide which jump $\mathcal{J}_k$ happens according to the weights $\langle J_k^\dagger J_k\rangle_{\psi(t)}\delta t/p_{\text{jump}}(t)$.

Exercise 4.19 Make sure you understand eqn (4.100). Can you repeat the same steps that we used for the simple master equation (4.80) also for the general case?

Stochastic thermodynamics along quantum jump trajectories

All that we have found so far applies to any CPTP master equation. Now, we consider again the BMS equation, but this time for a slowly driven system:

$$\begin{aligned}\frac{\partial}{\partial t}\rho_S(t) =& -\frac{i}{\hbar}[H_S(\lambda_t), \rho_S(t)] \\ &+ \sum_{\omega_t} r(\omega_t)\left[S(\omega_t)\rho_S(t)S^\dagger(\omega_t) - \frac{1}{2}\{S^\dagger(\omega_t)S(\omega_t), \rho_S(t)\}\right].\end{aligned} \tag{4.103}$$

We have met this situation already in Section 3.4: $H_S(\lambda_t)$ is the time-dependent system Hamiltonian, which varies slowly compared with the decay of the bath correlation function, and $\omega_t = \omega(\lambda_t)$ denotes its instantaneous transition frequencies. Furthermore, to focus on the essential elements, we assume no additional index α in contrast to eqn (4.47). The generalization is easily carried out at the end. Finally, the notation above is still not compact enough for our purposes. In the following, we therefore write H_t, ω, r_ω, S_ω and $dn_\omega(t)$ instead of $H_S(\lambda_t)$, ω_t, $r(\omega_t)$, $S(\omega_t)$ and $dn(\omega_t, t)$.

The evolution of the stochastic density matrix $\sigma(t) = |\psi(t)\rangle\langle\psi(t)|$ along a quantum jump trajectory is then obtained by generalizing eqn (4.94) appropriately:

$$\begin{aligned}d\sigma(t) =& -dt\frac{i}{\hbar}[H_t, \sigma(t)] + \sum_\omega dn_\omega(t)\left[\frac{S_\omega\sigma(t)S_\omega^\dagger}{\langle S_\omega^\dagger S_\omega\rangle_{\sigma(t)}} - \sigma(t)\right] \\ &+ dt\sum_\omega r_\omega\left[\langle S_\omega^\dagger S_\omega\rangle_{\sigma(t)}\sigma(t) - \frac{1}{2}\{S_\omega^\dagger S_\omega, \sigma(t)\}\right].\end{aligned} \tag{4.104}$$

The stochastic increments satisfy the two rules $dn_\omega(t)dn_{\omega'}(t) = \delta_{\omega,\omega'}dn_\omega(t)$ and $\mathbb{E}_{\psi(t)}[dn_\omega(t)] = r_\omega \mathrm{tr}_S\{S_\omega^\dagger S_\omega\sigma(t)\}dt$. The goal is now to set up a stochastic thermodynamics framework for this equation.

We start with the **stochastic first law**. Based on our previous experience with the weak coupling regime, there is little reason to doubt why our definition of internal energy should not work along a single trajectory when the system state is pure. Thus, we define the **stochastic internal energy**

$$u_S[\psi(t), \lambda_t] \equiv \mathrm{tr}_S\{H_t\sigma(t)\} = \langle\psi(t)|H_t|\psi(t)\rangle. \tag{4.105}$$

The change in stochastic internal energy gives rise to the stochastic first law,

$$du_S[\psi(t), \lambda_t] = \mathrm{tr}_S\{[dH_t]\sigma(t)\} + \mathrm{tr}_S\{H_t[d\sigma(t)]\}, \tag{4.106}$$

after appropriately identifying stochastic heat and work. In Section 3.1 we saw that the definition of mechanical work is independent of the system state. Thus, we define the **stochastic work**

$$đw(t) \equiv \mathrm{tr}_S\{[dH_t]\sigma(t)\}. \tag{4.107}$$

The term $đq(t) \equiv \mathrm{tr}_S\{H_t[d\sigma(t)]\}$ is consequently identified as **stochastic heat**. A close look at the heat along a quantum jump trajectory reveals that interesting insights can be gained by splitting it into two parts, *classical* heat and **quantum heat**,

$$\boxed{đq(t) = đq_{\rm cl}(t) + đq_{\rm qu}(t).} \tag{4.108}$$

We start with the intuitive definition of classical heat, which reads

$$đq_{\rm cl}(t) = -\sum_\omega \hbar\omega dn_\omega(t). \tag{4.109}$$

Thus, whenever a jump labelled by ω happens, the energy of the system decreases by $\hbar\omega$, which adds a contribution $-\hbar\omega$ to the stochastic heat. Remember that at any time step at most one stochastic increment satisfies $dn_\omega(t) = 1$. The histogram of $đq_{\rm cl}(t)$ constructed from simulating many quantum jump trajectories therefore coincides with the heat exchange statistics obtained from the counting field method developed in Section 4.4. Note that analysing the heat along a stochastic trajectory of a *classical* rate master equation as considered in Section 2.4 would give rise to the same expression (4.109), hence the terminology 'classical' heat. In fact, one can show that coherences play no role in the evaluation of $đq_{\rm cl}(t)$ because coherences do not change the probability of a jump happening:

$$p_{\text{jump }\omega}(t) = r_\omega \text{tr}_S\{S_\omega^\dagger S_\omega \rho_S(t)\}dt = r_\omega \text{tr}_S\{S_\omega^\dagger S_\omega \mathcal{D}_{H_t}\rho_S(t)\}dt. \tag{4.110}$$

Here, $\mathcal{D}_{H_t}$ denotes the dephasing operation with respect to the eigenbasis of H_t.

Exercise 4.20 Derive eqn (4.110).

In general, it turns out that $đq(t) \neq đq_{\rm cl}(t)$ and the missing part is called quantum heat, which is a novel feature that does not become transparent in the full counting statistics framework. The explicit definition of quantum heat looks a bit cumbersome:

$$\begin{aligned} đq_{\rm qu}(t) \equiv & \sum_\omega dn_\omega(t)\text{tr}_S\left\{\frac{S_\omega[H_t\sigma(t) + \sigma(t)H_t]S_\omega^\dagger}{2\langle S_\omega^\dagger S_\omega\rangle_{\sigma(t)}} - H_t\sigma(t)\right\} \\ & + dt\sum_\omega r_\omega \text{tr}_S\left\{\langle S_\omega^\dagger S_\omega\rangle_{\sigma(t)} H_t\sigma(t) - H_t S_\omega^\dagger S_\omega\sigma(t)\right\}. \end{aligned} \tag{4.111}$$

Its derivation, which is straightforward, is relegated to an exercise.

Exercise 4.21 Derive eqn (4.111) starting from $đq_{\rm qu}(t) = \text{tr}_S\{H_t[d\sigma(t)]\} - đq_{\rm cl}(t)$ and using eqn (4.104). *Hint:* It could be useful to first express the classical heat as

$$đq_{\rm cl}(t) \equiv \sum_\omega dn_\omega(t)\text{tr}_S\left\{H_t\frac{S_\omega\sigma(t)S_\omega^\dagger}{\langle S_\omega^\dagger S_\omega\rangle_{\sigma(t)}} - \frac{S_\omega[H_t\sigma(t) + \sigma(t)H_t]S_\omega^\dagger}{2\langle S_\omega^\dagger S_\omega\rangle_{\sigma(t)}}\right\}, \tag{4.112}$$

which can be derived by using $[H_t, S_\omega^\dagger S_\omega] = 0$ and $[S_\omega, H_t] = \hbar\omega S_\omega$.

The first crucial property of quantum heat is that it vanishes whenever the state along a quantum jump trajectory is classical, which means that it has no coherence in the energy eigenbasis. Since $\sigma(t)$ is pure, such a state can always be written as $\sigma(t) = |\epsilon_t\rangle\langle\epsilon_t|$, where $|\epsilon_t\rangle$ labels a state with eigenenergy ϵ_t. Note that, if ϵ_t labels a degenerate eigenenergy subspace, there is a freedom to rotate the energy eigenbasis to justify the desired form. Based on this, it is not too hard to confirm that

$$đq_{\text{qu}}(t) = 0 \quad \text{if } \sigma(t) \text{ is classical.} \tag{4.113}$$

To give an example where $đq_{\text{qu}}(t)$ does not vanish, consider a two-level system with time-independent Hamiltonian $H_S = \frac{\hbar\Omega}{2}(|1\rangle\langle 1| - |0\rangle\langle 0|)$. Suppose the system starts in the maximally coherent state $|+\rangle\langle +|$ with $|+\rangle = (|1\rangle + |0\rangle)/\sqrt{2}$ and suddenly jumps at the next time step to, say, the ground state $|0\rangle$. According to eqn (4.109), the classical heat is $đq_{\text{cl}}(t) = -\hbar\Omega$. If that were the only contribution to heat, we would encounter an odd situation because the internal energy is initially $u_S(|+\rangle) = 0$ and finally $u_S(|0\rangle) = -\hbar\Omega/2$. Hence, $\Delta u_S = -\hbar\Omega/2 \neq đq_{\text{cl}}(t)$. The missing part of the first law is provided by the quantum heat.

A second crucial property of the quantum heat is that it vanishes after taking the ensemble average:

$$\mathbb{E}[đq_{\text{qu}}(t)] = 0. \tag{4.114}$$

Of course, this must be the case because we showed in the previous section that the counting field formalism, which takes into account only the classical part of the heat, coincides with the conventional definition of heat obtained in Chapter 3; compare with Exercise 4.10. Showing eqn (4.114) is left as an exercise.

Exercise 4.22 Show eqn (4.114) based on the explicit definition (4.111).

Finally, we ask whether we can also analyse the second law along a quantum jump trajectory generated by eqn (4.104). This requires us to introduce the notion of **stochastic system entropy** along a quantum jump trajectory. It turns out that this problem has not yet been satisfactorily resolved. It is only if the system state stays diagonal in the energy eigenbasis that there is a known solution, which, of course, already follows from the framework of classical stochastic thermodynamics.

To exemplify this problem, we consider again the case of the time-independent two-level system, but this time we consider the initial state $\rho_S(0) = \lambda_+|+\rangle\langle +| + \lambda_-|-\rangle\langle -|$ with $|-\rangle = (|0\rangle - |1\rangle)/\sqrt{2}$. First, according to Section 4.3, the stochastic entropy of the system is defined as minus the logarithm of the probabilities of finding the system in an eigenstate of $\rho_S(t)$, i.e. in the case of $\rho_S(0)$ this is either $-k_B \ln \lambda_+$ or $-k_B \ln \lambda_-$. This is the interpretation within the two-point measurement scheme, which gives rise to the entropy production fluctuation theorem (4.32). Yet, how should this stochastic entropy be computed along a quantum jump trajectory? For instance, assume that in a particular run of the simulation the system jumps at the first time step to either the state $|0\rangle$ or the state $|1\rangle$. Which of the two stochastic entropies $-k_B \ln \lambda_+$ and $-k_B \ln \lambda_-$ should be associated with the state $|0\rangle$ or $|1\rangle$? The problem persists for any time t as long as the average state $\rho_S(t)$ contains coherences in the energy eigenbasis.

Of course, one can use quantum jump trajectories to compute the change in $k_B S_{\mathrm{vN}}[\rho_S(t)]$ with $\rho_S(t) = \mathbb{E}[\sigma(t)]$, which—as we know from Chapter 3—yields a second law of the form $k_B \Delta S_{\mathrm{vN}}[\rho_S(t)] - Q(t)/T \geq 0$. The only problem is that it is not possible to establish a unique and general definition of stochastic entropy $s_S[\psi(t)]$ along a quantum jump trajectory such that $\mathbb{E}\{s_S[\psi(t)]\} = k_B S_{\mathrm{vN}}[\rho_S(t)]$.

We remark that we have so far viewed the quantum jump trajectory method (as well as the counting field approach) as a convenient theoretical and numerical tool to set up a framework of stochastic thermodynamics for weakly coupled open quantum systems. Perhaps the above problems diminish if we view it from an *experimental* perspective? It turns out that the opposite is the case.

First, let us ask what happens if we try to infer stochastic trajectories of a quantum system as in classical stochastic thermodynamics by monitoring the *system*. For instance, one could repeatedly perform measurements of the system energy at small time steps. Unfortunately, whenever the system has coherences in the energy eigenbasis, this measurement strategy destroys them. Thus, this changes the quantum heat and the system entropy even *on average*. Furthermore, measurments of observables other than energy certainly change the (thermo)dynamics too, and there is no reason to expect that previously derived fluctuation theorems continue to hold.

A theoretical framework that can take into account measurement *backaction* and *disturbance* will be developed in Chapter 5. Here, we only summarize that there is no apparent measurement strategy of the system with which we can confirm the theoretical picture above—unless coherences are negligible, i.e. the dynamics is classical.

To circumvent this problem, the quantum jump trajectory method is often interpreted in a different experimental context, which is related to our initial example of a damped cavity. Suppose we have a perfect photodetector in front of our cavity and detect every emitted photon. If we assume that the initial state of the cavity is known and pure, the dynamics of the system is described by the stochastic Schrödinger equation (4.90) if $n(t)$ is the number of experimentally counted photons. Thus, there is a one-to-one correspondence between the stochastic increments $dn(t)$ and detection events in the environment, which is remarkable.

Unfortuntely, there are two problems. First, for the general case of a system described by the BMS equation, this procedure requires *all* energy emissions *and* absorptions of the system to be experimentally detected. This appears to be experimentally difficult for realistic baths. In fact, this picture is equivalent to the two-point measurement scheme as it requires all degrees of freedom in the environment to be monitored.

The second problem is more subtle. In classical stochastic thermodynamics, the average of the stochastic heat equals the total heat flow into the bath. However, focusing again on the damped cavity case for illustration, all emitted photons are *absorbed* by the detectors. Thus, on average, there is actually no longer a real heat flow in this experiment as all excitations have been converted into digits stored in a computer. Thus, this interpretation of a quantum jump trajectory, as well as the interpretation of the two-point measurement scheme, changes the *ontological* status of a bath. A 'bath' should by definition be an object about which we have very *limited* microscopic information. This is reversed in the present picture by assuming precise microscopic information about the bath.

To conclude, while there exist powerful theoretical tools to study fluctuations in quantum systems, which are in agreement with classical fluctuation theorems, problems appear when trying to confirm them experimentally. Coming back to the single-molecule-pulling experiments from Section 2.9, it seems unlikely that we can use the quantum work fluctuation theorem in the same way, i.e. to infer free energy differences of complex open systems by local measurements of genuinely quantum non-equilibrium processes. Nevertheless, in the rest of this section we will indeed see that quantum fluctuation theorems do have important practical consequences.

4.6 Thermodynamic Uncertainty Relations

We used the previous sections to generalize classical fluctuation theorems to the quantum regime. Apart from the classical fluctuation theorem (2.166), we were indeed able to generalize all of them and we also discussed fluctuation theorems that we left out in Chapter 2 (such as the exchange and steady-state fluctuation theorems). Obviously, by application of the inequality $e^x \geq 1+x$, we can use these quantum fluctuation theorems to derive almost all second laws that we found by different means in Chapter 3, an exception being eqn (3.106). In this section we discuss another important implication of the fluctuation theorem: the thermodynamic uncertainty relation. In general, thermodynamic uncertainty relations summarize a family of inequalities, where the mean entropy production bounds fluctuations of other relevant thermodynamic quantities such as currents, work or the entropy productions themselves.

Mathematical preliminaries: convex and concave functions

Before we can start, we need to briefly talk about convex and concave functions. A function $f = f(x)$ is called **convex** on an interval I if for all $x_1, x_2 \in I$ and $\lambda \in [0, 1]$

$$f[(1-\lambda)x_1 + \lambda x_2] \leq (1-\lambda)f(x_1) + \lambda f(x_2). \tag{4.115}$$

In contrast, the function $f(x)$ is called **concave** if

$$f[(1-\lambda)x_1 + \lambda x_2] \geq (1-\lambda)f(x_1) + \lambda f(x_2). \tag{4.116}$$

Equivalently, f is concave if $-f$ is convex. Note that a function can be convex on one interval, but concave on another. If f is twice differentiable, then f is convex (concave) if and only if $f''(x) \geq 0$ ($f''(x) \leq 0$) for all $x \in I$. Thus, when plotting a convex (concave) function $y = f(x)$, the graph is 'bent upwards' ('downwards'). Finally, if f is invertible and convex (concave) on an interval I, then its inverse f^{-1} is concave (convex) on $f(I)$. This becomes transparent by recalling that the plot of $x = f^{-1}(y)$ can be obtained from the plot $y = f(x)$ by reflecting it across the line $y = x$.

Exercise 4.23 Make sketches to convince yourself of the statements above. Furthermore, consider the binary Shannon entropy $S(p) \equiv -p\ln p - (1-p)\ln(1-p)$ for some probability $p \in [0, 1]$ and show that $S(p)$ is concave. This result holds in greater generality; compare with eqn (A.9). Viewing Shannon or von Neumann entropy as a measure about our uncertainty, what is the interpretation of this result?

Knowing whether a function is convex or concave allows one to apply **Jensen's inequality**. To state it, consider a probability distribution $p(x)$ of some random variable x and let us denote the expectation value of some function f by $\mathbb{E}[f(x)] = \sum_x p(x)f(x)$. Then, Jensen's inequality states that

$$\boxed{f(\mathbb{E}[x]) \leq \mathbb{E}[f(x)] \quad \text{if } f \text{ is convex.}} \tag{4.117}$$

The opposite inequality holds if f is concave. The equality holds if $p(x)$ is 'pure', i.e. if there exists an x_0 such that $p(x) = \delta_{x,x_0}$, or in the typically irrelevant cases where f is constant or affine, i.e. $f(x) = ax + b$, for all x with $p(x) > 0$.

Exercise 4.24 Consider the integral fluctuation theorem $\langle e^{-\sigma}\rangle = 1$. Show that application of Jensen's inequality implies the second law $\langle\sigma\rangle \geq 0$.

Thermodynamic uncertainty relation for entropy production

We now turn to applications of physical relevance. Central to our investigations in this section is the detailed fluctuation theorem of the form

$$\frac{P(\sigma)}{P(-\sigma)} = e^{\sigma/k_B}. \tag{4.118}$$

Despite not enjoying universal validity, this fluctuation theorem still captures a multitude of applications such as the exchange and steady-state fluctuation relations (4.35) and (4.46). Another example is provided by the work fluctuation theorem (4.23) for a time-symmetric driving protocol $\lambda_t = \lambda_{\tau-t}$ and a Hamiltonian obeying time-reversal symmetry: $H_\Theta(\lambda_t) = H(\lambda_t)$. Thus, the random variable σ in eqn (4.118) does not necessarily need to be the stochastic entropy production, but it can also be, for example, the stochastic work or some other measure of dissipation. In our applications below, however, we only focus on the case where σ is the stochastic entropy production.

Next, we introduce a new probability distribution,

$$Q(\sigma) \equiv (1 + e^{-\sigma/k_B})P(\sigma), \tag{4.119}$$

which is defined *only* on the interval $[0, \infty)$. It is straightforward to see that $Q(\sigma)$ is non-negative and normalization is a consequence of the detailed fluctuation theorem (4.118): $\int_0^\infty d\sigma Q(\sigma) = 1$. To avoid confusion, we denote expectation values of any function $f(\sigma)$ with respect to Q by $\langle f(\sigma)\rangle_Q$, whereas expectation values with respect to P get *no* subscript. Furthermore, the first and second moments of σ are connected:

$$\Sigma \equiv \langle\sigma\rangle = \left\langle \sigma \tanh\left(\frac{\sigma}{2k_B}\right)\right\rangle_Q, \quad \langle\sigma^2\rangle = \langle\sigma^2\rangle_Q. \tag{4.120}$$

Here, we used the upper case Greek letter Σ to denote the mean value of σ and $\tanh(x) \equiv (e^x - e^{-x})/(e^x + e^{-x})$ denotes the hyperbolic tangent.

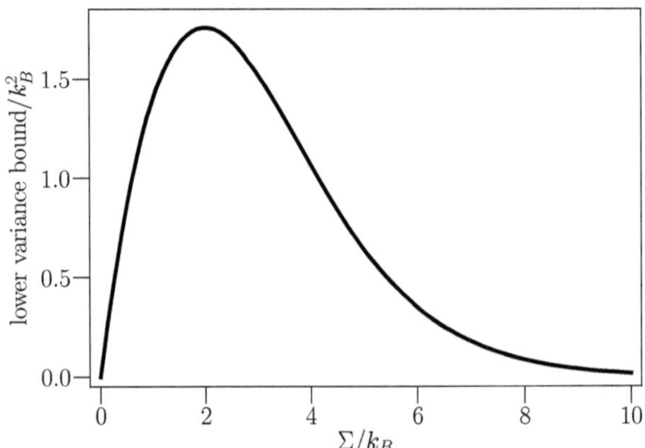

Fig. 4.4 Plot of the bound (4.122).

Exercise 4.25 Derive relations (4.120).

We now introduce the function $g(x) \equiv x \tanh(x/2)$, such that $\Sigma/k_B = \langle g(\sigma/k_B)\rangle_Q$. For $x \geq 0$, $g(x)$ is a monotonically increasing function and, hence, invertible. Furthermore, $g(\sqrt{x})$ is also monotonically increasing and, moreover, concave for $x \geq 0$, which can be checked by looking at the first and second derivatives. Now, if we denote by $h(y) \equiv g^{-1}(y)$ the inverse of g, it follows that the function $h(y)^2 = [g^{-1}(y)]^2$ is the inverse of $g(\sqrt{x})$ and that $h^2(y)$ is convex for $y \geq 0$. Taken together, this allows us to conclude

$$\langle \sigma^2 \rangle = \langle \sigma^2 \rangle_Q = k_B^2 \langle h^2[g(\sigma/k_B)] \rangle_Q \geq k_B^2 h^2 \left(\langle g(\sigma/k_B) \rangle_Q \right) = k_B^2 h^2(\Sigma/k_B), \quad (4.121)$$

where we used Jensen's inequality and eqn (4.120). Substracting Σ^2 on both sides, we can conclude that the variance in the entropy production can be upper bounded by a function depending only on the mean value Σ of the entropy production:

$$\mathrm{Var}(\sigma) \geq k_B^2 h^2(\Sigma/k_B) - \Sigma^2. \quad (4.122)$$

In Fig. 4.4 we plot the right-hand side of eqn (4.122). For large mean entropy production, $\Sigma/k_B \gg 1$, the bound becomes uninformative. For a small mean entropy production, it gives, however, a strong bound. From $g(x) = x^2/2 + \mathcal{O}(x^3)$ we can conclude for small x that $h(y) = \sqrt{2y}$. Then, eqn (4.122) implies $\langle \sigma^2 \rangle \geq 2k_B\Sigma$. Thus, the variance in stochastic entropy production has to grow faster than the mean entropy production (times $2k_B$): *nature does not allow a dissipative process* ($\Sigma > 0$) *without fluctuations* ($\langle \sigma^2 \rangle = 0$). This is at least true for small dissipations and it will play a crucial role in the next section.

Furthermore, let us consider the integral fluctuation theorem for $\sigma/k_B \ll 1$. A direct expansion of it yields $1 = \langle e^{-\sigma/k_B} \rangle = 1 - \Sigma/k_B + \langle \sigma^2 \rangle / 2k_B^2 + \mathcal{O}[(\sigma/k_B)^3]$, which predicts $2k_B\Sigma \approx \langle \sigma^2 \rangle$, i.e. the bound (4.122) becomes *tight* for small mean entropy production. Notice that the relation $2k_B\Sigma \approx \langle \sigma^2 \rangle$ is nothing but a manifestation of the *fluctuation–dissipation theorem*, which we met previously in Exercise 2.17.

Thermodynamic uncertainty relation for current-type observables

We now extend our analysis above to derive bounds for quantities other than the variance in entropy production. To be as general as possible, we abstractly denote by γ some stochastic trajectory such that $P(\sigma) = \sum_\gamma \delta[\sigma - \sigma(\gamma)]p(\gamma)$, where $p(\gamma)$ is the probability of observing trajectory γ. Examples for γ include the stochastic trajectory of a classical system (as in Section 2.4) or a trajectory obtained within the two-point measurement scheme introduced in Section 4.1, where $\gamma = (x_\tau, x_0)$. We further denote by $\gamma^\dagger$ the conjugate trajectory to γ, which arises by considering a 'time-reversed' process, e.g. $(x_\tau, x_0)^\dagger = (x_0, x_\tau)$. Note that $(\gamma^\dagger)^\dagger = \gamma$. Furthermore, for the time-symmetric fluctuation theorems considered in this section, it turns out that

$$e^{\sigma(\gamma)/k_B} = \frac{p(\gamma)}{p(\gamma^\dagger)}, \tag{4.123}$$

which implies eqn (4.118). Equation (4.123) is a consequence of the fundamental symmetry (4.7), when applied to a time-symmetric probability $p_{\rm tr}(\gamma) = p(\gamma)$ with appropriately chosen boundary conditions.

Exercise 4.26 Recall the exchange fluctuation theorem (4.35) and define the stochastic entropy production $\sigma \equiv k_B \sum_\nu \beta_\nu(\Delta\epsilon_\nu - \mu_\nu \Delta n_\nu)$. Which measurement results define the trajectory γ in this case? Show that eqn (4.123) is identical to eqn (4.35) and that the detailed fluctuation theorem in the form $P(\sigma)/P(-\sigma) = e^{\sigma/k_B}$ holds.

Next, consider an arbitrary functional $\phi(\gamma)$ of the stochastic trajectory γ and assume that ϕ is *anti-symmetric* under time-reversal, i.e.

$$\phi(\gamma^\dagger) = -\phi(\gamma). \tag{4.124}$$

A simple example of such a functional is provided by the entropy production itself, $\sigma(\gamma^\dagger) = -\sigma(\gamma)$, which follows from eqn (4.123).

Exercise 4.27 Consider the example from the previous exercise. Show that the change of energy $\Delta\epsilon_\nu$ or particle number Δn_ν of bath ν, as well as any linear combination of them, satisfies eqn (4.124).

Next, we consider the joint distribution to observe a stochastic entropy production σ and some value ϕ:

$$P(\sigma, \phi) = \sum_\gamma \delta[\sigma - \sigma(\gamma)]\delta[\phi - \phi(\gamma)]p(\gamma). \tag{4.125}$$

It follows that

$$\begin{aligned} P(\sigma, \phi) &= e^{\sigma/k_B} \sum_\gamma \delta[\sigma - \sigma(\gamma)]\delta[\phi - \phi(\gamma)]p(\gamma^\dagger) \\ &= e^{\sigma/k_B} \sum_\gamma \delta[\sigma - \sigma(\gamma^\dagger)]\delta[\phi - \phi(\gamma^\dagger)]p(\gamma) \\ &= e^{\sigma/k_B} \sum_\gamma \delta[\sigma + \sigma(\gamma)]\delta[\phi + \phi(\gamma)]p(\gamma) = e^{\sigma/k_B} P(-\sigma, -\phi), \end{aligned} \tag{4.126}$$

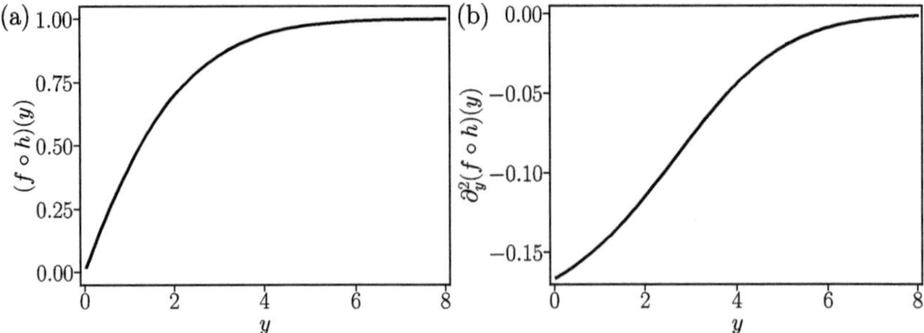

Fig. 4.5 Plot of $(f \circ h)(y)$ (a) and its second derivative (b) to demonstrate concavity.

where we used eqn (4.123), $\sum_\gamma f(\gamma) = \sum_{\gamma^\dagger} f(\gamma)$ and the anti-symmetry of $\sigma(\gamma)$ and $\phi(\gamma)$. In analogy to eqn (4.119), we introduce the probability distribution

$$Q(\sigma, \phi) = (1 + e^{\sigma/k_B})P(\sigma, \phi), \tag{4.127}$$

which, again, is only defined on $\sigma \in [0, \infty)$. Furthermore, in analogy to eqn (4.120), it is straightforward to confirm that

$$\langle \phi \rangle = \left\langle \phi \tanh\left(\frac{\sigma}{2k_B}\right)\right\rangle_Q, \quad \langle \phi^2 \rangle = \langle \phi^2 \rangle_Q. \tag{4.128}$$

The previous relation allows us to conclude that

$$\langle \phi \rangle^2 \le \langle \phi^2 \rangle_Q \left\langle \tanh^2\left(\frac{\sigma}{2k_B}\right)\right\rangle_Q = \langle \phi^2 \rangle \left\langle \tanh^2\left(\frac{\sigma}{2k_B}\right)\right\rangle_Q, \tag{4.129}$$

where we used the **Cauchy–Schwarz inequality**.

Exercise 4.28 The Cauchy–Schwarz inequality states that for any inner product space with scalar product $(\cdot,\cdot)$ it is true that $|(u,v)|^2 \le (u,u)(v,v) = \|u\|^2\|v\|^2$ for any two vectors u and v of that space. Use this result to confirm eqn (4.129).

Next, we return to our function $g(x) = x\tanh(x/2)$ and its inverse $h(y) = g^{-1}(y)$ introduced above. Furthermore, let us also introduce $f(x) \equiv \tanh^2(x/2)$. By construction, we can then write

$$\left\langle \tanh^2\left(\frac{\sigma}{2k_B}\right)\right\rangle_Q = \langle f(\sigma/k_B)\rangle_Q = \langle (f \circ h \circ g)(\sigma/k_B)\rangle_Q, \tag{4.130}$$

where, for clarity, we introduced the concatenation $\circ$ of two functions $(f \circ g)(x) \equiv f[g(x)]$. The claim is now that $(f \circ h)(y) = (f \circ g^{-1})(y)$ is concave, which allows us to apply Jensen's inequality:

$$\left\langle \tanh^2\left(\frac{\sigma}{2k_B}\right)\right\rangle_Q \le (f \circ h)[\langle g(\sigma/k_B)\rangle_Q] = (f \circ h)\left(\frac{\Sigma}{k_B}\right), \tag{4.131}$$

where we used eqn (4.120) at the end. The 'proof' that $f \circ h$ is concave is done by plotting it and its second derivative in Fig. 4.5.

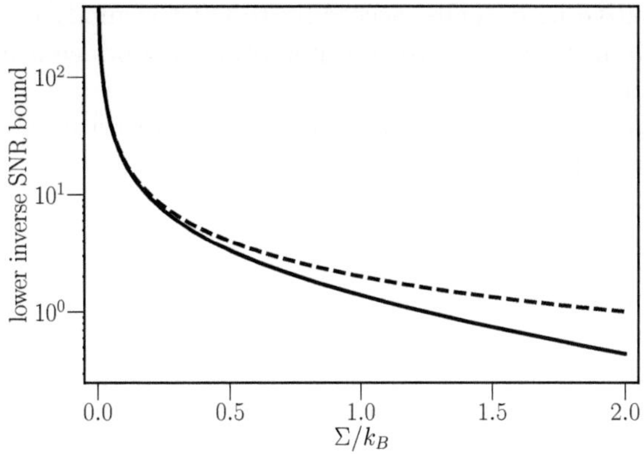

Fig. 4.6 Plot of the right-hand side of eqn (4.133) (solid line) and the right-hand side of eqn (4.134) (dashed line). Note the logarithmic scale. SNR, signal-to-noise ratio.

Finally, combining the inequalities (4.129) and (4.131), we obtain

$$\langle \phi \rangle^2 \leq \langle \phi^2 \rangle (f \circ h) \left(\frac{\Sigma}{k_B} \right). \tag{4.132}$$

After some rearrangement and after introducing the hyperbolic cosecant $\text{csch}(x) \equiv 2/(e^x - e^{-x})$, we find

$$\boxed{\frac{\text{Var}(\phi)}{\langle \phi \rangle^2} \geq \text{csch}^2 \left[\frac{h(\Sigma/k_B)}{2} \right].} \tag{4.133}$$

We call this central relation the **thermodynamic uncertainty relation**. It bounds the variance of some anti-symmetric observable divided by its mean square by a function of the mean entropy production Σ. This function is plotted in Fig. 4.6.

To interpret the thermodynamic uncertainty relation, we notice first that the left-hand side of eqn (4.133) is the *inverse signal-to-noise ratio.* If $(\langle \phi^2 \rangle - \langle \phi \rangle^2)/\langle \phi \rangle^2$ is large, then the stochastic quantity $\phi(\gamma)$ is dominated by its noise. In contrast, if $(\langle \phi^2 \rangle - \langle \phi \rangle^2)/\langle \phi \rangle^2$ is small, then the stochastic process generates a very 'precise signal'. Thus, the bound (4.133) allows us to conclude (see also Fig. 4.6) the following.

Thermodynamic uncertainty relation. *A small signal-to-noise ratio of any current-type observable (including entropy production itself) requires a large mean entropy production.*

At equilibrium, when $\Sigma = 0$, the bound diverges, which is related to the fact that all currents vanish at equilibrium. This is consistent because $\langle \phi \rangle$, or better $\langle \phi \rangle / t$, typically corresponds to a current (recall Exercise 4.27).

In Fig. 4.6 we also plot the function $2k_B/\Sigma$, which approximates well the right-hand side of eqn (4.133), in particular for small Σ. In fact, for a *Markov process at steady state*, which obeys local detailed balance, one finds the simpler bound

$$\boxed{\frac{\text{Var}(\phi)}{\langle \phi \rangle^2} \geq \frac{2k_B}{\Sigma},} \tag{4.134}$$

which we do not derive here. Thus, according to the thermodynamic uncertainty relation, non-Markovian processes have the potential to produce a more precise signal than Markov processes.

An important application of the thermodynamic uncertainty relation follows in Section 4.7. We therefore close this section with a short exercise.

Exercise 4.29 We come back to the example of the single-electron transistor from Exercise 4.13, where you were asked to compute the cumulant generating function in the long time limit. Use this result to confirm eqn (4.134). Choose as the observable ϕ the number of electrons n flowing from left to right at steady state. Convince yourself of the fact that eqn (4.134) can be expressed as

$$\frac{\mathrm{Var}(I)}{I^2} \geq \frac{2k_B}{\dot{\Sigma}}, \tag{4.135}$$

where $I = \langle n \rangle(t)/t$ is the current, $\mathrm{Var}(I) = [\langle n^2 \rangle(t) - \langle n \rangle^2(t)]/t$ is the current variance and $\dot{\Sigma}$ is the entropy production rate, all evaluated at steady state.

4.7 (Impossibility of) Carnot Efficiency at Finite Power

The Carnot cycle demonstrates that it is possible to reach *Carnot efficiency* $\eta_C = 1 - T_C/T_H$, which is the highest possible efficiency for an engine operating between a hot and a cold heat bath at fixed temperatures T_H and T_C, respectively. In theory, the Carnot cycle reaches Carnot efficiency only in the ideal limit of an infinite cycle time $\tau \to \infty$ such that its power output vanishes: $-P \equiv -W/\tau \to 0$. All real-world engines, which necessarily operate in finite time to produce a finite power output, were observed to have an efficiency smaller than Carnot. Therefore, the following question has significant fundamental and practical relevance: Is it possible to construct a heat engine with Carnot efficiency *and* finite power output?

Consider a system in contact with two large thermal baths, the temperature of which can be assumed to remain constant throughout the process. The first and second laws are

$$\Delta U_S = Q_C + Q_H + W, \quad \Sigma = \Delta S_S - \frac{Q_C}{T_C} - \frac{Q_H}{T_H} \geq 0. \tag{4.136}$$

We are interested in the limit where the system has reached a (periodic) steady state such that $\Delta U_S = 0$ and $\Delta S_S = 0$. Then, after using the first law, the second law can be written as $T_C\Sigma = W + \eta_C Q_H \geq 0$. Since we are interested in extracting work, $W < 0$, we infer $Q_H > 0$. Thus, the following efficiency is always positive:

$$\eta \equiv \frac{-W}{Q_H} = \eta_C - \frac{T_C\Sigma}{Q_H} \leq \eta_C. \tag{4.137}$$

As expected, the efficiency is bounded by the Carnot efficiency, but more importantly this result also tells us that $\eta = \eta_C$ if and only if $\Sigma = 0$ (excluding cases causing divergences such as $T_C = 0$, to which we return in Section (4.8), and $Q_H = \infty$). Thus, the question of attaining Carnot efficiency at finite power comes down to the question of whether there are finite-time processes that extract work *reversibly*, i.e. with zero entropy production. Importantly, this question is not answered by the basic laws of thermodynamics itself. To obtain insights into it, we therefore need to go *beyond* the phenomenological laws of non-equilibrium thermodynamics.

An important insight into this question can be obtained from the thermodynamic uncertainty relation, assuming that the heat engine obeys the time-reversal symmetric fluctuation theorem (4.118). Since we just found that $\eta \to \eta_C$ happens only in the limit $\Sigma \to 0$, it suffices to consider the simpler thermodynamic uncertainty relation (4.134) because eqn (4.133) reduces to it for small Σ (see Fig. 4.6). The thermodynamic uncertainty relation applies to any 'current-type' observable that is anti-symmetric under time-reversal that includes the stochastic work done on the system:

$$\frac{\mathrm{Var}(w)}{\langle w\rangle^2} \geq \frac{2k_B}{\Sigma}. \tag{4.138}$$

Exercise 4.30 Return to Section 4.2 and confirm that the stochastic work introduced in the two-point measurement scheme is anti-symmetric under time-reversal, as required by eqn (4.124): $w(\gamma^\dagger) = -w(\gamma)$. Convince yourself that this does not change if one applies this definition to the setting of Section 4.3, where we treated multiple baths. Thus, as long as $P_{\mathrm{tr}}(\sigma) = P(\sigma)$, eqn (4.138) holds for small Σ. Furthermore, show that chemical work, introduced in eqn (3.166), is also anti-symmetric under time-reversal. Thus, all conclusions in this section also apply to thermoelectric devices in the absence of magnetic fields or other mechanisms breaking time-reversal symmetry.

Since we are interested in the steady-state thermodynamics of our heat engine, eqn (4.138) is not yet written in a useful form since all thermodynamic quantities appearing in it, $\langle w\rangle$, $\mathrm{Var}(w)$ and Σ, diverge in the long time limit as they are proportional to time t; compare also with the discussion around eqn (4.65). Therefore, we introduce the power $P = \langle w\rangle/t = W/t$, the fluctuations in power

$$\delta P^2 \equiv \lim_{t\to\infty} \frac{\mathrm{Var}(w)}{t} \tag{4.139}$$

and the entropy production rate $\dot{\Sigma} = \Sigma/t$. Now, all these quantities have a finite steady-state value and, in terms of them, the thermodynamic uncertainty relation (4.138) reduces to

$$\frac{\delta P^2}{P^2} \geq \frac{2k_B}{\dot{\Sigma}}. \tag{4.140}$$

Next, we use $T_C\dot{\Sigma} = P + \eta_C\dot{Q}_H$, where $\dot{Q}_H$ is the steady-state heat flux from the hot bath. This yields the uncertainty relation $\delta P^2/P^2 \geq 2k_BT_C/(P + \eta_C\dot{Q}_H)$. After rearrangement we find the trade–off

$$\boxed{\frac{1}{2} \geq k_BT_C\frac{-P}{\delta P^2}\frac{\eta}{\eta_C - \eta},} \tag{4.141}$$

where we introduced the efficiency $\eta = -P/\dot{Q}_H$.

Equation (4.141) is the first central result of this section and its interpretation is the following. Suppose we want to have a heat engine with an efficiency arbitrary close to Carnot efficiency: $\eta \to \eta_C$. Then, the last factor on the right-hand side of eqn (4.141) diverges, but the overall right-hand side has to be *smaller* than $1/2$. Thus, this divergence must be compensated for by the factor $-T_C P/\delta P^2$, which has to go to zero at least as fast as η approaches η_C. Thus, eqn (4.141) implies one of the following three scenarios: an engine operating at Carnot efficiency must either have (i) zero power output ($-P \searrow 0$), (ii) diverging fluctuations in the output power ($\delta P^2 \to \infty$) or (iii) access to a zero temperature heat bath ($T_C \searrow 0$).

The possibility of case (i) is well known owing to the Carnot cycle, which was our starting point. Case (ii) is a novel finding, but it seems highly undesirable to have a heat engine with *vanishing precision* (or signal-to-noise ratio) $P^2/\delta P^2 \to 0$. Such an engine would be completely unreliable. Finally, case (iii) is more subtle and will be treated in detail only in Section 4.8. There, we will see that it is possible to exclude this case based on an intuitive physical assumption. Remarkably, it turns out that this assumption simultaneously excludes the possibility of case (ii).

Before treating the case of low-temperature environments, we present another argument, which shows that reversible processes extracting a finite amount of work in finite time cannot exist. This argument, which also implies the impossibility of Carnot efficiency at finite power, is more general than the argument above, but also weaker in its prediction than eqn (4.141).

We consider our conventional scenario of a quantum system coupled to multiple heat baths described by the Hamiltonian $H_{SB}(\lambda_t) = H_S(\lambda_t) + \sum_\nu [V_{SB}^{(\nu)}(\lambda_t) + H_B^{(\nu)}]$. Furthermore, we consider a process lasting a time τ and we demand that $V_{SB_\nu}(\lambda_0) = V_{SB_\nu}(\lambda_\tau) = 0$ because we do not want any (free) energy to be hidden in the coupling. The initial state is taken to be $\rho_{SB}(0) = \rho_S(0) \bigotimes_\nu \pi_\nu(\beta_\nu)$.

Now, the claim is the following. Suppose that the total Hilbert space $\mathcal{H}_{SB}$ is finite (or can be approximated as being finite). Then, extracting a finite amount of work ($W < 0$) in a reversible process ($\Sigma = 0$) is only possible if the internal energy of the system is lowered without changing its entropy. Moreover, the extracted work is bounded by $-W(\tau) \leq -\Delta U_S(\tau)$. Thus, reversible work extraction in finite time is only possible by using the system as a 'battery', i.e. by consuming resources contained in the initial system state $\rho_S(0)$. If one additionally demands that after the process the system is returned to its initial state, then no work extraction is possible, i.e. $\Sigma = 0$ implies $W \geq 0$.

To derive this statement, we recall the notion of entropy production from Section 3.7, which can be applied as long as the temperatures of the baths do not change significantly. We found that

$$\begin{aligned}\Sigma(\tau) &= \Delta S_S(\tau) - \sum_\nu \frac{Q_\nu(\tau)}{T_\nu} \\ &= k_B \sum_\nu D[\rho_\nu(\tau)|\pi_\nu(\beta_\nu)] + k_B I_{\text{tot}}[\rho_{SB}(\tau)] \geq 0.\end{aligned} \tag{4.142}$$

Here, $\Delta S_S(\tau)$ is the change in von Neumann entropy of the system (times k_B), $Q_\nu(\tau)$ is minus the change in energy of bath ν and $I_{\text{tot}}[\rho_{SB}(\tau)] = \Delta S_{\text{vN}}[\rho_S(\tau)] + \sum_\nu \Delta S_{\text{vN}}[\rho_{B_\nu}(\tau)]$ is the non-negative total information, which here equals the change in von Neumann entropy of the marginal system and bath states owing to the initial product state assumption. Furthermore, the first law reads $W(\tau) = \Delta U_S(\tau) - \sum_\nu Q_\nu(\tau)$ and we demand $W < 0$.

Next, we make use of Theorem A.5, which states that the relative entropy can be lower bounded as $D(\rho|\sigma) \geq [S_{\text{vN}}(\rho) - S_{\text{vN}}(\sigma)]^2/(3\ln^2 d)$, where d is the dimension of the Hilbert space. Therefore, since $\rho_\nu(0) = \pi_\nu(\beta_\nu)$, we can lower bound the entropy production as

$$\Sigma(\tau) \geq k_B I_{\text{tot}}(\tau) + k_B \sum_\nu \frac{\Delta S_{\text{vN}}[\rho_\nu(\tau)]^2}{3\ln^2 d_\nu} \geq 0, \tag{4.143}$$

where $d_\nu = \dim \mathcal{H}_\nu$ is the dimension of the Hilbert space of bath ν, which is finite by assumption. This result shows that, if $\Delta S_{\text{vN}}[\rho_\nu(\tau)]^2 > 0$ for only one ν, then $\Sigma > 0$. Thus, for a reversible process with $\Sigma = 0$ we need $\Delta S_{\text{vN}}[\rho_\nu(\tau)]^2 = 0$ for all ν. Consequently, we find $I_{\text{tot}}[\rho_{SB}(\tau)] = \Delta S_{\text{vN}}[\rho_S(\tau)]$. Hence, the condition $\Sigma = 0$ implies $\Delta S_S(\tau) = 0$.

Finally, recall that, for a fixed energy, the Gibbs state maximizes the von Neumann entropy. Conversely, for a fixed von Neumann entropy, the Gibbs state *minimizes* the energy. Since we just found that $S_{\text{vN}}[\rho_\nu(\tau)] = S_{\text{vN}}[\pi_\nu(\beta_\nu)]$, this implies that $Q_\nu(\tau) \leq 0$ for all ν. Then, from the first law we infer $W(\tau) = \Delta U_S(\tau) - \sum_\nu Q_\nu(\tau) \geq \Delta U_S(\tau)$. Thus, the only possibility to extract work is provided by lowering the energy of the system *without* changing its von Neumann entropy, i.e. the system acts as a battery. If we exclude these hidden resources, for instance by demanding that the system returns to its initial state or starts from a passive state (see Exercise 3.13), we find

$$\boxed{\Sigma(\tau) = 0 \quad \Rightarrow \quad W(\tau) \geq 0.} \tag{4.144}$$

Remarkably, we did not need to assume any time-reversal symmetry here. Everything followed from the definition of entropy production in eqn (4.142). Although this definition assumes large baths, the bath Hilbert space dimension d_ν can still be finite (we discuss this point further in Section 4.8). However, in the end, this also reveals a weakness in the argument above: typically, d_ν is enormously large and the bound in eqn (4.143) becomes uninformative.

To conclude, while there are good reasons to expect that no process can extract finite power reversibly, i.e. strict Carnot efficiency at finite power is impossible, it seems that it is possible to *approximate such a process arbitrarily well* if time-reversal symmetry is broken. Put differently, accepting finite measurement resolution as an experimental fact, Carnot efficiency at finite power seems—in principle—possible. If time-reversal symmetry is obeyed, we know from eqn (4.141) that additional constraints play a role that indicate that Carnot efficiency at finite power cannot be reached, not even approximately.

4.8 Third Law of Thermodynamics

The third law of thermodynamics aims to describe systems at very low temperature. In contrast to the zeroth, first and second laws, the third law has a less fundamental status. Many different formulations exist that are not entirely equivalent and also offer different physical perspectives. In some sense, this is not surprising. When the theory of thermodynamics was conceived, it was unknown that systems can exhibit strange low-temperature behaviour. The idea that thermodynamic systems are composed of many reversibly interacting atomistic particles was not universally accepted before the early 20th century. In fact, arguments about the third law of thermodynamics—much exchanged between Nernst, Einstein and Planck—contributed a lot to the initial acceptance of quantum physics.

This is one of the reasons why we postponed a discussion of the third law, which has little influence on the remaining part of this book, to the end of Chapter 4. Another reason is that we found in Section 4.7 that Carnot efficiency at finite power seems possible in the limit of a vanishing temperature $T \to 0$ of the cold bath. The goal of this section is to show that it is likely to be impossible to prepare any physical system at zero temperature. In this sense, the present section connects well to the previous section, albeit it has relatively little to do with the general topic of this chapter ('quantum fluctuation theorems').

Historical development

It is beneficial to start with a brief historical account of the development of the third law of thermodynamics together with its different formulations. We therefore recall some equilibrium thermodynamics. Let $\mathcal{F}(T, \lambda)$ be the equilibrium free energy (often called the *Helmholtz free energy* in textbooks on thermodynamics) of some thermodynamic system at temperature T with further control parameters λ specifying its state (e.g. volume, chemical concentrations, external fields). By definition, $\mathcal{F}(T,\lambda) = \mathcal{U}(T,\lambda) - T\mathcal{S}(T,\lambda)$ with the internal energy $\mathcal{U}$ and thermodynamic entropy $\mathcal{S}$. Consequently, the differential of $\mathcal{F}$ reads $d\mathcal{F} = d\mathcal{U} - Td\mathcal{S} - \mathcal{S}dT$. As in conventional textbook thermodynamics, let us consider the first law $d\mathcal{U} = đQ - pdV + \sum_\nu \mu_\nu dN_\nu$, where Q, p, V, μ_ν and N_ν are the heat, pressure, volume, chemical potential and particle number of species ν, respectively. The second law states that $Td\mathcal{S} = đQ$ for a reversible process at equilibrium. Using both laws, we obtain

$$d\mathcal{F} = -pdV - \mathcal{S}dT + \sum_\nu \mu_\nu dN_\nu, \tag{4.145}$$

which shows that the free energy $\mathcal{F}(T, V, N_\nu)$ is a function of T, V and N_ν. In particular, the previous relation implies

$$\mathcal{S} = -\left(\frac{\partial \mathcal{F}}{\partial T}\right)_{V,N_\nu}. \tag{4.146}$$

Now, for a reversible isothermal process where the temperature T of the system remains constant, we deduce that

$$\Delta\mathcal{F} = \Delta\mathcal{U} - T\Delta\mathcal{S} = \Delta\mathcal{U} + T\left(\frac{\partial \Delta\mathcal{F}}{\partial T}\right)_{V,N_\nu}. \tag{4.147}$$

We remark that the notation of the partial derivative is confusing at this point. While V and N_ν (or any other parameter λ) are *fixed* while taking the partial derivative with respect to T, these parameters can *change* during the isothermal process.

Now, we take a look at this equation in the limit $T \to 0$. We make the mild assumption that $\Delta\mathcal{S}$ remains finite for $T \to 0$, which is sometimes called *Einstein's statement of the third law.* We then find that the change in free energy is completely determined by the change in internal energy: $\Delta\mathcal{F} = \Delta\mathcal{U}$ for $T \to 0$.

The question is now whether it is always possible to express $\Delta\mathcal{F}$ as a function of $\Delta\mathcal{U}$. This had important consequences as $\Delta\mathcal{F}$ is a crucial quantity determining whether or not a process in the presence of a heat bath happens spontaneously (if $\Delta\mathcal{F} < 0$), whereas $\Delta\mathcal{U}$ is typically an experimentally accessible quantity. In fact, one could hope that the differential equation (4.147) fixes $\Delta\mathcal{F}$ as a function of $\Delta\mathcal{U}$. Unfortunately, this is not the case. If $\Delta\mathcal{F} = f(T)$ is a solution of eqn (4.147), then so is $\Delta\mathcal{F} = f(T) + bT$ for an arbitrary constant b. This constant can only be determined by fixing the first partial derivative of $\Delta\mathcal{F}$ with respect to T at some particular point $T = T^*$. This ambiguity was a known and important practical problem in chemistry around 1900. Based on empirical observations, Nernst postulated in 1906 that

$$\lim_{T\to 0}\left(\frac{\partial\Delta\mathcal{F}}{\partial T}\right)_{V,N_\nu} = \lim_{T\to 0}\left(\frac{\partial\Delta\mathcal{U}}{\partial T}\right)_{V,N_\nu}. \tag{4.148}$$

Let us consider the consequences of eqn (4.148). First of all, by differentiating eqn (4.147) with respect to temperature T and assuming that the second partial derivative of $\Delta\mathcal{F}$ with respect to T is also well behaved (i.e. finite) for $T \to 0$, eqn (4.147) directly implies that

$$\lim_{T\to 0}\left(\frac{\partial\Delta\mathcal{U}}{\partial T}\right)_{V,N_\nu} = 0. \tag{4.149}$$

Recalling that $\partial\mathcal{U}/\partial T = C$ is the heat capacity, we deduce that the heat capacity is constant at zero temperature. Now, let us additionally consider eqn (4.148). Combining it with eqn (4.149), we obtain from eqn (4.146) the striking consequence

$$\lim_{T\to 0}\mathcal{S}(T,\lambda) = \lim_{T\to 0}\mathcal{S}(T,\lambda'). \tag{4.150}$$

Thus, at zero temperature thermodynamic entropy becomes a constant independent of λ. Notice that this does not fix the absolute value of entropy and, somewhat remarkably, eqn (4.150) was not explicitly noticed by Nernst in his early papers. Nevertheless, the third law in the form of eqn (4.150) is often referred to as *Nernst's heat theorem.*

Clearly, Nernst's heat theorem is much stronger than Einstein's version of the third law, which only requires the difference $\Delta\mathcal{S}$ to remain finite for $T \to 0$. For this reason, eqn (4.150) was refuted by Einstein and others. In fact, there are examples violating Nernst's heat theorem; for instance, a change in the groundstate degeneracy of a quantum many-body system as a function of λ implies a change in the equilibrium entropy at $T = 0$.

Planck decided to replace Nernst's heat theorem by demanding that the entropy of any chemically homogeneous system (i.e. a system consisting of the same kind of molecule whose thermodynamic state—in addition to its chemical nature—is completely determined by its temperature T, volume V and mass M) approaches a universal constant, which he took to be zero, for $T \to 0$. This is known as *Planck's formulation* of the third law, which is weaker than eqn (4.150) and does not completely resolve the problem of ambiguity mentioned above.

Finally, another formulation of the third law is known as *Nernst's unattainability principle*, which was put forward by Nernst in later publications. It states: *No thermodynamic process can reach zero temperature by a finite number of steps or within a finite time.* Below, we exclusively focus on this formulation, which seems to be the most plausible version of the third law based on arguments from quantum (statistical) mechanics. We note that the unattainability principle was put forward by Nernst to support his heat theorem. Strictly speaking, both are, however, not equivalent unless further assumptions are used.

In the end, the controversy about the third law is related to the fact that the theory of thermodynamics stands on shaky ground at zero temperature. It is not clear, for instance, whether the notion of an isothermal process even makes sense for $T \to 0$. Microscopically, the problem is rooted in the fact that at very low temperatures quantum dynamics happens effectively in a Hilbert space of very small dimension. For sufficiently small T, we can truncate the Hilbert space dimension, say, to $d = 100$, even if the system contains $N = 10^{23}$ particles. At this scale, there are few reasons to expect that thermodynamic principles and standard methods of statistical mechanics continue to hold.

Microscopic explanations of the unattainability principle

Let us now investigate the unattainability principle in detail from a microscopic perspective. Indeed, from an experimental point of view there are good reasons to believe in it as it is very challenging to prepare very-low-temperature states, somewhat similar to approaching the speed of light: any massive object can be further accelerated by putting more energy into it, but it will never reach the speed of light. Furthermore, we make the following agreement: we say that a quantum system is at zero temperature if it is in a *pure* state. Indeed, once a quantum system is in some pure state, we can—at least in principle—use a unitary transformation to map it to the ground state.

The claim is now that under reasonable assumptions it is impossible to purify a mixed quantum state. To this end, we consider our standard set-up of a system coupled to a bath with Hamiltonian and initial state

$$H_{SB}(\lambda_t) = H_S(\lambda_t) + H_B(\lambda_t) + V_{SB}(\lambda_t), \quad \rho_{SB}(0) = \rho_S(0) \otimes \pi_B(\beta). \tag{4.151}$$

We allow all parts of the system–bath Hamiltonian to be time dependent; our only assumption—echoing the above-mentioned fact that at low temperatures most degrees of freedom of a quantum system are 'frozen out'—is that the bath Hilbert space is of *finite dimension:* $\dim \mathcal{H}_B \equiv d_B < \infty$. In this case, the rank of $\pi_B(\beta)$ (seen as a matrix) is maximal: $\text{rank}[\pi_B(\beta)] = d_B$. The rank of the system state is assumed to be larger than 1 (otherwise we started in a pure state) but finite too: $1 < \text{rank}[\rho_S(0)] < \infty$.

Clearly, a necessary condition to map a mixed state to a pure state is that the rank of the state decreases. However, unitary dynamics, independent of the Hamiltonian, always *preserves the rank.* This can be easily seen by recalling, first, that the rank of a matrix coincides with its number of non-zero eigenvalues and, second, that a unitary evolution leaves the spectrum unchanged. Thus, we obtain

$$\begin{aligned} \text{rank}[\rho_S(0)]d_B &= \text{rank}[\rho_{SB}(0)] = \text{rank}[\rho_{SB}(t)] \\ &\leq \text{rank}[\rho_S(t)]\text{rank}[\rho_B(t)] \leq \text{rank}[\rho_S(t)]d_B. \end{aligned} \tag{4.152}$$

Here, the first equality is a consequence of $\text{rank}(A \otimes B) = \text{rank}(A) \cdot \text{rank}(B)$; the third inequality $\text{rank}(\rho_{SB}) \leq \text{rank}(\rho_S)\text{rank}(\rho_B)$ is derived in the exercise below; and, finally, we used the elementary fact that the rank of any matrix equals at most the dimension of the vector space it is acting on. Rewritten, eqn (4.152) reads

$$\text{rank}[\rho_S(0)] \leq \text{rank}[\rho_S(t)]. \tag{4.153}$$

Hence, it is impossible to decrease the rank of the system state.

Exercise 4.31 The goal is to prove $\text{rank}(\rho_{SB}) \leq \text{rank}(\rho_S)\text{rank}(\rho_B)$. For that purpose let $\rho_{SB} = \sum_\alpha \mu_\alpha |\psi_\alpha\rangle\langle\psi_\alpha|_{AB}$ with $\mu_\alpha > 0$. We define the *support* of ρ_{SB} as the linear subspace spanned by the vectors $|\psi_\alpha\rangle\langle\psi_\alpha|_{AB}$ (seen as elements of a matrix vector space), i.e. $\text{supp}(\rho_{AB}) = \text{Span}\{|\psi_\alpha\rangle\langle\psi_\alpha|_{AB}\}_\alpha$. Convince yourself of the fact that $\text{rank}(\rho_{SB}) = \dim[\text{supp}(\rho_{AB})]$ and that the desired inequality follows from

$$\text{supp}(\rho_{AB}) \subseteq \text{supp}(\rho_A \otimes \rho_B) = \text{supp}(\rho_A) \otimes \text{supp}(\rho_B). \tag{4.154}$$

To derive eqn (4.154), start with a pure bipartite state $\rho_{SB,\alpha} \equiv |\psi_\alpha\rangle\langle\psi_\alpha|_{AB}$ such that $\rho_{SB} = \sum_\alpha \mu_\alpha \rho_{SB,\alpha}$. Now, show that $\text{supp}(\rho_{AB,\alpha}) \subseteq \text{supp}(\rho_{A,\alpha}) \otimes \text{supp}(\rho_{B,\alpha})$ by using the **Schmidt decomposition**, which says that one can always write $|\psi_\alpha\rangle = \sum_i c_{i,\alpha}|i_\alpha\rangle_A \otimes |i_\alpha\rangle_B$ for some orthonormal set of vectors $\{|i_\alpha\rangle_{A/B}\} \in \mathcal{H}_{A/B}$ and some *positive, real-valued* coefficients $c_{i,\alpha}$. Next, show that eqn (4.154) follows from convex combination and the fact that if $|i_\alpha\rangle\langle i_\alpha|_A \in \text{supp}(\rho_{A,\alpha})$, then also $|i_\alpha\rangle\langle i_\alpha|_A \in \text{supp}(\rho_A)$.

It should be obvious that the result (4.153) can be generalized to include the presence of multiple heat baths, arbitrary initial bath states as long as they have full rank and arbitrary system–bath correlations as long as the initial rank is not smaller than $\text{rank}[\rho_S(0)]d_B$. Result (4.153) is therefore fairly general and hinges only on the two assumptions that the bath can be approximated to be finite dimensional and the initial bath state has full rank. The second assumption is necessary to exclude any purity as a hidden resource in the bath because that would only entail the question of how the bath was initially prepared in a state with low rank. The first assumption is more subtle and relates to a fundamental assumption about the world around us, namely that it is *computable*. Without delving into details, let us call a process 'computable' if there exists a digital computer that can *simulate* it up to a fixed but arbitrary accuracy. Note that we do not demand that the digital computer meets current technological standards. In fact, the simulation of many physical processes requires computational capacities exceeding by far all human capabilities, now and in the future. Nevertheless, to the best of the author's knowledge, there is no evidence in nature that there are processes that cannot be simulated—at least in principle—to arbitrary accuracy by finite-size computers. Applied to quantum mechanics, this means that every infinite-dimensional Hilbert space can be truncated to finite dimensions for all practical purposes.

In view of the last statement, it is interesting to return to our findings of Section 4.7. There, we found that eqn (4.141) allows for Carnot efficiency at finite power either if the cold bath is at zero temperature or if the fluctuations in the power diverge. Now, what we have just seen is that zero temperature is not reachable in a finite-dimensional setting. Furthermore, a finite-dimensional setting also excludes any divergences in the power fluctuations. Thus, it seems as if the assumption of having finite dimension forbids both reaching zero temperature and having Carnot efficiency at finite power. As the status and precise formulation of the third law are in any case debated, this motivates us to suggest a novel formulation of the third law. Expressed in a memorable way:

Third law of thermodynamics. *We live in a computable universe.*

We conclude this section by three further comments. First of all, in our proof above based on eqn (4.153) one might wonder at which place the argument of time enters the discussion. After all, the unattainability principle asserted that it is impossible to reach zero temperature in *finite time* whereas our proof above holds for any unitary evolution, independent of how we scale time. It turns out, however, that there is a profound connection between time and our assumption about a finite-dimensional Hilbert space. Namely, provided the work injected remains finite, any quantum system can only explore a finite region of the Hilbert space in finite time. In some sense, the Schrödinger equation tells us that the wavefunction $|\psi(t)\rangle$ can explore the Hilbert space only with a *finite speed* $\partial_t|\psi(t)\rangle$. Thus, the fact that we have only finite work resources and do not have infinite time provides another argument in favour of the conjecture that all processes in nature are computable.

Second, one might wonder whether there are alternative proofs of the unattainability principle. In fact, there are and another simple argument is the following. If it is possible to prepare a quantum system S in a pure state $|\psi\rangle\langle\psi|_S$, then this implies that it has zero correlations with any other part of the world. In equations, if E denotes the environment of the system, then having a pure system state implies a global state of the form

$$\rho_{SE} = |\psi\rangle\langle\psi|_S \otimes \rho_E, \tag{4.155}$$

where ρ_E is left unspecified. Indeed, while we have used a state of the form (4.155) repeatedly throughout this book as a *good approximation* (in particular, for an initial state), having a strict equality in eqn (4.155) seems impossible if one recalls that the fundamental forces in nature decay to zero only for an infinite distance between two objects. Although the decay to zero is extremely quick for the weak and strong forces, and also the Coulomb force tends to quickly cancel out over large distances since most objects have a negligible net electric charge, at very low temperatures even the tiniest external influences can become important.

Exercise 4.32 Show that $\mathrm{tr}_E\{\rho_{SE}\} = |\psi\rangle\langle\psi|_S$ for some $|\psi\rangle_S$ implies eqn (4.155) for any bipartite state ρ_{SE}.

Finally, let us comment on the question of whether quantum measurements can help. At first sight, it seems as if a projective quantum measurement can prepare a system in a pure state. However, recall the way we have modelled quantum measurements in Section 1.4, where we let an external ancilla interact with the system such that the system–ancilla dynamics is unitary. Within this picture it becomes impossible again to cool down to zero. The final exercise explores this further.

Exercise 4.33 Show that for an initially mixed ancilla state an ideal projective measurement (see eqns (1.26) and (1.27)) becomes impossible within the framework of Section 1.4. Furthermore, show that $\text{rank}[\rho_S] \cdot \text{rank}[\rho_A] \leq \text{rank}[\rho'_S] \cdot \text{rank}[\rho'_A]$, where $\rho_{S/A}$ ($\rho'_{S/A}$) denotes the marginal state of the system/ancilla before (after) the interaction. Thus, the rank of ρ_S can only be decreased by 'trading in' at least an equal rank increase for the ancilla. Again, if the initial ancilla state has full rank, it becomes impossible to decrease the system rank.

Further reading

4.1. After the discovery of classical fluctuation theorems, it was not immediately obvious how to define fluctuations of thermodynamic quantities in the quantum regime such that formally the same fluctuation theorems ensue. Piechocinska (2000) wrote the first paper achieving this successfully in the context of bit erasure (Landauer's principle); since then the two-point measurement scheme has become a widespread tool.
4.2. The first general derivation of the quantum work fluctuation theorem was provided by Kurchan (2000) and Tasaki (2000). The content of Exercise 4.4 is also due to Tasaki (2000), while the content of Exercise 4.5 is due to Campisi *et al.* (2009).
4.3. The first entropy production fluctuation theorem in the quantum regime was derived very early by Piechocinska (2000). The terminology 'exchange fluctuation theorem' was coined in a paper by Jarzynski and Wójcik (2004), where eqn (4.41) was derived for the first time. Afterwards, exchange fluctuation theorems were derived in a broader context (Saito and Utsumi, 2008; Andrieux *et al.*, 2009; Esposito *et al.*, 2009). Experimental confirmations of the exchange fluctuation theorem were achieved in *double quantum dot* set-ups, where two quantum dots are tunnel-coupled to each other and placed in between two electron reservoirs (Utsumi *et al.*, 2010; Küng *et al.*, 2012).
4.4. Counting field methods have a long tradition in quantum transport and mesoscopic physics (Nazarov and Blanter, 2009) and have been used to derive quantum fluctuation theorems, in particular for model-specific studies, in parallel with the development of the two-point measurement scheme. That the two methods give identical statistics was shown by Esposito *et al.* (2009). Furthermore, counting field methods offer a rigorous way to derive not only steady-state exchange fluctuation theorems, but also other fluctuation theorems (Liu and Xi, 2016)
4.5. Whereas counting field methods are preferably used in quantum transport and mesoscopic physics, the quantum jump trajectory approach has its origin in quantum optics. The books of Carmichael (1993) and Wiseman and Milburn (2010), among many others, provide good introductions. Publications using quantum jump trajectories to derive fluctuation theorems in agreement with the counting field and two-point measurement approach are numerous. The first paper pointing to the presence of genuine quantum features in the energetics of a quantum jump trajectory was written by Elouard *et al.* (2017), who also coined the terminology 'quantum heat'. However, the terminology was used therein in multiple contexts, which partially contradict our findings of Chapter 5.
4.6. The terminology 'thermodynamic uncertainty relation' was coined by Barato and Seifert (2015), who first derived eqn (4.134) for special cases and conjectured that it is true for all Markov processes. Only 5 years later, thermodynamic uncertainty relations have already become a well-established tool in classical stochastic thermodynamics (Horowitz and Gingrich, 2020). However, in the non-Markovian regime, which is particularly relevant for small quantum systems, it became clear that eqn (4.134) does not share universal validity (Agarwalla

and Segal, 2018). Here, we followed Zhang (2019), who extended the proof of Hasegawa and Van Vu (2019) to derive eqn (4.133), which was (in a somewhat more complicated way) first shown by Timpanaro *et al.* (2019) to be the tightest and most saturable thermodynamic uncertainty relation. Essential for our proof was the probability distribution (4.119), which was introduced by Merhav and Kafri (2010), who also derived the bound (4.122). Another essential element was the time-reversal symmetric form of the detailed fluctuation theorem. For broken time-reversal symmetry, thermodynamic uncertainty relation no longer have the same simple form, but useful generalizations exist (Macieszczak *et al.*, 2018; Proesmans and Horowitz, 2019; Potts and Samuelsson, 2019).

4.7. The question of whether Carnot efficiency can be reached at finite power became an important topic in quantum and stochastic thermodynamics, in particular since results by Benenti *et al.* (2011) indicated that this is possible if time-reversal symmetry is broken. The bound (4.141) was derived by Pietzonka and Seifert (2018). For our second argument we extended the work of Mohammady and Romito (2019). Specific heat engines approaching Carnot efficiency at finite power were constructed based on diverging power fluctuations (Campisi and Fazio, 2016), or for cyclic processes (Holubec and Ryabov, 2018), for which the bound (4.141) does not apply.

4.8. Many textbooks on thermodynamics remain surprisingly quiet about the third law. For my historical survey, I relied on the works by Kox (2006) and Klimenko (2012), and I also consulted the paper by Nernst (1906), which initiated the discussion, and the book by Planck (1927). General proofs showing that it is impossible to reach a pure state in a finite-dimensional setting were given, for example, by Allahverdyan *et al.* (2011) and Wu *et al.* (2013); for an alternative method to deal with low-temperature problems see the recent work by Uzdin and Rahav (2021). Inequalities about the rank or Schmidt number of (marginals of) bipartite quantum states play an important role in quantum information theory (Cadney *et al.*, 2014). A review about *quantum speed limits* was given by Deffner and Campbell (2017). More precise considerations about resource costs of quantum measurements were investigated by Guryanova et al. (2020). Readers interested in a timeline of experimentally reached low temperatures can find useful results in the Wikipedia article 'Timeline of low-temperature technology' (accessed 2 December 2020) or in the article by Tuoriniemi (2016).

5
Operational Quantum Stochastic Thermodynamics

Summary. The last two chapters built on the assumption that central thermodynamic quantities can be defined without disturbing the dynamics of the system. This assumption cannot be retained in light of real experiments, where quantum measurements are disturbing. This chapter starts by discussing why it is necessary to overcome the semi-classical two-point measurement scheme. Then, consistent notions of internal energy, heat, work and system entropy are defined for a (Markovian and non-Markovian) quantum stochastic process as introduced in Chapter 1, which only relies on interventions performed on the system. The thermodynamic description of quantum measurements and Maxwell's demon is studied in detail. The chapter concludes with applying these 'operational' definitions to a Nobel-prize-winning experiment.

5.1 Classicality of Quantum Fluctuation Theorems

In Chapter 4 we described a successful theoretical approach, the two-point measurement scheme, to study fluctuations of work, heat, energy and entropy in quantum systems such that formally the same fluctuation theorems as known from classical stochastic thermodynamics ensue. Unfortunately, we also had to point out that the experimental implementation of this theoretical approach faces severe challenges as soon as the bath has a reasonable size such that it is actually justified to call it a 'bath'. The main experimental difficulty is measuring the precise microstate of the bath, which is assumed to be possible in the two-point measurement scheme. We summarize this important finding as follows.

Observation. *No measurement strategy of work, heat, energy or entropy fluctuations exists that simultaneously satisfies the following three criteria: (i) it is local (i.e. done only on the system), (ii) it implies fluctuation theorems as in classical stochastic thermodynamics and (iii) it cannot be generated by a classical stochastic process.*

The previous observation has the status of a *conjecture*: within 20 years of research on thermodynamic fluctuations in quantum systems, nobody succeeded in finding a measurement strategy satisfying points (i), (ii) and (iii). However, it has not yet been proven that this is impossible. The goal of this chapter is to first elucidate this observation further by pointing out facts that we did not discuss in Chapter 4. Afterwards, we will introduce a framework that satisfies points (i) and (iii), but violates point (ii). This 'operational' framework is therefore contrary to the two-point measurement scheme, which satisfies (ii) but violates (i) and (iii).

Quantum Stochastic Thermodynamics. Philipp Strasberg, Oxford University Press.
 DOI: 10.1093/oso/9780192895585.003.0005

The topic of this section is to show that the two-point measurement scheme and related approaches violate point (iii): their measurement statistics *can* be generated by a classical stochastic process. Recall the definition of a classical stochastic process from Section 1.6. It is described by a probability distribution $p(r_n, \dots, r_0)$, which satisfies the Kolmogorov consistency condition

$$p(r_n, \dots, \cancel{r_\ell}, \dots, r_0) = \sum_{r_\ell} p(r_n, \dots, r_\ell, \dots, r_0). \tag{5.1}$$

Whenever a process is described by such a probability distribution, it is indistinguishable from a classical stochastic process *based on the available measurement statistics.* This does *not* imply that the process has no quantum features in general; it only states that one could emulate the process with a classical device that reproduces the same outcomes. Non-classicality in the quantum world has many facets, and violation of eqn (5.1) is only one of them.

To prove that the two-point measurement scheme has classical measurement statistics, recall the definition of the two-point probability:

$$p(x_\tau, x_0) = \text{tr}\{\Pi(x_\tau)U(\tau, 0)\Pi(x_0)\rho(0)\Pi(x_0)U^\dagger(\tau, 0)\}. \tag{5.2}$$

Furthermore, to derive fluctuation theorems, we assumed the initial condition $\rho(0) = \sum_{x_0} \mu(x_0)\Pi(x_0)$. This implies eqn (5.1), as the next exercise shows.

Exercise 5.1 Show that, independent of the initial state, $\sum_{x_\tau} p(x_\tau, x_0) = p(\cancel{x_\tau}, x_0)$ always holds. Instead, confirm that the initial state assumption is important to conclude that $\sum_{x_0} p(x_\tau, x_0) = p(x_\tau, \cancel{x_0})$.

One might object to the above argument that quantum fluctuation theorems continue to hold even if the initial state $\rho(0)$ contains coherences in the initial measurement basis as long as its diagonal elements have the correct weights $\Pi(x_0)\rho(0)\Pi(x_0) = \mu(x_0)\Pi(x_0)$ (e.g. the weights of a Gibbs ensemble in the case of the work fluctuation theorem). Indeed, since all these coherences get destroyed during the initial measurement, this is certainly true on a formal level. However, it is also true that the two-point measurement scheme does not say how fluctuations of work, heat, energy or entropy should be defined in the case that the first measurement is *not* carried out: What is, for example, fluctuating work based on $p(x_\tau, \cancel{x_0})$ or $p(\cancel{x_\tau}, x_0)$? We will see that it is possible to define work fluctuations in this case, but they no longer obey a fluctuation theorem in general.

One might speculate whether the problem gets simpler in the much studied weak coupling and Markovian regime. At least there one could hope that it is possible to define fluctuating thermodynamic quantities satisfying points (i), (ii) and (iii). Unfortunately, this also does not seem to be the case. In fact, throughout the book we have already collected ample evidence for classical measurement statistics building on the general exposition from Section 1.9. Let us briefly summarize the main insights.

First, in Exercise 3.9, we saw that the BMS equation gives rise to classical measurement statistics if they are performed in the energy eigenbasis, which clearly is the most important basis to talk about internal energy, heat and work. Furthermore, in the absence of degeneracies, the BMS equation predicts that the population dynamics in the energy eigenbasis obeys a classical rate master equation (Exercise 3.12). In Exercise 4.20 we concluded that the probability of jumping between different energies is independent of the coherences and we also recall that the energy exchange statistics obtained from the full counting statistics method are identical to the two-point measurement scheme. Moreover, while we have identified a quantum part to the stochastic heat along quantum jump trajectories in Section 4.5, we also saw that this contribution vanishes on average (Exercise 4.22). In the same section we also identified severe problems in the identification of stochastic entropy along a quantum jump trajectory whenever coherences are initially present.

Finally, there is another point supporting our claim. If we average quantum jump trajectories, we obtain the evolution of the density matrix described by the BMS equation. In equations, we expressed this relation in Section 4.5 as $\rho_S(t) = \mathbb{E}[|\psi(t)\rangle\langle\psi(t)|]$. Thus, in order to confirm any result from the previous chapter based on local measurements of the system only, these measurements must be tuned such that their averages coincide with the (unmeasured) evolution predicted by the BMS equation. This puts severe restrictions on any attempt to experimentally measure energetic fluctuations in systems with quantum coherence if one wants them to obey fluctuation theorems.

5.2 A No-Go Theorem

In Section 5.1, we collected general evidence that approaches able to confirm quantum fluctuation theorems are characterized by classical measurement statistics. In this section, we are a bit more specific in our set-up with the benefit that we are able to draw sharper conclusions. These conclusions confirm that it is impossible to find a general characterization of work fluctuations with the same properties as in classical stochastic thermodynamics. Put differently, we can conclude that quantum stochastic thermodynamics is *more* than a mere extension of classical stochastic thermodynamics—it is a theory with completely new aspects because of measurement backaction. Of course, we remark that classical measurements can also be disturbing and as soon as these are considered one ends up having similar trouble.

Before we can formalize our main conclusion of this section, we introduce a neat trick that in some sense shows that the *two*-point measurement scheme can be regarded as a *one*-point measurement scheme. Consider an isolated system with Hamiltonian $H(\lambda_t)$, which we assume to be non-degenerate for the moment. The work probability distribution according to the two-point measurement scheme then becomes

$$p_{\text{TPMS}}(w) = \sum_{\epsilon_\tau,\epsilon_0} \delta[w - (\epsilon_\tau - \epsilon_0)]|\langle\epsilon_\tau|U(t,0)|\epsilon_0\rangle|^2\langle\epsilon_0|\rho(0)|\epsilon_0\rangle, \tag{5.3}$$

where ϵ_0 and ϵ_τ are eigenenergies of $H(\lambda_0)$ and $H(\lambda_\tau)$, respectively. We now introduce the operator

$$M_{\text{TPMS}}(w) \equiv \sum_{\epsilon_\tau,\epsilon_0} \delta[w - (\epsilon_\tau - \epsilon_0)]|\langle\epsilon_\tau|U(\tau,0)|\epsilon_0\rangle|^2|\epsilon_0\rangle\langle\epsilon_0|. \tag{5.4}$$

Obviously, this operator is constructed such that a measurement of it on the initial state equals

$$\mathrm{tr}\{M_{\mathrm{TPMS}}(w)\rho(0)\} = p_{\mathrm{TPMS}}(w). \tag{5.5}$$

Furthermore, $M_{\mathrm{TPMS}}(w)$ is evidently positive and satisfies the completeness relation

$$\int đw M_{\mathrm{TPMS}}(w) = I. \tag{5.6}$$

Using the terminology of quantum measurement theory from Section 1.4, we see that the set of operators $\{M_{\mathrm{TPMS}}(w)\}$ constitutes a POVM.

Exercise 5.2 If $H(\lambda_t)$ contains degeneracies, the above procedure no longer works directly. A way around it is the following. Let $|\epsilon_t, g_t\rangle$ denote an eigenstate of $H(\lambda_t)$ with eigenvalue ϵ_t and g_t labels eigenvectors in the corresponding degenerate subspace. Then, show that

$$M_{\mathrm{TPMS}}(w) = \sum_{\epsilon_\tau, g_\tau, \epsilon_0, g_0} \delta[w - (\epsilon_\tau - \epsilon_0)] |\langle \epsilon_\tau, g_\tau | U(\tau, 0) | \epsilon_0, g_0 \rangle|^2 |\epsilon_0, g_0\rangle\langle \epsilon_0, g_0| \tag{5.7}$$

is still normalized and positive, i.e. it constitutes a POVM. Show that $\mathrm{tr}\{M_{\mathrm{TPMS}}(w)\rho(0)\}$ coincides with the work probability distribution from the two-point measurement approach if the initial state obeys $\langle \epsilon_0, g_0 | \rho(0) | \epsilon_0, g_0' \rangle \sim \delta_{g_0, g_0'}$.

The above procedure shows that it is possible to measure stochastic work at a single time using a rather complicated POVM, which requires knowledge of $U(\tau, 0)$. In fact, no conventional quantum measurement of any system observable can suffice to determine $p_{\mathrm{TPMS}}(w)$ because the number of measurement outcomes w is in general larger than the Hilbert space dimension (and hence, the possible number of eigenvalues any system observable can have). Nevertheless, we know that every POVM can be implemented with an ancilla and projective measurements. The construction is as follows.

The ancilla is assumed to be a particle moving in one dimension described by position and momentum operators X and P and it is initialized in the position eigenstate $|x = 0\rangle$. We assume that its own Hamiltonian can be neglected during the experiment, which consists of four steps. First, we entangle the system and ancilla by applying the unitary $U_0 = \exp[-iH(\lambda_0) \otimes P/\hbar]$. Then, the system is evolved as usual with respect to the time-dependent Hamiltonian $H(\lambda_t)$, whereas the ancilla remains untouched. The unitary over this driving period is $U(\tau, 0)$. This step is followed by applying another entangling unitary to the system and ancilla of the form $U_\tau = \exp[-iH(\lambda_\tau) \otimes P/\hbar]$. Thus, the total sequence of unitaries is $U_\tau U(\tau, 0) U_0$. Finally, we projectively measure the position of the ancilla with projectors $|x\rangle\langle x|$. The claim is now that for any initial system state $\rho(0)$ the work probability distribution (5.3) coincides with the probability of obtaining outcome $x = w$

$$p_{\mathrm{TPMS}}(x = w) = \mathrm{tr}\left\{ |x\rangle\langle x| U_\tau U(\tau, 0) U_0 [\rho(0) \otimes |0\rangle\langle 0|] U_0^\dagger U^\dagger(\tau, 0) U_\tau^\dagger \right\}. \tag{5.8}$$

The proof is left as an exercise.

Exercise 5.3 Prove eqn (5.8). *Hint:* Use $e^{i\alpha P}|x\rangle = |x + \hbar\alpha\rangle$ for any $\alpha \in \mathbb{R}$.

We are now in a position to prove an interesting 'no-go theorem'. To that end, we denote by $\{M(w)\}$ an arbitrary POVM, i.e. we have $M(w) \geq 0$, $\sum_w M(w) = I$ and $p(w) \equiv \mathrm{tr}\{M(w)\rho(0)\}$ is the probability of measuring w given an initial state $\rho(0)$. Now, we want $M(w)$ to model the measurement of work fluctuations. Therefore, we impose two reasonable requirements on $M(w)$, which are certainly satisfied in the classical case.

(i) For an isolated system, the change in internal energy in the absence of measurements equals the first moment of the work distribution $p(w)$. In equations,

$$\mathrm{tr}\{H(\lambda_\tau)\rho(\tau)\} - \mathrm{tr}\{H(\lambda_0)\rho(0)\} = \Delta U = \sum_w wp(w) \quad \text{for all } \rho(0), \tag{5.9}$$

where $\rho(\tau) = U(\tau,0)\rho(0)U^\dagger(\tau,0)$ is the time-evolved state.

(ii) When applied to an incoherent initial state $\rho(0) = \mathcal{D}_{H(\lambda_0)}\rho(0)$, where $\mathcal{D}_X$ denotes the dephasing operation with respect to the eigenbasis of X, the distribution $p(w)$ coincides with the distribution $p_{\mathrm{TPMS}}(w)$ from the two-point measurement scheme. In equations,

$$\mathrm{tr}\{M(w)\rho(0)\} = \mathrm{tr}\{M_{\mathrm{TPMS}}(w)\rho(0)\} \quad \text{for all } \rho(0) = \mathcal{D}_{H(\lambda_0)}\rho(0). \tag{5.10}$$

The first requirement is very natural and follows by imposing the first law of thermodynamics to an isolated system. The second requirement demands that for classical initial states the two-point measurement approach gives the correct work distribution. In particular, for an initial thermal state, this guarantees that $p(w) = \mathrm{tr}\{M(w)\rho(0)\}$ satisfies the work fluctuation theorem. We then obtain the following remarkable result.

No-go theorem. *No measurement strategy $\{M(w)\}$ of work exists that simultaneously satisfies requirements (i) and (ii) for all initial states.*

Notice that for initial diagonal states the two-point measurement scheme satisfies point (i): $\Delta U = \sum_w wp_{\mathrm{TPMS}}(w)$. But if the initial state has coherences in the energy eigenbasis, the two-point measurement scheme no longer satisfies point (i). Now, the no-go theorem implies that, if we search for a measurement strategy overcoming this problem, we necessarily have to violate requirement (ii).

The proof of the no-go theorem goes as follows. First, requirement (i) can be written more explicitly as

$$\sum_w \mathrm{tr}\{wM(w)\rho(0)\} = \mathrm{tr}\{[U^\dagger(\tau,0)H(\lambda_\tau)U(\tau,0) - H(\lambda_0)]\rho(0)\}. \tag{5.11}$$

Since this relation has to hold for all initial states $\rho(0)$, we obtain the operator identity

$$\sum_w wM(w) = U^\dagger(\tau,0)H(\lambda_\tau)U(\tau,0) - H(\lambda_0), \tag{5.12}$$

which we use at the end.

Next, let us consider requirement (ii) for an initial energy eigenstate $\rho(0) = |\epsilon_0\rangle\langle\epsilon_0|$. We obtain that $\langle\epsilon_0|M(w)|\epsilon_0\rangle = \mathrm{tr}\{M_{\mathrm{TPMS}}(w)|\epsilon_0\rangle\langle\epsilon_0|\}$ has to hold for all $|\epsilon_0\rangle$ and all w. Clearly, the right-hand side is zero whenever $w + \epsilon_0 \neq \epsilon_\tau$ for some eigenvalue ϵ_τ of the final Hamiltonian. This implies that, for all $w \neq \epsilon_\tau - \epsilon_0$, the measurement operator $M(w)$ is equal to zero. The proof of this statement is left as an exercise.

Exercise 5.4 Consider an arbitrary non-negative operator $A \geq 0$. Show that if $\langle n|A|n\rangle = 0$ for all elements of some basis set $\{|n\rangle\}$, then $A = 0$ identically. *Hint:* Convince yourself of the fact that $(\phi, \psi) \equiv \langle\phi|(A + \epsilon)|\psi\rangle$ defines a scalar product for any $\epsilon > 0$. Then, in the limit $\epsilon \searrow 0$ the Cauchy–Schwarz inequality implies $|\langle m|A|n\rangle|^2 \leq \langle m|A|m\rangle\langle n|A|n\rangle$. Use this to derive the desired result.

The previous point implies that the difference $\epsilon_\tau - \epsilon_0$ generates all possible work values w we need to consider because $M(w) = 0$ otherwise. Now, consider the case where the work values $w = \epsilon_\tau - \epsilon_0$ are *non-degenerate*. For an initial energy eigenstate $|\epsilon_0'\rangle$ we obtain by virtue of requirement (ii)

$$\langle\epsilon_0'|M(\epsilon_\tau - \epsilon_0)|\epsilon_0'\rangle = \mathrm{tr}\{M_{\mathrm{TPMS}}(\epsilon_\tau - \epsilon_0)|\epsilon_0'\rangle\langle\epsilon_0'|\} = \delta_{\epsilon_0',\epsilon_0}|\langle\epsilon_\tau|U(\tau,0)|\epsilon_0\rangle|^2. \quad (5.13)$$

Thus, $M(\epsilon_\tau - \epsilon_0)$ seen as a matrix has only a single non-zero element on the diagonal. Using the Cauchy–Schwarz inequality (see Exercise 5.4), we infer that $|\langle\epsilon_0'|M(\epsilon_\tau - \epsilon_0)|\epsilon_0''\rangle|^2 \leq \langle\epsilon_0'|M(\epsilon_\tau - \epsilon_0)|\epsilon_0'\rangle\langle\epsilon_0''|M(\epsilon_\tau - \epsilon_0)|\epsilon_0''\rangle = 0$ for all $\epsilon_0' \neq \epsilon_0''$. Hence, all off-diagonal elements of $M(w)$ for $w = \epsilon_\tau - \epsilon_0$ vanish. Therefore, we reach the conclusion that, for all measurements with non-degenerate work values, requirement (ii) implies that the operator $M(w)$ is *fixed and identical* to the two-point measurement scheme: $M(w) = |\langle\epsilon_\tau|U(t,0)|\epsilon_0\rangle|^2|\epsilon_0\rangle\langle\epsilon_0|$ for $w = \epsilon_\tau - \epsilon_0$.

Finally, we show that this finding is incompatible with requirement (i), for which it suffices to find a counterexample. Consider a two-level system with initial and final Hamiltonian $H(\lambda_0) = \epsilon|1\rangle\langle 1|$ and $H(\lambda_\tau) = \epsilon'|1\rangle\langle 1|$, respectively (the energy of the ground state $|0\rangle$ is set to zero). We consider $\epsilon' \neq \epsilon$, which implies the work values $w \in \{0, -\epsilon, \epsilon', \epsilon' - \epsilon\}$. Furthermore, let the unitary time evolution operator be $U(\tau, 0) = |0\rangle\langle +| + |1\rangle\langle -|$. It follows that eqn (5.12) reduces to

$$\frac{\epsilon'}{2}|0\rangle\langle 0| + \frac{\epsilon' - 2\epsilon}{2}|1\rangle\langle 1| = \epsilon'|-\rangle\langle -| - \epsilon|1\rangle\langle 1|. \quad (5.14)$$

This gives the desired contradiction and concludes the proof.

To summarize, this section and Section 5.1 directly call into question the idea that the two-point measurement scheme is the one and only meaningful approach to characterizing work, heat, energy and entropy fluctuations in quantum systems. In addition, the two-point measurement scheme and related approaches from Chapter 4 remain unsatisfactory from an experimental point of view as they assume perfect knowledge about all degrees of freedom. Thus, quantum stochastic thermodynamics differs radically from classical stochastic thermodynamics. This makes it interesting, but on the downside one has to admit that experimental insights in the quantum case are much harder to obtain than in the classical case, where experimental insights have already been obtained, as discussed in Section 2.9.

Nevertheless, quantum mechanics allows us to perform a plethora of other interesting experiments. In many quantum experiments, however, measurements are disturbing and knowledge is only available about the system (and sometimes even that knowledge is incomplete). In the remainder of the book, we develop a framework that is able to cope with these challenges.

5.3 Thermodynamics of Quantum Measurements

A particularly challenging problem when formulating a theory of quantum stochastic thermodynamics is to understand the energetic and entropic impact of a quantum measurement. In this section, we exlusively focus on the thermodynamics of a measurement (and, later, also a more general intervention) happening at a *single time*. There is a lot we can learn from this simple situation, but there is also some ambiguity left in the assignment of stochastic heat and work during the measurement, which cannot be removed without knowing the precise experimental context.

Thermodynamics of projective measurements

To model the measurement of some system observable $R_S = \sum_{r=1}^n \lambda(r)\Pi_S(r)$ with n distinct eigenvalues, we briefly review the framework discussed in Section 1.4. The basic idea was to make use of an external ancilla system to implement the measurement. This was motivated by the fact that one typically has precise control about the measurement equipment in a laboratory (detectors, external fields, etc.) but not the system itself. For the moment, let us call the external ancilla a *memory* M and let us assume that it is an **ideal memory**, characterized by the following properties.

(i) The dimension of the memory Hilbert space equals the number of distinct measurement results, i.e. $\dim \mathcal{H}_M = n$ in our case.

(ii) The states of the memory are energetically degenerate, i.e. $H_M \sim I_M$, where I_M is the identity operator in the memory Hibert space.

(iii) Initially, the memory is decorrelated from the system and prepared in the state $\rho_M^{(0)} = |1\rangle\langle 1|_M$, where $|1\rangle_M$ denotes some standard reference state.

These properties are certainly unrealistic for most currently used memories, but we remark that we are only interested in finding a *minimal* thermodynamic description for a quantum measurement. It is always possible to add further thermodynamic costs resulting from realistic implementations at the end.

As discussed in greater detail in Section 1.4, our ideal measurement is implemented by letting the system and memory interact such that the resulting unitary operator is $U_{SM} = \sum_{r=1}^n \Pi_S(r) \otimes \sum_{r'=1}^n |r+r'-1\rangle\langle r'|_M$ (with $r+r'-1$ modulo n whenever $r+r'-1 > n$). The state after the unitary interaction is $\rho_{SM}^{(1)} \equiv \mathcal{U}_{SM}\rho_{SM}^{(0)} \equiv U_{SM}\rho_{SM}^{(0)}U_{SM}^\dagger$ with the initial system–memory state $\rho_{SM}^{(0)} = \rho_S^{(0)} \otimes |1\rangle\langle 1|_M$. The marginal system state $\rho_S^{(1)} = \sum_r \mathcal{P}_S(r)\rho_S^{(0)} \equiv \sum_r \Pi_S(r)\rho_S^{(0)}\Pi_S(r)$ describes the *average* (or unconditional) effect of the projective measurement. To obtain the system state conditioned on result r, we perform a measurement of the memory observable $R_M = \sum_{r=1}^n r|r\rangle\langle r|_M$. If this measurement reveals result r, the normalized post-measurement state of the system and memory is $\rho_{SM}^{(2)}(r) \equiv \mathcal{P}_S(r)\rho_S^{(0)}/p(r) \otimes |r\rangle\langle r|_M$ with $p(r) = \mathrm{tr}_S\{\Pi_S(r)\rho_S^{(0)}\}$. We denote the averaged post-measurement state by $\rho_{SM}^{(2)} \equiv \sum_r p(r)\rho_{SM}^{(2)}(r)$. This scheme and part of the notation is summarized in the circuit diagram of Fig. 5.1.

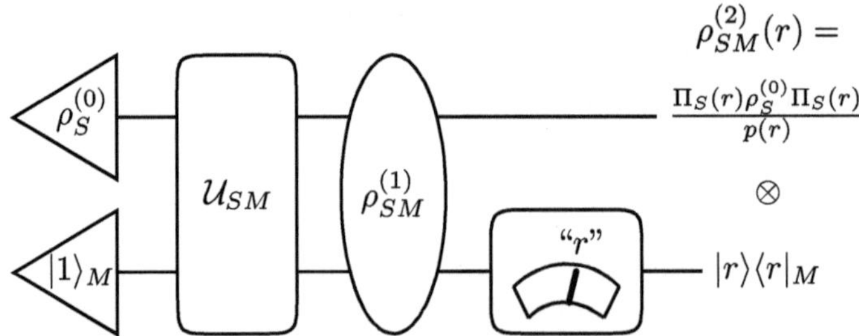

Fig. 5.1 Circuit diagram to illustrate the implementation of a measurement with time running from left to right.

Let us now analyse this situation from a thermodynamic perspective, starting with energetic considerations. Since the energies of the memory are degenerate, they always cancel out when computing changes in energy. We can therefore ignore the energy of the memory and define the internal energy as

$$U_j \equiv \mathrm{tr}_S\{H_S\rho_S^{(j)}\} \tag{5.15}$$

with $j \in \{0, 1, 2\}$. Here, we assumed the system Hamiltonian H_S to be fixed. We also introduce the conditional or stochastic internal energy

$$u_2(r) \equiv \mathrm{tr}_S\{H_S\rho_S^{(2)}(r)\}, \tag{5.16}$$

which depends on the measurement outcome r and satisfies $\sum_r p(r)u_2(r) = U_2$. Note that $u_j(r) = U_j$ for $j \in \{0, 1\}$ since there is not yet any measurement result to take into account. To study the first law, we split the change in stochastic internal energy into two parts,

$$\Delta u_{20}(r) \equiv u_2(r) - U_0 = \Delta u_{21}(r) + \Delta U_{10}, \tag{5.17}$$

where $\Delta u_{21}(r) \equiv u_2(r) - U_1$ and $\Delta U_{10} \equiv U_1 - U_0$. It is instructive to investigate the two contributions in eqn (5.17) separately.

First, the term ΔU_{10} describes a change in energy due to a global unitary evolution of the isolated system and memory. From Section 3.1 we know that this change in energy should be counted as work. Thus, we set $\Delta U_{10} = W^{\mathrm{meas}}$.

The second term $\Delta u_{21}(r)$ is more subtle, but at least we can easily confirm one important property. Namely, since $\sum_r p(r)\rho_S^{(2)}(r) = \rho_S^{(1)}$, we obtain $\sum_r p(r)\Delta u_{21}(r) = 0$. This allows us to conclude that a quantum measurement is always a work source *on average*:

$$\boxed{\Delta U_{20} \equiv \sum_r p(r)\Delta u_{20}(r) = W^{\mathrm{meas}}.} \tag{5.18}$$

This makes sense if one takes into account that measurements do not happen 'spontaneously' like a thermal fluctuation, but require an *active* intervention from the outside. Furthermore, if we measure an observable R_S, which commutes with the Hamiltonian, we find $\Delta U_{20} = W^{\mathrm{meas}} = 0$. This is also the case considered in classical stochastic thermodynamics, where ideal measurements are assumed to have no work cost.

Exercise 5.5 Show $\Delta U_{20} = W^{\text{meas}} = 0$ if $[R_S, H_S] = 0$.

Unfortunately, the previous insight does not fix how we should interpret the *stochastic* change in internal energy $\Delta u_{21}(r)$ due to a measurement: Is this work, heat, both or neither? As it turns out, nobody has yet succeeded in giving a general and satisfactory answer to this question. In the following, by considering two special cases, we give arguments as to *why* this question is hard to answer.

First, we recall the two-point measurement scheme and the work fluctuation theorems from Section 4.2. There, we discussed below eqn (4.22) the interpretation of the non-energy-conserving collapse of the wavefunction. We decided to interpret the change in energy due to the second measurement of an *isolated* system in a *pure* state as work, which ensured the validity of the work fluctuation theorem. On the other hand, the change in energy due to the first measurement in eqn (4.20) was *not* counted as fluctuating work. Does this imply that it should be counted as heat?

At least in some cases it seems better to count energy changes due to measurements as heat. Consider the example of a *classical* system *coupled* to a thermal bath and for simplicity we ignore any driving. We assume that the system has two states and is prepared at the initial time $t = 0$ in the excited state with energy ϵ (we set the ground state energy to zero) and at some later time τ a measurement reveals the system to be in the ground state (we assume *no* further measurements in between $t = 0$ and $t = \tau$). What is the interpretation of the observed system energy change $-\epsilon$ in this case? Clearly, we would identify it as a heat flux because we would conclude that at some time $t \in (0, \tau)$ the system must have jumped from the excited state to the ground state (perhaps the system was jumping multiple times, but the last jump must be from the excited state to the ground state). Let us now consider a step-by-step analysis of the energetics along the process. Prior to the measurement at $t = \tau$, the energy of the system is $\epsilon p_e(\tau)$, where $p_e(\tau)$ is the probability of finding the system in the excited state. Since the system started in the excited state, the energy change so far is $\epsilon p_e(\tau) - \epsilon = Q$, which must be heat (no driving). Then, at time τ the measurement reveals the system in the ground state and the associated stochastic change in energy (using our notation above) is $\Delta u_{21} = -\epsilon p_e(\tau)$, such that the total change in energy between 0 and τ is $\Delta u_{21} + Q = -\epsilon$. But we said initially that this must be interpreted as heat. Hence, $\Delta u_{21} \equiv q_{\text{meas}}$ must be fluctuating heat, which here has a purely classical contribution originating from an update of our state of knowledge. In fact, in Section 5.4 we will find that this choice matches the heat defined in classical stochastic thermodynamics in the proper limit.

Thus, we have reached the point where we have two opposing interpretations for Δu_{21} *depending on the physical context.* Moreover, it is interesting to compare our findings above with the concept of *quantum heat* introduced in Section 4.5. In the presence of a bath, quantum heat identifies contributions to the change in internal energy originating from a coherent superposition of energy eigenstates. In contrast to the two-point measurement scheme, where we considered an active measurement intervening the system dynamics, the interpretation of quantum heat arose from a fictitious thought experiment used to unravel the system dynamics because an actual measurement would *disturb* the dynamics.

To conclude, we need to confess that the correct interpretation of $\Delta u_{21}(r)$ remains an open question and perhaps it is fundamentally not answerable. Therefore, for the rest of this section, we simply label the change in internal energy due to the final read-out of the memory by $\Delta u^{\text{meas}}(r) \equiv \Delta u_{21}(r)$. Luckily, the interpretation of this term has no consequences on average, as demonstrated by eqn (5.18). Further labelling the total change in internal energy by $\Delta u(r) \equiv \Delta u_{20}(r)$, the stochastic **first law for a quantum measurement** becomes

$$\boxed{\Delta u(r) = \Delta u^{\text{meas}}(r) + W^{\text{meas}}} \tag{5.19}$$

with the property $\sum_r p(r)\Delta u^{\text{meas}}(r) = 0$.

After this long discussion about the energetics of quantum measurements, we finally turn to the second law. Surprisingly, the situation is simpler here. For the set-up above the initial thermodynamic entropy is $S_0 \equiv k_B S_{\text{vN}}[\rho_{SM}^{(0)}] = k_B S_{\text{vN}}[\rho_S^{(0)}]$ (remember that the memory is prepared in a pure state with zero entropy). Furthermore, after applying the unitary U_{SM}, the entropy is still the same: $S_0 = S_1 = k_B S_{\text{vN}}[\rho_{SM}^{(1)}]$. Only the definition of the final entropy requires more care. To that end, suppose that we obtain measurement outcome r. A naive guess to define the conditional (or stochastic) entropy of the system is to set $s_2(r) = k_B S_{\text{vN}}[\rho_{SM}^{(2)}(r)]$. However, for a projective measurement of a rank-1 observable R_S, we get $\rho_{SM}^{(2)}(r) = |r\rangle\langle r|_S \otimes |r\rangle\langle r|_M$ and, hence, $s_2(r) = 0$. This implies $s_2(r) - S_0 \leq 0$ for all r and all initial states in apparent violation of the second law of thermodynamics.

The second law is saved by taking into account that the classical measurement result r also has to be stored somewhere. This has to be done in some physical device and Landauer's principle (see Section 2.2) tells us that this entails an entropic cost of at least $k_B S_{\text{Sh}}[p(r)]$. Therefore, we define the conditional entropy

$$\boxed{s_2(r) \equiv -k_B \ln p(r) + k_B S_{\text{vN}}\left[\rho_S^{(2)}(r)\right],} \tag{5.20}$$

where we used $S_{\text{vN}}[\rho_{SM}^{(2)}(r)] = S_{\text{vN}}[\rho_S^{(2)}(r)]$. Note that eqn (5.20) coincides with the classical definition of stochastic entropy, eqn (2.56), if the system state is pure. On average, the entropy after the measurement is

$$S_2 = \sum_r p(r)s_2(r) = k_B S_{\text{Sh}}(\boldsymbol{p}) + k_B \sum_r p(r) S_{\text{vN}}\left[\rho_S^{(2)}(r)\right], \tag{5.21}$$

where $\boldsymbol{p}$ is the probability vector corresponding to $p(r)$.

Then, the **second law for a quantum measurement** follows as

$$\boxed{S_2 - S_0 \geq 0.} \tag{5.22}$$

This tells us that a quantum measurement also cannot reduce the entropy once one takes into account all participating degrees of freedom. Furthermore, for a measurement of a rank-1 observable, the second law is equivalent to

$$k_B S_{\text{Sh}}(\mathbf{p}) \geq k_B S_{\text{vN}}\left[\rho_S^{(0)}\right] \tag{5.23}$$

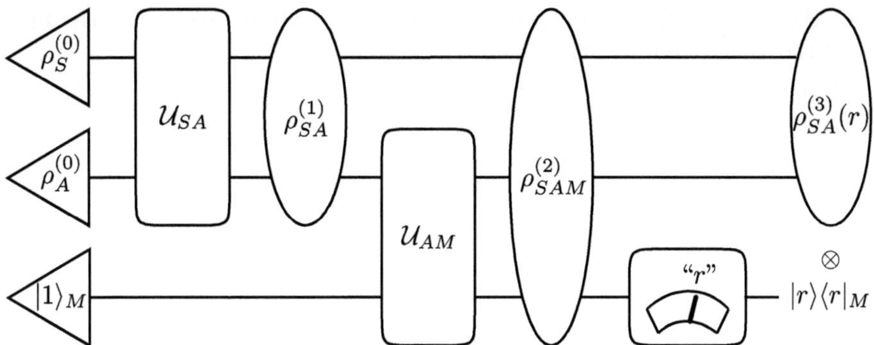

Fig. 5.2 Circuit diagram to illustrate the implementation of a general control operation.

with equality if and only if $[R_S, \rho_S^{(0)}] = 0$. The proof of the last two inequalities is left as an exercise.

Exercise 5.6 Prove eqns (5.22) and (5.23). *Hint:* Use Theorems A.1 and A.2.

Thermodynamics of a general control operation

We consider the general control operation happening at a single time. Based on our previous findings, its thermodynamics is straightforward. First, recall the content of Section 1.5 and, in particular, the unitary dilation theorem, eqn (1.48). Every control operation (or intervention) can be modelled by an instrument, i.e. a set of CP maps $\{\mathcal{C}(r)\}$, which add up to a CPTP map $\mathcal{C} = \sum_r \mathcal{C}(r)$. The action of each CP map can be written as

$$\tilde{\rho}_S(r) = \mathcal{C}(r)\rho_S^{(0)} = \mathrm{tr}_A\left\{\mathcal{P}_A(r)\mathcal{U}_{SA}\left[\rho_S^{(0)} \otimes \rho_A^{(0)}\right]\right\}. \tag{5.24}$$

Here, A denotes an ancilla degree of freedom, $\mathcal{U}_{SA}$ the superoperator corresponding to a unitary U_{SA} acting on the system–ancilla space and $\mathcal{P}_A(r)$ the superoperator corresponding to a projector $\Pi_A(r)$ acting on the ancilla only. The state after applying the map $\mathcal{C}(r)$ is not normalized and its norm $p(r) = \mathrm{tr}_S\{\tilde{\rho}_S(r)\}$ equals the probability of the map $\mathcal{C}(r)$ happening. Note that, in contrast to the case of a projective measurement studied before, the initial ancilla state $\rho_A^{(0)}$ and the unitary U_{SA} can be arbitrary.

To analyse the situation from a thermodynamic point of view, we assume that the final measurement of the ancilla, which is described by the projectors $\Pi_A(r)$, is carried out by an ideal memory M as described before. Thus, we now have a tripartite system SAM. A corresponding circuit diagram is shown in Fig. 5.2, which now divides the control operation into four stages and the corresponding states are labelled $\rho_{SAM}^{(j)}$ with $j \in \{0, 1, 2, 3\}$. Furthermore, we leave the ancilla Hamiltonian H_A arbitrary, whereas the memory Hamiltonian $H_M \sim I_M$ is assumed to be trivial. This allows us to include energetic changes of the ancilla in the description that are typically not negligible in actual experiments (compare, for example, with the micromaser studied in Section 3.9 and its extension studied in Section 5.7).

Now, the first law follows by extending our results above. We define the internal energy of the system and ancilla at stage j

$$U_j \equiv \mathrm{tr}_{SA}\left\{(H_S + H_S)\rho_{SA}^{(j)}\right\}. \tag{5.25}$$

The change in internal energy during the first two steps is due to work, which we denote as

$$\Delta U_{10} = U_1 - U_0 = W^{\mathrm{ctrl}}, \quad \Delta U_{21} = U_2 - U_1 = W^{\mathrm{meas}}. \tag{5.26}$$

Here, the superscript 'ctrl' stands for control. If the measurement of the ancilla happens in the energy eigenbasis, then $W^{\mathrm{meas}} = 0$ always (Exercise 5.5). Furthermore, during the final step, we learn something about the system–ancilla state. Therefore, we introduce the conditional or stochastic system–ancilla internal energy

$$u_3(r) = \mathrm{tr}_{SA}\left\{(H_S + H_S)\rho_{SA}^{(3)}(r)\right\}, \tag{5.27}$$

which satisfies $\sum_r p(r)u_3(r) = U_3$. Similar to eqn (5.19), the first law for a control operation reads

$$\boxed{\Delta u(r) = \Delta u^{\mathrm{meas}}(r) + W^{\mathrm{meas}} + W^{\mathrm{ctrl}},} \tag{5.28}$$

where $\Delta u(r) = u_3(r) - U_0$ and $\Delta u^{\mathrm{meas}}(r) = u_3(r) - U_2$. Again, the interpretation of $\Delta u^{\mathrm{meas}}(r)$ depends on the situation, but it is straightforward to confirm that $\sum_r p(r)\Delta u^{\mathrm{meas}}(r) = 0$. Thus, every control operation corresponds on average to a work source. This makes sense if one takes into account that a control operation has to be actively implemented in a laboratory. Furthermore, this finding is also in unison with the framework of repeated interactions studied in Section 3.9. There, we also found that there is a work cost W^{ctrl} resulting from the interaction of the external ancilla with the system, see eqn (3.141). Clearly, since there were no explicit measurements, we had $W^{\mathrm{meas}} = \Delta u^{\mathrm{meas}}(r) = 0$ in Section 3.9.

Next, we turn to the second law. The definition of entropy prior to the measurement of the ancilla is again simple: we have $S_0 = k_B S_{\mathrm{vN}}[\rho_{SA}^{(0)}] = S_1 = S_2$. Furthermore, inspired by eqn (5.20), we define the stochastic entropy at stage 3 by

$$s_3(r) = -k_B \ln p(r) + k_B S_{\mathrm{vN}}\left[\rho_{SA}^{(2)}(r)\right]. \tag{5.29}$$

On average, we then find again that entropy increases during a control operation:

$$\boxed{S_3 - S_0 = \sum_r p(r)s_3(r) - S_0 \geq 0.} \tag{5.30}$$

If we relabel SA by S, this result follows immediately by application of our previous second law (5.22). Importantly, while entropy increases on average, it can decrease along a single trajectory, as you are asked to verify in the next exercise.

Exercise 5.7 Find an example where $s_3(r) - S_0 < 0$ for some r, but $S_3 - S_0 \geq 0$. *Hint:* It suffices to consider a simple measurement.

To conclude, we established a first and second law for an arbitrary single control operation and, in particular, we defined stochastic internal energy and entropy conditioned on receiving a certain outcome r. So far, however, our results have been quite formal and the character of the second law was purely information-theoretic. In the following sections, we use the results above to construct thermodynamic quantities along stochastic trajectories for a system, which is driven, in contact with a heat bath and interrupted by *multiple* measurements or control operations.

5.4 Stochastic Thermodynamics of Quantum Markov Processes

The goal of this section is to equip a quantum Markov process as defined in Section 1.8 with a consistent thermodynamic interpretation along a single trajectory of measurement results recorded in a laboratory. Applications of that formalism will be treated in Sections 5.5 and 5.7. The extension to the non-Markovian regime is discussed in Section 5.6.

Preliminary considerations

We start by briefly recalling the definition of a quantum Markov process. Consider an open quantum system interrupted at arbitrary times $t_n > \cdots > t_0 = 0$ by arbitrary control operations. Each control operation at time t_k is characterized by a set of CP maps $\{\mathcal{C}_k(r_k)\}$, which add up to a CPTP map $\mathcal{C}_k = \sum_{r_k} \mathcal{C}_k(r_k)$. Here, the subscript k labels the time t_k and r_k labels measurement results. Now, for a quantum Markov process the conditional system state at time t_n *after* the nth control operation is

$$\tilde{\rho}_S(t_n|\mathbf{r}_n) = \mathcal{C}_n(r_n)\mathcal{E}_{n,n-1}\dots\mathcal{C}_1(r_1)\mathcal{E}_{1,0}\mathcal{C}_0(r_0)\rho_S(0). \tag{5.31}$$

Here, the dynamical maps $\mathcal{E}_{\ell,k} \equiv \mathcal{E}(t_\ell, t_k)$ are CPTP and propagate the system state forwards in time from t_k to t_ℓ. Furthermore, the conditional system state $\tilde{\rho}_S(t_n|\mathbf{r}_n)$ is not normalized and its norm equals the probability of obtaining the measurement results $\mathbf{r}_n \equiv (r_n, \dots, r_1, r_0)$: $p(\mathbf{r}_n) = \text{tr}_S\{\tilde{\rho}_S(t_n|\mathbf{r}_n)\}$. The normalized system state is denoted $\rho_S(t_n|\mathbf{r}_n) = \tilde{\rho}_S(t_n|\mathbf{r}_n)/p(\mathbf{r}_n)$.

Now, we are in a position to formulate precisely the goal of this section. First of all, whenever we speak about a *stochastic trajectory*, we mean the trajectory of measurement results $\mathbf{r}_n$. It is precisely this trajectory of measurement results that is recorded in a laboratory. We do *not* assume any fictitious unravelling process of the dynamical maps $\mathcal{E}_{\ell,k}$ (e.g. some stochastic Schrödinger equation) nor do we assume any detailed knowledge about the bath degrees of freedom. *All* accessible information is encoded in the control operations $\mathcal{C}_k(r_k)$ giving rise to $\mathbf{r}_n$. This philosophy distinguishes the present approach from the ones in Chapter 4 and justifies the terminology *operational* quantum stochastic thermodynamics. The goal is then to find for a given trajectory $\mathbf{r}_n$ definitions of stochastic internal energy $u(\mathbf{r}_n)$, heat $q(\mathbf{r}_n)$, work $w(\mathbf{r}_n)$ and entropy $s(\mathbf{r}_n)$. These definitions shall satisfy a stochastic first law and a second law on average. Moreover, we demand that the definitions reduce to previously established results in their respective limit of validity.

Furthermore, we have to make an agreement on the dynamical maps $\mathcal{E}_{\ell,k}$. These maps cannot be arbitrary, but they should have a consistent thermodynamic interpretation. For instance, consider a driven system with Hamiltonian $H_S(\lambda_t)$ weakly coupled to an ideal heat bath at temperature T. In the absence of any interventions, the first and second laws for the time interval (t_{n-1}, t_n) read

$$\Delta U_S^{(n)} = Q^{(n)} + W^{(n)}, \quad \Delta S_S^{(n)} - Q^{(n)}/T \geq 0. \tag{5.32}$$

Here, we denoted $\Delta U_S^{(n)} \equiv U_S(t_n) - U_S(t_{n-1})$ and $\Delta S_S^{(n)} \equiv S_S(t_n) - S_S(t_{n-1})$ with corresponding definitions $U_S(t) = \text{tr}_S\{H_S(\lambda_t)\rho_S(t)\}$ and $S_S(t) = k_B S_{\text{vN}}[\rho_S(t)]$. Furthermore, heat and work are defined as

$$Q^{(n)} = \int_{t_{n-1}}^{t_n} dt \text{tr}_S\{H_S(\lambda_t)\partial_t\rho_S(t)\}, \quad W^{(n)} = \int_{t_{n-1}}^{t_n} dt \text{tr}_S\{\rho_S(t)\partial_t H_S(\lambda_t)\}. \tag{5.33}$$

Note that these thermodynamic identities hold *for any state* $\rho_S(t)$; in particular, $\rho_S(t) = \rho_S(t|\mathbf{r}_{n-1})$ could also be conditioned on previous measurement results.

The standard example is the BMS equation with generator $\mathcal{L}(\lambda_t)$, which describes the dynamics of the system in the absence of interventions. In this case, the corresponding dynamical maps are given by the time-ordered exponential

$$\mathcal{E}_{n,n-1} = \exp_+ \left[\int_{t_{n-1}}^{t_n} dt \mathcal{L}(\lambda_t)\right]. \tag{5.34}$$

Because of its familiarity and to focus on the essential new arguments, we use this description throughout this section as a guiding example. We remark, however, that the analysis below can also be applied to situations beyond the validity of the BMS approximation (see also Section 5.6). It is only important to start from a consistent thermodynamic framework *in the absence* of any interventions and control operations.

To cover a larger class of experimental situations, we also allow that the external agent can perform *feedback control*. If the feedback control is classical, i.e. conditional on the measurement results $\mathbf{r}_n$, this means that eqn (5.31) is generalized to

$$\tilde{\rho}_S(t_n|\mathbf{r}_n) = \mathcal{C}_n(r_n|\mathbf{r}_{n-1})\mathcal{E}_{n,n-1}(\mathbf{r}_{n-1})\dots\mathcal{C}_1(r_1|r_0)\mathcal{E}_{1,0}(r_0)\mathcal{C}_0(r_0)\rho_S(0). \tag{5.35}$$

Now, the control operations performed at time t_n can depend on previous measurement results $\mathbf{r}_{n-1}$, which we denoted by $\mathcal{C}_n(r_n|\mathbf{r}_{n-1})$. Furthermore, the external agent can change the driving protocol $\lambda_t = \lambda_t(\mathbf{r}_n)$ (with $t \geq t_n$), such that the dynamical maps depend on previous measurement results too. This description captures all possible *classical* control scenarios. Quantum (or coherent) control scenarios are briefly discussed in Section 5.6.

As done multiple times before, here we also model the control operations by using an external stream of ancillas and an ideal memory. The physical picture that emerges from this mathematical description is sketched in Fig. 5.3 (sketches of real experiments are displayed in Figs. 3.8 and 5.7). The picture is essentially equivalent to the repeated interaction scheme from Section 3.9 with two important differences. First, we include here the effect of an ancilla measurement in the description. Second, we assume that the interaction with the ancillas happens *instantaneously*. In contrast, the repeated interaction scheme allows for finite system–ancilla interaction time. The reason for

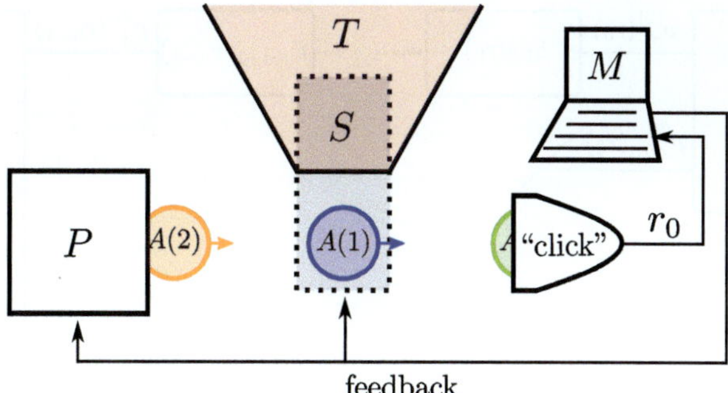

Fig. 5.3 A sketch showing a system S in contact with a heat bath at temperature T interacting with a stream of ancillas $A(0), A(1), A(2), \ldots$ (note that we start counting at $n = 0$). From left to right: ancilla $A(2)$ is prepared in a *preparation apparatus* P in some initial state; ancilla $A(1)$ is in close proximity to the system (grey area) and interacts via a sudden interaction with it; and the first ancilla $A(0)$ just gets detected to give rise to result r_0. Finally, the result r_0 is stored in a memory M. As indicated by the arrow with the label 'feedback', the initial ancilla state (sketched in different colors) or the system–ancilla interaction can be manipulated based on the state of the memory. The figure is taken from Strasberg (2020).

considering instantaneous interactions is rather pragmatic: it clearly separates the effect of the control operations (described by $\mathcal{C}_n(r_n|\mathbf{r}_{n-1})$) from the effect of the bath (described by $\mathcal{E}_{n,n-1}(\mathbf{r}_{n-1})$). Recall that this is a necessary assumption in the theory of (quantum) stochastic processes: the interventions of the external agent are assumed to be fully controllable. From a thermodynamic point of view, this assumption is not necessary.

We also extend the notation from the previous section to multiple control operations $\mathcal{C}_n(r_n|\mathbf{r}_{n-1})$, using a new ancilla at each step. The unitary dilation theorem from eqn (5.24) is then expressed as

$$\mathcal{C}_n(r_n|\mathbf{r}_{n-1})\rho_S^{(0)} = \mathrm{tr}_{A(n)}\left\{\mathcal{P}_{A(n)}(r_n)\mathcal{U}_{SA(n)}(\mathbf{r}_{n-1})\left[\rho_S^{(0)} \otimes \rho_{A(n)}^{(0)}(\mathbf{r}_{n-1})\right]\right\} \tag{5.36}$$

for any state $\rho_S^{(0)}$. All that we changed here compared with eqn (5.24) is to label the ancilla by $A(n)$ instead of A and to allow the initial ancilla state $\rho_{A(n)}^{(0)}$ and the system–ancilla unitary $U_{SA(n)}$ to depend on previous measurement results $\mathbf{r}_{n-1}$. This is necessary if we want to implement all kinds of classical feedback control. Furthermore, for simplicity we assume in the following that $\Pi_{A(n)}(r_n)$ describes a projective measurement that commutes with the ancilla Hamiltonian:

$$[\Pi_{A(n)}(r_n), H_{A(n)}] = 0. \tag{5.37}$$

In unison with Section 5.3, we also denote the system state prior to the nth control operation by $\rho_S^{(0)}(\mathbf{r}_{n-1})$ and introduce the notation

$$\rho_{SA(n)}^{(1)}(\mathbf{r}_{n-1}) \equiv \mathcal{U}_{SA(n)}(\mathbf{r}_{n-1})[\rho_S^{(0)}(\mathbf{r}_{n-1}) \otimes \rho_{A(n)}^{(0)}(\mathbf{r}_{n-1})], \tag{5.38}$$

$$\rho_{SA(n)}^{(2)}(\mathbf{r}_n) \equiv \frac{1}{p(r_n|\mathbf{r}_{n-1})}\mathcal{P}_{A(n)}(r_n)\rho_{SA(n)}^{(1)}(\mathbf{r}_{n-1}). \tag{5.39}$$

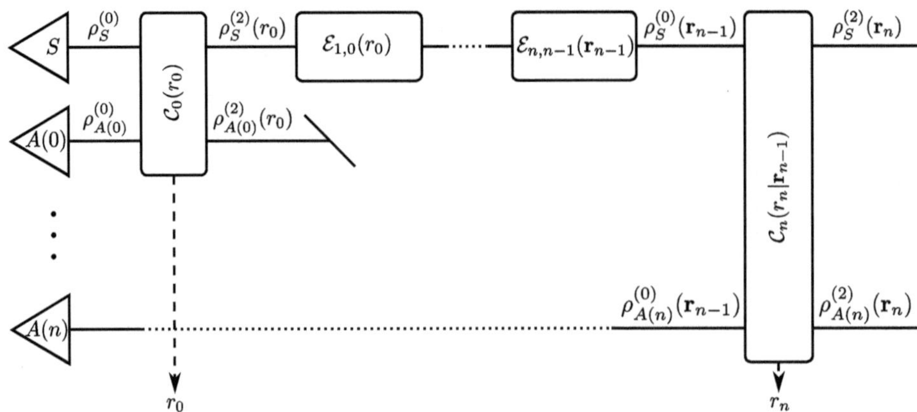

Fig. 5.4 Circuit diagram of a quantum Markov process where the external interventions are implemented by ancillas. The system (nth ancilla) state before the nth intervention is denoted by $\rho_S^{(0)}(\mathbf{r}_{n-1})$ ($\rho_{A(n)}^{(0)}(\mathbf{r}_{n-1})$) and after the intervention by $\rho_S^{(2)}(\mathbf{r}_n)$ ($\rho_{A(n)}^{(2)}(\mathbf{r}_n)$), given result r_n. We do not display 'intermediate states' during a control operation; these are labelled with a superscript '(1)' , such as that in eqn (5.38). The slash at the end of the $A(0)$ line should simply indicate that we discard the ancilla and no longer care about its dynamics after the first control operation. The framework presented here, however, continues to hold if we 're-use' ancilla $A(0)$; see also Section 5.6.

Here, $p(r_n|\mathbf{r}_{n-1}) = p(\mathbf{r}_n)/p(\mathbf{r}_{n-1})$ denotes the probability of obtaining result r_n conditioned on $\mathbf{r}_{n-1}$. Note that the superscript (0) and (1) ((2)) together with $\mathbf{r}_{n-1}$ ($\mathbf{r}_n$) uniquely specify that the states refer to the control operation happening at time t_n, which allows us to drop the time argument in the notation. We confirm for the marginal system state the relation

$$\sum_{r_n} p(r_n|\mathbf{r}_{n-1})\rho_S^{(2)}(\mathbf{r}_n) = \rho_S^{(1)}(\mathbf{r}_{n-1}). \tag{5.40}$$

Part of the notation above is summarized in Fig. 5.4.

Finally, we denote the state of the system and all ancillas at time $t > t_n$ (assuming $t < t_{n+1}$ if there is another intervention at time t_{n+1}) conditioned on the results $\mathbf{r}_n$ by $\rho_{SA}(t|\mathbf{r}_n)$. Here and in the following, we write $A \equiv A(0)A(1)\dots A(n)$ to denote the entire stream of ancillas. Furthermore, for $t > t_n$ the Hamiltonian of the system and all ancillas reads

$$H_{SA}[\lambda_t(\mathbf{r}_n)] = H_S[\lambda_t(\mathbf{r}_n)] + \sum_{j=0}^{n} H_{A(j)}, \tag{5.41}$$

which depends on the previous measurement results $\mathbf{r}_n$. Note that the ancillas are assumed to be non-interacting. Furthermore, the instantaneous interaction between the nth ancilla and the system at time t_n is assumed to be absorbed in the unitary $U_{SA(n)}$ in eqn (5.36) and therefore is not further specified in eqn (5.41).

Operational stochastic thermodynamics of a quantum Markov process

We now turn to the thermodynamic description and start with the definition of thermodynamic state functions. Based on the Hamiltonian (5.41), it follows that the **stochastic internal energy** of the system and all ancillas splits as

$$\boxed{u_{SA}(\mathbf{r}_n,t) \equiv \mathrm{tr}_{SA}\{H_{SA}[\lambda_t(\mathbf{r}_n)]\rho_{SA}(t|\mathbf{r}_n)\} = u_S(\mathbf{r}_n,t) + \sum_{j=0}^{n} u_{A(j)}(\mathbf{r}_n,t).} \tag{5.42}$$

This extends our previous definition (5.27) to the case of multiple ancillas. Likewise, in analogy with the stochastic entropy (5.29) for a single intervention, we define the **stochastic entropy** of the system and all ancillas as

$$\boxed{s_{SA}(\mathbf{r}_n,t) \equiv -k_B \ln p(\mathbf{r}_n) + k_B S_{\mathrm{vN}}[\rho_{SA}(t|\mathbf{r}_n)].} \tag{5.43}$$

It remains for us to establish the first law along a single trajectory and the second law on average. We aim at establishing them during a time interval, which starts shortly after the $(n-1)$th intervention and finishes after the nth intervention. Since n is arbitrary and the dynamics Markovian, this is sufficient for all general purposes. We denote this time interval by $(n] \equiv \lim_{\epsilon\searrow 0}(t_{n-1}+\epsilon, t_n+\epsilon]$. The interval *excluding* the nth control operation is denoted $(n) \equiv \lim_{\epsilon\searrow 0}(t_{n-1}+\epsilon, t_n-\epsilon)$.

During the time interval (n), the first law is simple. First of all, since the state of the ancillas does not change, their internal energy does not change either:

$$\Delta u_A^{(n)}(\mathbf{r}_{n-1}) \equiv \lim_{\epsilon\searrow 0}[u_A(\mathbf{r}_{n-1},t_n-\epsilon) - u_A(\mathbf{r}_{n-1},t_{n-1}+\epsilon)] = 0. \tag{5.44}$$

Exercise 5.8 Convince yourself of eqn (5.44).

The first law then follows directly from eqns (5.32) and (5.33):

$$\Delta u_S^{(n)}(\mathbf{r}_{n-1}) = q^{(n)}(\mathbf{r}_{n-1}) + w^{(n)}(\mathbf{r}_{n-1}). \tag{5.45}$$

Here, the heat and work flow conditioned on the previous outcomes $\mathbf{r}_{n-1}$ are

$$q^{(n)}(\mathbf{r}_{n-1}) = \int_{t_{n-1}}^{t_n} dt \mathrm{tr}_S\{H_S[\lambda_t(\mathbf{r}_{n-1})]\partial_t\rho_S(t|\mathbf{r}_{n-1})\}, \tag{5.46}$$

$$w^{(n)}(\mathbf{r}_{n-1}) = \int_{t_{n-1}}^{t_n} dt \mathrm{tr}_S\{\rho_S(t|\mathbf{r}_{n-1})\partial_t H_S[\lambda_t(\mathbf{r}_{n-1})]\}. \tag{5.47}$$

Next, if the nth control operation happens, the ancillas can also change their internal energy. In fact, not only can the internal energy of the ancilla $A(n)$ change, but also the internal energy of previous ancillas.

Exercise 5.9 Construct an example with two ancillas $A(0)$ and $A(1)$, where the internal energy of $A(0)$ changes after receiving result r_1. Show that this is not a quantum effect in general. *Hint:* It suffices to consider trivial dynamics with map $\mathcal{E}_{1,0} = \mathcal{I}_S$.

The first law during the control operation then follows by properly generalizing eqn (5.28) to

$$\begin{aligned}\Delta u_{SA}^{\text{ctrl}}(\mathbf{r}_n) &\equiv \lim_{\epsilon\searrow 0}[u_{SA}(\mathbf{r}_n, t_n+\epsilon) - u_{SA}(\mathbf{r}_{n-1}, t_n-\epsilon)] \\ &= \Delta u^{\text{meas}}(\mathbf{r}_n) + w^{\text{ctrl}}(\mathbf{r}_{n-1}).\end{aligned} \tag{5.48}$$

Note that our assumption (5.37) implies that $W^{\text{meas}} = 0$ in eqn (5.28). Therefore, we find that the work cost of the nth control operation reads

$$\begin{aligned}w^{\text{ctrl}}(\mathbf{r}_{n-1}) &= \text{tr}_S\left\{H_S[\lambda_n(\mathbf{r}_{n-1})]\left[\rho_S^{(1)}(\mathbf{r}_{n-1}) - \rho_S^{(0)}(\mathbf{r}_{n-1})\right]\right\} \\ &\quad + \text{tr}_{A(n)}\left\{H_{A(n)}\left[\rho_A^{(1)}(\mathbf{r}_{n-1}) - \rho_A^{(0)}(\mathbf{r}_{n-1})\right]\right\} \\ &\equiv w_S^{\text{ctrl}}(\mathbf{r}_{n-1}) + w_{A(n)}^{\text{ctrl}}(\mathbf{r}_{n-1}),\end{aligned}$$

which we split into parts affecting only the system and only the ancilla $A(n)$, respectively. Furthermore, λ_n is shorthand for λ_t evaluated at time $t = t_n$. Next, we consider $\Delta u^{\text{meas}}(\mathbf{r}_n)$ in eqn (5.48), which describes the random change in energy conditioned on receiving the outcome r_n. We have already discussed in the previous section that, depending on the situation, $\Delta u^{\text{meas}}(\mathbf{r}_n)$ can have a heat- or work-like character. Here, we split it as

$$\Delta u^{\text{meas}}(\mathbf{r}_n) = q_S^{\text{meas}}(\mathbf{r}_n) + w_A^{\text{meas}}(\mathbf{r}_n), \tag{5.49}$$

i.e. the change in energy of the system after receiving result r_n is interpreted as heat, whereas the corresponding change in energy of the ancillas is interpreted as work:

$$q_S^{\text{meas}}(\mathbf{r}_n) \equiv \text{tr}_S\left\{H_S[\lambda_n(\mathbf{r}_{n-1})]\left[\rho_S^{(2)}(\mathbf{r}_n) - \rho_S^{(1)}(\mathbf{r}_{n-1})\right]\right\}, \tag{5.50}$$

$$w_A^{\text{meas}}(\mathbf{r}_n) \equiv \text{tr}_A\left\{H_A\left[\rho_A^{(2)}(\mathbf{r}_n) - \rho_A^{(1)}(\mathbf{r}_{n-1})\right]\right\}. \tag{5.51}$$

This splitting is motivated by the fact that the system is in contact with a heat bath whereas the ancillas are not. Further justification for this identification is also given below and in Section 5.6, but—as emphasized in Section 5.3—an unambiguous identification does not seem possible. Moreover, similar to Section 5.3, it is possible to show that

$$\sum_{r_n} p(r_n|\mathbf{r}_{n-1})q_S^{\text{meas}}(\mathbf{r}_n) = 0, \quad \sum_{r_n} p(r_n|\mathbf{r}_{n-1})w_A^{\text{meas}}(\mathbf{r}_n) = 0. \tag{5.52}$$

Exercise 5.10 Derive eqn (5.52).

To summarize the **first law**, we denote by $\Delta u_{SA}^{(n]}(\mathbf{r}_n)$ the total change in energy during the time interval $(n]$. Collecting our previous results, the stochastic first law can be split into a system part

$$\boxed{\Delta u_S^{(n]}(\mathbf{r}_n) = q_S^{(n]}(\mathbf{r}_n) + w_S^{(n]}(\mathbf{r}_{n-1}),} \tag{5.53}$$

where $q_S^{(n]}(\mathbf{r}_n) \equiv q^{(n)}(\mathbf{r}_{n-1}) + q_S^{\text{meas}}(\mathbf{r}_n)$ and $w_S^{(n]}(\mathbf{r}_{n-1}) \equiv w^{(n)}(\mathbf{r}_{n-1}) + w_S^{\text{ctrl}}(\mathbf{r}_{n-1})$, and an ancilla part

$$\boxed{\Delta u_A^{(n)}(\mathbf{r}_n) = 0, \quad \Delta u_A^{\text{ctrl}}(\mathbf{r}_n) = w_A^{\text{meas}}(\mathbf{r}_n) + w_{A(n)}^{\text{ctrl}}(\mathbf{r}_{n-1}).} \tag{5.54}$$

Next, we turn to the second law. We introduce the **stochastic entropy production**

$$\sigma^{(n)}(\mathbf{r}_{n-1}) = \Delta s_{SA}^{(n)}(\mathbf{r}_{n-1}) - \frac{q^{(n)}(\mathbf{r}_{n-1})}{T}, \tag{5.55}$$

$$\sigma^{\text{ctrl}}(\mathbf{r}_n) = \Delta s_{SA}^{\text{ctrl}}(\mathbf{r}_n) - \frac{q_S^{\text{meas}}(\mathbf{r}_n)}{T}, \tag{5.56}$$

such that $\sigma^{(n]}(\mathbf{r}_n) = \sigma^{(n)}(\mathbf{r}_{n-1}) + \sigma^{\text{ctrl}}(\mathbf{r}_n)$ denotes the total stochastic entropy production during the time interval $(n]$. The claim is now that $\sigma^{(n)}(\mathbf{r}_{n-1})$ is always non-negative, whereas non-negativity of $\sigma^{\text{ctrl}}(\mathbf{r}_n)$ follows only on average.

We start by explicity writing down the stochastic entropy production in the absence of control operations,

$$\sigma^{(n)}(\mathbf{r}_{n-1}) = k_B\left\{S_{\text{vN}}\left[\rho_{SA}^{(0)}(\mathbf{r}_{n-1})\right] - S_{\text{vN}}\left[\rho_{SA}^{(2)}(\mathbf{r}_{n-1})\right]\right\} - \frac{q^{(n)}(\mathbf{r}_{n-1})}{T}. \tag{5.57}$$

Here, the joint system ancilla state shortly before the nth control operation is linked to the one shortly after the $(n-1)$th control operation by

$$\rho_{SA}^{(0)}(\mathbf{r}_{n-1}) = [\mathcal{E}_{n,n-1}(\mathbf{r}_{n-1}) \otimes \mathcal{I}_A]\rho_{SA}^{(2)}(\mathbf{r}_{n-1}) \equiv \mathcal{E}_{n,n-1}(\mathbf{r}_{n-1})\rho_{SA}^{(2)}(\mathbf{r}_{n-1}). \tag{5.58}$$

By splitting $S_{\text{vN}}(\rho_{SA}) = S_{\text{vN}}(\rho_S) + S_{\text{vN}}(\rho_A) - I_{S:A}(\rho_{SA})$, we can infer that

$$\sigma^{(n)}(\mathbf{r}_{n-1}) \geq I_{S:A}\left[\rho_{SA}^{(2)}(\mathbf{r}_{n-1})\right] - I_{S:A}\left[\mathcal{E}_{n,n-1}(\mathbf{r}_{n-1})\rho_{SA}^{(2)}(\mathbf{r}_{n-1})\right], \tag{5.59}$$

where we used the second law from eqn (5.32) together with the fact that the marginal states of the ancillas do not change. Thus, the stochastic entropy production in the absence of control operations is lower bounded by the change in correlations between the system and all ancillas. The next exercise shows that these correlations cannot increase; hence,

$$\boxed{\sigma^{(n)}(\mathbf{r}_{n-1}) \geq 0.} \tag{5.60}$$

Exercise 5.11 Show the positivity of the right-hand side of eqn (5.59). *Hint:* Using monotonicity of relative entropy could be helpful.

Next, we show that $\sigma^{\text{ctrl}}(\mathbf{r}_n)$ is non-negative on average. A sufficient condition for this is

$$\boxed{\sum_{r_n} p(r_n|\mathbf{r}_{n-1})\sigma^{\text{ctrl}}(\mathbf{r}_n) \geq 0.} \tag{5.61}$$

More explicitly and after taking into account eqn (5.52), this becomes

$$\begin{aligned}\sum_{r_n} p(r_n|\mathbf{r}_{n-1})\frac{\sigma^{\text{ctrl}}(\mathbf{r}_n)}{k_B} &= S_{\text{Sh}}[p(r_n|\mathbf{r}_{n-1})] + \sum_{r_n} p(r_n|\mathbf{r}_{n-1})S_{\text{vN}}\left[\rho_{SA}^{(2)}(\mathbf{r}_n)\right] \\ &\quad - S_{\text{vN}}\left[\rho_{SA}^{(0)}(\mathbf{r}_{n-1})\right].\end{aligned} \tag{5.62}$$

This expression is identical to

$$S_{\text{Sh}}[p(r_n|\mathbf{r}_{n-1})] + \sum_{r_n} p(r_n|\mathbf{r}_{n-1})S_{\text{vN}}\left[\frac{\mathcal{P}(r_n)\rho_{SA}^{(1)}(\mathbf{r}_{n-1})}{p(r_n|\mathbf{r}_{n-1})}\right] - S_{\text{vN}}\left[\rho_{SA}^{(1)}(\mathbf{r}_{n-1})\right],$$

where we used that the von Neumann entropy is invariant under unitary transformations. Finally, the non-negativity of eqn (5.61) can be verified by recalling Theorem A.1.

This concludes our derivation of the laws of thermodynamics within the operational framework for a quantum Markov process. For later convenience, we introduce the averages

$$X(t) \equiv \sum_{\mathbf{r}_n} p(\mathbf{r}_n)x(\mathbf{r}_n), \tag{5.63}$$

where X or x is a placeholder for internal energy, heat, work, entropy or entropy production. Thus, we use capital letters for averages in unison with our convention in classical stochastic thermodynamics.

Special cases

In the remainder, we connect the present framework to previous results and discuss the case of multiple heat baths.

First, the definitions given above reduce to the repeated interaction framework if one does not perform any ancilla measurement, i.e. in the case that $\mathcal{P}_{A(n)}(r_n)$ is the identity operation. Second, if one performs *no* control operation at all, the definitions above reduce to the standard quantum thermodynamics framework based on the BMS equation as described in Section 3.4. This is an important distinctive feature. In standard classical stochastic thermodynamics, this case is described by *averaging* over all trajectories. However, in quantum mechanics each measurement strategy can influence the dynamics. Thus, the traditional thermodynamic picture, such as the one in Section 3.4, is not obtained by averaging, but by *not measuring.*

Exercise 5.12 Verify the statements of the previous paragraph.

It is also instructive to compare the operational approach with the case traditionally considered in classical stochastic thermodynamics, which is described by continuous projective measurements in the system's energy eigenbasis. Of course, by 'continuous' we mean *very frequent* measurements. They can be modelled by performing measurements every time step δt such that the dynamical map can be approximated as

$$\mathcal{E}(t+\delta t, t) \approx \mathcal{I}_S + \delta t \mathcal{L}(\lambda_t). \tag{5.64}$$

Furthermore, the conventional framework of classical stochastic thermodynamics neglects any sort of feedback control and, hence, the dynamical map does not depend on previous measurement results. Since the system Hamiltonian commutes at different times in the classical limit, we can write their spectral decomposition as $H_S(\lambda_t) = \sum_s \epsilon(s, \lambda_t)|s\rangle\langle s|$ with a time-independent eigenbasis. The control operation at time $t_n = n\delta t$ is

$$\mathcal{C}_n(s_n)\rho_S^{(0)}(\mathbf{s}_{n-1}) = |s_n\rangle\langle s_n|\rho_S^{(0)}(\mathbf{s}_{n-1})|s_n\rangle\langle s_n|, \tag{5.65}$$

where we labelled the measured results r_n by s_n. Finally, any energetic cost associated with the dynamics of the ancillas is neglected in classical stochastic thermodynamics. Put differently, one considers the limit of an ideal memory measuring the system as defined at the beginning of Section 5.3.

We now ask what happens to our definitions in this case. We focus on one time step from $t_{n-1} = (n-1)\delta t$ to $t_n = n\delta t$, starting as usual with the state of the system shortly after the measurement at t_{n-1} and ending shortly after the measurement t_n. Let the measurement results be s_{n-1} and s_n, respectively. The associated internal energies are then $\epsilon(s_{n-1}, \lambda_{n-1})$ and $\epsilon(s_n, \lambda_n)$, respectively. These results are clearly in unison with classical stochastic thermodynamics.

Next, we turn to work and heat. First, by discretizing the integral in eqn (5.47), which is justified since δt is very small, it follows that the work in between the measurements reads

$$w^{(n)} = \epsilon(s_{n-1}, \lambda_n) - \epsilon(s_{n-1}, \lambda_{n-1}). \tag{5.66}$$

Furthermore, since we measure in the energy eigenbasis, our framework above predicts that there is no work cost associated with the measurement. For the heat it is different. First, discretizing again the integral in eqn (5.46), we find that

$$q^{(n)}(\mathbf{s}_{n-1}) = \mathrm{tr}_S\{H_S(\lambda_n)\mathcal{L}(\lambda_{n-1})|s_{n-1}\rangle\langle s_{n-1}|\}\delta t. \tag{5.67}$$

Next, we look at the heat associated with the measurement. According to eqn (5.50), this heat is given by comparing the internal energy after the measurement with the expectation value of the internal energy prior to the measurement. Since the state prior to the measurement is $[\mathcal{I}_S + \delta t\mathcal{L}(\lambda_t)]|s_{n-1}\rangle\langle s_{n-1}|$, we obtain

$$q_S^{\mathrm{meas}}(\mathbf{s}_n) = \epsilon(s_n, \lambda_n) - \mathrm{tr}_S\left\{H_S(\lambda_n)[\mathcal{I}_S + \delta t\mathcal{L}(\lambda_t)]|s_{n-1}\rangle\langle s_{n-1}|\right\}. \tag{5.68}$$

Taken together, the total heat flow becomes

$$q_S^{(n]}(\mathbf{s}_n) = q^{(n)}(\mathbf{s}_{n-1}) + q_S^{\text{meas}}(\mathbf{s}_n) = \epsilon(s_n, \lambda_n) - \epsilon(s_{n-1}, \lambda_n). \tag{5.69}$$

Finally, by comparing eqns (5.66) and (5.69) with the classical case (eqns (2.47) and (2.49)), we find that they agree. Thus, the stochastic energetics of the operational approach is consistent with classical stochastic thermodynamics.

What about stochastic entropy? A quick calculation reveals that the change in stochastic entropy in the present case is

$$\Delta s_S^{(n]}(\mathbf{s}_n) = -k_B \ln p(s_n|\mathbf{s}_{n-1}) = -k_B \ln p(s_n|s_{n-1}). \tag{5.70}$$

Here, we used the Markov property $p(s_n|\mathbf{s}_{n-1}) = p(s_n|s_{n-1})$, which holds because the control operations are rank-1 measurements (i.e. causal breaks in the terminology of Section 1.8).

Exercise 5.13 Derive eqn (5.70) starting from eqn (5.43).

Thus, the average entropy production becomes in the operational approach

$$đ\Sigma^{(n]} = k_B \sum_{s_{n-1}} p(s_{n-1}) S_{\text{Sh}}[p(s_n|s_{n-1})] - \frac{đQ_S^{(n]}}{T} \geq 0, \tag{5.71}$$

which does *not* coincide with the result from classical stochastic thermodynamics. Instead of containing the change in Shannon entropy of the system distribution, $S_{\text{Sh}}[p(s_n)] - S_{\text{Sh}}[p(s_{n-1})]$, it contains the average of $S_{\text{Sh}}[p(s_n|s_{n-1})]$, which is known as the *entropy rate* of the stochastic process. This discrepancy is *not* an inconsistency since our operational framework is designed to *include* the thermodynamic costs associated with the external interventions. In the present situation, the term proportional to $S_{\text{Sh}}[p(s_n|s_{n-1})]$ describes the cost of generating information in our memory. This becomes particularly transparent if one assumes the system to be at equilibrium. Standard classical stochastic thermodynamics predicts zero entropy production in this case, whereas the present approach predicts $đ\Sigma^{(n]} \sim S_{\text{Sh}}[p(s_n|s_{n-1})] > 0$ at equilibrium. This is a consequence of Landauer's principle if we assume that an experimentalist still continues to measure the system at equilibrium. Moreover, the operational approach allows all kinds of feedback control to be included in the description, whereas the standard description of classical stochastic thermodynamics breaks down in that case. The thermodynamics of feedback control will be the topic of Section 5.5.

Before we come to that, we briefly comment on the case of multiple heat baths. In principle, everything above can be extended to multiple baths. The only problem now is that it is no longer possible to associate a unique temperature with the measurement heat q_S^{meas}, which causes some ambiguity for the definition of stochastic entropy production. Note that the average entropy production does not suffer from that problem since q_S^{meas} vanishes on average. Furthermore, we also met the same problem in the classical case. For a system coupled to multiple baths, the observation of a change in the system state does, in general, not allow a unique heat bath to be associated with this transition.

5.5 Maxwell's Demon and the Thermodynamics of Feedback Control

We mentioned in Section 5.4 that the operational framework can take into account all kinds of feedback control. In the present section we study the case of classical feedback control in detail. Note that "classical" feedback control means that the information processed by the external agent is treated classically, but the system to be controlled can be quantum in general. In the following, we start with the general picture and afterwards treat a simple example. Another example is treated in Section 5.7.

General results

We start by considering the operational framework of quantum stochastic thermodynamics for a subclass of processes, that considerably simplify the thermodynamic treatment. This subclass of processes is created by ancillas that implement an ideal memory as defined at the beginning of Section 5.3. In contrast to Section 5.3, we here do not restrict the control operations $\mathcal{C}_k(r_k|\mathbf{r}_{k-1})$ to be projective measurements. However, we assume that the final measurement $\mathcal{P}_{A(n)}(r_n)$ of the ancilla is performed with rank-1 projectors such that the post-measurement state of the ancilla is pure.

Under these circumstances, it turns out that the influence of the ancillas on the thermodynamic description can be completely removed. This is clear for the first law since the ancilla Hamiltonian is energetically degenerate now. To investigate the impact on the second law, we start by denoting the initial pure ancilla state by $\rho_{A(n)}^{(0)} = |1\rangle\langle 1|_{A(n)}$. Then, we find for the initial state of the system and all ancillas

$$\rho_{SA}(0) = \rho_S(0) \otimes |1\rangle\langle 1|_{A(0)} \otimes |1\rangle\langle 1|_{A(1)} \otimes \cdots \otimes |1\rangle\langle 1|_{A(n)}, \tag{5.72}$$

where $\rho_S(0)$ is arbitrary. Furthermore, since the final ancilla state is $\rho_{A(n)}^{(2)} = |r_n\rangle\langle r_n|$, the system–ancilla state at time $t > t_n$ after the nth measurement reads

$$\rho_{SA}(t|\mathbf{r}_n) = \rho_S(t|\mathbf{r}_n) \otimes |r_0\rangle\langle r_0|_{A(0)} \otimes |r_1\rangle\langle r_1|_{A(1)} \otimes \cdots \otimes |r_n\rangle\langle r_n|_{A(n)}, \tag{5.73}$$

where $\rho_S(t|\mathbf{r}_n)$ is the system state conditioned on $\mathbf{r}_n$, which is not further specified here. From these considerations it follows that the change in stochastic entropy is

$$\begin{aligned} \Delta s_{SA}(\mathbf{r}_n, t) &= s_{SA}(\mathbf{r}_n, t) - s_{SA}(0) \\ &= -k_B \ln p(\mathbf{r}_n) + k_B S_{\text{vN}}[\rho_S(t|\mathbf{r}_n)] - k_B S_{\text{vN}}[\rho_S(0)]. \end{aligned} \tag{5.74}$$

Note that $s_{SA}(0) = k_B S_{\text{vN}}[\rho_S(0)]$ contains no term involving the probabilities $p(\mathbf{r}_n)$ stored in the memory. This is because the memory is supposed to be initially prepared in the blank state $|11\ldots1\rangle_M$, its initial probability distribution is $p(\mathbf{r}_n, 0) = \delta_{r_0,1}\ldots\delta_{r_n,1}$. To be consistent with the requirement that this has zero average entropy $S_{\text{Sh}}[p(\mathbf{r}_n, 0)] = 0$, we need to set $\ln p(\mathbf{r}_n, 0) = 0$ for all sequences $\mathbf{r}_n$.

The average second law becomes

$$\Sigma(t) = k_B S_{\text{Sh}}[p(\mathbf{r}_n)] + k_B \sum_{\mathbf{r}_n} p(\mathbf{r}_n) S_{\text{vN}}[\rho_S(t|\mathbf{r}_n)] - k_B S_{\text{vN}}[\rho_S(0)] - \frac{Q(t)}{T} \geq 0. \tag{5.75}$$

Here, the average heat flow contains both the contribution (5.46) in the absense of control operations and the measurement heat (5.50),

$$Q(t) = \sum_{\mathbf{r}_n} p(\mathbf{r}_n) \sum_{j=0}^{n} \left[q^{(j)}(\mathbf{r}_{j-1}) + q_S^{\text{meas}}(\mathbf{r}_j) \right]. \tag{5.76}$$

This heat flow also appears in the averaged first law,

$$\Delta U_S(t) = U_S(t) - U_S(0) = Q(t) + W(t), \tag{5.77}$$

where the average internal energy is $U_S(t) = \sum_{\mathbf{r}_n} p(\mathbf{r}_n)\text{tr}_S\{H_S[\lambda_t(\mathbf{r}_n)]\rho_S(t|\mathbf{r}_n)\}$.

The above laws of thermodynamics include the case of arbitrary ideal measurements performed on the system followed by an arbitrary classical feedback control loop utilizing the previous measurement results. For simplicity, let us consider the case where the system at the end of the process is returned to its initial state. For instance, if the initial state $\rho_S(0)$ is thermal, this can be achieved by allowing the system to relax back to the thermal state after all control operations have finished: $\rho_S(t|\mathbf{r}_n) \to \rho_S(0)$. This assumption simplifies the treatment because $\Delta U_S(t) = 0$ and $S_{\text{vN}}[\rho_S(t|\mathbf{r}_n)] - S_{\text{vN}}[\rho_S(0)] = 0$ in this case. Then, we can use the first law $Q(t) = -W(t)$ to express the second law as

$$\Sigma(t) = k_B S_{\text{Sh}}[p(\mathbf{r}_n)] + \frac{W(t)}{T} \geq 0. \tag{5.78}$$

This can be rearranged to

$$\boxed{-W(t) \leq k_B T S_{\text{Sh}}[p(\mathbf{r}_n)],} \tag{5.79}$$

which we call the **second law for feedback control**. Recalling that work is defined to be positive if it is done on the system, the second law for feedback control suggests that it is possible to *extract work*, $W(t) < 0$, *in the presence of a single heat bath.*

Exercise 5.14 Assume that the initial ancilla state is *mixed*, which can be used to model faulty measurements. Show that this always reduces the amount of extractable work.

To gain further insights about the second law for feedback control, we consider only a single measurement with result r followed by a work extraction (feedback) process modelled by a dynamical map $\mathcal{E}(r)$. We start by writing down the *average* first and second laws during the measurement step (see eqns (5.18) and (5.22)):

$$\Delta U_S^{\text{meas}} = W^{\text{meas}}, \quad S_{\text{Sh}}[p(r)] + \sum_r p(r) S_{\text{vN}}[\rho_S^{(2)}(r)] - S_{\text{vN}}(\rho_S^{(0)}) \geq 0. \tag{5.80}$$

Next, we consider the work extraction process. The average first law reads $\Delta U_S^{\text{fb}} = W^{\text{fb}} + Q$, where we used a superscript 'fb' to indicate that this is the process where the feedback happens. The second law reads

$$\Delta S_{SA}^{\text{fb}} - \frac{Q}{T} = k_B \sum_r p(r) \left\{ S_{\text{vN}} \left[\mathcal{E}(r)\rho_S^{(2)}(r) \right] - S_{\text{vN}} \left[\rho_S^{(2)}(r) \right] \right\} - \frac{Q}{T} \geq 0. \tag{5.81}$$

We can get a more transparent interpretation of this second law by considering the mutual information between the system and the ancilla after the measurement:

$$I_{S:A}^{(2)} \equiv I_{S:A}^{(2)}\left[\sum_r p(r)\rho_S^{(2)}(r) \otimes |r\rangle\langle r|_A\right] = S_{\rm vN}\left[\rho_S^{(1)}\right] - \sum_r p(r) S_{\rm vN}\left[\rho_S^{(2)}(r)\right], \quad (5.82)$$

where $\rho_S^{(1)} = \sum_r p(r)\rho_S^{(2)}(r)$ denotes the average post-measurement state; compare with eqn (5.40). Using the first and second laws, we can then bound the extractable work during the feedback process as

$$-W^{\rm fb} \leq -\Delta U_S^{\rm fb} + k_B T\left(I_{S:A}^{(2)} + \sum_r p(r) S_{\rm vN}\left[\mathcal{E}(r)\rho_S^{(2)}(r)\right] - S_{\rm vN}\left[\rho_S^{(1)}\right]\right). \quad (5.83)$$

As above, we focus on the case where the system is returned to its initial state after the work extraction process has finished: $\rho_S^{(0)} = \mathcal{E}(r)\rho_S^{(2)}$. In addition, we assume that the initial measurement is *classical*. Mathematically, if we model the measurement as $\mathcal{C}(r)\rho_S = P_r\rho_S P_r$ for some POVM $\{P_r^2\}$, a classical measurement is defined by demanding $[P_r, \rho_S] = 0$. This implies $\rho_S^{(1)} = \rho_S^{(0)}$. Consequently, we obtain $\Delta U_S^{\rm meas} = W^{\rm meas} = 0$ and, since the system is returned to its initial state after the feedback process, also $\Delta U_S^{\rm fb} = 0$. The bound on the extractable work then simplifies to

$$\boxed{-W^{\rm fb} \leq k_B T I_{S:A}^{(2)}.} \quad (5.84)$$

In contrast to eqn (5.79), this bound conveys the important message that the ability to extract work with feedback crucially relies on the *correlations* built up between the system and the ancilla (or memory). If the measurement establishes no correlations, we get $-W^{\rm fb} \leq 0$ and work extraction becomes impossible. The inequality (5.84) can be called the *second law for classical single-time feedback control.*

We remark that at first sight both eqn (5.79) and eqn (5.84) appear to violate the Kelvin–Planck formulation of the second law. A close look reveals, however, that the Kelvin–Planck formulation states that it is impossible to devise a *cyclically* operating heat engine that absorbs energy from a single heat bath and turns it into useful work. While we made sure with our assumption above that the system operates cyclically, i.e. returns to its initial state, this is not yet so for the memory. In order to return the memory to its initial state, we have to erase the information stored in it and need to spend at least an amount of work equal to $k_B T S_{\rm Sh}[p(\mathbf{r}_n)]$ according to Landauer's principle (see Section 2.2). This amount of work compensates for the work that we can possibly extract according to eqn (5.79). Thus, the Kelvin–Planck formulation of the second law is also safe.

Still, the possibility of extracting work from a single heat bath by some measurement and feedback loop remains intriguing. It offers a fascinating alternative to the conventional paradigm of heat engines that operate between two thermal baths at different temperatures. However, we have not yet demonstrated *how* to design a device that is capable of extracting work based on measurement and feedback. After all, it could be that the second law (5.79) is too loose and for some—not yet known—reason we always have $W(t) \geq 0$. As it turns out, this is not the case and it is indeed possible to extract work based on measurement and feedback control. For historical reasons, the external agent that implements the measurement and feedback loop is called **Maxwell's demon** in this case.

To construct such a 'Maxwell's demon device', we first recall Maxwell's original idea, which we quote here as formulated in his own words:

> ... if we conceive of a being whose faculties are so sharpened that he can follow every molecule in its course, such a being, whose attributes are as essentially finite as our own, would be able to do what is impossible to us. For we have seen that molecules in a vessel full of air at uniform temperature are moving with velocities by no means uniform, though the mean velocity of any great number of them, arbitrarily selected, is almost exactly uniform. Now let us suppose that such a vessel is divided into two portions, A and B, by a division in which there is a small hole, and that a being, who can see the individual molecules, opens and closes this hole, so as to allow only the swifter molecules to pass from A to B, and only the slower molecules to pass from B to A. He will thus, without expenditure of work, raise the temperature of B and lower that of A, in contradiction to the second law of thermodynamics.

Thus, in our modern language, the demon measures the position and velocity of molecules in a gas and the feedback control law says that the demon opens the hole whenever a molecule that is faster (slower) than the average has a chance to pass from A to B (B to A); otherwise, the hole remains closed. Over the course of time, this creates a temperature imbalance even if A and B were initially at the same temperature. If the work needed to open and close the hole can be assumed to be negligibly small, this *appears* to violate the second law if one neglects the information generation in the demon's memory.

Example: The electronic Maxwell demon

We now explicitly construct a device in close analogy to Maxwell's original idea and related to an actual experimental implementation of it. We call it an 'electronic Maxwell demon' because it is based on the idea of inducing a flow of electrons *against* the voltage bias by sorting electrons instead of gas molecules. The following exposition requires knowledge about the single-electron transistor as introduced in Section 3.10.

We recall the rate master equation describing the single-electron transistor,

$$\frac{d}{dt}\begin{pmatrix} p_F(t) \\ p_E(t) \end{pmatrix} = \sum_\nu \Gamma_\nu \begin{pmatrix} -[1-f_\nu(\epsilon_0)] & f_\nu(\epsilon_0) \\ 1-f_\nu(\epsilon_0) & -f_\nu(\epsilon_0) \end{pmatrix} \begin{pmatrix} p_F(t) \\ p_E(t) \end{pmatrix}, \tag{5.85}$$

and assume that we (or the demon) can control the bare tunnelling rates Γ_ν by switching them between two values: either $\Gamma_\nu = 0$ or $\Gamma_\nu \equiv \Gamma_0 > 0$. We assume weak coupling such that the energy cost associated with switching the tunnelling rate from zero to Γ_0 is negligible, similar to the original idea that the demon can open and close the hole with negligible energy expenditure. Furthermore, we set the chemical potentials to $\mu_L = \epsilon_0 + eV/2$ and $\mu_R = \epsilon_0 - eV/2$, where V denotes the voltage bias and e the

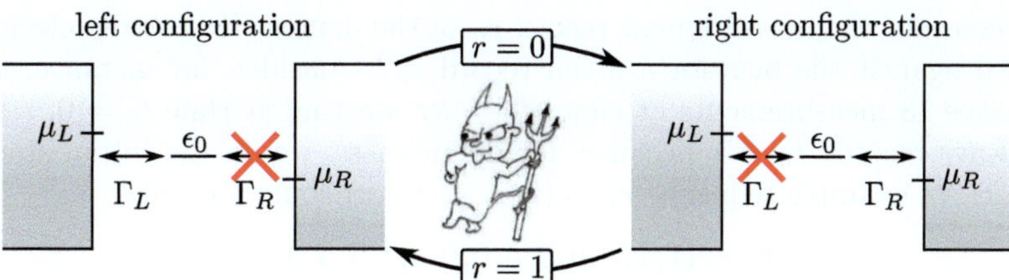

Fig. 5.5 Sketch of an electronic Maxwell demon, where the demon measures the state of the dot at every time τ and switches between a 'left' and 'right' configuration corresponding to $(\Gamma_L, \Gamma_R) = (\Gamma_0, 0)$ and $(\Gamma_L, \Gamma_R) = (0, \Gamma_0)$, respectively. Whether the configuration gets switched depends on the measurement result $r \in \{0, 1\}$ and the current configuration.

electron charge. Then, it turns out that the only parameter entering the Fermi functions is the dimensionless variable $\alpha \equiv \beta eV/2$ and for definiteness we choose $V > 0$ in the following. Without any feedback control, this implies that electrons have the tendency to flow from left to right.

Now, consider the following measurement and feedback loop. Suppose we start at time $t_0 = 0$ with a filled dot and tunnelling rates switched to the value $(\Gamma_L, \Gamma_R) = (\Gamma_0, 0)$. We call this the 'left configuration' because the dot is only coupled to the left bath. Then, we wait for a time τ before we measure the system state, finding either no electron ($r_1 = 0$) or one electron ($r_1 = 1$) on the dot. If the result is $r_1 = 0$, we switch the tunnelling rates to $(\Gamma_L, \Gamma_R) = (0, \Gamma_0)$, which we call the 'right configuration' for obvious reasons. If $r_1 = 1$, we keep the left configuration. In the latter case, we continue measuring the state of the dot at every time $t_\ell = \ell\tau$ until we find $r_\ell = 0$, whereupon we switch to the right configuration. Also in the right configuration we continue to measure the state of the dot at every time τ, but the feedback control law is now *reversed.* If at some time t_ℓ we find $r_\ell = 0$ for the right configuration, we keep the right configuration. If we find $r_\ell = 1$ instead, we switch back to the left configuration. Then, we continue measuring at every time τ applying the rules above. Finally, we stop the control protocol after some time $t_n = n\tau$. A sketch of the control loop is shown in Fig. 5.5.

Exercise 5.15 Convince yourself of the fact that for any finite α and sufficiently large n this control protocol transports electrons from the right to the left *against* the voltage bias.

To describe the dynamics of the system and the demon, let us introduce the probabilities $p_{j,\nu}$ with $j \in \{0, 1\}$ denoting the number of electrons on the dot and $\nu \in \{L, R\}$ the configuration of the single-electron transistor. To simplify the treatment, we assume that $\tau \gg \Gamma_0^{-1}$, which implies that the system has approximately reached the steady state before each measurement. Since the dot is at each time coupled to only one bath, this steady state is in fact an equilibrium state with respect to either the left or right bath. Denoting by $\pi_{j|\nu}$ the equilibrium probability of finding the dot with j electrons given that we are in configuration ν, we find

$$\pi_{0|L} = \pi_{1|R} = \frac{1}{e^{\alpha} + 1}, \quad \pi_{1|L} = \pi_{0|R} = \frac{1}{e^{-\alpha} + 1}. \tag{5.86}$$

Next, consider the measurement record $\mathbf{r}_n$ of the demon. How many electrons are transported against the bias for a given record $\mathbf{r}_n$? Consider, for instance, that the demon makes 13 measurements at times $t_k = k\tau$ starting at time $t_0 = 0$, where the demon always records $r_0 = 1$ because, for definiteness, we fix the initial state to be $p_{0,L}(0) = 1$. An example sequence $\mathbf{r}_n = (r_n, \dots, r_1, r_0)$ could look like

$$\mathbf{r}_n = (1,1,1,0,0,0,0,0,0,1,1,1,1). \tag{5.87}$$

In this case, the dot was filled until t_3 before the demon found it empty at time t_4. Then, the demon switched from the left to the right configuration and subsequently found the dot empty from t_5 until t_9. At time t_{10} the demon found the dot filled again and switched back to the left configuration. Afterwards, the dot remained filled at t_{11} and t_{12}. Thus, we know that at time t_4 there was one net electron transfer from the system to the left bath. Furthermore, we know that at time t_{10} there was another net transfer of one electron from the right bath to the system. Thus, from t_0 to t_{12}, the sequence (5.87) describes the transfer of one electron from the right to the left bath. The sequence (5.87) can be written in a more convenient way by making clear when a transition from 1 to 0 or 0 to 1 happens:

$$\mathbf{r}_n = (1\ 1\ 1|0\ 0\ 0\ 0\ 0\ 0|1\ 1\ 1\ 1). \tag{5.88}$$

The number of bars $|$ in the sequence $\mathbf{r}_n$, denoted $\#(\mathbf{r}_n)$, has an important meaning. If $\#(\mathbf{r}_n) = 2m$, there has been a net number of m electrons transferred from the right to the left bath and the final state of the system at t_n is a filled dot in the left configuration. If $\#(\mathbf{r}_n) = 2m+1$, there has been $m+1$ electrons transferred to the left bath, but only m electrons were transferred from the right bath, which means that the final system state is an empty dot in the right configuration.

Exercise 5.16 Convince yourself of the previous statements.

We can go on even further by grouping all sequences $\mathbf{r}_n$ with $k \equiv \#(\mathbf{r}_n)$ bars into the set R_k. The probability of observing such a sequence $\mathbf{r}_n$ is

$$p(\mathbf{r}_n) = \pi_{1|L}^{n-k}\pi_{0|L}^{k} \quad \text{if} \quad \mathbf{r}_n \in R_k, \tag{5.89}$$

where we used $\pi_{0|L} = \pi_{1|R}$ and $\pi_{1|L} = \pi_{0|R}$. All sequences in R_k have the same probability, which is a consequence of the fact that we are working with the steady-state probabilities (5.86). The number of sequences in R_k is given by the binomial coefficient:

$$\#R_k = \binom{n}{k} = \frac{n!}{k!(n-k)!}. \tag{5.90}$$

The previous considerations allow us to compute the Shannon entropy of $p(\mathbf{r}_n)$:

$$S_{\text{Sh}}[p(\mathbf{r}_n)] = -\sum_{k=0}^{n}\sum_{\mathbf{r}_n \in R_k} p(\mathbf{r}_n)\ln p(\mathbf{r}_n) = nS_{\text{Sh}}(\pi_{0|L}, \pi_{1|L}). \tag{5.91}$$

Alternatively, this result could have been derived by taking into account that each measurement copies the state of the system, which has entropy $S_{\text{Sh}}(\pi_{0|L}, \pi_{1|L}) = S_{\text{Sh}}(\pi_{0|R}, \pi_{1|R})$ and which is independent of the state measured previously.

Exercise 5.17 Derive eqn (5.91) by using $\sum_{k=0}^{n}\binom{n}{k}p^{n-k}q^k = (p+q)^n$.

Now, we can analyse our electronic Maxwell demon from a thermodynamic perspective. In general, the integrated first law after the nth measurement is

$$\Delta U_S(t_n) = Q_L(t_n) + Q_R(t_n) + W_{\text{chem}}(t_n). \tag{5.92}$$

If we denote by $N_{L/R}(t_n)$ the average number of electrons transferred from the left/right bath to the system, we can express the heat flows as $Q_\nu(t_n) = (\epsilon_0 - \mu_\nu)N_\nu(t_n)$ and the chemical work as $W_{\text{chem}}(t_n) = \mu_L N_L(t_n) - \mu_R N_R(t_n)$. Furthermore, let us focus on the limit where the number of measurement cycles n is very large. Since $\Delta U_S(t_n) \in [-\epsilon_0, \epsilon_0]$ for any n, we can neglect this boundary term for large n since the term $N_\nu(t_n) \sim n$ dominates. Then, we also have approximately $N_L(t_n) \approx -N_R(t_n)$.

Next, we focus on the change in entropy of the dot state only. Since the initial state of the dot is assumed to be pure, we have $S_{\text{vN}}[\rho_S(0)] = 0$. Likewise, since the final state of the dot immediately after the nth measurement is also pure, we get $S_{\text{vN}}[\rho_S(t_n|\mathbf{r}_n)] = 0$ too. Thus, and in unison with eqn (5.78), the second law becomes

$$\Sigma(t) = k_B S_{\text{Sh}}[p(\mathbf{r}_n)] - \frac{Q_L(t) + Q_R(t)}{T} = k_B S_{\text{Sh}}[p(\mathbf{r}_n)] + \frac{W_{\text{chem}}(t_n)}{T} \geq 0, \tag{5.93}$$

where we used the first law to get the second equality.

It remains for us to compute the chemical work. Using $N_L(t_n) \approx -N_R(t_n)$, we get

$$W_{\text{chem}}(t_n) = (\mu_L - \mu_R)\sum_{\mathbf{r}_n} p(\mathbf{r}_n) n_L(\mathbf{r}_n, t_n), \tag{5.94}$$

where $n_L(\mathbf{r}_n, t_n)$ is the number of electrons transferred from the left bath conditioned on the measurement sequence $\mathbf{r}_n$. If $\mathbf{r}_n \in R_k$, we have $n_L(\mathbf{r}_n, t_n) = -k/2$ or $n_L(\mathbf{r}_n, t_n) = -(k+1)/2$ depending on the question of whether k is even or odd, respectively. However, since we ignore such boundary effects, we set $n_L(\mathbf{r}_n, t_n) = -k/2$ for $\mathbf{r}_n \in R_k$ and obtain

$$W_{\text{chem}}(t_n) = -\frac{\mu_L - \mu_R}{2}\sum_{k=0}^{n}\sum_{\mathbf{r}_n \in R_k} k p(\mathbf{r}_n) = -\frac{\mu_L - \mu_R}{2} n\pi_{0|L}. \tag{5.95}$$

Putting everything together, the entropy production *per* measurement and feedback cycle becomes

$$\Sigma^{(n]} = \frac{\Sigma(t_n)}{n} = k_B\left[S_{\text{Sh}}(\pi_{0|L}, \pi_{1|L}) - \alpha\pi_{0|L}\right] \geq 0. \tag{5.96}$$

This expression allows us to define an efficiency for the Maxwell demon quantifying how much information needs to be generated in the memory to extract a certain amount of work:

$$\eta \equiv \frac{-\beta W_{\text{chem}}(t_n)}{S_{\text{Sh}}[p(\mathbf{r}_n)]} = \frac{\alpha\pi_{0|L}}{S_{\text{Sh}}(\pi_{0|L}, \pi_{1|L})} \leq 1. \tag{5.97}$$

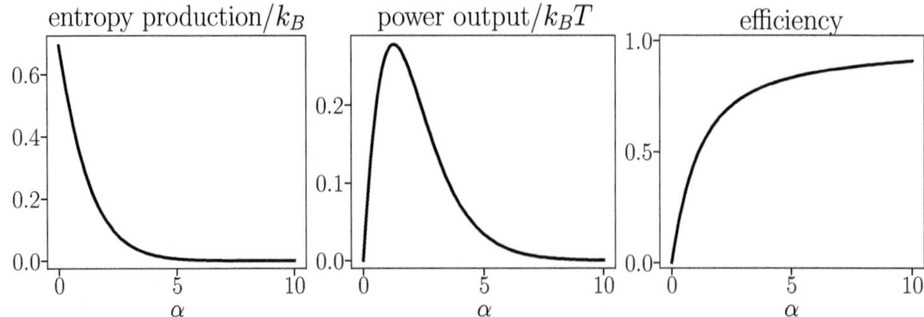

Fig. 5.6 Plot of the entropy production per cycle, the power output per cycle and the efficiency as a function of the dimensionless parameter $\alpha = \beta eV/2$.

Numerical results for the entropy production, the power output and the efficiency as a function of α are shown in Fig. 5.6. If one recalls that α is proportional to the voltage bias V, the results appear surprising at first sight. The entropy production is *maximal* for $V = 0$ and decays to *zero* in the infinite bias regime $V \to \infty$. The thermodynamics of the single-electron transistor *without* any demon is exactly *opposite*. Thus, the demon significantly changes the thermodynamic interpretation.

To understand this, consider first the case $V = 0$. Then, there is no thermodynamic cost associated with the movement of an electron from one bath to the other. However, for $V = 0$ the entropy (5.91) generated in the memory becomes maximal since $S_{\text{Sh}}(\pi_{0|L}, \pi_{1|L}) = \ln 2$. In contrast, for $V \to \infty$ the thermodynamic cost associated with the electron transport diverges. However, in that limit we also find that $(\pi_{0|L}, \pi_{1|L}) \to (0, 1)$. Thus, since there is never an electron on the dot in the left configuration, the demon never switches to the right configuration. The transport is completely blocked. Furthermore, the entropy generated in the memory is zero because the only possible measurement sequence is $\mathbf{r}_n = (1, \ldots, 1, 1)$.

Apart from this peculiarity, we also see similarities to conventional heat engines. The power output is maximized for some intermediate value $V \in (0, \infty)$ and the maximal efficiency $\eta \to 1$ is reached in the limit of zero power output. Thus, a Maxwell demon also faces the power–efficiency dilemma (see Section 4.7) and has to decide on high power at moderate efficiency *or* high efficiency at moderate power.

5.6 Autonomous Approach and Strong Coupling Corrections

The goal of this section is twofold. First, we extend the operational approach to the strong coupling and non-Markovian regime, thereby demonstrating that it has a wide range of applicability. Second, we derive the thermodynamic definitions appearing in the operational approach from a different perspective, thereby providing an independent justification for them. To that end, we approach the problem from an *autonomous* perspective modelling the external agent fully quantum mechanically. Everything evolves unitarily and no explicit measurements appear in that description.

Therefore, we can rely on the results from Chapter 3, which were derived *in the absence* of measurements, to draw conclusions about the case *in the presence* of measurements.

Autonomous implementation of the process tensor

Let us return to Fig. 5.3, which shows a sketch of the operational approach for a system S coupled to a bath B. It involves a preparation apparatus P, ancillas A, a detector, a memory M and a feedback loop. All these parts are modelled explicitly in this section, albeit it turns out that the detector and feedback loop do not require additional physical degrees of freedom. The global Hamiltonian of the problem therefore reads

$$H_{SBPAM}(\lambda_t) = H_{SB}(\lambda_t) + H_{PA}(\lambda_t) + V_{SA}(\lambda_t) + H_{AM}(\lambda_t). \tag{5.98}$$

Here, $H_{SB}(\lambda_t) = H_S(\lambda_t) + H_B + V_{SB}$ is the conventional system–bath Hamiltonian (which is left arbitrary in this section), $H_{PA}(\lambda_t)$ describes the preparation of the ancillas, $V_{SA}(\lambda_t)$ is the system–ancilla interaction and $H_{AM}(\lambda_t)$ is the Hamiltonian describing the memory and its interaction with the ancilla. Furthermore, the present scenario does not yet include feedback control, but this is easy to add later on.

The unitary time evolution with respect to the Hamiltonian (5.98) of the global system–bath–preparation apparatus–ancilla–memory state $\rho_{SBPAM}(t)$ is written as

$$\rho_{SBPAM}(t) = U_{SBPAM}(t,0)\rho_{SBPAM}(0)U^\dagger_{SBPAM}(t,0). \tag{5.99}$$

The goal is now the following. Suppose that the memory M stores *all possible* measurement results $\mathbf{r}_n$. Then, given a *hypothetical* measurement of the memory yielding result $\mathbf{r}_n$, we want that the *conditional* system state reads

$$\tilde{\rho}_S(t|\mathbf{r}_n) = \mathrm{tr}_{BPA}\{\langle \mathbf{r}_n|\rho_{SBPAM}(t)|\mathbf{r}_n\rangle\} = \mathfrak{T}[\boldsymbol{C}_{n:0}(\mathbf{r}_n)], \tag{5.100}$$

where $\mathfrak{T}$ is the process tensor introduced in Section 1.7 and $\boldsymbol{C}_{n:0}(\mathbf{r}_n)$ is an n-time control operation, which we are free to choose. If eqn (5.100) holds, we say that the Hamiltonian (5.98) together with a suitable initial state $\rho_{SBPAM}(0)$ 'simulates' or 'autonomously implements' the process tensor or the quantum stochastic process.

Equation (5.100) should be compared with Fig. 1.6, where we stated that there exists a unitary dilation theorem for the entire process tensor. The goal here is to study in greater detail the role played by $H_{PA}(\lambda_t)$, $V_{SA}(\lambda_t)$ and $H_{AM}(\lambda_t)$, in particular from a thermodynamic perspective. To keep the exposition tractable, this involves a number of idealizations. Since we are not interested in practical realizations of our autonomous model, but in the theoretical foundations of quantum stochastic thermodynamics, we take the freedom to use as many idealizations as we want.

We start by finding a way of implementing a sequence of control operations $\mathcal{C}_j = \sum_{r_j} \mathcal{C}_j(r_j)$ at times t_j with $\mathcal{C}_j(r_j)$ given by eqn (5.36) (if one replaces n by j in that equation and ignores the effect of feedback control).

To this end, we first consider $H_{PA}(\lambda_t) = H_P + H_A + V_{PA}(\lambda_t)$, where H_P and H_A are the Hamiltonians of the preparation apparatus and the ancillas alone, assuming them to be time independent for simplicity. The interaction $V_{PA}(\lambda_t)$ between the preparation apparatus and the ancillas is designed such that it prepares the ancillas in the desired initial states $\rho^{(0)}_{A(0)}, \rho^{(0)}_{A(1)}, \ldots, \rho^{(0)}_{A(n)}$ required for the implementation of the control operation according to eqn (5.36). The reader should by now be sufficiently familiar with the unitary dilation theorem to know that, by choosing an appropriate initial state $\rho_P(0)$ for the preparation apparatus and an appropriate interaction $V_{PA}(\lambda_t)$, we can prepare any ancilla state we like. Therefore, we refrain from specifying the details of H_P, $\rho_P(0)$ and $V_{PA}(\lambda_t)$ further.

Second, we consider the system–ancilla interaction $V_{SA}(\lambda_t)$. Again, ignoring the situation involving feedback control for the moment, $V_{SA}(\lambda_t)$ is responsible for implementing the unitary operations $\mathcal{U}_{SA(j)}$ in eqn (5.36). Thus, we set

$$V_{SA}(\lambda_t) = \sum_j V_{SA(j)}(\lambda_t), \quad V_{SA(j)}(\lambda_t) = \hbar\delta(t - t_j) i \ln(U_{SA(j)}), \tag{5.101}$$

where t_j is the time at which the control operation $\mathcal{C}_j(r_j)$ happens.

Exercise 5.18 Show that $V_{SA(j)}(\lambda_t)$ is Hermitian and convince yourself of the fact that the time evolution according to $V_{SA(j)}(\lambda_t)$ implements the desired unitary operator $U_{SA(j)}$ at time t_j. *Hint:* One possibility is to discretize time into steps dt and to represent the Dirac delta function as $\delta(t - t_j) \approx dt^{-1}\Theta(t - t_j)\Theta(t_j + dt - t)$ and to let $dt \to 0$ afterwards.

Finally, we consider $H_{AM}(\lambda_t) = \sum_j V_{A(j)M(j)}(\lambda_t)$, where $V_{A(j)M(j)}(\lambda_t)$ describes the interaction of the jth ancilla with the jth part of the memory, which is assumed to be ideal (see Section 5.3). The interaction is assumed to be of the same form as in eqn (5.101):

$$V_{A(j)M(j)}(\lambda_t) = \hbar\delta(t - t_j^+) i \ln(U_{A(j)M(j)}). \tag{5.102}$$

Here, $t_j^+ = t_j + \epsilon$ denotes a time *shortly after* t_j since the system–ancilla interaction has to happen before the ancilla–memory interaction (see Fig. 5.3). After tracing out the memory, the action of $V_{A(j)M(j)}(\lambda_t)$ implements the CPTP map $\sum_{r_j} \mathcal{P}_{A(j)}(r_j)$, which describes the average effect of the measurement on the ancilla. How this can be done was described in detail in Sections 1.4 and 5.3 and is not repeated here.

Exercise 5.19 Convince yourself of the fact that the above description implements a set of control operations $\mathcal{C}_j = \sum_{r_j} \mathcal{C}_j(r_j)$ at times t_j.

We continue by adding one more assumption, which makes our exposition easier and also more realistic; namely, we assume that the memory quickly dephases. Expressed in equations, this means the following. If we label by $|\mathbf{r}_n\rangle$ the basis of the memory used to encode the measurement results, the assumption of quick dephasing implies that the memory state can be approximated for practically all times as

$$\rho_M(t) = \sum_{\mathbf{r}_n} p(\mathbf{r}_n; t) |\mathbf{r}_n\rangle\langle\mathbf{r}_n|_M. \tag{5.103}$$

Here, $p(\mathbf{r}_n; t)$ is the probability of finding the memory in the state $|\mathbf{r}_n\rangle$ at time t. For instance, if $t < 0$, $p(\mathbf{r}_n; t) = \delta_{r_0,1} \ldots \delta_{r_n,1}$ describes the initial memory state, where each register is set to its standard reference state $|1\rangle$. On the other hand, if $t > t_n$, then $p(\mathbf{r}_n; t) = p(\mathbf{r}_n)$ is the probability of obtaining measurement results $\mathbf{r}_n$. In general,

$$p(\mathbf{r}_n; t) = p(\mathbf{r}_{k-1})\delta_{r_k,1} \dots \delta_{r_n,1} \quad \text{for} \quad t \in (t_{k-1}, t_k). \tag{5.104}$$

Since a dephasing operation is not a unitary operation, this requires us to include an additional 'dephasing bath' B' in the description, which causes the memory states to quickly and irreversibly decohere (note that B' could also be a part of B). After tracing out B', the system–preparation apparatus–ancilla–memory state at any time t can be written as

$$\rho_{SPAM}(t) = \sum_{\mathbf{r}_n} \tilde{\rho}_{SPA}(t|\mathbf{r}_n) \otimes |\mathbf{r}_n\rangle\langle\mathbf{r}_n|_M, \tag{5.105}$$

with $p(\mathbf{r}_n; t) = \text{tr}_{SPA}\{\tilde{\rho}_{SPA}(t|\mathbf{r}_n)\}$. The state $\tilde{\rho}_{SPA}(t|\mathbf{r}_n)$ can be interpreted as the (non-normalized) conditional state of the system, preparation apparatus and all ancillas in the case that somebody (e.g. some external *superobserver*) makes a projective measurement of the memory and obtains as a result the sequence $\mathbf{r}_n$. Note that this interpretation does not depend on any dephasing bath B'. Even in the absence of B', a measurement of $\mathbf{r}_n$ would destroy any coherent superposition of the measurement results: $\tilde{\rho}_{SPA}(t|\mathbf{r}_n) = \text{tr}_{BM}\{|\mathbf{r}_n\rangle\langle\mathbf{r}_n|\rho_{SBPAM}(t)\}$. While it is useful to think about $\tilde{\rho}_{SPA}(t|\mathbf{r}_n)$ as a *hypothetical* post-measurement state, we keep in mind that we never assume that any actual measurement is performed in this section.

Based on eqn (5.105), it also becomes transparent how to include feedback control in the description. For that purpose, we generalize the Hamiltonian (5.98) to

$$H_{SBPAM}(\lambda_t) = \sum_{\mathbf{r}_n} H_{SBPA}[\lambda_t(\mathbf{r}_n)] \otimes |\mathbf{r}_n\rangle\langle\mathbf{r}_n|_M + H_{AM}(\lambda_t). \tag{5.106}$$

Here, $H_{SBPA}[\lambda_t(\mathbf{r}_n)]$ describes the same physical situation as before with the only difference being that every term appearing in it is allowed to depend on the state of the memory $\mathbf{r}_n$. Now, consider the time evolution from t_n after the nth control operation to some later time t. For simplicity in the presentation, let us ignore the short time delay $t_n^+ - t_n$ between the system–ancilla interaction (5.101) and the ancilla–memory interaction (5.102). Then, during this time $H_{AM}(\lambda_t)$ acts only trivially on the memory degrees of freedom and the unitary time evolution is

$$\rho_{SBPAM}(t) = \sum_{\mathbf{r}_n} U_{SBPA}(\mathbf{r}_n)\tilde{\rho}_{SBPA}(t_n|\mathbf{r}_n)U_{SBPA}^\dagger(\mathbf{r}_n) \otimes |\mathbf{r}_n\rangle\langle\mathbf{r}_n|_M, \tag{5.107}$$

where

$$U_{SBPA}(\mathbf{r}_n) = \exp_+\left[-\frac{i}{\hbar}\int_{t_n}^t H_{SBPA}[\lambda_s(\mathbf{r}_n)]ds\right]. \tag{5.108}$$

Exercise 5.20 Derive eqns (5.107) and (5.108).

The construction above allows us to consider all kinds of conceivable feedback scenarios, whether they are classical or quantum (where, for instance, one could entangle an ancilla with the system and afterwards *re-use* it to implement a further control operation). More generally speaking, the above protocol allows us to implement all kinds of n-time control operations $\boldsymbol{C}_{n:0}(\mathbf{r}_n)$ as long as they are CPTP. Thus, we can confirm eqn (5.100): $\text{tr}_{BPA}\{\langle\mathbf{r}_n|\rho_{SBPAM}(t)|\mathbf{r}_n\rangle\} = \mathfrak{T}[\boldsymbol{C}_{n:0}(\mathbf{r}_n)]$. This concludes our dynamical description of an autonomous model that is able to implement an arbitary quantum stochastic process. In the end, we merely used the unitary dilation theorem many times.

Exercise 5.21 Show that eqn (5.100) reduces to eqn (5.35) if the dynamics is Markovian and we consider only classical feedback control. Convince yourself of the fact that the construction above continues to hold for non-Markovian dynamics. Finally, recall Fig. 1.6, which shows the most general strategy to generate $\mathfrak{T}[\boldsymbol{C}_{n:0}(\mathbf{r}_n)]$. Verify that this case is also included in our description.

Thermodynamic description

Before we delve into the details, we make the agreement that the preparation apparatus is *ideal.* This means that we assume all thermodynamic costs associated with the preparation of the ancillas to be *negligible.* In view of our agreement above to focus on foundational instead of practical issues, we could indeed argue that the preparation apparatus is a sophisticated machine implementing all thermodynamic processes in a *reversible* manner (i.e. with zero entropy production). Furthermore, our goal is to precisely understand the thermodynamic impact of non-equilibrium resources such as a stream of ancillas prepared in an arbitrary state. Adding the potential costs for their preparation is not offering any deeper insights into the present discussion.

Next, we fix the initial state to be

$$\rho_{SBPAMB'}(0) = \pi_{SB}(\lambda_0) \otimes \rho_P(0) \otimes \rho_A(0) \otimes \rho_M(0) \otimes \pi_{B'}, \tag{5.109}$$

which we explain from right to left. First, $\pi_{B'}$ describes the dephasing bath prepared in a canonical ensemble at temperature T; this is assumed to be an ideal weakly coupled thermal bath. Next, $\rho_M(0) = |1\rangle\langle 1|_{M(0)} \otimes \cdots \otimes |1\rangle\langle 1|_{M(n)}$ describes the initial state of the memory with each register j prepared in the standard reference state $|1\rangle$. The initial state $\rho_A(0)$ of the ancillas is assumed to be decorrelated from the rest, but otherwise it can be arbitrary as the desired initial state is implemented via the preparation apparatus. For that purpose, we need to prepare the preparation apparatus adequately, for instance by supplying it with enough free energy resources. However, as stated above, the precise state $\rho_P(0)$ is unimportant for our purposes and it is neglected in the remainder. Finally, the system and bath are assumed to be prepared in a canonical ensemble $\pi_{SB}(\lambda_0)$ at temperature T. One might wonder why we do not consider more general initial system states, but, in fact, we can view the first control operation at $t_0 = 0$ as a preparation of *any* system state we like.

Now, we recall results from Section 3.7, where we derived a general thermodynamic framework using the Hamiltonian of mean force, which was—at least formally—valid for any system–bath Hamiltonian (to be thermodynamically consistent, we needed to require the bath to be large such that its temperature barely changes from a macroscopic point of view). Remember that the 'system' in our philosophy is the part of the universe about which we have precise control. In the present framework, this certainly does not only include the system S itself, but also the ancillas and the memory. We thus define the *supersystem* $S' \equiv SAM$. The Hamiltonian of mean force $H^*_{S'}(\lambda_t)$ for S' can be indirectly defined through the relation

$$\pi^*_{S'}(\lambda_t) = \text{tr}_{BB'}\{\pi_{S'BB'}(\lambda_t)\} = \frac{e^{-\beta H^*_{S'}(\lambda_t)}}{\mathcal{Z}^*_{S'}(\lambda_t)}, \quad \mathcal{Z}^*_{S'}(\lambda_t) \equiv \frac{\mathcal{Z}_{S'BB'}(\lambda_t)}{\mathcal{Z}_{BB'}(\lambda_t)}. \tag{5.110}$$

Our first result is that $H^*_{S'}(\lambda_t)$ can be written at all times as

$$H^*_{S'}(\lambda_t) = \sum_{\mathbf{r}_n} \{H^*_S[\lambda_t(\mathbf{r}_n)] + H_A(\mathbf{r}_n)\} \otimes |\mathbf{r}_n\rangle\langle\mathbf{r}_n|_M, \tag{5.111}$$

where $H^*_S[\lambda_t(\mathbf{r}_n)]$ is indirectly defined via the relation $\pi^*_S[\lambda_t(\mathbf{r}_n)] = \text{tr}_B\{\pi_{SB}[\lambda_t(\mathbf{r}_n)]\}$. Implicitly contained in the result (5.111) is the assumption that we are not interested in a thermodynamic description 'during' the control operation since the control operation is *instantaneous*. Strictly speaking, eqn (5.111) is therefore valid for all times $t \notin \{t_0, t_1, \ldots, t_n\}$.

Exercise 5.22 Consider a tripartite system XYB with Hamiltonian $H_X + H_Y + H_B + V_{XB}$. Show that the Hamiltonian of mean force H^*_{XY} for XY can be written as $H^*_X + H_Y$ and deduce from this insight eqn (5.111).

Since we are considering a large isolated system evolving unitarily in time, we can now use insights from Chapter 3 to define basic thermodynamic quantities. First of all, the total work done on S' during the process is

$$W(t) = \int_0^t ds \text{tr}_{S'}\left\{\frac{\partial H_{S'}(\lambda_s)}{\partial s}\rho_{S'}(s)\right\}. \tag{5.112}$$

Next, we list the definitions for the strong coupling internal energy and entropy for S' (recall that the Hamiltonian of mean force depends on temperature)

$$U^*_{S'}(t) = \text{tr}_{S'}\{\rho_{S'}(t)\left[H^*_{S'}(\lambda_t) + \beta\partial_\beta H^*_{S'}(\lambda_t)\right]\}, \tag{5.113}$$
$$S^*_{S'}(t) = k_B\text{tr}_{S'}\left\{\rho_{S'}(t)\left[-\ln\rho_{S'}(t) + \beta^2\partial_\beta H^*_{S'}(\lambda_t)\right]\right\}. \tag{5.114}$$

Furthermore, heat is indirectly defined via the first law: $Q^*(t) = \Delta U^*_{S'}(t) - W(t)$. Note that all these definitions follow from Section 3.7 by replacing S by S'.

Next, we consider the second law, which stipulates

$$\Sigma^*(t) = \Delta S^*_{S'}(t) - \frac{Q^*(t)}{T} \geq 0. \tag{5.115}$$

Positivity of the entropy production follows from monotonicity of the relative entropy because

$$\Sigma^*(t) = k_B D[\rho_{S'BB'}(t)|\pi_{S'B}(\lambda_t) \otimes \pi_{B'}] - k_B D[\rho_{S'}(t)|\pi^*_{S'}(\lambda_t)]. \tag{5.116}$$

In fact, this expression can be seen as a combination of the two second laws (eqns (3.98) and (3.107)) applied to a weakly coupled dephasing bath B' and an arbitrarily strongly coupled bath B, respectively.

Exercise 5.23 Derive eqn (5.116).

So far, the above definitions have provided a consistent thermodynamic framework for a large supersystem S' in contact with the thermal baths B and B'. The above framework holds because there are *no* explicit measurements. The goal is now to rewrite the expressions in a way that the role played by the memory in our autonomous description becomes transparent.

We start with the internal energy and system entropy. Using eqn (5.105), which, after tracing out P, becomes $\rho_{SAM}(t) = \sum_{\mathbf{r}_n} p(\mathbf{r}_n;t)\rho_{SA}(t|\mathbf{r}_n) \otimes |\mathbf{r}_n\rangle\langle\mathbf{r}_n|_M$, and eqn (5.111), we find for the internal energy

$$U^*_{S'}(t) = \sum_{\mathbf{r}_n} p(\mathbf{r}_n;t)\mathrm{tr}_{SA}\left\{\rho_{SA}(t|\mathbf{r}_n)\big(H^*_S[\lambda_t(\mathbf{r}_n)] + \beta\partial_\beta H^*_S[\lambda_t(\mathbf{r}_n)] + H_A(\mathbf{r}_n)\big)\right\}. \tag{5.117}$$

To lighten the notation, we drop for now the dependence on $\mathbf{r}_n$ in any Hamiltonian, keeping in mind that the present description includes the possibility of performing feedback control. In addition, we also drop any subscript on the trace operation $\mathrm{tr}\{\cdot\}$, which become clear from context. Thus, we write the previous equation as

$$U^*_{S'}(t) = \sum_{\mathbf{r}_n} p(\mathbf{r}_n;t)\mathrm{tr}\left\{\rho_{SA}(t|\mathbf{r}_n)[H^*_S(\lambda_t) + \beta\partial_\beta H^*_S(\lambda_t) + H_A]\right\}. \tag{5.118}$$

Likewise, we find for the entropy the expression

$$S^*_{S'}(t) = k_B\sum_{\mathbf{r}_n} p(\mathbf{r}_n;t)\left[S_{\mathrm{vN}}[\rho_{SA}(t|\mathbf{r}_n)] - \ln p(\mathbf{r}_n;t) + \beta^2\mathrm{tr}\{\rho_S(t|\mathbf{r}_n)\partial_\beta H^*_S(\lambda_t)\}\right]. \tag{5.119}$$

Next, we consider the work. Our Hamiltonian contains three time-dependent terms: $H_S(\lambda_t)$, $V_{SA}(\lambda_t)$ and $V_{AM}(\lambda_t)$. Consequently, we split the work into three parts: $W(t) = W_S(t) + W_{SA}(t) + W_{AM}(t)$. The first directly refers to the change in system energy due to the external driving and reads

$$W_S(t) = \sum_{\mathbf{r}_n}\int_0^t ds p(\mathbf{r}_n;s)\mathrm{tr}\left\{\frac{\partial H_S(\lambda_s)}{\partial s}\rho_S(s|\mathbf{r}_n)\right\}, \tag{5.120}$$

where we used again eqn (5.105). The other two parts describe energy changes due to sudden unitary operations and have to be considered with greater care.

The work associated with the system–ancilla interaction is written

$$W_{SA}(t) = \sum_{\mathbf{r}_n}\int_0^t ds p(\mathbf{r}_n;s)\mathrm{tr}\left\{\frac{\partial V_{SA}(\lambda_s)}{\partial s}\rho_{SA}(s|\mathbf{r}_n)\right\}. \tag{5.121}$$

Now, recall that $V_{SA}(\lambda_t)$ describes a train of delta-kicks at times t_k, implementing a unitary $U_{SA(k)}$ affecting the system and the kth ancilla. Let us consider first the case of a weakly coupled bath B such that the energy stored in V_{SB} is negligible. Then, we can write $W_{SA}(t)$ as a sum over changes in the expectation value of $H_S(\lambda_k) + H_{A(k)}$ happening at times t_k,

$$W_{SA}(t) = \sum_{\mathbf{r}_n} \sum_{k=0}^{n} p(\mathbf{r}_n; t_k^-) \text{tr} \left\{ [H_S(\lambda_k) + H_{A(k)}] \left[\rho_{SA}^{(1)}(t_k|\mathbf{r}_n) - \rho_{SA}^{(0)}(t_k|\mathbf{r}_n) \right] \right\}. \tag{5.122}$$

Here, we used the notation from Sections 5.3 and 5.4, where $\rho_{SA}^{(0)}(t_k|\mathbf{r}_n)$ denotes the system–ancilla state prior to the control operation and $\rho_{SA}^{(1)}(t_k|\mathbf{r}_n) = \mathcal{U}_{SA(k)}\rho_{SA}^{(0)}(t_k|\mathbf{r}_n)$ denotes the state after the system–ancilla interaction but before the ancilla–memory interaction. Furthermore, we introduced the notation $p(\mathbf{r}_n; t_k^-)$ to describe the probability that the memory is found in $\mathbf{r}_n$ *before* the ancilla–memory interaction at time t_k. Thus, $p(\mathbf{r}_n; t_k^-) = p(\mathbf{r}_{k-1})\delta_{r_k,1} \dots \delta_{r_n,1}$; compare with eqn (5.104). Now, let us return to the strong coupling case, where V_{SB} is no longer negligible. Then, although $U_{SA(k)}$ does not directly act on the bath degrees of freedom, the expectation value of V_{SB} can change because of the change in the system state. The proper generalization of eqn (5.122) needs to take into account the bath degrees of freedom and reads

$$\begin{aligned} W_{SA}(t) = & \sum_{\mathbf{r}_n} \sum_{k=0}^{n} p(\mathbf{r}_n; t_k^-) \\ & \times \text{tr} \left\{ [H_S(\lambda_k) + H_{A(k)} + V_{SB}] \left[\rho_{SBA}^{(1)}(t_k|\mathbf{r}_n) - \rho_{SBA}^{(0)}(t_k|\mathbf{r}_n) \right] \right\}. \end{aligned} \tag{5.123}$$

Finally, we apply the same logic to $W_{AM}(t)$, which describes the sudden changes in the ancilla–memory energy due to the unitaries $U_{A(k)M}$ at times t_k. This time, however, we do not have to worry about strong coupling corrections because the marginal state of S does not change during this operation. Taking further into account that the memory Hamiltonian is degenerate, we find

$$W_{AM}(t) = \sum_{\mathbf{r}_n} \sum_{k=0}^{n} \text{tr} \left\{ H_A \left[p(\mathbf{r}_n; t_k^+)\rho_A^{(2)}(t_k|\mathbf{r}_n) - p(\mathbf{r}_n; t_k^-)\rho_A^{(1)}(t_k|\mathbf{r}_n) \right] \right\}. \tag{5.124}$$

Here, $\rho_A^{(2)}(t_k|\mathbf{r}_n)$ denotes the state of the ancillas after the measurement of $A(k)$ and $p(\mathbf{r}_n; t_k^+)$ describes the resulting probability distribution of the memory states, which differs from $p(\mathbf{r}_n; t_k^-)$ exactly at the kth entry. Moreover, since the marginal state of S does not change at this step, we actually have alternative options to decompose $W_{AM}(t)$ by taking into account an arbitrary fraction μ of the system energy $H_S(\lambda_t)$:

$$\begin{aligned} W_{AM}(t) = & \mu \sum_{\mathbf{r}_n} \sum_{k=0}^{n} \text{tr} \left\{ H_S(\lambda_k) \left[p(\mathbf{r}_n; t_k^+)\rho_S^{(2)}(t_k|\mathbf{r}_n) - p(\mathbf{r}_n; t_k^-)\rho_S^{(1)}(t_k|\mathbf{r}_n) \right] \right\} \\ & + \sum_{\mathbf{r}_n} \sum_{k=0}^{n} \text{tr} \left\{ H_A \left[p(\mathbf{r}_n; t_k^+)\rho_A^{(2)}(t_k|\mathbf{r}_n) - p(\mathbf{r}_n; t_k^-)\rho_A^{(1)}(t_k|\mathbf{r}_n) \right] \right\}. \end{aligned} \tag{5.125}$$

We come back to this equation at the end.

Exercise 5.24 Show that eqn (5.124) and (5.125) are equal.

We have now reached a position where we can draw interesting conclusions by asking: What is the internal energy, system entropy, work done and heat flow conditioned on a memory in state $|\mathbf{r}_n\rangle$? To motivate these questions, imagine that they are asked by an external superobserver who projectively measures the state of the memory at the final time t. Yet, these measurements do not have to be implemented in practice—they are a fictitious concept. Alternatively, we could also imagine 'us' as an inside observer described as a part of the global Hamiltonian (5.98) asking these questions.

Above, we have already written all state functions in the suggestive form

$$X(t) = \sum_{\mathbf{r}_n} p(\mathbf{r}_n; t) x(\mathbf{r}_n, t), \tag{5.126}$$

where X refers to either the internal energy or the entropy of the supersystem. The interpretation of eqn (5.126) is clear: if the external superobserver measured result $\mathbf{r}_n$, then the corresponding internal energy or entropy is given by $x(\mathbf{r}_n, t)$. Therefore, $x(\mathbf{r}_n, t)$ should describe the associated stochastic thermodynamic quantity at the trajectory level. Now, by looking at eqns (5.118) and (5.119), we infer

$$\boxed{\begin{aligned} u^*_{S'}(\mathbf{r}_n, t) &= \operatorname{tr}\left\{\rho_{SA}(t|\mathbf{r}_n)[H^*_S(\lambda_t) + \beta\partial_\beta H^*_S(\lambda_t) + H_A]\right\}, \\ s^*_{S'}(\mathbf{r}_n, t) &= k_B\left[S_{\rm vN}[\rho_{SA}(t|\mathbf{r}_n)] - \ln p(\mathbf{r}_n; t) + \beta^2 \operatorname{tr}\{\rho_S(t|\mathbf{r}_n)\partial_\beta H^*_S(\lambda_t)\}\right]. \end{aligned}} \tag{5.127}$$

In the weak coupling limit, we have $H^*_S(\lambda_t) = H_S(\lambda_t)$ and $\partial_\beta H^*_S(\lambda_t) = 0$ and the definitions above become *identical* to the previous definition of stochastic internal energy and system entropy; compare with eqns (5.42) and (5.43), respectively.

Now, we consider the work. First, the work (5.120) done on the system from t_{n-1} to t_n becomes

$$W_S(t_n) - W_S(t_{n-1}) = \sum_{\mathbf{r}_n} \int_{t_{n-1}}^{t_n} ds p(\mathbf{r}_n; s) \operatorname{tr}\left\{\frac{\partial H_S(\lambda_s)}{\partial s}\rho_S(s|\mathbf{r}_n)\right\}. \tag{5.128}$$

Now, notice that during this time interval there is no control operation. Hence, $p(\mathbf{r}_n; s) = p(\mathbf{r}_{n-1})\delta_{r_n,1}$ does not change in time and we can write

$$W_S(t_n) - W_S(t_{n-1}) = \sum_{\mathbf{r}_{n-1}} p(\mathbf{r}_{n-1}) \int_{t_{n-1}}^{t_n} ds \operatorname{tr}\left\{\frac{\partial H_S(\lambda_s)}{\partial s}\rho_S(s|\mathbf{r}_{n-1})\right\}. \tag{5.129}$$

Again, given that we knew that the memory is in state $\mathbf{r}_n = (\mathbf{r}_{n-1}, 1)$, the work done on the system in the time interval (t_{n-1}, t_n) becomes

$$w_S^{(n)}(\mathbf{r}_{n-1}) = \int_{t_{n-1}}^{t_n} ds \operatorname{tr}\left\{\frac{\partial H_S(\lambda_s)}{\partial s}\rho_S(s|\mathbf{r}_{n-1})\right\}, \tag{5.130}$$

which exactly *coincides* with our previous result (5.47). Next, we turn to the work due to the system–ancilla interaction. From eqn (5.123) we infer that the work $W_{SA}(t_n)$ done during the nth control operation is

$$W_{SA}(t_n) = \sum_{\mathbf{r}_{n-1}} p(\mathbf{r}_{n-1}) \tag{5.131}$$
$$\times \operatorname{tr}\left\{[H_S(\lambda_n) + H_{A(n)} + V_{SB}]\left[\rho^{(1)}_{SBA}(t_n|\mathbf{r}_{n-1}) - \rho^{(0)}_{SBA}(t_n|\mathbf{r}_{n-1})\right]\right\},$$

where we used $p(\mathbf{r}_n; t_n^-) = p(\mathbf{r}_{n-1})\delta_{r_n,1}$. This suggests identifying

$$w_{SA}(\mathbf{r}_{n-1}, t_n) = \operatorname{tr}\left\{[H_S(\lambda_n) + H_{A(n)} + V_{SB}]\left[\rho^{(1)}_{SBA}(t_n|\mathbf{r}_{n-1}) - \rho^{(0)}_{SBA}(t_n|\mathbf{r}_{n-1})\right]\right\} \tag{5.132}$$

as the work done during the system–ancilla interaction conditioned on the results $\mathbf{r}_{n-1}$. Again, it is straightforward to see that in the weak coupling regime, $V_{SB} \approx 0$, this definition *coincides* with what we have previously called the work due to the control operation, $w^{\text{ctrl}}(\mathbf{r}_{n-1})$; compare with eqn (5.49). Finally, we turn to the work due to the ancilla–memory interaction. From eqn (5.124) we infer that the work $W_{AM}(t_n)$ spent during the nth measurement is

$$\begin{aligned} W_{AM}(t_n) &= \sum_{\mathbf{r}_n} \operatorname{tr}\left\{H_A\left[p(\mathbf{r}_n; t_n^+)\rho^{(2)}_A(t_n|\mathbf{r}_n) - p(\mathbf{r}_n; t_n^-)\rho^{(1)}_A(t_n|\mathbf{r}_n)\right]\right\} \\ &= \sum_{\mathbf{r}_n} p(\mathbf{r}_n)\operatorname{tr}\left\{H_A\left[\rho^{(2)}_A(t_n|\mathbf{r}_n) - \rho^{(1)}_A(t_n|\mathbf{r}_{n-1})\right]\right\}. \end{aligned} \tag{5.133}$$

Here, we used $p(\mathbf{r}_n; t_n^-) = p(\mathbf{r}_{n-1})\delta_{r_n,1}$, we set $p(\mathbf{r}_n; t_n^+) = p(\mathbf{r}_n)$ and we denoted $\rho^{(1)}_A(t_n|\mathbf{r}_n) = \rho^{(1)}_A(t_n|\mathbf{r}_{n-1}, 1) \equiv \rho^{(1)}_A(t_n|\mathbf{r}_{n-1})$ to indicate that the ancilla state prior to the measurement does not depend on the result r_n. The previous expression now suggests using the following definition at the trajectory level:

$$w_{AM}(\mathbf{r}_n, t_n) = \operatorname{tr}\left\{H_A\left[\rho^{(2)}_A(t_n|\mathbf{r}_n) - \rho^{(1)}_A(t_n|\mathbf{r}_{n-1})\right]\right\}. \tag{5.134}$$

By now, the reader might not be too surprised to find out that this expression coincides with what we previously called the measurement work, which we denoted $w_A^{\text{meas}}(\mathbf{r}_n)$ and defined in eqn (5.51).

Conclusions

We constructed a giant 'black box', which simulates a quantum stochastic process in a unitary way. Using results from Chapter 3, we equipped this box with a consistent thermodynamic interpretation. Furthermore, this black box had a memory that was designed to store a 'measurement record' $\mathbf{r}_n$ that we used in Section 5.4 to define a stochastic trajectory. We then asked: Given that the memory is found in state $|\mathbf{r}_n\rangle$, what are the internal energy, system entropy and work done *conditional* on $\mathbf{r}_n$?

As we explicitly demonstrated above, all these quantities coincide with the definitions of Section 5.4 at weak coupling. This provides an *independent* and *solid* justification for the definitions used in operational quantum stochastic thermodynamics. Clearly, by using the first law at the trajectory level, the heat defined in the autonomous approach conditioned on $\mathbf{r}_n$ also coincides with the definition of stochastic heat from Section 5.4. Moreover, we established the positivity of the entropy production *on average*, which follows by noting that we can write eqn (5.115) as

$$\Sigma^*(t) = \sum_{\mathbf{r}_n} p(\mathbf{r}_n;t)\sigma^*(\mathbf{r}_n,t) \geq 0 \tag{5.135}$$

with $\sigma^*(\mathbf{r}_n,t) \equiv \Delta s^*_{S'}(\mathbf{r}_n,t) - q(\mathbf{r}_n,t)/T$, where $q(\mathbf{r}_n,t)$ is defined by the stochastic first law. As a consequence, the present findings also reveal that the second law of thermodynamics holds for arbitrary quantum stochastic processes, independent of the Markov assumption. Our sole requirement was that the initial system–bath state is modelled by a global canonical ensemble.

Finally, we return to our alternative expression (5.125) for the work done during the ancilla–memory interaction. Setting, for example, $\mu = 1$ and following the same steps as above, we find

$$W_{AM}(t_n) = \sum_{\mathbf{r}_n} p(\mathbf{r}_n)\mathrm{tr}\left\{[H_S(\lambda_n) + H_A]\left[\rho^{(2)}_{SA}(t_n|\mathbf{r}_n) - \rho^{(1)}_{SA}(t_n|\mathbf{r}_{n-1})\right]\right\}. \tag{5.136}$$

From here, we would conclude that the stochastic work conditioned on $\mathbf{r}_n$ is

$$\tilde{w}_{AM}(\mathbf{r}_n,t_n) = \mathrm{tr}\left\{[H_S(\lambda_n) + H_A]\left[\rho^{(2)}_{SA}(t_n|\mathbf{r}_n) - \rho^{(1)}_{SA}(t_n|\mathbf{r}_{n-1})\right]\right\}. \tag{5.137}$$

Crucially, it is not hard to see that $w_{AM}(\mathbf{r}_n,t_n) \neq \tilde{w}_{AM}(\mathbf{r}_n,t_n)$, i.e. the two different proposals do *not* coincide. This echoes the conundrum discussed in Sections 4.2, 5.3 and 5.4. There is a small but non-negligible freedom to interpret the change in energy due to the ‘collapse’ of the wavefunction or the update of the observer’s state of knowledge as a heat- or work-like contribution. This ambiguity cannot be fixed by the present autonomous approach either. Therefore, this choice remains partially a matter of convention, albeit—as we have emphasized in previous sections—further consistency checks can remove at least part of the ambiguity.

The present autonomous approach is powerful and general, but it is also quite abstract. Therefore, Section 5.7 is devoted to an illustrating experimental example.

5.7 Application to Photon Number Stabilization Experiments

The last section of this book is devoted to applying our previous findings to a real-world experiment from quantum optics. This experiment demonstrates that we have reached an era where we have complete control about individual quantum systems—something which Schrödinger in 1952 believed to be as hard as ‘raising Ichthyosauria in the zoo’. The experiment involves real-time and time-delayed feedback control, the handling of incomplete information, among other complications, and therefore provides an ultimate test for the operational approach to quantum stochastic thermodynamics. The rest of this section requires some basic knowledge about cavity electrodynamics, which was provided in Exercises 3.1 and 3.28.

Experimental set-up and dynamical description

We start with a step-by-step analysis of the experimental set-up, which is an extension of the micromaser set-up considered in Section 3.9; see Fig. 5.7. To get a better feeling for the physics involved, we also list the experimental parameters in the following.

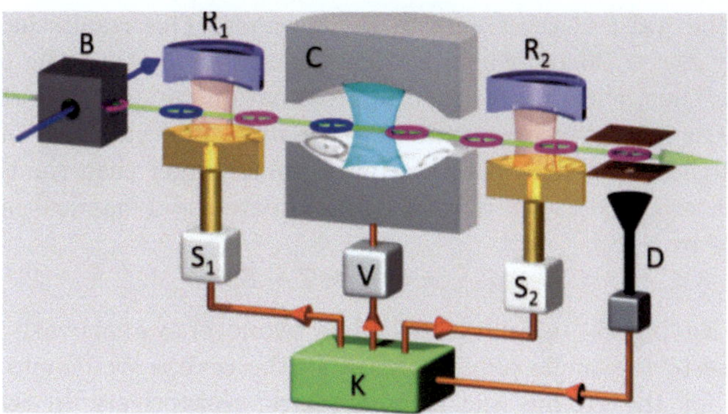

Fig. 5.7 Sketch of the experimental set-up taken from Zhou *et al.* (2012). The system we wish to control is the microwave cavity C in the centre. To do so, one uses a beam of atoms prepared in B, which can be manipulated by the Ramsey cavities R_1 and R_2 and are read out by the detector D. Then, the measurement results are sent to a controller K, which decides in real time whether to send a sensor atom to measure the state of the cavity (pink circles) or an emitter or absorber atom to manipulate the state of the cavity (blue circles). In the latter case the atoms are brought into exact resonance with the cavity by applying a voltage V.Reprinted figure with permission from X. Zhou *et al.*, *Phys. Rev. Lett.* **108**, 243602 (2012). Copyright (2021) by the American Physical Society.

First, the system we want to control is a superconducting Fabry–Perot cavity, denoted by C in Fig. 5.7. Its Hamiltonian is $\hbar\omega_c a^\dagger a$, where $a^\dagger$ and a denote photon creation and annihilation operators and $\omega_c/2\pi = 51.1$ GHz is the experimentally measured frequency of the cavity.

This cavity is an open system coupled to an environment at temperature $T = 0.8$ K, which implies a Bose–Einstein distribution $N_{\text{th}} = (e^{\beta\hbar\omega_c} - 1)^{-1} \approx 0.05$. The dynamics of the cavity is well described by the master equation (in a rotating frame)

$$\begin{aligned}\partial_t \rho_S(t) &= \mathcal{L}_0 \rho_S(t) \qquad (5.138)\\ &\equiv \frac{1+N_{\text{th}}}{2\tau_c}\left[a\rho_S(t)a^\dagger - \frac{1}{2}\{a^\dagger a, \rho_S(t)\}\right] + \frac{N_{\text{th}}}{2\tau_c}\left[a^\dagger\rho_S(t)a - \frac{1}{2}\{aa^\dagger, \rho_S(t)\}\right].\end{aligned}$$

The experimental cavity lifetime is $\tau_c = 65$ ms. We remark that strong coupling or non-Markovian effects play *no* role in this experiment. Therefore, we can later apply the thermodynamic framework as developed in Section 5.4.

Without any external interventions, the cavity will thermalize to the canonical ensemble π_S. For the present parameters, this means that the cavity is found with probability 0.95 in the vacuum state $|0\rangle$ with zero photons. The goal of the feedback loop is to reverse the effect of the dissipation and to stabilize a pure photon number (or *Fock*) state $|n\rangle = |n_T\rangle$, where $n_T > 0$ denotes the target number of photons, which can be adapted in the experiment. For the numerical simulations, we choose $n_T = 2$.

To reverse the effect of dissipation, a beam of atoms is created in B via velocity selection and laser excitation. The atoms are repeatedly prepared at regular intervals of duration $\tau = 82$ μs and they leave B with a velocity of $v = 250$ m/s. The interaction time of each atom with the cavity can be estimated as $t_{\text{int}} = w/v$, where

$w = 6$ mm is the waist of the Gaussian cavity mode. This results in an interaction time of roughly $t_{\text{int}} \approx 30\ \mu$s, which is 2000 times *smaller* than the cavity lifetime: $t_{\text{int}} \approx \tau_c/2000$. Thus, within very good approximation we can treat the interactions with the atoms as instantaneous, as done in the formal development of our theory in Section 5.4. Furthermore, the cavity lifetime is much larger than the time τ between two atoms ($\tau/\tau_c \approx 1/800$), such that we approximate the dynamical map in between two interactions by

$$\mathcal{E}_{k+1,k} = e^{\mathcal{L}_0\tau} \approx \mathcal{I} + \mathcal{L}_0\tau. \tag{5.139}$$

The atoms leaving B are well described as two-level systems with an energy gap $\hbar\omega_a \approx \hbar\omega_c$ close to the single photon energy in the cavity. We denote the two levels as $|g\rangle$ and $|e\rangle$ for the ground and excited states, respectively. In reality, however, both states correspond to excited Rydberg states of the atom where the orbit of the outer electron is far away from the nucleus, creating in turn a large dipole moment. Furthermore, the atoms leave B in the ground state $|g\rangle$.

To counteract the dissipation by feedback control, we first need to measure the number of photons in C, which is done with the pink 'sensor' atoms in Fig. 5.7. Remarkably, this happens in a non-destructive way *without* absorbing or emitting a single photon using a modified Ramsey interferometry scheme. If the initial state of the cavity contains no coherences in the Fock basis, this implements a *quantum non-demolition (QND) measurement*. The basic idea behind this remarkable measurement scheme is that the cavity can imprint a phase shift proportional to its energy on certain atomic levels if the parameters of the experiment are properly tuned. Since the energy of the cavity is proportional to the number n of photons in it, this imprints information about n in the phase of the atom without changing the photon number n. This phase shift can then be detected by interferometry. We will not describe the physical mechanism behind this phase shift here, which is known as the *dispersive atom–field coupling regime* in quantum optics. However, the next exercise reveals details about how the Ramsey interferometry scheme works in general.

Exercise 5.25 Before the atom reaches the detector D, it interacts with the three cavities R_1, C and R_2. In the first Ramsey cavity R_1, a $\pi/2$ microwave pulse is implemented, which induces the transformations $|g\rangle \mapsto (|g\rangle + |e\rangle)/\sqrt{2}$ and $|e\rangle \mapsto (-|g\rangle + |e\rangle)/\sqrt{2}$ on the atom. Afterwards, if the cavity C is in a Fock state with n photons, the dispersive interaction with the atom implements the phase-shift $|e\rangle \mapsto e^{-i\Phi_0 n}|e\rangle$ with Φ_0 the phase shift per photon. Importantly, *only* the excited state experiences a phase shift. Finally, the atom interacts with the second Ramsey cavity R_2, which implements a phase–shifted $\pi/2$ pulse such that $|g\rangle \mapsto (|g\rangle + e^{i\varphi_r}|e\rangle)/\sqrt{2}$ and $|e\rangle \mapsto (-e^{-i\varphi_r}|g\rangle + |e\rangle)/\sqrt{2}$. Here, φ_r is the adjustable phase of the Ramsey interferometer. In the end, the atom is detected in D by ionization due to an electric field. Since the ground and excited states of the atom have different ionization energies, the detection of a resulting electron implements a projective measurement in the basis $\{|g\rangle, |e\rangle\}$. Now, provided that the atoms are prepared in the ground state in B, show that the probability of detecting the atom in $|g\rangle$ or $|e\rangle$ is

$$p_s(g|n) = \frac{1}{2}\left[1 - \cos(\Phi_0 n + \varphi_r)\right], \quad p_s(e|n) = \frac{1}{2}\left[1 + \cos(\Phi_0 n + \varphi_r)\right], \tag{5.140}$$

which depends on the number of photons n.

Let us denote by $p_s(r|n)$ the probability of detecting the atom in state $r \in \{0,1\}$, where $r = 0$ ($r = 1$) labels the ground (excited) state, conditioned on having n photons in the cavity. Furthermore, the subscript s shall remind us that this probability is due to sending a sensor atom through the cavity. Then, this probability is

$$p_s(r|n) = \frac{1}{2}\left\{1 + (-1)^{1-r}\cos\left[\frac{\pi}{4}(n - n_T) + \frac{\pi}{2}\right]\right\}, \tag{5.141}$$

which can be obtained from eqn (5.140) by choosing the adjustable phase of the Ramsey interferometer as $\varphi_r = \pi/2 - \Phi_0 n_T$ and by tuning the phase shift per atom to $\Phi_0 = \pi/4$. In reality, there are, of course, imprecisions such that eqn (5.141) does not exactly match the experimentally observed probability. The following table shows the probabilities $p_s(r|n)$ for a target number of $n_T = 2$ photons:

n	0	1	2	3	4
$p_s(0\|n)$	0.0	0.15	0.5	0.85	1.0
$p_s(1\|n)$	0.0	0.85	0.5	0.15	1.0

We see that, if $n = n_T$, one expects to detect an equal number of atoms in the ground and excited states. If $n > n_T$ ($n < n_T$), one expects to detect more (fewer) atoms in the ground state than in the excited state. This allows us to distinguish between $n > n_T$, $n = n_T$ or $n < n_T$ photons in the cavity, but note that we are far away from an ideal projective measurement of the cavity. This has to be the case since each atom can carry only one bit of information, whereas we need to distinguish between many (in principle, infinite) different Fock states $|n\rangle$.

Now, we can combine the undisturbed evolution of the cavity, eqn (5.139), with the measurements described above. To simplify the treatment, let us assume that the cavity contains initially no coherences in the Fock basis such that we can describe its state with probability $P_n(0)$ of having initially n photons in the cavity. Then, let us define a matrix with elements $E_{n,n'} \equiv \langle n|(\mathcal{I} + \mathcal{L}_0\tau)|n'\rangle$. Furthermore, we introduce measurement matrices with elements $M_{n,n'}(r) = \delta_{n,n'}p_s(r|n)$. Given a measurement record $\mathbf{r}_n$, the non-normalized state of the cavity can then be obtained by simple matrix multiplication:

$$\tilde{\boldsymbol{P}}(t_n|\mathbf{r}_n) = M(r_n)E \ldots M(r_1)EM(r_0)\boldsymbol{P}(0). \tag{5.142}$$

Here, $\tilde{\boldsymbol{P}}(t|\mathbf{r}_n)$ denotes the non-normalized vector of probabilities $\tilde{P}_n(t|\mathbf{r}_n)$ to have n photons in the cavity at time t conditioned on the measurement record $\mathbf{r}_n$. In the case of no initial coherences, this completely describes the state of the cavity.

Exercise 5.26 Derive eqn (5.142).

Finally, we turn to the feedback step. Suppose that, after sending a bunch of sensor atoms through the cavity, we infer that $n > n_T$ ($n < n_T$), i.e. we have more (fewer) photons than desired in C. If we want to change this, we can send an emitter or absorber atom into the cavity, which is prepared in either the excited state or ground state, respectively. For this purpose, the energy gap of the atoms is brought into exact resonance with the cavity by applying an external voltage V (*Stark shift*). In this case, the atom–cavity interaction is described by a Jaynes–Cummings Hamiltonian (see Exercise 3.1), which allows us to compute the conditional probability $p(r,n|r',n')$ of detecting the atom in state r and of having n photons in C given that the atom was prepared in r' and the cavity had n' photons. By properly tuning the interaction time, this yields for the conditional probability $p_e(r,n|n') \equiv p(r,n|e,n')$ in the case of sending an emitter atom into the cavity

$$p_e(r,n|n') = \delta_{n+r-1,n'} \sin^2\left(\frac{\pi}{2}\frac{\sqrt{n+r}}{\sqrt{n_T}} + \frac{\pi}{2}r\right). \tag{5.143}$$

For the case of sending an absorber atom into the cavity, the conditional probability $p_a(r,n|n') \equiv p(r,n|g,n')$ becomes

$$p_a(r,n|n') = \delta_{n+r,n'} \cos^2\left(\frac{\pi}{2}\frac{\sqrt{n+r}}{\sqrt{n_T+1}} + \frac{\pi}{2}r\right). \tag{5.144}$$

Note that, depending on the initial number n' of photons in the cavity, an absorber (emitter) atom will not always absorb (emit) a photon.

Exercise 5.27 Derive eqns (5.143) and (5.144) by using eqn (3.10) and by choosing an interaction time $t_e = \pi/(2g\sqrt{n_T})$ for the emitter case and $t_a = \pi/(2g\sqrt{n_T+1})$ for the absorber case. Again, because of various imperfections and constraints, the experimental probabilities do not exactly match eqns (5.143) and (5.144).

It is important to point out that in the experiment the atoms that are detected are *time-delayed*, as also indicated in Fig. 5.7. This means that, before the kth atom is registered with outcome r_k at D, there have already been $d = 5$ atoms which have interacted or are about to interact with the cavity. The state of these can no longer be influenced. This point is important for the design of the feedback control law, which concludes the dynamical description of the experiment. In order to decide at time $t_k \equiv k\tau$ what kind of atom to send into the cavity, we can only use the state estimate $\boldsymbol{P}(t_{k-d}|\mathbf{r}_{k-d})$ at time $t_{k-d} = (k-d)\tau$. After taking this into account, we decide to send an absorber atom as soon as we estimate

$$\sum_{n>n_T} P_n(t_{k-d}|\mathbf{r}_{k-d}) > P_{n_T}(t_{k-d}|\mathbf{r}_{k-d}), \tag{5.145}$$

i.e. there is a higher probability of having more photons than n_T in the cavity than of having n_T photons. Conversely, we send an emitter atom if we estimate $\sum_{n<n_T} P_n(t_{k-d}|\mathbf{r}_{k-d}) > P_{n_T}(t_{k-d}|\mathbf{r}_{k-d})$. Furthermore, after each feedback operation we wait d time steps before we apply the feedback control law again. This feedback scheme is slightly different from the experiment, but it works well, as Fig. 5.8 demonstrates.

Operational quantum stochastic thermodynamics

After having explained the dynamics of the experiment, we can turn to its thermodynamic description. By comparing our general sketch in Fig. 5.3 with the experimental

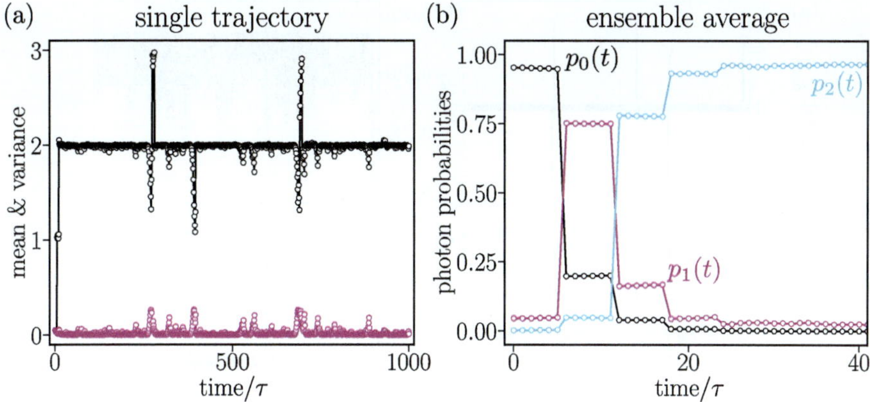

Fig. 5.8 Dynamics of the (numerical) photon number stabilization experiment versus time in units of τ, which equals the number of interventions. The initial cavity state is at thermal equilibrium. (a) Conditional mean photon number and variance for a single measurement record $\mathbf{r}_n$. The conditional mean is defined as $\sum_n nP_n(t|\mathbf{r}_n)$ and quickly stabilizes around the target number of $n_T = 2$ photons (black line). The rare fluctuations are due to photon emissions or absorptions or erroneous detection events. The conditional variance is defined as $\sum_n n^2 P_n(t|\mathbf{r}_n) - [\sum_n nP_n(t|\mathbf{r}_n)]^2$ and remains close to zero for all times (pink curve). This demonstrates that the purity of the cavity state is high. (b) Ensemble-averaged cavity state $P_n(t) = \sum_{\mathbf{r}_n} p(\mathbf{r}_n)P_n(t|\mathbf{r}_n)$ over 1000 trajectories. We plot the probability for $n = 0$ (black), 1 (pink) and 2 (turquoise) photons. The probability $p_{n_T}(t) \approx 0.96$ of having the desired number of photons is very high, which demonstrates the success of the feedback loop. Note that the effect of the time delay in the control loop is clearly visible in the plot.

set-up, Fig. 5.7, we see that both are strikingly similar. What we previously called a stream of ancillas corresponds to a stream of atoms in the experiment.

Since the cavity Hamiltonian is time independent, the first law in the time interval $(n\tau \quad \tau, n\tau]$ (including the control operation at time $n\tau$) for the cavity reads,

$$\Delta u_S^{(n]}(\mathbf{r}_n) = q_S^{(n)}(\mathbf{r}_{n-1}) + q_S^{\text{meas}}(\mathbf{r}_n) + w_S^{\text{ctrl}}(\mathbf{r}_{n-1}). \tag{5.146}$$

Here, $q_S^{(n)}(\mathbf{r}_{n-1})$ takes into account the heat flow due to photon exchanges with the environment and $q_S^{\text{meas}}(\mathbf{r}_n)$ is the heat associated with the update of the cavity state due to the measurement. Finally, $w_S^{\text{ctrl}}(\mathbf{r}_{n-1})$ is the work cost of the control operation. If we send a sensor atom into the cavity, this work is zero because we implement a quantum non-demolition measurement, which does not change the photon number. Non-zero values of $w_S^{\text{ctrl}}(\mathbf{r}_{n-1})$ therefore signify that we sent an absorber or emitter atom. The first law for the atoms is $\Delta u_A^{\text{ctrl}}(\mathbf{r}_n) = w_A^{\text{meas}}(\mathbf{r}_n) + w_A^{\text{ctrl}}(\mathbf{r}_{n-1})$, which is less interesting for our discussion here.

Turning to the second law, we first remark that the change in entropy of the atoms is zero because they are initially and finally (because of the projective measurement in D) in a pure state. The entropy production along a single trajectory becomes

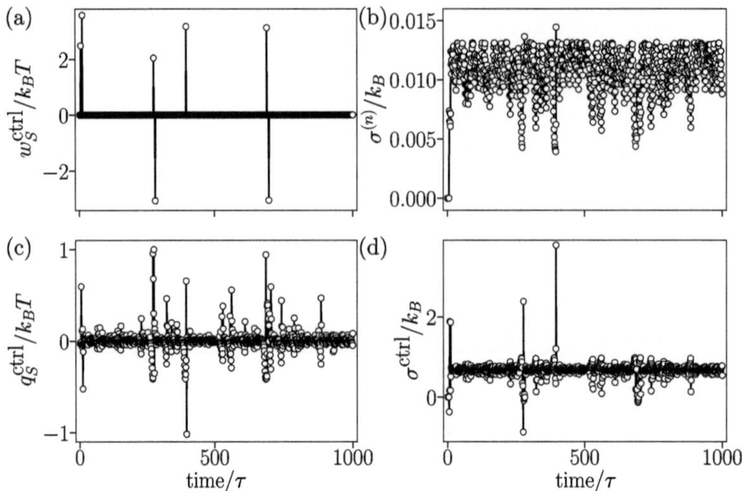

Fig. 5.9 Stochastic thermodynamics of the photon number stabilization experiment for a single measurement record starting with the cavity in equilibrium. We plot $w_S^{\rm ctrl}(\mathbf{r}_{n-1})$ (a), $\sigma^{(n)}(\mathbf{r}_n)$ (b), $q_S^{\rm meas}(\mathbf{r}_n)$ (c), and $\sigma^{\rm ctrl}(\mathbf{r}_n)$ (d) as a function of time in units of τ. For further explanation, see the main text.

$$\begin{aligned}\sigma^{(n]}(\mathbf{r}_n) &= \sigma^{(n)}(\mathbf{r}_n) + \sigma^{\rm ctrl}(\mathbf{r}_n) \\ &= -k_B \ln p(r_n|\mathbf{r}_{n-1}) + k_B S_{\rm Sh}[\boldsymbol{P}(t_n|\mathbf{r}_n)] - k_B S_{\rm Sh}[\boldsymbol{P}(t_{n-1}|\mathbf{r}_{n-1})] \\ &\quad - \frac{q_S^{(n)}(\mathbf{r}_{n-1}) + q_S^{\rm meas}(\mathbf{r}_n)}{T},\end{aligned} \tag{5.147}$$

where $\boldsymbol{P}(t_n|\mathbf{r}_n)$ denotes the cavity state immediately after the nth control operation.

The precise microscopic definitions of all quantities appearing in the first and second laws can be inferred from Section 5.4 and are not repeated here. In the following, we focus only on discussing the thermodynamic features of the experiment based on numerical results.

We start with the thermodynamics along a single trajectory of measurement records $\mathbf{r}_n$. The trajectory we consider is the same as that considered in Fig. 5.8a and various interesting thermodynamic quantities are plotted in Fig. 5.9.

First, we consider the work $w_S^{\rm ctrl}$ injected during the control operation (Fig. 5.9a). As explained above, this contribution is most of the time exactly zero because there is no work cost for the measurement with respect to changes in the cavity energy (there is, however, a work cost associated with the manipulations of the atoms, which we do not plot here). Thus, if $w_S^{\rm ctrl} \neq 0$, there is a feedback control operation and the sign of $w_S^{\rm ctrl}$ signifies whether we try to inject or extract a photon. Note that even during the feedback control operations $w_S^{\rm ctrl}$ fluctuates because our state of knowledge about the cavity fluctuates too. Therefore, we can never be certain that we emitted (or absorbed) exactly one photon into (from) the cavity and this uncertainty is reflected in the fluctuating $w_S^{\rm ctrl}$.

Second, we also plot $q_S^{\rm meas}(\mathbf{r}_n)$ (Fig. 5.9c), which fluctuates permanently because

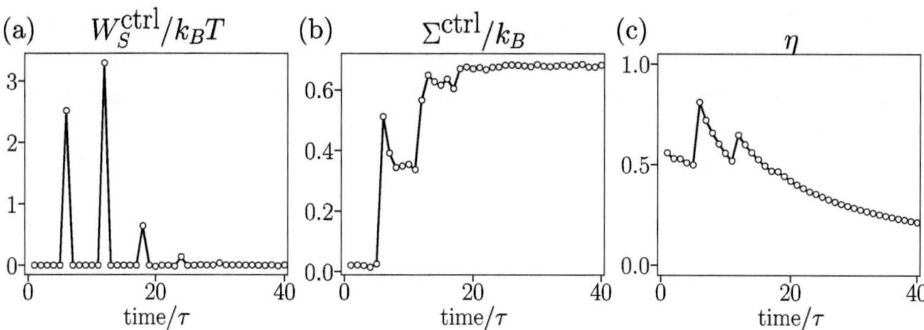

Fig. 5.10 Stochastic thermodynamics of the photon number stabilization experiment averaged over 1000 trajectories. We plot W_S^{ctrl} (a), Σ^{ctrl} (b) and η (c) as a function of time in units of τ. For further explanation, see the main text.

each measurement (and also each feedback control operation) is used to update our state of knowledge about the cavity. Again, since the cavity state is never pure, each detected atom reveals 'new' information (although this information could be erroneous), but compared with w_S^{ctrl} the magnitude of fluctuations is smaller.

Third, we plot the stochastic entropy production $\sigma^{(n)}(\mathbf{r}_n)$ in the absence of interventions (Fig. 5.9b). As predicted by our general theory from Section 5.4, it is always positive: $\sigma^{(n)}(\mathbf{r}_n) \geq 0$.

Fourth and finally, in Fig. 5.9d we display the stochastic entropy production $\sigma^{\text{ctrl}}(\mathbf{r}_n)$ due to the interventions. As we can see, it can happen that $\sigma^{\text{ctrl}}(\mathbf{r}_n)$ becomes negative. Furthermore, observe that $\sigma^{\text{ctrl}}(\mathbf{r}_n)$ is roughly two orders of magnitude larger than $\sigma^{(n)}(\mathbf{r}_n)$. Thus, from an entropic point of view, the most costly part of the experiment are the frequent measurements, whereas in between the measurements the cavity barely changes such that $\sigma^{(n)}(\mathbf{r}_n)$ remains very small.

The previous points are also confirmed by looking at the average thermodynamic behaviour, as demonstrated in Fig. 5.10. To that end, we averaged over 1000 trajectories, but we only considered a time interval with 40 interventions.

First, in Fig. 5.10a we plot the average work W_S^{ctrl} invested during the control operations. Since the initial state has with high probability zero photons, the transient dynamics is dominated by the injection of work. However, because of the time delay this happens only every 6τ. Since the emission of one photon into the cavity by an emitter atom has a high probability of success, we quickly reach the desired Fock state with $n_T = 2$ photons. Therefore, the work injected at 18τ and 24τ is small.

Second, we plot the average entropy production Σ^{ctrl} during the control operation in Fig. 5.10b. As predicted by our theory, we find $\Sigma^{\text{ctrl}} \geq 0$ in accordance with the second law of thermodynamics. Furthermore, after some transient dynamics, the entropy production Σ^{ctrl}/k_B per control operation saturates around $\ln 2$. This shows that the entropy production during the control step is dominated by the average of $-k_B \ln p(r_n|\mathbf{r}_{n-1})$, i.e. the first term in eqn (5.147). To understand the reason behind this, recall that the probability of measuring the sensor atom in the ground or excited state is 1/2 if there are $n = n_T$ many atoms in the cavity. Thus, we typically generate one bit of information per measurement if the cavity resides close to its target state.

Finally, Fig. 5.10c shows the *efficiency* of the experiment. In fact, from a simplistic thermodynamic point of view, the experiment does nothing other than use a non-equilibrium resource (the stream of atoms) to create and maintain a non-equilibrium state of the cavity, which would otherwise be at thermal equilibrium. To quantify this efficiency, we average the stochastic entropy production (5.147) and sum it up in each time step. After using the first law, we find

$$\sum_{j=1}^{N} \Sigma^{(j)} = \frac{W_S^{\rm ctrl} - \Delta F_S}{T} + k_B S_{\rm Sh}[p(\mathbf{r}_n)] \geq 0. \tag{5.148}$$

Here, $\Delta F_S = F_S(n\tau) + k_B T \ln \mathcal{Z}_S$ is the change in non-equilibrium free energy starting from a cavity state in equilibrium with partition function $\mathcal{Z}_S = \mathrm{tr}_S\{e^{-\beta\hbar\omega_c a^\dagger a}\}$. The average free energy after n time steps is computed by averaging the stochastic free energy $u_S(\mathbf{r}_n) - k_B T S_{\rm Sh}[\boldsymbol{P}(t_n|\mathbf{r}_n)]$ of the conditional cavity state $\boldsymbol{P}(t_n|\mathbf{r}_n)$:

$$F_S(n\tau) = \sum_{\mathbf{r}_n} p(\mathbf{r}_n)\{u_S(\mathbf{r}_n) - k_B T S_{\rm Sh}[\boldsymbol{P}(t_n|\mathbf{r}_n)]\}. \tag{5.149}$$

It follows from the second law (5.148) that the following efficiency is bounded by one:

$$\eta \equiv \frac{\Delta F_S}{W_S^{\rm ctrl} + k_B T S_{\rm Sh}[p(\mathbf{r}_n)]} \leq 1. \tag{5.150}$$

As Fig. 5.10c demonstrates, we can achieve remarkably high efficiencies peaked around values of 0.8 and 0.65 before they decay in the long run to zero. This decay is due to the fact that stabilizing the photon number state means that its free energy no longer changes while the measurement and feedback loop still consumes resources. Thus, the *preparation* of the photon number state is very efficient, but the *stabilization* of it has by definition an efficiency of zero.

One might wonder why only the work $W_S^{\rm ctrl}$ invested in the cavity enters the definition of the efficiency (5.150), but not the work $W_A^{\rm ctrl}$ invested in preparing the state of the atoms. In fact, the latter is non-negligible because the many sensor atoms leaving B in the ground state need to be rotated to a superposition of the excited and ground states, which costs work $W_A^{\rm ctrl} \approx \hbar\omega_a/2$. However, what we are interested in here is how efficiently we can use a given amount of non-equilibrium resources (atoms in a pure state) to perform some task (creation of Fock states). That efficiency should be the same independent of, for instance, the question of whether the atoms leaving B are in the ground or excited state (in the latter case we could even additionally extract work from the atoms). The second law of thermodynamics cares about *entropy changes*, which are zero for the incoming and outgoing stream of atoms. Indeed, the outgoing stream of atoms, which are all in a pure state, could be re-used for the next experiment.

This finishes the thermodynamic analysis of this remarkable experiment, which was clearly idealized. In reality, there are further costs resulting from, for example, the cooling of the environment down to less than 1 K, the laser preparation of the atoms in B or the electronics associated with the controller K. The goal here was to provide a 'minimal' thermodynamic description. Similar with other idealizations in thermodynamics, one could imagine that the hidden thermodynamic costs of running the laboratory equipment can be made arbitrarily small in an ideal world. The outcome of the present analysis is therefore that, in principle, thermodynamics can be applied in a consistent way to study the manipulation of single quantum systems. This is a remarkable outcome compared with the standard textbook conception of thermodynamics that prevailed during the 20th century.

Finally, we mention that the experiment evolves further complications. For instance, the number of atoms interacting with the cavity at a given time is not fixed to one, but Poisson distributed (with an average number of 0.6 atoms). In addition, the detector D can also miss detecting an atom. These and other small imperfections are the reason why the experimentally observed probability $P_{n_T}(t)$ is around 0.8. The important point is, however, that all of these imperfections can be taken into account within the operational framework of quantum stochastic thermodynamics.

Further reading

5.1. We started with the motivation of why the two-point measurement scheme is not the only way to define work fluctuations in the quantum regime. In fact, criticism of and alternatives to the two-point measurement scheme were communicated early on; for instance, by Allahverdyan and Nieuwenhuizen (2005). Various ideas of their approach are reflected in the following sections.

5.2. The 'one-point' measurement scheme for work that we considered was introduced by Roncaglia *et al.* (2014). The no-go theorem was derived by Perarnau-Llobet *et al.* (2017).

5.3. The thermodynamic description of quantum measurements has been discussed in many places, with different degrees of generality and emphasis put on other features. The exposition we provided follows as a special case from the multi-time treatment of Section 5.4 and, on average, is in unison with the repeated interaction framework in Section 3.9. We also tried to stress the fact that there is no unambiguous and general identification of stochastic heat and work during the measurement step.

5.4. The operational thermodynamic framework for a quantum Markov process was introduced by Strasberg (2019*a*), which—apart from the difference that $w_A^{\text{meas}}(\mathbf{r}_n)$ was considered as heat—agrees with our exposition.

5.5. The quote from Maxwell is taken from his book (Maxwell, 1871), where he first published the idea of the demon to illustrate the statistical nature of the second law. The paradox of Maxwell's demon then sparked a long historical debate about the question of how to save the second law. Today, it is widely accepted that the resolution to this paradox is provided by Landauer's principle. This was convincingly argued by Bennett (1982), but (without knowing Landauer's principle) Penrose (1970) had already reached the same conclusions. Because of recent theoretical and experimental progress, Maxwell's ancient idea has sparked another wave of research, with most interest focused on the special case of eqn (5.84), which we called the 'second law for classical single-time feedback control' (Parrondo *et al.*, 2015). Our example of the electronic Maxwell demon built on the theoretical idea by Schaller *et al.* (2011), which was experimentally realized by Chida *et al.* (2017) and which can also be realized autonomously using the Coulomb-coupled quantum dots studied at the end of Section 3.10 (Strasberg *et al.*, 2013). A related experiment based on pulling single molecules as described in Section 2.9 was designed by Ribezzi-Crivellari and Ritort (2019).

5.6. The autonomous approach to quantum stochastic thermodynamics was proposed by Strasberg (2020).

5.7. The initial quote about Ichthyosauria is from Schrödinger (1952). The experiments we considered were performed by the group of Haroche (Sayrin *et al.*, 2011; Zhou *et al.*, 2012). For these and earlier related experiments Haroche was awarded the Nobel Prize in Physics in 2012. The thermodynamic analysis of these experiments was performed by Strasberg (2019*a*). Readers interested in learning more about the dispersive coupling regime in quantum optics can consult Section 19.3 of Scully and Zubairy (1997).

6
Outlook

Congratulations! You have reached the end of this book. You should now be able to study internal energy, work, heat, entropy and entropy production (including its fluctuations!) within a consistent framework for a large variety of open quantum systems and experimental settings. The basics of this framework are not limited to the weak coupling or Markov assumption, and it can cope with various non-equilibrium resources, including quantum measurements and feedback control operations. Nevertheless, the content of every book is limited and I had to leave out many other interesting developments in the field. This last chapter gives a brief overview of these topics, in particular in connection with open questions.

First, the book did not review any techniques to compute the *dynamics* of open quantum systems in the strong coupling and non-Markovian regime. The quantum master equations studied here are restricted to the weak coupling and Markovian case. Clearly, the dynamical description of open quantum systems beyond that regime is much richer but also more complicated (de Vega and Alonso, 2017). Many different techniques are known, each has its advantages and drawbacks, and some seem to be better suited than others to compute thermodynamic quantities. Taking this into account, I believe that any (necessarily biased) attempt to list these techniques here can only fail. Thus, since there seems to be no good review article on strong coupling and non-Markovian thermodynamics, readers are encouraged to start browsing the Web on their own to delve deeper into this (still largely unexplored) territory.

Second, I need to mention that a part of the quantum thermodynamics community seems to be particularly motivated by the idea of revealing *quantum advantages* in thermodynamics. In this book, I purposely avoided any such claims, preferring to keep an eye on the overarching framework that thermodynamics and statistical mechanics provide in both the quantum and classical contexts. In fact, despite many claims about quantum advantages in thermodynamics, I have seen none that is convincing to me. In most works, quantum and classical resources are differently accounted for such that any comparison between them appears biased and, even if that were not the case, our currently employed theoretical methods appear too idealized to draw any final conclusions. In fact, designing, engineering and controlling quantum systems is very demanding, but the resources required to do so are often not accounted for. A windmill does not need to be cooled down to below 1 kelvin to start working. Certainly, I agree, however, that there is much space left for further exploration here.

Third, while the basic laws of thermodynamics hold and continue to be insightful at the nanoscale, some elements of macroscopic thermodynamics are lost in the microworld. In particular, the success of thermodynamics rests on the fact that the

Quantum Stochastic Thermodynamics. Philipp Strasberg, Oxford University Press.
 DOI: 10.1093/oso/9780192895585.003.0006

behaviour of macroscopic systems can be described by very few variables (e.g. temperature, volume, polarization, chemical concentrations and shear stress, among others, but seldomly more than a handful are necessary for applications). In contrast, the thermodynamic description of nanoscale systems requires more knowledge; for example, thermodynamic state functions of the open system depend on the entire system density matrix ρ_S and its dynamics depends in a complicated way on the details of the interaction Hamiltonian. Thus, if the system Hilbert space has dimension 10, the in principle 99 independent components of ρ_S have to be taken into account—which does not appear to be a very efficient description.

A recent theoretical framework, known as a thermodynamic *resource theory* (or the 'resource theory of thermal operations'), deserves credit for providing an alternative approach to quantum thermodynamics, which overcomes to some extent the above-mentioned problem. Instead of trying to derive the laws of thermodynamics for an arbitrary system–bath Hamiltonian, physically motivated restrictions on the resulting unitary evolution are *imposed* from the beginning, thereby greatly reducing the complexity from the start. Despite creating a lot of attention, I have excluded resource theories from the present book for two reasons. First, the mathematical machinery required to derive the final results is heavy and appears particularly unconventional for researchers with a traditional physics background, and I had to add an entire chapter to explain them (I admit that I am just not clever enough to understand the final mathematical details myself, so let me refer the reader to the review by Lostaglio (2019), who has understood them). However, the second more profound reason is that the resource theory of thermal operations has not yet produced any experimental insights because its basic premises are (at most) approximately met in reality. While being 'approximately true' is typically all that one wants from a physical theory, resource theories suffer from the fact that their mathematical results are very sensitive to slight changes in the initial premises. Thus, finding a truly useful resource theory for quantum thermodynamics is definitely a major challenge in the field.

Fourth, the book remained overall short on the application side because I gave priority to first settling a broad range of foundational questions in a (hopefully) transparent way. Furthermore, at the moment it is still not clear which practically useful applications (e.g. energy-efficient computation, sophisticated cooling strategies, powerful thermoelectrics and solar cells) could emerge from the field and which are the best experimental platforms to realize them. For me, the most promising candidates are thermoelectric devices in solid-state nanostructures, as reviewed in Section 3.10, which I would claim to be the first quantum heat machines ever realized. It would be amazing if we could use such devices as powerful nanoscale computers that recycle part of the heat they generate during computation and turn it into useful electric power again. Clearly, the race for interesting useful applications is open.

Fifth, an important and almost unexplored territory with profound consequences for all of the above-mentioned topics relates to the question about the size of the system and the bath; see Fig. 6.1. In traditional thermodynamics, both the system and the bath are macroscopically large. Throughout this book we explored the territory of a microscopically small system coupled to a large thermal bath which is well characterized by its initial temperature and chemical potential. While many of our

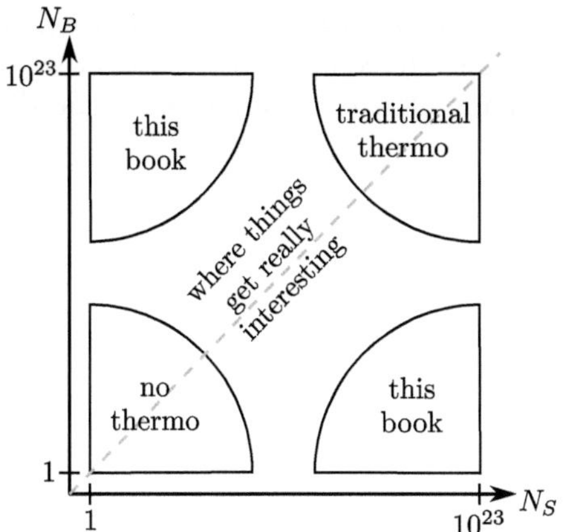

Fig. 6.1 Mental map of thermodynamics for a system coupled to a bath, whose sizes are characterized by their particle numbers N_S and N_B, respectively. If both are large, we are in the realm of traditional thermodynamics. If the system is small but the bath is large, we are in the 'traditional' realm of quantum and stochastic thermodynamics, as covered in this book. If both are small, things become simple again and there is no need to devise any sophisticated thermodynamic description. In between, things get really interesting. Note that I sketched the map to be symmetric with respect to exchanging the labels S and B: if the 'system' has 10^{23} particles and the 'bath' consists of a single particle, we need to relabel both. Finally, note that the exact meanings of 'large' and 'small' and the position of the boundaries in the sketch are only indicative and require further research.

algebraic manipulations remain *mathematically* correct for a system and a bath of any size, their *physical* meaning diminishes when applied outside the realm specified above. If the system and the bath both consist only of (say) a single spin, everything is reduced to simple unitary dynamics and there is no need to attempt any coarse-grained description, i.e. any thermodynamic concept (temperature, heat, entropy) becomes either trivial or meaningless then (compare also with our discussion about the zeroth law in Section 3.6). On the other hand, if the system becomes larger and consists of, say, 100 spins, there are already 2^{200} different observables characterizing its complete quantum state and any attempt to model its thermodynamics based on the entire system density matrix ρ_S is doomed to fail. In particular, the von Neumann entropy of ρ_S will no longer provide a useful concept for the thermodynamic entropy of the open system. Moreover, if the size of the bath gets reduced, finite-size effects become visible as characterized by time-dependent temperatures, chemical potentials or other (perhaps non-equilibrium) parameters and we laid down the basics for this situation in Section 3.8. A potentially much richer thermodynamic framework emerges therefore in the limit of larger (but not macroscopic) systems and smaller (but not microscopic) baths, where many fundamental questions also remain unanswered. For instance, it seems crucial to overcome the much used assumption of an initial Gibbs state of the bath and to focus on the more realistic microcanonical ensemble or even a pure state. Then, particularly interesting results could emerge when the equivalence of ensembles breaks down.

Finally, readers who are still seeking further inspiration should consult the book edited by Binder *et al.* (2018). It contains 41 chapters written by various experts in the field and it covers many alternative approaches and different applications not mentioned here.

Appendix A
Concepts from Information Theory

We briefly review some basic concepts of classical and quantum information theory and afterwards present some useful, advanced theorems without proof. Below, we denote by $d = \dim \mathcal{H}$ the dimension of the Hilbert space of the system or, in the classical case, d denotes the dimension of the space in which the vector of probabilities $\boldsymbol{p} \in \mathbb{R}^d$ lives. Furthermore, $d_{X,Y,\dots}$ refers to the dimension of a subsystem $X, Y, \dots$.

A.1 Basic Concepts

Classical information theory

We define the **Shannon entropy** of a probability distribution $\{p_x\}$ with probabillity vector $\boldsymbol{p}$ as

$$S_{\text{Sh}}(\{p_x\}) = S_{\text{Sh}}(\boldsymbol{p}) \equiv -\sum_x p_x \ln p_x. \tag{A.1}$$

Note that we use the natural logarithm in the definition (measuring entropy in 'nats'), whereas in information theory it is customary to use the logarithm $\log_2$ with respect to base 2 (measuring entropy in 'bits' $= \ln 2$ nats ≈ 0.69 nats).

If the state of a physical system X is described by probabilities $\{p_x\}$, then the Shannon entropy can be interpreted as the *information content* that the system has to offer to an external observer upon a measurement of the system. Switching perspective, from the observer's point of view the Shannon entropy can be interpreted as the *uncertainty* one has about a system X *before* measuring it. There is no uncertainty about a system (or the system contains no information for the observer, respectively) if the system is for sure in the state x^*: $p_x = \delta_{x,x^*}$. Then, it follows from the convention $0 \ln 0 \equiv 0$, which can be justified by the limit $\lim_{x \searrow 0} x \ln x = 0$, that $S_{\text{Sh}}(\{\delta_{x,x^*}\}) = 0$. Furthermore, one can verify the following four properties.

(i) The Shannon entropy is bounded from above and below: $0 \leq S_{\text{Sh}}(\boldsymbol{p}) \leq \ln d$.
(ii) The Shannon entropy $S_{\text{Sh}}(\{p_x\})$ is continuous in each p_x.
(iii) If $S_{\text{Sh}}(d)$ denotes the Shannon entropy of the uniform distribution $p_x = 1/d$ for all x, then $S_{\text{Sh}}(d) < S_{\text{Sh}}(d+1)$. Thus, the uncertainty of a system always increases if it can be in more different states about which we have no prior information.
(iv) Let us divide the system X into 'subsystems' $X = \bigoplus_\alpha X_\alpha$ and decompose the probabilities as $p_x = p_{x|\alpha} P_\alpha$, where P_α is the probability of being in subsystem X_α and $p_{x|\alpha}$ is the conditional probability of being in x given α. Then, the Shannon entropy 'additively' decomposes as

$$S_{\rm Sh}(\{p_x\}) = S_{\rm Sh}(\{P_\alpha\}) + \sum_\alpha P_\alpha S_{\rm Sh}(\{p_{x|\alpha}\}). \tag{A.2}$$

The above properties can be regarded as natural properties that a measure of uncertainty or information should satisfy, but, in fact, Shannon *deduced* eqn (A.1) (up to an overall positive factor) from properties (ii), (iii) and (iv).

Furthermore, there are a number of important information theory concepts related to Shannon entropy. First, we define the **classical relative entropy** between two probability vectors $\boldsymbol{p}$ and $\boldsymbol{q}$ of the same system X as

$$D(\boldsymbol{p}|\boldsymbol{q}) \equiv \sum_x p_x \ln \frac{p_x}{q_x} \geq 0. \tag{A.3}$$

The positivity of relative entropy can be derived from properties that we discuss at the beginning of Section 4.6; namely, that the convex function $f(x) \equiv x \ln x$ satisfies Jensen's inequality (4.117). Furthermore, eqn (A.3) can diverge whenever $q_x = 0$ but $p_x \neq 0$. Since $D(\boldsymbol{p}|\boldsymbol{q}) = 0$ if and only if $\boldsymbol{p} = \boldsymbol{q}$, relative entropy can be interpreted as measuring the 'distance' or 'difference' between the two probability vectors $\boldsymbol{p}$ and $\boldsymbol{q}$, i.e., the larger $D(\boldsymbol{p}|\boldsymbol{q})$ the easier it is to tell them apart. Nevertheless, relative entropy is not a metric in the mathematical sense because it is not symmetric: $D(\boldsymbol{p}|\boldsymbol{q}) \neq D(\boldsymbol{q}|\boldsymbol{p})$.

Another important concept is the **classical mutual information**, which is defined for a bipartite system XY with joint probability vector $\boldsymbol{p}_{XY}$ as

$$I(\boldsymbol{p}_{XY}) \equiv S_{\rm Sh}(\boldsymbol{p}_X) + S_{\rm Sh}(\boldsymbol{p}_Y) - S_{\rm Sh}(\boldsymbol{p}_{XY}) = \sum_{x,y} p_{x,y} \ln \frac{p_{x,y}}{p_x p_y}, \tag{A.4}$$

where $\boldsymbol{p}_X$ and $\boldsymbol{p}_Y$ describe the marginal distributions of $\boldsymbol{p}_{XY}$. We remark that we often write $I_{X:Y} \equiv I(\boldsymbol{p}_{XY})$ if the probability vector $\boldsymbol{p}_{XY}$ is clear from context. Classical mutual information is bounded from above and below by

$$0 \leq I(\boldsymbol{p}_{XY}) \leq \ln \min\{d_X, d_Y\}. \tag{A.5}$$

The first inequality follows after noting that $I(\boldsymbol{p}_{XY}) = D(\boldsymbol{p}_{XY}|\boldsymbol{p}_X \otimes \boldsymbol{p}_Y)$, which also establishes the fact that mutual information quantifies the *correlations* contained in $\boldsymbol{p}_{XY}$. To derive the second inequality, we suppose that $d_X \leq d_Y$ without loss of generality. Next, we write

$$I(\boldsymbol{p}_{XY}) = S_{\rm Sh}(\{p_x\}) - \sum_y p_y S_{\rm Sh}(\{p_{x|y}\}). \tag{A.6}$$

The claim is now that there exists a legitimate probability distribution $p_{x,y}$, which (simultaneously) maximizes the first term and minimizes the second term in eqn (A.6). To see this, consider any injective function $f : X \to Y$ (note that this step requires $d_X \leq d_Y$) and define $\tilde{p}_{x,y} \equiv \delta_{y,f(x)}/d_X$. This gives $\tilde{p}_{x|y} = \delta_{x,f^{-1}(y)}$ for all x, y such that $f(x) = y$. For all y for which no x exists such that $f(x) = y$, we have $\tilde{p}_y = 0$. In either case, the second term in eqn (A.6) vanishs. Thus, this distribution satisfies $I(\tilde{\boldsymbol{p}}_{XY}) = \ln d_X$.

Quantum information theory

The subsequent quantum concepts in information theory follow from the classical case by analogy. We start with the **von Neumann entropy**, which is defined as

$$S_{\text{vN}}(\rho) \equiv -\text{tr}\{\rho \ln \rho\} = -\sum_k \lambda_k \ln \lambda_k = S_{\text{Sh}}(\{\lambda_k\}), \tag{A.7}$$

where we used the spectral decomposition of $\rho = \sum_k \lambda_k |k\rangle\langle k|$ at the end. The von Neumann entropy of a quantum system therefore measures the *classical* uncertainty an external observer has about the state of the system. Equivalently, the classical information content of a system in state ρ equals its von Neumann entropy. The von Neumann entropy has some elementary properties, which we will occasionally use in this book and which we state here without proof.

(i) The von Neumann entropy is bounded from above and below: $0 \leq S_{\text{vN}}(\rho) \leq \ln d$.

(ii) The von Neumann entropy is invariant under any unitary transformation U: $S_{\text{vN}}(U\rho U^\dagger) = S_{\text{vN}}(\rho)$.

(iii) For a bipartite quantum system in a state ρ_{XY} with marginal states $\rho_{X/Y} = \text{tr}_{Y/X}\{\rho_{XY}\}$ we have

$$S_{\text{vN}}(\rho_{XY}) \leq S_{\text{vN}}(\rho_X) + S_{\text{vN}}(\rho_Y) \tag{A.8}$$

with equality if and only if $\rho_{XY} = \rho_X \otimes \rho_Y$. Equation (A.8) is known as **subadditivity** of von Neumann entropy.

(iv) The von Neumann entropy is concave, i.e. for any convex combination $\sum_j \lambda_j \rho_j$ of density matrices ρ_j with weights $\lambda_j \geq 0$ such that $\sum_j \lambda_j = 1$,

$$\sum_j \lambda_j S_{\text{vN}}(\rho_j) \leq S_{\text{vN}}\left(\sum_j \lambda_j \rho_j\right). \tag{A.9}$$

Further concepts are directly related to the von Neumann entropy. First, the **quantum relative entropy** between two states ρ and σ is defined as

$$D(\rho|\sigma) \equiv \text{tr}\{\rho(\ln \rho - \ln \sigma)\} \geq 0. \tag{A.10}$$

Its positivity follows from *Klein's inequality*, which we do not discuss or derive here. Furthermore, $D(\rho|\sigma) = 0$ if and only if $\rho = \sigma$. As in the classical case, relative entropy measures the statistical difference between two quantum states. For instance, let $\sigma = I/d$ be the maximally mixed state. Then,

$$D\left(\rho\middle|\frac{I}{d}\right) = \ln d - S_{\text{vN}}(\rho), \tag{A.11}$$

which is intuitively appealing: the larger the uncertainty about ρ, the less it can be distinguished from a maximally mixed or uninformative state. However, since the relative entropy is not symmetric, $D(\rho|\sigma) \neq D(\sigma|\rho)$, it does not define a metric.

Furthermore, it diverges whenever the support of ρ is not contained in the support of σ as in the classical case. Like the von Neumman entropy, relative entropy is invariant under any unitary transformation U: $D(U\rho U^\dagger|U\sigma U^\dagger) = D(\rho|\sigma)$. In contrast to von Neumann entropy, it is not concave but convex: $D(\rho|\sigma) \le \sum_j \lambda_j D(\rho_j|\sigma_j)$ for convex combinations $\rho = \sum_j \lambda_j \rho_j$ and $\sigma = \sum_j \lambda_j \sigma_j$.

A second important concept is the **quantum mutual information** of a bipartite state ρ_{XY}:

$$I(\rho_{XY}) \equiv S_{\text{vN}}(\rho_X) + S_{\text{vN}}(\rho_Y) - S_{\text{vN}}(\rho_{XY}) \ge 0. \tag{A.12}$$

We also write $I_{X:Y} \equiv I(\rho_{XY})$ if the state ρ_{XY} is clear from the context. Furthermore, positivity follows from subadditivity, eqn (A.8), or after noting that $I(\rho_{XY}) = D(\rho_{XY}|\rho_X \otimes \rho_Y)$, which shows that quantum mutual information measures the correlations between X and Y as in the classical case. However, in contrast to the classical case (see eqn (A.5)), quantum mutual information is upper bounded by (note the factor 2)

$$I(\rho_{XY}) \le 2 \ln \min\{d_X, d_Y\}. \tag{A.13}$$

This implies that quantum correlations can be twice as strong as classical correlations. To prove the bound, we need to find again a state that simultaneously maximizes the first two terms and minimizes the last term in eqn (A.12). At least in the case that $d_X = d_Y$, it is easy to see that these conditions are satisfied by a *maximally entangled state* $\rho_{XY} = |\psi\rangle\langle\psi|_{XY}$ with $|\psi\rangle_{XY} = \sum_{k=1}^{d_X} |k\rangle_X |k\rangle_Y / \sqrt{d_X}$. Here, $|k\rangle_X$ ($|k\rangle_Y$) denotes an arbitrary set of orthonormal states in $\mathcal{H}_X$ ($\mathcal{H}_Y$).

It is interesting to note that we used a *pure* state to maximize the mutual information in the quantum case, whereas in the classical case we used a *mixed* state. In fact, the classical mutual information of any classical pure state, which is always of the form $p_{x,y} = \delta_{x,x^*}\delta_{y,y^*}$ for some $x^* \in X$ and $y^* \in Y$, is *zero*—in strong contrast to the quantum case. The origin of this striking difference between the quantum and classical world comes from entanglement: the reduced state $\rho_X = \text{tr}_Y\{|\psi\rangle\langle\psi|_{XY}\}$ of a joint pure state is in general not pure anymore.

Finally, we will sometimes also need to quantify correlations of an N-partite state $\rho_{12\ldots N}$ with $N > 2$. This can be done in many different ways and in this book we will sometimes encounter the **total information**

$$I_{\text{tot}}(\rho_{12\ldots N}) \equiv \sum_{i=1}^{N} S_{\text{vN}}(\rho_i) - S_{\text{vN}}(\rho_{12\ldots N}). \tag{A.14}$$

This quantity also has a variety of properties. A few of them are explored in the next exercise.

Exercise A.1 Show that the total information can be written in the two alternative forms

$$\begin{aligned} I_{\text{tot}}(\rho_{12\ldots N}) &= D(\rho_{12\ldots N}|\rho_1 \otimes \rho_2 \otimes \cdots \otimes \rho_N) \\ &= I_{1:2} + I_{12:3} + \cdots + I_{12\ldots N-1:N}. \end{aligned} \tag{A.15}$$

Notice that eqn (A.15) implies that the total information is positive.

A.2 Advanced Inequalities

We here state a couple of useful inequalities without proof, which will be used at certain points in this book and which are good to know in general. Since not all of these inequalities have a (simple memorable) name, we label them as 'Theorem A.1', 'Theorem A.2', etc. for later reference in the main text.

The first inequality bounds the entropy of a mixture of quantum states and should be compared with eqn (A.9).

Theorem A.1 *Let $\rho = \sum_n \lambda_n \rho_n$ be a convex combination of density matrices. Then,*

$$S_{\mathrm{vN}}(\rho) \leq \sum_n \lambda_n S_{\mathrm{vN}}(\rho_n) + S_{\mathrm{Sh}}(\boldsymbol{\lambda}). \tag{A.16}$$

The equality holds if and only if the ρ_n have support on mutually orthogonal subspaces.

The previous theorem is closely related to the next one. To state it, we need the notion of a positive operator-valued measure (POVM) from Section 1.4.

Theorem A.2 *Let $\{P_n\}$ be a set of positive operators such that $\{P_n^2\}$ is a POVM (i.e. $\sum_n P_n^2 = I$). Then, for any state ρ,*

$$S_{\mathrm{vN}}(\rho) \leq S_{\mathrm{vN}}\left(\sum_n P_n \rho P_n\right). \tag{A.17}$$

The content of the last two theorems can be neatly summarized in one equation. Defining $\rho_n \equiv P_n \rho P_n / \lambda_n$ with $\lambda_n \equiv \mathrm{tr}\{P_n^2 \rho\}$, we obtain

$$S_{\mathrm{vN}}(\rho) \leq S_{\mathrm{vN}}\left(\sum_n \lambda_n \rho_n\right) \leq \sum_n \lambda_n S_{\mathrm{vN}}(\rho_n) + S_{\mathrm{Sh}}(\boldsymbol{\lambda}). \tag{A.18}$$

Exercise A.2 Show that projective measurements increase the von Neumann entropy of the *average* post-measurement state.

The next result is known as **monotonicity of relative entropy**. If you are not yet familiar with the notion of CPTP maps, postpone the rest of this section until you are familiar with the content of Section 1.5.

Theorem A.3 *For any CPTP map $\mathcal{C}$ and all states ρ, σ we have*

$$\boxed{D(\mathcal{C}\rho|\mathcal{C}\sigma) \leq D(\rho|\sigma).} \tag{A.19}$$

This theorem is so important that we put a box around it: it appears in many microscopic derivations of the second law in the main text. Some insights into the derivation and the use of it can be gained from the following exercise.

Exercise A.3 Consider two arbitrary bipartite states ρ_{AB} and σ_{AB} and show that eqn (A.19) implies

$$D(\rho_A|\sigma_A) \leq D(\rho_{AB}|\sigma_{AB}). \tag{A.20}$$

Then, use eqn (A.20) to derive eqn (A.19) in general. *Hint:* Remember the unitary dilation theorem, eqn (1.49).

We also use the classical counterpart of Theorem A.3 in this book, for which we need the notion of a stochastic matrix T defined by the requirement that $T_{x,x'} \geq 0$ for all x, x' and $\sum_x T_{x,x'} = 1$ for all x'.

Theorem A.4 *For any stochastic matrix T and any two probability distributions $\mathbf{p}$ and $\mathbf{q}$ we have*

$$D(T\mathbf{p}|T\mathbf{q}) \leq D(\mathbf{p}|\mathbf{q}). \tag{A.21}$$

Finally, a useful theorem to bound the quantum relative entropy is the following.

Theorem A.5 *For arbitrary states ρ and σ acting on a Hilbert space with dimension d we have*

$$D(\rho|\sigma) \geq \frac{[S_{\rm vN}(\rho) - S_{\rm vN}(\sigma)]^2}{3(\ln d)^2}. \tag{A.22}$$

Further reading

The fields of classical and quantum information theory are enormous and better covered elsewhere (Cover and Thomas, 1991; Nielsen and Chuang, 2000). That they are intimately linked to the field of statistical mechanics was, in some sense, even noticed before the advent of information theory. The origin of the word entropy goes back to Clausius and some of the definitions given above were already used by Boltzmann, Gibbs and others.

The proof of Theorem A.1 can be looked up in the book by Nielsen and Chuang (2000) (Theorem 11.10 therein) and the proof of Theorem A.2 can be found in the book by Jacobs (2014) (Theorem 11 therein). Whereas the proofs of the previous theorems are not too complicated, the proof of Theorem A.3 is involved (Lindblad, 1975; Ruskai, 2002). The last theorem, Theorem A.5, was discovered by Reeb and Wolf (2015).

Appendix B
Superoperators

Maps that map matrices onto matrices (or operators onto operators) are often called **superoperators** and we denote them by calligraphic letters such as $\mathcal{A}$, $\mathcal{B}$, $\mathcal{C}\ldots$ throughout this book. The goal of this appendix is to show that superoperators are simply 'big matrices' and to provide some background information useful for practical calculations and numerical implementations.

Recall that, whenever you have a vector space V and a map $\Phi : V \to V$ acting linearly on it, this map can be represented by a matrix. To see this, consider an arbitrary vector $\mathbf{v} = \sum_i v_i \mathbf{e}_i \in V$ decomposed in an arbitrary basis $\{\mathbf{e}_i\}$ of that vector space. Since Φ acts linearly, we have

$$\Phi(\mathbf{v}) = \sum_i v_i \Phi(\mathbf{e}_i). \tag{B.1}$$

The components of $\Phi(\mathbf{v}) \in V$ can be read off by taking the scalar product

$$\Phi(\mathbf{v})_j \equiv \langle \mathbf{e}_j | \Phi(\mathbf{v}) \rangle = \sum_i v_i \langle \mathbf{e}_j | \Phi(\mathbf{e}_i) \rangle \equiv \sum_i \hat{\Phi}_{ji} v_i. \tag{B.2}$$

In the last step we introduced the matrix $\hat{\Phi}$ with components $\hat{\Phi}_{ji} = \langle \mathbf{e}_j | \Phi(\mathbf{e}_i) \rangle$. In order to distinguish it from the abstract map Φ and in order to emphasize that the representation of $\hat{\Phi}$ depends on the chosen basis, we put a 'hat' on it. In that basis the action of Φ is given by ordinary matrix multiplication and we identify $\Phi(\mathbf{v}) \leftrightarrow \hat{\Phi}\mathbf{v}$.

Now, the same recipe can also be applied to superoperators since the set of complex $d \times d$ matrices (where d equals the dimension of the underlying Hilbert space in quantum mechanical applications) forms a vector space. All that we have to do is to fix a basis and a scalar product for the vector space of complex $d \times d$ matrices. Below, we will review two popular choices: one is typically used for numerical manipulations, whereas the other has additional mathematical and structural advantages.

B.1 Numerically Convenient Superoperator Mapping

For the first choice consider an arbitrary matrix ρ written in some basis $\{|k\rangle\}$ as

$$\rho = \sum_{k,l} \rho_{kl} |k\rangle\langle l|. \tag{B.3}$$

Now, we wish to map the set of matrix coefficients $\{\rho_{kl}\}$ to a vector, a procedure also known as **vectorization**. One way to do so is by identifying the basis element $|k\rangle\langle l|$ in the space of matrices with the tensor product of the two vectors $|k\rangle$ and $|l\rangle^*$. Here, the star denotes complex conjugation in the case that $|l\rangle$ has complex coefficients. This choice turns out to be convenient, but other choices are possible too. Henceforth, we identify $|k\rangle\langle l| \leftrightarrow |k\rangle \otimes |l\rangle^* \equiv |kl\rangle\rangle$, where we used a 'double-ket' notation to denote vectorized operators. Thus, we identify

$$\rho \leftrightarrow |\rho\rangle\rangle = \sum_{k,l} \rho_{kl}|kl\rangle\rangle = \sum_{k,l} \rho_{kl}|k\rangle \otimes |l\rangle^*. \tag{B.4}$$

The dual vectors $\langle\langle kl| = \langle k| \otimes \langle l|^* = (|k\rangle \otimes |l\rangle^*)^\dagger$, denoted by a 'double-bra', are constructed as usual by taking the conjugate transpose: $\langle\langle \rho| = \sum_{k,l} \rho^*_{kl}\langle k| \otimes \langle l|^*$.

Next, we introduce a scalar product $(\rho|\sigma)$ between two matrices ρ and σ (assumed to be of equal size) by defining

$$(\rho|\sigma) \equiv \mathrm{tr}\{\rho^\dagger\sigma\}. \tag{B.5}$$

This scalar product is known as the Frobenius or Hilbert–Schmidt scalar product.

Exercise B.1 Show that $(\rho|\sigma) = \langle\langle\rho|\sigma\rangle\rangle$, where $|\rho\rangle\rangle$ and $|\sigma\rangle\rangle$ are the vectorized matrices introduced above and $\langle\langle\rho|\sigma\rangle\rangle$ denotes the usual scalar product of two complex vectors. Next, introduce the vector $|I\rangle\rangle \equiv \sum_k |k\rangle \otimes |k\rangle^*$. Show that $|I\rangle\rangle$ is the vectorization of the identity matrix I. Show also that the trace can be written in superoperator space as $\langle\langle I|A\rangle\rangle = \mathrm{tr}\{A\}$, where A is an arbitrary matrix.

In order to use ordinary matrix calculus for superoperators, we need to know the matrix representation $\hat{\mathcal{A}}$ of an arbitrary superoperator $\mathcal{A}$ with respect to the above chosen basis. For this purpose we use the fact that the action of every superoperator $\mathcal{A}$ can be written as

$$\mathcal{A}\rho = \sum_k X_k\rho Y_k \tag{B.6}$$

for some set of (not further specified) matrices $\{X_k\}$ and $\{Y_k\}$. The proof of this statement follows from our considerations below, as we discuss later. Consequently, in order to write $\mathcal{A}$ as a matrix $\hat{\mathcal{A}}$, we need to know how to represent the operation $X\rho Y$ for arbitrary X and Y as a matrix. To this end, we look at its matrix elements:

$$\begin{aligned}(X\rho Y)_{mn} &= \sum_{k,l} \rho_{kl}\langle m|(X|k\rangle)\langle l|(Y|n\rangle) = \sum_{k,l} \rho_{kl}\langle m|(X|k\rangle)[\langle n|(Y^\dagger|l\rangle)]^* \\ &= \langle m| \otimes \langle n|^* X \otimes Y^T \sum_{k,l} \rho_{kl}|k\rangle \otimes |l\rangle^* = \langle\langle mn|\big(X \otimes Y^T|\rho\rangle\rangle\big).\end{aligned} \tag{B.7}$$

Here, a superscript T denotes the transpose of a matrix. Note that X and Y need not be Hermitian, which is the reason why we were extra careful and wrote, for instance, $\langle m|(X|k\rangle)$ to make clear on which side of the scalar product X is acting. From eqn (B.7) we then infer the following matrix representation of the superoperator $\mathcal{A}$:

$$\mathcal{A}\rho = X\rho Y \leftrightarrow \hat{\mathcal{A}} = X \otimes Y^T. \tag{B.8}$$

Obviously, the matrix representation of the superoperator in eqn (B.6) follows as $\hat{\mathcal{A}} = \sum_k X_k \otimes Y_k^T$.

Finally, we show that the concatenation of two superoperators $\mathcal{A}_2 \circ \mathcal{A}_1$ corresponds to ordinary matrix multiplication $\hat{\mathcal{A}}_2 \cdot \hat{\mathcal{A}}_1$, here explicitly denoted with a dot. We do so by considering the superoperators $\mathcal{A}_1\rho = X_1\rho Y_1$ and $\mathcal{A}_2\rho = X_2\rho Y_2$. The more general case of eqn (B.6) follows from linearity. Then, to see this, we define $\mathcal{B}\rho \equiv \mathcal{A}_2 \circ \mathcal{A}_1\rho = X_2X_1\rho Y_1Y_2$ and from eqn (B.8) it follows that

$$\hat{\mathcal{B}} = X_2X_1 \otimes (Y_1Y_2)^T = X_2X_1 \otimes Y_2^T Y_1^T = (X_2 \otimes Y_2^T)(X_1 \otimes Y_1^T) = \hat{\mathcal{A}}_2 \cdot \hat{\mathcal{A}}_1. \quad \text{(B.9)}$$

Another direct but tedious way is to confirm that for any ρ it holds that $\langle i|(\mathcal{A}_2 \circ \mathcal{A}_1\rho)|j\rangle = \sum_{k,l} \langle\langle ij|\hat{\mathcal{A}}_2|kl\rangle\rangle\langle\langle kl|\hat{\mathcal{A}}_1|\rho\rangle\rangle$.

Let us now return to eqn (B.6) by going in the reverse direction. We saw that any superoperator $\mathcal{A}$ can be represented by a matrix $\hat{\mathcal{A}}$. If we denote by $\{B_k\}$ a basis for the vector space of $d \times d$ matrices on which $\mathcal{A}$ acts, then the space of $d^2 \times d^2$ matrices $\hat{\mathcal{A}}$ is spanned by the basis $\{B_k \otimes B_l^T\}$. In particular, any matrix representation of the superoperator $\mathcal{A}$ can be expanded as

$$\hat{\mathcal{A}} = \sum_{k,l} a_{kl} B_k \otimes B_l^T \quad \text{(B.10)}$$

for some complex coefficients a_{kl}. It follows from our representation (B.8) that the corresponding superoperator $\mathcal{A}$ reads

$$\mathcal{A}\rho = \sum_{k,l} a_{kl} B_k \rho B_l, \quad \text{(B.11)}$$

which is of the general form (B.6) if we define $X_k \equiv B_k$ and $Y_k \equiv \sum_l a_{kl} B_l$.

While the above exposition provides all the necessary tools to deal with superoperators numerically, the devil is (as usual) in the detail and simply requires further practice. As a first lesson the reader is therefore asked to confirm the following results.

Exercise B.2 An important superoperator in quantum mechanics results from considering the time evolution of an isolated system, which is given by $\mathcal{U}\rho \equiv U\rho U^\dagger$ for some unitary matrix U. According to the above construction, the matrix representation of $\mathcal{U}$ is given by $\hat{\mathcal{U}} = U \otimes U^*$. Show that $\hat{\mathcal{U}}$ is unitary in the usual sense of $\hat{\mathcal{U}} \cdot \hat{\mathcal{U}}^\dagger = \hat{\mathcal{U}}^\dagger \cdot \hat{\mathcal{U}} = \hat{\mathcal{I}}$, where $\hat{\mathcal{I}} = I \otimes I$ is the identity matrix in superoperator space. On the other hand, according to the definitions (i) and (i′) of Section 1.5 the map $\mathcal{U}\rho \equiv U\rho U^\dagger$ is CP. Show that this does *not* imply that the matrix $\hat{\mathcal{U}}$ is positive.

B.2 The Choi–Jamiokowski Isomorphism

We now consider another way to represent superoperators by matrices, which somehow reverses the properties found in the foregoing exercise. This representation plays an outstanding role in mathematical physics and is typically refered to as the Choi–Jamiołkowski isomorphism.

To introduce it, let $\mathcal{A}$ be a superoperator acting on quantum states ρ_1 defined over a Hilbert space $\mathcal{H}_1$. Next, we consider the tensor product space $\mathcal{H}_1 \otimes \mathcal{H}_0$, where $\mathcal{H}_0 = \mathcal{H}_1$ is just a copy of the original Hilbert space. It turns out, however, to be convenient to use a different index '0' instead of '1' in the notation. A central role in the following is played by the non-normalized maximally entangled state

$$|\psi^+\rangle \equiv \sum_i |i\rangle \otimes |i\rangle \in \mathcal{H}_1 \otimes \mathcal{H}_0, \tag{B.12}$$

where $\{|i\rangle\}$ is an arbitrary orthonormal basis in $\mathcal{H}_1$ or $\mathcal{H}_0$, respectively. The vectorization procedure for an arbitrary operator X_1 acting on $\mathcal{H}_1$ in the Choi–Jamiołkowski isomorphism is accomplished via the mapping

$$|X\rangle\rangle \equiv (X_1 \otimes I_0)|\psi^+\rangle = \sum_i (X|i\rangle) \otimes |i\rangle \in \mathcal{H}_1 \otimes \mathcal{H}_0. \tag{B.13}$$

Similarly, a superoperator $\mathcal{A}$ is mapped to a matrix A via the prescription

$$\mathsf{A} \equiv (\mathcal{A}_1 \otimes \mathcal{I}_0)|\psi^+\rangle\langle\psi^+| = \sum_{i,j} \mathcal{A}(|i\rangle\langle j|) \otimes |i\rangle\langle j|, \tag{B.14}$$

which is a matrix acting on the vector space $\mathcal{H}_1 \otimes \mathcal{H}_0$. In this context, A is known as the *Choi matrix*. To recover the action of $\mathcal{A}$ on a state ρ, one uses the relation

$$\mathcal{A}\rho = \mathrm{tr}_0\{(I_1 \otimes \rho_0^T)\mathsf{A}\}, \tag{B.15}$$

which is easy to confirm by direct calculation.

Working with the Choi matrix (B.14) has an important advantage: the superoperator $\mathcal{A}$ is CP if and only if its Choi matrix is positive: $\mathsf{A} \geq 0$. In fact, the direction '$\Rightarrow$' is easy to see, but the converse direction requires more work. To get further acquainted with Choi matrices, the reader is asked to do the following exercise.

Exercise B.3 Show that, if $\mathcal{A}$ is trace-preserving, the trace of the Choi matrix is $\mathrm{tr}\{\mathsf{A}\} = d$, where $d = \dim \mathcal{H}_1$. Thus, recalling that $\mathsf{A} \geq 0$, we can view A/d as a quantum state of a bipartite system, which is in one-to-one correspondence with the map $\mathcal{A}$. However, in contrast to our earlier construction, the Choi matrix of a unitary time evolution map $\mathcal{U}\rho = U\rho U^\dagger$ is no longer unitary. Convince yourself of this.

We proceed by looking at the concatenation of superoperators, which can be described using the **link product** $*$. To this end, it turns out to be convenient to explicitly include the spaces on which the Choi matrices are defined in the notation. Thus, let A_{10} act on two spaces labeled 1 and 0 and B_{21} acts on two spaces labeled 2 and 1. Their link product is defined as

$$\mathsf{B}_{21} * \mathsf{A}_{10} = \mathrm{tr}_1\{(I_2 \otimes \mathsf{A}_{10}^{T_1})(\mathsf{B}_{21} \otimes I_0)\}, \tag{B.16}$$

where T_1 means transpose on space 1 only, i.e., $\mathsf{A}_{10}^{T_1} = \sum_{i,j} \mathcal{A}(|i\rangle\langle j|)_1^T \otimes |i\rangle\langle j|_0$. In words, the link product takes two Choi matrices, puts them in reverse order, appends them with identity matrices to make them act on the same joint space (here, the space 210), performs a partial transpose of one matrix with respect to the common space (here, the space 1), and finally traces them over the common space.

The link product satisfies a number of elementary properties, which can be checked by direct calculation. For instance, it is associative if the three matrices share *no* common space, which we write as

$$\mathsf{C}_{32} * \mathsf{B}_{21} * \mathsf{A}_{10} = (\mathsf{C}_{32} * \mathsf{B}_{21}) * \mathsf{A}_{10} = \mathsf{C}_{32} * (\mathsf{B}_{21} * \mathsf{A}_{10}). \tag{B.17}$$

Moreover, if two matrices have no space in common, the link product reduces to the tensor product:

$$\mathsf{B}_2 * \mathsf{A}_0 = \mathsf{B}_2 \otimes \mathsf{A}_0. \tag{B.18}$$

If they act on the same space instead, one obtains the Hilbert–Schmidt scalar product

$$\mathsf{B}_0 * \mathsf{A}_0 = \mathrm{tr}_0\{\mathsf{B}_0^T \mathsf{A}_0\}. \tag{B.19}$$

Finally, the link product is commutative up to relabeling of Hilbert spaces,

$$\mathsf{B}_{21} * \mathsf{A}_{10} = S_{2,0}\mathsf{A}_{10} * \mathsf{B}_{21} S_{2,0}, \tag{B.20}$$

where $S_{2,0} = S_{2,0}^\dagger$ is the unitary swap operator between spaces 2 and 0, defined as $S_{2,0}|j\rangle_2 \otimes |k\rangle_0 = |k\rangle_0 \otimes |j\rangle_2$. In particular, the link product is commutative if the participating spaces are traced over in the final expression. For instance, for arbitary different spaces 0, 1, 2, 3, 4 and 5 it is true that

$$\mathrm{tr}_3\{\mathsf{D}_{54} * \mathsf{C}_{432} * \mathsf{B}_{21} * \mathsf{A}_{10}\} = \mathrm{tr}_3\{\mathsf{D}_{54} * \mathsf{B}_{21} * \mathsf{C}_{432} * \mathsf{A}_{10}\}. \tag{B.21}$$

To connect the link product to the concatenation of superoperators acting on a state ρ it is useful to introduce some conventions. First of all, we identify the Choi matrix of ρ with ρ itself. Equation (B.15) then implies $\mathcal{A}\rho = \mathrm{tr}_0\{(I_1 \otimes \rho_0^T)\mathsf{A}_{10}\} = \mathsf{A}_{10} * \rho_0$. Moreover, although the state space of ρ is always the same, it is useful to denote state spaces at different times with different labels, i.e., we formally associate to each time a different space. Thus, instead of writing $\mathcal{C} \circ \mathcal{B} \circ \mathcal{A}\rho$, we write $\mathcal{C}_{32} \circ \mathcal{B}_{21} \circ \mathcal{A}_{10}\rho_0$, and we call, e.g., space 1 the *output space* of $\mathcal{A}$ and the *input space* of $\mathcal{B}$. With this identification we have

$$\mathcal{C}_{32} \circ \mathcal{B}_{21} \circ \mathcal{A}_{10}\rho_0 = \mathsf{C}_{32} * \mathsf{B}_{21} * \mathsf{A}_{10} * \rho. \tag{B.22}$$

B.3 Matrix Representations of the Process Tensor

We return to the process tensor introduced in Section 1.7, which we can now express in a neat form. To this end, we label the input space of the nth control operation $\mathcal{C}(r_n)$ with S_n and its output space with S_n' (remember that the control operations only act on the system). The bath space during the nth control operation is labeled B_n. This convention is illustrated in Fig. B.1. The process tensor defined in eqn (1.61) is then written as

$$\begin{aligned} &\mathfrak{T}[\mathcal{C}(r_n), \ldots, \mathcal{C}(r_1), \mathcal{C}(r_0)] = \\ &\quad \mathrm{tr}_{B_n}\left\{\mathcal{C}_{S_n',S_n}\mathcal{U}_{S_nB_n,S_{n-1}'B_{n-1}} \cdots \mathcal{C}_{S_1',S_1}\mathcal{U}_{S_1B_1,S_0'B_0}\mathcal{C}_{S_0',S_0}\rho_{S_0B_0}\right\}. \end{aligned} \tag{B.23}$$

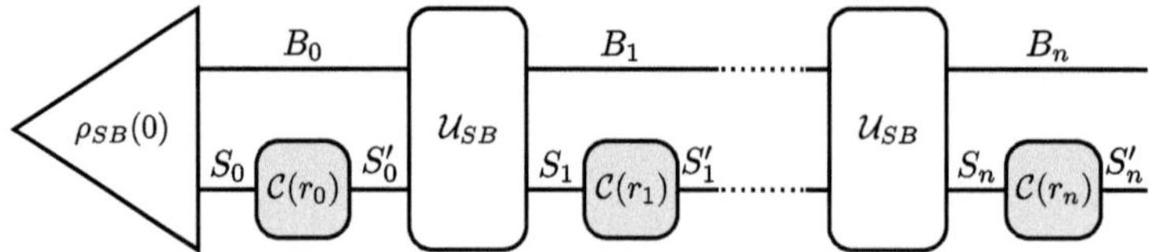

Fig. B.1 Process tensor with input and output spaces labeled according to our convention.

where we dropped any dependence on time or the measurement results r_n for notational simplicity. Now, observe that each space in eqn (B.23) appears exactly twice (except of S_n', which appears only once). Generalizing eqn (B.22), we find

$$\mathfrak{T}[\mathcal{C}(r_n),\ldots,\mathcal{C}(r_1),\mathcal{C}(r_0)] = \tag{B.24}$$
$$\mathrm{tr}_{B_n}\left\{\mathsf{C}_{S_n',S_n} * \mathsf{U}_{S_nB_n,S_{n-1}'B_{n-1}} * \cdots * \mathsf{C}_{S_1',S_1} * \mathsf{U}_{S_1B_1,S_0'B_0} * \mathsf{C}_{S_0',S_0} * \rho_{S_0B_0}\right\}.$$

Next, observe that we can apply eqn (B.21) to rearrange terms into

$$\mathfrak{T}[\mathcal{C}(r_n),\ldots,\mathcal{C}(r_1),\mathcal{C}(r_0)] = \tag{B.25}$$
$$\mathsf{C}_{S_n',S_n} * \cdots * \mathsf{C}_{S_1',S_1} * \mathsf{C}_{S_0',S_0} * \mathrm{tr}_{B_n}\left\{\mathsf{U}_{S_nB_n,S_{n-1}'B_{n-1}} * \cdots * \mathsf{U}_{S_1B_1,S_0'B_0} * \rho_{S_0B_0}\right\}.$$

We can write this concisely as

$$\mathfrak{T}[\mathcal{C}(r_n),\ldots,\mathcal{C}(r_1),\mathcal{C}(r_0)] = \mathsf{C}(\mathbf{r}_n) * \mathsf{T} \tag{B.26}$$

with $\mathsf{C}(\mathbf{r}_n) = \mathsf{C}(r_n) \otimes \cdots \otimes \mathsf{C}(r_0)$, where we used eqn (B.18) and introduced the dependence on the measurement results again, and $\mathsf{T} = \mathrm{tr}_B\{\mathsf{U}_{n-1} * \cdots * \mathsf{U}_0 * \rho_0\}$, where the subscripts now denote the time steps, tacitly assuming the association of input and output spaces as illustrated in Fig. B.1.

Taking the trace over the final output space and using eqn (B.19), we can compactly write

$$p(\mathbf{r}_n) = \mathrm{tr}_S\{\mathfrak{T}[\mathcal{C}(r_n),\ldots,\mathcal{C}(r_1),\mathcal{C}(r_0)]\} = \mathrm{tr}\{\mathsf{C}(\mathbf{r}_n)^T\mathsf{T}\}, \tag{B.27}$$

where the final trace is over the spaces $S_n', S_n, \ldots, S_0', S_0$. Equation (B.27) can be interpreted as a generalization of the Born rule to multiple time steps. Indeed, at a single time the Born rule says that $p(r) = \mathrm{tr}\{P(r)\rho\}$ for some POVM element $P(r)$ and state ρ. By analogy, $\mathsf{C}(\mathbf{r}_n)^T$ corresponds to a multitime POVM element and T corresponds to a multitime 'state.' In particular, we note that both objects are positive.

Thus, we can conclude that the link product defined on Choi states acting on multiple input/output spaces allows for a convenient separation of the process T itself, which captures all the effects of the environment, and the external control operations $\mathsf{C}(\mathbf{r}_n)$. The probability to get a sequence of outcomes $\mathbf{r}_n$ can be obtained by matrix multiplication of the two objects followed by a trace. This separation is, of course, not a particular consequence of the Choi–Jamiołkowski isomorphism—it follows from any representation as we illustrate now, which further elucidates the meaning of the link product and of eqn (B.25).

For this purpose we consider an explicit matrix representation of the state and superoperators. Let $\rho(0) = \sum \rho_{a_0,b_0}^{\alpha_0,\beta_0}|\alpha_0 a_0\rangle\langle\beta_0 b_0|$ be the initial state, where Greek (Latin) indices label the bath (system) degrees of freedom (thus, the index b_0 does *not* label bath degrees of freedom), and summation symbols $\sum$ without subscripts correspond to sums over all indices in the following. Note that we do not specify a particular basis for the representation. The unitary time evolution operator in the first time step from 0 to t_1 is written as $U_0 = \sum U_{a_1,a_0}^{\alpha_1,\alpha_0} |\alpha_1 a_1\rangle\langle\alpha_0 a_0|$ and the action of the time evolution superoperator on the initial state is written as $\mathcal{U}_0\rho(0) = \sum \mathcal{U}_{a_1 a_0,b_1 b_0}^{\alpha_1\alpha_0,\beta_1\beta_0} \rho_{a_0,b_0}^{\alpha_0,\beta_0} |\alpha_1 a_1\rangle\langle\beta_1 b_1|$.

For simplicity, we consider only a three–step process in the following; the generalization to an n–step process is mathematically straightforward but cumbersome to write down. We find (again dropping the dependence on $\mathbf{r}_n$ in the notation)

$$\begin{aligned}
&\mathcal{C}_2\mathcal{U}_1\mathcal{C}_1\mathcal{U}_0\mathcal{C}_0\rho(0) = \\
&\quad \sum \mathcal{C}_{a_2'a_2,b_2'b_2}\mathcal{U}_{a_2a_1',b_2b_1'}^{\alpha_2\alpha_1,\beta_2\beta_1}\mathcal{C}_{a_1'a_1,b_1'b_1}\mathcal{U}_{a_1a_0',b_1b_0'}^{\alpha_1\alpha_0,\beta_1\beta_0}\mathcal{C}_{a_0'a_0,b_0'b_0}\rho_{a_0,b_0}^{\alpha_0,\beta_0}|\alpha_2a_2'\rangle\langle\beta_2b_2'|.
\end{aligned} \tag{B.28}$$

Tracing over the system and bath degrees of freedom and regrouping terms, we find

$$\begin{aligned}
p(r_2,r_1,r_0) &= \sum_{\text{Latin}}\left(\sum_{\text{Greek}}\mathcal{U}_{a_2a_1',b_2b_1'}^{\epsilon\alpha_1,\epsilon\beta_1}\mathcal{U}_{a_1a_0',b_1b_0'}^{\alpha_1\alpha_0,\beta_1\beta_0}\rho_{a_0,b_0}^{\alpha_0,\beta_0}\right)\mathcal{C}_{ea_2,eb_2}\mathcal{C}_{a_1'a_1,b_1'b_1}\mathcal{C}_{a_0'a_0,b_0'b_0} \\
&\equiv \sum_{\text{Latin}}\mathfrak{T}_{a_2a_1'a_1a_0'a_0,b_2b_1'b_1b_0'b_0}\mathfrak{C}_{a_2a_1'a_1a_0'a_0,b_2b_1'b_1b_0'b_0} = \mathrm{tr}\{\mathfrak{T}\mathfrak{C}^T\}.
\end{aligned} \tag{B.29}$$

Here, we have again separated the influence of the environment from the control operations. In fact, the trace over the Latin indices corresponds precisely to the trace on the right hand side of eqn (B.27). Furthermore, the term in the brackets presents an explicit matrix representation of the process tensor $\mathfrak{T}$, which can be associated to $\mathrm{tr}_B\{\mathcal{U}_1 * \mathcal{U}_0 * \rho(0)\}$ if one interprets the link product as a matrix product in the environment space and a tensor product in the system space. Likewise, the term $\mathfrak{C}$ containing the control operations can be associated to $\mathcal{C}_2 * \mathcal{C}_1 * \mathcal{C}_0 = \mathcal{C}_2 \otimes \mathcal{C}_1 \otimes \mathcal{C}_0$. This analogy can be even further strengthened by tracing out only the bath degrees of freedom in eqn (B.28) and by noting that

$$\begin{aligned}
&\mathrm{tr}_B\{\mathcal{C}_2\mathcal{U}_1\mathcal{C}_1\mathcal{U}_0\mathcal{C}_0\rho(0)\} = \\
&\quad \mathrm{tr}_{S_2S_1'S_1S_0'S_0}\Bigg\{I_{S_2'}\otimes\sum_{\text{Latin}}\sum_{\text{Greek}}\mathcal{U}_{a_2a_1',b_2,b_1'}^{\epsilon\alpha_1,\epsilon\beta_1}\mathcal{U}_{a_1a_0',b_1,b_0'}^{\alpha_1\alpha_0,\beta_1\beta_0}\rho_{a_0,b_0}^{\alpha_0,\beta_0}|a_2a_1'a_1a_0'a_0\rangle\langle b_2b_1'b_1b_0'b_0| \\
&\qquad \times\sum_{\text{Latin}}\mathcal{C}_{\bar{a}_2'\bar{a}_2,\bar{b}_2'\bar{b}_2}\mathcal{C}_{\bar{a}_1'\bar{a}_1,\bar{b}_1'\bar{b}_1}\mathcal{C}_{\bar{a}_0'\bar{a}_0,\bar{b}_0'\bar{b}_0}|\bar{a}_2'\bar{a}_2\bar{a}_1'\bar{a}_1\bar{a}_0'\bar{a}_0\rangle\langle\bar{b}_2'\bar{b}_2\bar{b}_1'\bar{b}_1\bar{b}_0'\bar{b}_0|\Bigg\}
\end{aligned} \tag{B.30}$$

Exercise B.4 Show that eqn (B.30) is identical to eqn (B.25) if we use the Choi representation for superoperators, i.e., if we identify $\mathcal{U} \leftrightarrow \mathsf{U}$ and $\mathcal{C} \leftrightarrow \mathsf{C}$.

To complete our exposition, we turn to another equivalent representation of the process tensor, which shows that the process tensor can be seen as a state of a suitable quantum many–body system. Specifically, we convert an n–step process to a state living on $\mathcal{H}_S^{\otimes(2n+1)}$, which is isomorphic to our previous representations. To motivate the

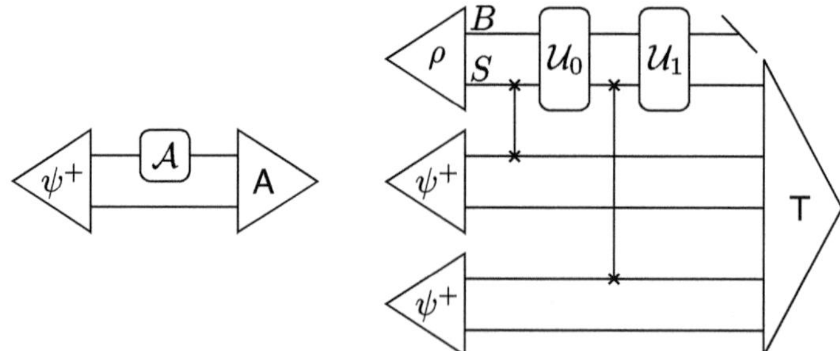

Fig. B.2 Left: Circuit diagram of the Choi matrix representation of a superoperator $\mathcal{A}$. Right: Circuit diagram to construct the Choi matrix of the process tensor according to a generalized Choi–Jamiołkowski isomorphism. The vertical lines denote a swap between state spaces marked with a cross.

following, we return to the Choi matrix A corresponding to a superoperator $\mathcal{A}$, which was constructed by letting $\mathcal{A}$ act on one half of an (unnormalized) maximally entangled state, see eqn (B.14). Furthermore, recall that the Choi matrix can be naturally interpreted as the (unnormalized) state of a bipartite quantum system (Exercise B.3). This is graphically depicted in Fig. B.2.

Similarly, to turn the process tensor into its corresponding Choi matrix, we let it interact at each intervention time t_j with one half of a maximally entangled state. This is achieved by *swapping* at each time the current system state with a fresh state, which forms one half of a maximally entangled ancilla state A_j. Precisely, let $\psi_j^+ = \sum |a_j' a_j'\rangle\langle b_j' b_j'|$ denote the (unnormalized) maximally entangled state of A_j and let the unitary swap superoperator be

$$\mathcal{S}_{S,A_j}(\rho_S \otimes \psi_j^+) = \mathcal{S}_{S,A_j} \sum \rho_{a,b} |a a_j' a_j'\rangle\langle b b_j' b_j'| \equiv \sum \rho_{a,b} |a_j' a a_j'\rangle\langle b_j' b b_j'|, \tag{B.31}$$

which exchanges the system with the first half of the ancilla. Then, the final claim of this section is

$$\begin{aligned} \mathsf{T} &= \mathrm{tr}_{B_2}\{\mathsf{U}_{S_2B_2,S_1'B_1} * \mathsf{U}_{S_1B_1,S_0'B_0} * \rho_{S_0B_0}\} \\ &\cong \mathrm{tr}_B\{\mathcal{U}_1\mathcal{S}_{SA_1}\mathcal{U}_0\mathcal{S}_{SA_0}\rho_{SB}(0) \otimes \psi_0^+ \otimes \psi_1^+\}, \end{aligned} \tag{B.32}$$

where $\cong$ means that the result is identical up to reordering of the Hilbert spaces and if the superoperators are given in Choi representation. Furthermore, we have restricted ourselves again to the first three control operations, but the generalization to n interventions is straightforward.

Exercise B.5 Derive eqn (B.32). Furthermore, show that the Choi matrix corresponding to a quantum Markov process defined in Section 1.8 is isomorphic to $\mathcal{E}(t_n, t_{n-1}) \otimes \cdots \mathcal{E}(t_1, 0) \otimes \rho_S(0)$, i.e., a many-body state where correlations only exist between a preparation and its subsequent measurement or, in view of the terminology of this appendix, between an output state S_{j-1}' and an input state S_j.

Further reading

Pedagogical accounts of how to represent quantum operations by matrices can be found in many places. As the name suggests, the Choi–Jamiołkowski isomorphism is due to the work of Jamiołkowski (1972) and Choi (1975). The link product was introduced by Chiribella *et al.* (2009) and the process tensor representation in terms of a many–body state was given by Pollock *et al.* (2018a); see also Milz and Modi (2021) and references therein for a detailed exposition.

Appendix C
Time-Reversal Symmetry

We show that the equations of motion of classical and quantum mechanics have a remarkable property, which we call *time-reversal symmetry* or *microreversibility.* Its consequences for the arrow of time and the second law of thermodynamics are also discussed at the end.

In both cases, classically and quantum mechanically, the notion of time-reversal symmetry will be introduced with respect to the following abstract thought experiment, which is also depicted as a diagram in Fig. C.1. It is important to note here that this thought experiment can be carried out in the real world without the need to actually 'reverse' the time. The reader therefore should *not* take the terminology 'time-reversal' literally (see the discussion in Section C.3). As usual in physics, there are two main players involved. Initial states, here abstractly denoted by S_0, and some dynamical law, here abstractly denoted by $\mathcal{E}(\tau)$, which maps the initial states to some final states $S_\tau = \mathcal{E}(\tau)S_0$ after a time τ has elapsed. Now, suppose that we find an invertible map Θ, which we call the time-reversal operator, and some conjugate dynamics $\mathcal{E}_\Theta(\tau)$ such that

$$S_0 = \Theta^{-1}\mathcal{E}_\Theta(\tau)\Theta\mathcal{E}(\tau)S_0. \tag{C.1}$$

In words: we get back to the initial state S_0 if we let the system evolve in time for a duration τ, then time-reverse the state followed by an evolution for a duration τ with respect to the conjugate dynamics and, finally, apply the inverse of the time-reversal operator. We now define the following.

Time-reversal symmetry. *A physical system defined by some dynamical law $\mathcal{E}(\tau)$ is said to possess time-reversal symmetry if there exists an invertible time-reversal operator Θ and a conjugate dynamical law $\mathcal{E}_\Theta(\tau)$ such that eqn (C.1) holds under the following three conditions:*

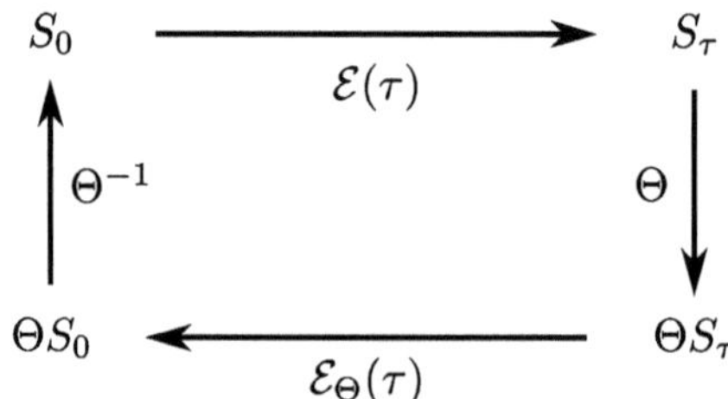

Fig. C.1 Abstract diagram of the thought experiment related to the notion of time-reversal symmetry.

(i) Equation (C.1) holds for all legitimate initial conditions S_0.
(ii) The states S_τ, ΘS_τ *and* ΘS_0 *are legitimate physical states.*
(iii) The conjugate dynamical law $\mathcal{E}_\Theta(\tau)$ *is a legitimate physical evolution law, which can (in principle) be implemented in a laboratory.*

Here, the word 'legitimate' has to be defined by the physical context at the end, e.g. legitimate states or evolution laws should not give rise to negative probabilities. Furthermore, the conjugate dynamics $\mathcal{E}_\Theta(\tau)$ must be realizable in a laboratory in principle (e.g. generated by a legitimate Hamiltonian). As we see below, this excludes the choice $\Theta = I$ and $\mathcal{E}_\Theta(\tau) = \mathcal{E}(\tau)^{-1}$, which would trivially satisfy eqn (C.1).

C.1 Time-Reversal Symmetry in Classical Mechanics

The state of a classical system with N particles and $2f$ degrees of freedom per particle is described by its phase-space coordinates $(\mathbf{q}, \mathbf{p}) \in \mathbb{R}^{2Nf}$, where $\mathbf{q} = (q_1, \ldots, q_{Nf})$ are the generalized coordinates and $\mathbf{p} = (p_1, \ldots, p_{Nf})$ are the conjugate momenta. Note that we here restrict the discussion to 'pure' states, which are represented by a single point in phase space. Since mixtures evolve via Liouville's equation linearly in classical mechanics, this assumption does not entail any loss of generality. In addition, to focus on the essential ingredients of time-reversal symmetry, we set $N = 1$ and $f = 1$ in the following (i.e. a single particle moving in one dimension). The generalization to many particles and multiple degrees of freedom is indeed just a matter of notation. Thus, we denote the Hamiltonian of the system by $H(q,p)$ and Hamilton's equations of motion become

$$\dot{q} = \frac{\partial H(q,p)}{\partial p}, \quad \dot{p} = -\frac{\partial H(q,p)}{\partial q}. \tag{C.2}$$

To be particularly cautious and in view of what follows, we write Hamilton's equations in difference form as

$$q_{t+dt} = q_t + dt \left.\frac{\partial H}{\partial p}\right|_{(q_t,p_t)}, \quad p_{t+dt} = p_t - dt \left.\frac{\partial H}{\partial q}\right|_{(q_t,p_t)}, \tag{C.3}$$

assuming dt to be small enough such that terms of order $\mathcal{O}(dt^2)$ are negligible throughout. Note that we explicitly keep the information at which phase-space point the partial derivative is evaluated.

We now define the time-reversal operator Θ via its action on a phase-space point (q,p) as

$$\Theta(q,p) \equiv (q,-p), \tag{C.4}$$

i.e. we leave the coordinate unchanged but flip the momentum. We see that this definition implies that Θ is an *involution*: $\Theta^2 = I$. Put differently, the time-reversal operator is its own inverse: $\Theta = \Theta^{-1}$. This will be true for the rest of this section, but it is no longer the case for quantum systems in general. Figure C.2 explains the concept of time-reversal for the case of a simple harmonic oscillator.

Furthermore, we start with the simplest situation and assume a Hamiltonian, which obeys the symmetry $H(q,-p) = H(q,p)$. This is often satisfied, for example, when the Hamiltonian $H(q,p) = p^2/2m + V(q)$ describes the motion of a particle in some

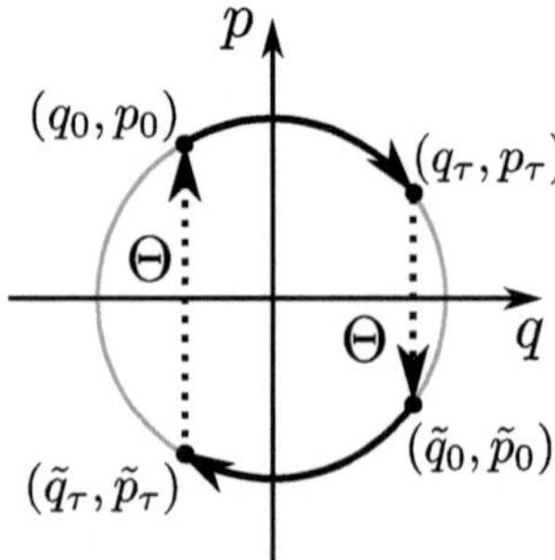

Fig. C.2 Time-reversal symmetry illustrated in the phase space of a harmonic oscillator. The grey circle defines a 'surface' of constant energy. Time-reversal symmetry of the dynamics means that one can go back from (q_τ, p_τ) to (q_0, p_0) by, first, mapping (q_τ, p_τ) to its time-reversed image $(\tilde{q}_0, \tilde{p}_0) = (q_\tau, -p_\tau)$, then letting the system evolve for a time τ and, finally, mapping $(\tilde{q}_\tau, \tilde{p}_\tau)$ back to $(q_0, p_0) = (\tilde{q}_\tau, -\tilde{p}_\tau)$.

external potential $V(q)$, but we treat exceptions soon. We then postulate the conjugate dynamics to also be described by Hamilton's equations with respect to the *same* Hamiltonian

$$H_\Theta(q, p) \equiv H(q, p). \tag{C.5}$$

We now check the validity of the diagram in Fig. C.1 for a small time step dt from $t = 0$ to dt. We introduce the notation

$$(\tilde{q}_0, \tilde{p}_0) \equiv \Theta(q_{dt}, p_{dt}) = (q_{dt}, -p_{dt}). \tag{C.6}$$

This choice might seem unconventional, but it has the advantage that the time parameter also increases in the time-reversed experiment, as it would in any actual experiment. Our goal is now to show that the time-reversed state, evolved for a small time step dt with respect to the Hamiltonian (C.5), obeys $(\tilde{q}_{dt}, \tilde{p}_{dt}) = \Theta(q_0, p_0) = (q_0, -p_0)$. To see this, we start by writing down Hamilton's equation for the time-reversed evolution with $H_\Theta = H$

$$\tilde{q}_{dt} = \tilde{q}_0 + dt\, \frac{\partial H}{\partial p}\bigg|_{(\tilde{q}_0, \tilde{p}_0)}, \quad \tilde{p}_{dt} = \tilde{p}_0 - dt\, \frac{\partial H}{\partial q}\bigg|_{(\tilde{q}_0, \tilde{p}_0)}. \tag{C.7}$$

Next, we use definition (C.6) and the fact that any differentiable function obeying $f(x) = f(-x)$ satisfies $f'(-x) = -f'(x)$. Applied to $H(q, p) = H(q, -p)$, this implies

$$\tilde{q}_{dt} = q_{dt} - dt\, \frac{\partial H}{\partial p}\bigg|_{(q_{dt}, p_{dt})}, \quad \tilde{p}_{dt} = -p_{dt} - dt\, \frac{\partial H}{\partial q}\bigg|_{(q_{dt}, p_{dt})}. \tag{C.8}$$

Using, furthermore, our assumption that terms of order $\mathcal{O}(dt^2)$ are negligible allows us to evaluate the partial derivatives at (q_0, p_0) instead of (q_{dt}, p_{dt}). Multiplying the second equation by -1, we end up with

$$\tilde{q}_{dt} = q_{dt} - dt\, \frac{\partial H}{\partial p}\bigg|_{(q_0, p_0)}, \quad -\tilde{p}_{dt} = p_{dt} + \frac{\partial H}{\partial q}\bigg|_{(q_0, p_0)}. \tag{C.9}$$

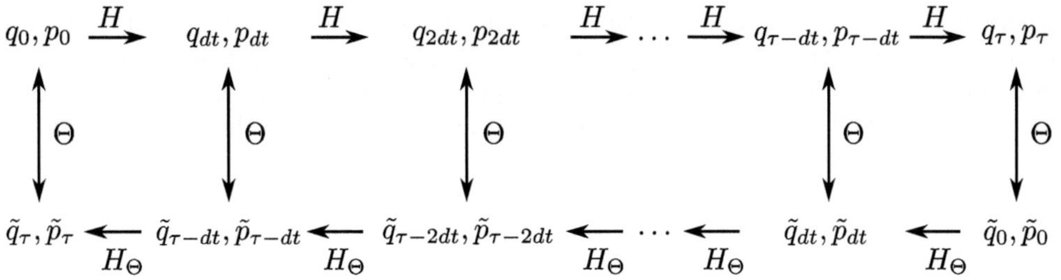

$$
\begin{array}{ccccccccc}
q_0, p_0 & \xrightarrow{H} & q_{dt}, p_{dt} & \xrightarrow{H} & q_{2dt}, p_{2dt} & \xrightarrow{H} \cdots \xrightarrow{H} & q_{\tau-dt}, p_{\tau-dt} & \xrightarrow{H} & q_\tau, p_\tau \\
\updownarrow \Theta & & \updownarrow \Theta & & \updownarrow \Theta & & \updownarrow \Theta & & \updownarrow \Theta \\
\tilde{q}_\tau, \tilde{p}_\tau & \xleftarrow[H_\Theta]{} & \tilde{q}_{\tau-dt}, \tilde{p}_{\tau-dt} & \xleftarrow[H_\Theta]{} & \tilde{q}_{\tau-2dt}, \tilde{p}_{\tau-2dt} & \xleftarrow[H_\Theta]{} \cdots \xleftarrow[H_\Theta]{} & \tilde{q}_{dt}, \tilde{p}_{dt} & \xleftarrow[H_\Theta]{} & \tilde{q}_0, \tilde{p}_0
\end{array}
$$

Fig. C.3 Commutative diagram describing time-reversal symmetry for a finite interval $[0, \tau]$.

But this equation is identical—after some rearrangement—to eqn (C.3) at $t = 0$ if we identify $(\tilde{q}_{dt}, \tilde{p}_{dt}) = (q_0, -p_0)$. To conclude, we have shown that we end up at the initial phase-space coordinate (q_0, p_0) if we start a time step dt later with (q_{dt}, p_{dt}), apply the time-reversal operator Θ, let the time-reversed state evolve forwards in time for a step dt with respect to the Hamiltonian (C.5) and finally apply Θ again.

Is the above argument sufficient to conclude that time-reversal symmetry applies also to a *finite* interval from 0 to τ? It is if we recall that we can always discretize any trajectory (q_t, p_t), $t \in [0, \tau]$, with a sufficiently small time step dt and if we use the fact that the time-reversal operator obeys $\Theta^2 = I$. This means that each step shown in the diagram in Fig. C.3, where we used the notation

$$(\tilde{q}_t, \tilde{p}_t) \equiv \Theta(q_{\tau-t}, p_{\tau-t}) = (q_{\tau-t}, -p_{\tau-t}), \tag{C.10}$$

obeys the mapping described above. Thus, the process obeys time-reversal symmetry.

It is interesting to wonder whether the time-reversal symmetry relation worked out above is unique or whether other choices for Θ and H_Θ are possible as well. As the next exercise shows, other choices are indeed possible.

Exercise C.1 Assume that the Hamiltonian obeys the symmetry $H(q,p) = H(-q,p)$, which is, for instance, the case for a harmonic oscillator. Now, consider the time-reversal operator $\Theta'(q,p) \equiv (-q,p)$, which flips the coordinate but not the momentum, and define the reversed dynamics via $H_{\Theta'}(q,p) \equiv H(q,p)$. By following the same steps as above, show that these transformations also lead to the notion of time-reversal symmetry. Flipping the coordinates and not the momenta might seem awkward at first sight. However, recall the flexibility that is offered by Hamilton's framework of classical mechanics. In particular, remember that any canonical transformation preserves the form of Hamilton's equations and describes the same physics in a different coordinate system. Convince yourself of the fact that the mapping $(Q,P) \equiv (p,-q)$ is a canonical transformation.

We have so far considered time-independent Hamiltonians obeying the symmetry $H(q,p) = H(q,-p)$. Two generalizations are important. First, it might be that $H(q,p) \neq H(q,-p)$. A prominent example is a particle with charge q in an external electromagnetic field described by an electric potential ϕ and a vector potential $\boldsymbol{A}$, which creates a magnetic field via $\boldsymbol{B} = \nabla \times \boldsymbol{A}$. Such a system is described by the Hamiltonian

$$H(\boldsymbol{q}, \boldsymbol{p}; \boldsymbol{B}) = \frac{1}{2m}\left(\boldsymbol{p} - \frac{q}{c}\boldsymbol{A}\right)^2 + q\phi(\boldsymbol{q}), \tag{C.11}$$

where $\boldsymbol{p} \in \mathbb{R}^3$ ($\boldsymbol{q} \in \mathbb{R}^3$) describes the three-dimensional momentum (position) and c is the speed of light. Clearly, $H(\boldsymbol{q}, -\boldsymbol{p}; \boldsymbol{B}) \neq H(\boldsymbol{q}, \boldsymbol{p}; \boldsymbol{B})$. However, if we recall that any magnetic field is created by moving charges and if we would include these in our description as well, then the time-reversal operator Θ would flip the velocity of these charges and, hence, the direction of the magnetic field. Therefore, eqn (C.11) obeys the symmetry $H(\boldsymbol{q}, -\boldsymbol{p}; -\boldsymbol{B}) = H(\boldsymbol{q}, \boldsymbol{p}; \boldsymbol{B})$. To describe time-reversal symmetry for the situation where we treat the magnetic field as an *external* parameter, we can keep the definition of Θ from eqn (C.4) and define the conjugate dynamics by applying Θ to the Hamiltonian (C.11):

$$H_\Theta(q, p; B) \equiv H(q, -p; B) = H(q, p; -B). \tag{C.12}$$

Here, we returned to a scalar notation (i.e. $\boldsymbol{q} \to q$, etc.) for simplicity. Because of the symmetry $H(q, p; B) = H(q, -p; -B)$, the conjugate dynamics is therefore generated by a Hamiltonian, which describes the same system but with an inverted magnetic field. Also because of that symmetry, we can confirm that

$$\left.\frac{\partial H_\Theta(q, p; B)}{\partial p}\right|_{(q_t, -p_t)} = -\left.\frac{\partial H(q, p; B)}{\partial p}\right|_{(q_t, p_t)}. \tag{C.13}$$

This relation is sufficient to repeat the same steps as above and to show the time-reversal symmetry of the dynamics.

The second generalization concerns explicitly time-dependent Hamiltonians of the form $H(q, p; B, \lambda_t)$, where λ_t is some externally specified driving protocol. To include this scenario in our description, remember that our calculation above holds for any Hamiltonian at a fixed time t. In particular, we could replace $H(q, p)$ by $H(q, p; \lambda_t)$ in eqns (C.5)–(C.9) without invalidating any argument. For a finite time interval $[0, \tau]$ and with respect to the diagram in Fig. C.3 this means that the time argument of the Hamiltonian $H(q, p; B, \lambda_t)$ in the forward process must match the time argument of the Hamiltonian $H_\Theta(q, p; B, \lambda_t)$ in the backward process. Together with the convention (C.10) that the time-reversed trajectory starts at $t = 0$ and ends at $t = \tau$ this implies that the time-reversed Hamiltonian is given by

$$H_\Theta(q, p; B, \lambda_t) \equiv H(q, p; -B, \lambda_{\tau - t}), \tag{C.14}$$

i.e. the protocol is executed in the reverse order, starting with λ_τ and ending with λ_0.

We end this section with a remark and a small exercise. The remark concerns the observations that the notion of time-reversal symmetry is closely linked to Liouville's theorem, which states that the flow of points in phase space generated by Hamilton's equations is divergence-free and preserves the volume. Since this fact is used in the main text, we state it here without proof (which can be looked up in any standard textbook on classical mechanics).

Liouville's theorem. *Consider a classical Hamiltonian system of N particles with $2f$ degrees of freedom per particle. Let $\Gamma = (\boldsymbol{q}, \boldsymbol{p}) \in \mathbb{R}^{2Nf}$ denote a point in the phase space. Furthermore, let $\Gamma^t = \phi^t(\Gamma^0)$ denote the time-evolved point in phase space given* *initial condition* Γ^0*. Then, the Jacobian of this transformation equals one, i.e. using a compact notation*

$$J = \det\left(\frac{\partial\Gamma^t}{\partial\Gamma^0}\right) = 1. \tag{C.15}$$

In particular, this implies for any function $f(\Gamma)$

$$\boxed{\int d\Gamma^0 f[\phi^t(\Gamma^0)] = \int d\Gamma^t f(\Gamma^t)} \tag{C.16}$$

Roughly speaking, Liouville's theorem states that *no information is lost* during the evolution of a Hamiltonian system. This idea is closely related to time-reversal symmetry since the latter principle states that we can find for each 'forward' trajectory a conjugated 'reversed' trajectory, which also obeys Hamiltonian dynamics. The fact that conservation of information plays a crucial role here is exemplified by the following exercise.

Exercise C.2 Consider a master equation $d_t\boldsymbol{p}(t) = R\boldsymbol{p}(t)$ described by a time-independent rate matrix R. The solution of the dynamics is given by the transition matrix e^{Rt}. Thus, the dynamics is clearly *invertible* as we can associate to each final state $\boldsymbol{p}(t)$ a unique initial state $\boldsymbol{p}(0)$ via $\boldsymbol{p}(0) = e^{-Rt}\boldsymbol{p}(t)$. Now, it is tempting to conclude that the dynamics of a master equation obeys time-reversal symmetry by setting $\Theta = I$ and by postulating that the time-reversed dynamics obeys the 'master equation' $d_t\boldsymbol{p}(t) = -R\boldsymbol{p}(t)$. While the above steps are mathematically correct, show that the above construction does not satisfy the requirements of time-reversal symmetry. What is wrong with the time-reversed master equation? What is the difference between Hamiltonian dynamics and master equation dynamics (compare, for example, the physical insight obtained from Liouville's theorem with respect to Theorem A.4)?

C.2 Time-Reversal Symmetry in Quantum Mechanics

We introduce time-reversal symmetry in quantum mechanics by postulating that there exists a time-reversal operator Θ such that $|\psi(0)\rangle = \Theta^{-1}U_\Theta(\tau,0)\Theta U(\tau,0)|\psi(0)\rangle$ holds for any initial state $|\psi(0)\rangle$. Here, $U(\tau,0)$ is the unitary time evolution for the forward dynamics and $U_\Theta(\tau,0)$ is the unitary time evolution of a suitably defined conjugate or backward dynamics. In words, the relation $|\psi(0)\rangle = \Theta^{-1}U_\Theta(\tau,0)\Theta U(\tau,0)|\psi(0)\rangle$ says that, if we let an arbitary state evolve for a time τ, time-reverse it, let it evolve for a time τ using the conjugate dynamics and apply the inverse time-reversal, then we end up with the same state. Since this relation is supposed to hold for any initial state, we can also write it as an operator identity:

$$\boxed{\Theta^{-1}U_\Theta(\tau,0)\Theta U(\tau,0) = I.} \tag{C.17}$$

How to define Θ and U_Θ remains, of course, the open question.

To approach it, we consider the simplest situation first and rely on some classical intuition. Consider a single particle with mass m moving in a potential V with Hamiltonian $H = P^2/2m + V(X)$, where X and P are the position and momentum operators, respectively. For this case we found classically that the time-reversal operator is an involution, $\Theta^{-1} = \Theta$, and we have good reasons to believe that the conjugate dynamics is identical to the forward dynamics, i.e. $U_\Theta(\tau) = U(\tau) = e^{-iH\tau/\hbar}$. We use this as our hypothesis now and investigate its consequences. Expanding eqn (C.17) for small τ, we arrive at

$$I + \Theta(-iH\tau\Theta - \Theta iH\tau) + \mathcal{O}(\tau^2) = I, \tag{C.18}$$

where we used $\Theta^2 = I$. From this equation we obtain the condition $-iH\tau\Theta = \Theta iH\tau$. Now, if Θ was a linear operator, the last condition reduces to $-H\Theta = \Theta H$, i.e. the time-reversal operator anti-commutes with the Hamiltonian. This conclusion, however, causes trouble because one immediately confirms that for any eigenstate $|E\rangle$ of the Hamiltonian with eigenenergy $E > 0$ there exists another eigenstate $\Theta|E\rangle$ with negative eigenenergy $-E$. This is problematic because we expect the spectrum of every reasonable physical Hamiltonian to be *bounded from below* (otherwise it would be thermodynamically unstable and we could draw an infinite amount of energy from it). Thus, our assumption that Θ is a linear operator must have been wrong. Therefore, we consider the alternative possibility that Θ is *anti-linear*, by which we mean that $\Theta i = -i\Theta$. With this assumption the condition $-iH\tau\Theta = \Theta iH\tau$ reduces to $H\Theta = \Theta H$, i.e. the time-reversal operator commutes with the Hamiltonian. This, indeed, does not give rise to any paradoxical situations for the spectrum of the Hamiltonian.

Anti-linear operators are somewhat bizarre objects—they cannot be represented by a matrix and typically one does not encounter them in quantum mechanics. To become familiar with them, we consider a simple example, namely the complex conjugation operator denoted by K. It is clear that $K^2 = I$, but K is not yet well defined if we do not specify in which basis it causes complex conjugation. Within the context of time-reversal symmetry and for the example considered above of a particle in a potential V, the most reasonable choice turns out to be complex conjugation with respect to the coordinate representation of the wavefunction. Thus, we set $\Theta = K$ and, by expanding any state in the coordinate representation, $|\psi\rangle = \int dx\langle x|\psi\rangle|x\rangle = \int dx\psi(x)|x\rangle$, we define K as $K|\psi\rangle = \int dx\psi(x)^*|x\rangle$. The motivation for this choice comes from the fact that the position and momentum operators X and P then transform as expected from the classical case (C.4):

$$\Theta X\Theta = X, \quad \Theta P\Theta = -P. \tag{C.19}$$

In this context, X is said to be an *even* observable and P an *odd* observable. Proving $\Theta X\Theta = X$ is simple. To prove $\Theta P\Theta = -P$, consider a momentum eigenstate $\psi_k(x) = e^{ikx/\hbar}$ with eigenvalue k in the coordinate representation (this is a plane wave). We then obtain the chain of equalities

$$\Theta P\Theta\psi_k(x) = \Theta Pe^{-ikx/\hbar} = \Theta(-k)e^{-ikx/\hbar} = -k\psi_k(x) = -p\psi_k(x), \tag{C.20}$$

where we used that the momentum operator has the coordinate representation $P = -i\hbar\frac{\partial}{\partial x}$. Since the $\psi_k(x)$ form an (over)complete set of basis vectors, we can conclude from eqn (C.20) that $\Theta P\Theta = -P$. Furthermore, we see that a Hamiltonian of the form

$H = P^2/2m + V(X)$ obeys $\Theta H \Theta = H$ and, hence, it is invariant under time-reversal. This also allows us to easily prove eqn (C.17) for $U_\Theta(\tau) = U(\tau)$ and $\Theta^{-1} = \Theta$:

$$\Theta e^{-iH\tau/\hbar} \Theta e^{-iH\tau/\hbar} = \Theta^2 e^{iH\tau/\hbar} e^{-iH\tau/\hbar} = I. \tag{C.21}$$

Unfortunately, the time-reversal operator Θ is not always given by complex conjugation in the position representation. It has, however, a couple of general features. The first is related to anti-linearity and is called *anti-unitarity*. It is defined by the requirement that for any two states $|\psi\rangle$ and $|\phi\rangle$

$$\langle \Theta\psi | \Theta\phi \rangle = \langle \psi | \phi \rangle^* = \langle \phi | \psi \rangle. \tag{C.22}$$

This property ensures that $|\langle \Theta\psi | \Theta\phi \rangle| = |\langle \psi | \phi \rangle|$, i.e. all probabilities are left unchanged. To become further acquainted with the 'bizarreness' of anti-unitarity, the reader is asked to do the following three small exercises.

Exercise C.3 Show that anti-unitarity implies anti-linearity.

Exercise C.4 Show for any operator O that $\mathrm{tr}\{\Theta O \Theta^{-1}\} = \mathrm{tr}\{O\}^* = \mathrm{tr}\{O^\dagger\}$. *Hint:* You can use that if Θ is anti-unitary then Θ^{-1} is also.

Exercise C.5 Show for any observable O that the time-reversed observable $\Theta O \Theta^{-1}$ is also an observable, i.e. it is Hermitian. *Hint:* Show this directly by confirming $\langle \psi | \Theta O \Theta^{-1} \phi \rangle = \langle \Theta O \Theta^{-1} \psi | \phi \rangle$ for any $|\psi\rangle$ and $|\phi\rangle$. We remark that it is unclear how and never necessary to define the Hermitian conjugate of the anti-unitary operator Θ.

Furthermore, show for any observable O that $\Theta O \Theta^{-1}$ has the same spectrum as O, i.e. the same eigenvalues but not necessarily the same eigenvectors.

It is clear from definition (C.22) that the product of two anti-unitary operators is unitary. This insight makes the following result intuitively appealing. Namely, every anti-unitary operator can be written in the standard form

$$\Theta = UK, \tag{C.23}$$

where K denotes as above complex conjugation in some fixed basis and U is a unitary operator. We will, however, not prove eqn (C.23) here. Furthermore, we demand that applying Θ twice to a wave function should give back the same wave function apart from a phase factor: $\Theta^2|\psi\rangle = e^{i\varphi}|\psi\rangle$. This requirement leads to an interesting conclusion. First, from $\Theta^2 = UKUK = e^{i\varphi}$ one derives $U^* = KUK = e^{i\varphi}U^\dagger = e^{i\varphi}(U^*)^T$. Iterating the last relation once, we obtain $U^* = e^{i\varphi}[e^{i\varphi}(U^*)^T]^T = e^{2i\varphi}U^*$. Hence, $e^{2i\varphi} = 1$, which implies $e^{i\varphi} = \pm 1$. Thus, we arrive at the conclusion

$$\Theta^2 = \pm I. \tag{C.24}$$

The case $\Theta^2 = I$ has already appeared above, but as we will see now the case $\Theta^2 = -I$ becomes important to describe time-reversal of a spin 1/2 particle.

From eqn (C.19) we infer that angular momentum $\mathbf{J} = \mathbf{X} \times \mathbf{P}$ of a particle moving in three dimensions is odd under time-reversal: $\Theta \mathbf{J} \Theta = -\mathbf{J}$. By analogy, we postulate the same for the spin: $\Theta\boldsymbol{\sigma}\Theta = -\boldsymbol{\sigma}$, where $\boldsymbol{\sigma} = (\sigma_x, \sigma_y, \sigma_z)$ is the vector of Pauli matrices. If $\Theta = K$, this is not possible independent of the basis that we choose for K. We therefore set $\Theta = UK$ and assume that K denotes complex conjugation in the coordinate representation *and* in the σ_z representation, where the Pauli matrices take on the standard form:

$$\sigma_x = \begin{pmatrix} 0 & 1 \\ 1 & 0 \end{pmatrix}, \quad \sigma_y = \begin{pmatrix} 0 & -i \\ i & 0 \end{pmatrix}, \quad \sigma_z = \begin{pmatrix} 1 & 0 \\ 0 & -1 \end{pmatrix}. \tag{C.25}$$

In this representation the condition $\Theta\boldsymbol{\sigma}\Theta = -\boldsymbol{\sigma}$ yields three equations:

$$\Theta\sigma_x\Theta^{-1} = U\sigma_x U^{-1} \quad = -\sigma_x \quad \Leftrightarrow \quad \{U, \sigma_x\} = 0, \tag{C.26}$$

$$\Theta\sigma_y\Theta^{-1} = -U\sigma_y U^{-1} = -\sigma_y \quad \Leftrightarrow \quad [U, \sigma_y] = 0, \tag{C.27}$$

$$\Theta\sigma_z\Theta^{-1} = U\sigma_z U^{-1} \quad = -\sigma_z \quad \Leftrightarrow \quad \{U, \sigma_z\} = 0. \tag{C.28}$$

Now, every complex 2×2 matrix can be written as $U = \alpha\sigma_x + \beta\sigma_y + \gamma\sigma_z + \delta I$ with $\alpha, \beta, \gamma, \delta \in \mathbb{C}$. Inserting this ansatz in the above three equations and evaluating the (anti-)commutators reveals that $\alpha = \delta = 0$ (from the first equation) and $\gamma = 0$ (from the second equation). The third equation is then automatically satisfied and we are left with $U = \beta\sigma_y$. The parameter β can be chosen freely as long as U is unitary, which implies $|\beta| = 1$. The conventional choice is $\beta = i$, such that

$$\Theta = i\sigma_y K = e^{i\pi\sigma_y/2} K. \tag{C.29}$$

One easily confirms that $\Theta^2 = -I$. Moreover, for a system with N spin 1/2 particles the time-reversal operator becomes $\Theta = \exp[i\pi(\sigma_y^{(1)} + \cdots + \sigma_y^{(N)})/2]K$ and we have $\Theta^2 = I$ if N is even and $\Theta^2 = -I$ if N is odd.

It is instructive to consider the distinction between systems with $\Theta^2 = I$ and $\Theta^2 = -I$ a little more in detail. We start with $\Theta^2 = I$ and assume that $[\Theta, H] = 0$. As the next exercise shows, it is then possible to always write the Hamiltonian as a real-valued matrix *without* explicit knowledge of the energy eigenbasis.

Exercise C.6 Show that any Hamiltonian that obeys $[\Theta, H] = 0$ for an anti-unitary operator Θ with $\Theta^2 = I$ can be given a real matrix representation without knowing the eigenbasis (remember that, in general, a Hamiltonian matrix has *complex* entries although its eigenvalues are always real since $H = H^\dagger$). For this purpose start with an arbitrary vector $|\phi_1\rangle$ and complex number a_1 and set $|\psi_1\rangle = a_1|\phi_1\rangle + \Theta a_1|\phi_1\rangle$, which is clearly Θ-invariant: $\Theta|\psi_1\rangle = |\psi_1\rangle$. Next, take a second vector $|\phi_2\rangle$ *orthogonal to* $|\psi_1\rangle$ and set $|\psi_2\rangle = a_2|\phi_2\rangle + \Theta a_2|\phi_2\rangle$. Show that $\langle\psi_2|\psi_1\rangle = 0$. Continued application of this recipe results in a set of basis vectors $\{|\psi_n\rangle\}$ that obey $\langle\psi_m|\psi_n\rangle = \delta_{m,n}$ where an appropriate choice of a_n guarantees normalization. Finally, show that the Hamiltonian is real in that basis by proving that $H_{mn} = \langle\psi_m|H|\psi_n\rangle = H_{mn}^*$. Hamiltonians that can be given such a real-valued matrix representation are said to possess *time-reversal invariance* or *time-reversal symmetry.* This notion should not be confused with our notion of time-reversal symmetry: a Hamiltonian that does not obey $[\Theta, H] = 0$ for an anti-unitary operator Θ with $\Theta^2 = I$ also gives rise to a time-reversal symmetry of the *dynamics*, as defined at the beginning of this appendix.

As a consequence of the previous exercise, we can choose without loss of generality $|E\rangle = \Theta|E\rangle$ for any energy eigenstate $|E\rangle$ if $[\Theta, H] = 0$ and $\Theta^2 = I$. In contrast, let us now consider the second option: $[\Theta, H] = 0$ with $\Theta^2 = -I$. We then confirm

$$\langle E|\Theta E\rangle = \langle \Theta E|\Theta^2 E\rangle^* = -\langle \Theta E|E\rangle^* = -\langle E|\Theta E\rangle. \tag{C.30}$$

Hence, $\langle E|\Theta E\rangle = 0$, which implies that $|E\rangle$ and $\Theta|E\rangle$ are two *orthogonal* states with the same energy eigenvalue. Thus, all eigenvalues of a Hamiltonian obeying $[\Theta, H] = 0$ with $\Theta^2 = -I$ are doubly degenerate. This result is known as *Kramer's degeneracy.*

After this excursion into how to define the time-reversal operator Θ for quantum systems, let us return to the dynamical picture. Consider a Hamiltonian $H(\lambda_t)$ with some driving protocol λ_t, $t \in [0, \tau]$. The time evolution operator for the forward time evolution can be approximated as

$$U(\tau, 0) \approx e^{-iH(\lambda_{N-1})\delta t/\hbar} \dots e^{-iH(\lambda_0)\delta t/\hbar}, \tag{C.31}$$

where we divided the time interval into steps of size $\delta t = \tau/N$ and implicitly kept in mind the limit $N \to \infty$ in which eqn (C.31) becomes exact. From our basic equation (C.17) we can then infer that the conjugate dynamics is described by the time evolution operator

$$\begin{aligned} U_\Theta(\tau, 0) &= \Theta U^\dagger(\tau, 0)\Theta^{-1} \\ &= \Theta e^{iH(\lambda_0)\delta t/\hbar} \dots e^{iH(\lambda_{N-1})\delta t/\hbar}\Theta^{-1} \\ &= e^{-iH_\Theta(\lambda_0)\delta t/\hbar} \dots e^{-iH_\Theta(\lambda_{N-1})\delta t/\hbar}. \end{aligned} \tag{C.32}$$

In the last equation, we defined $H_\Theta(\lambda_t) \equiv \Theta H(\lambda_t)\Theta^{-1}$. Thus, as expected from the classical case, the conjugate dynamics is defined by changing the protocol backwards in time from λ_τ to λ_0 with respect to the time-reversed Hamiltonian. Again in unison with the classical case, the time-reversed Hamiltonian describing a particle in an external magnetic field B can be obtained by simply flipping the magnetic field: $\Theta H(B, \lambda_t)\Theta^{-1} = H(-B, \lambda_t)$. Note, however, that it is *not* true that $\Theta B \Theta^{-1} = -B$. Treated as an *external* field, B is simply a real-valued parameter that remains unaffected by the time-reversal operator Θ.

Finally, in the last exercise we prove and generalize a statement that we used in Section 2.3 to derive the important relation of *local detailed balance.*

Exercise C.7 Consider first two observables $X = \sum_x x\Pi(x)$ and $Y = \sum_y y\Pi(y)$ and their time-reversal $\Theta X\Theta^{-1} = \sum_x x\Pi_\Theta(x)$ and $\Theta Y \Theta^{-1} = \sum_y y\Pi_\Theta(y)$. Show the validity of the following identity, which is also known as **microreversibility**:

$$\mathrm{tr}\{\Pi(y)U(t,0)\Pi(x)U^\dagger(t,0)\} = \mathrm{tr}\{\Pi_\Theta(x)U_\Theta(t,0)\Pi_\Theta(y)U_\Theta^\dagger(t,0)\}. \tag{C.33}$$

Now, recall that the rate to jump from a coarse-grained state x' to x under the assumption of time-scale separation was computed in eqn (2.41) and, reads:

$$R_{x,x'} = \frac{1}{\delta t}\frac{1}{V_{E,x'}}\mathrm{tr}\{\Pi(E,x)U(\delta t)\Pi(E,x')U^\dagger(\delta t)\}. \tag{C.34}$$

We now consider the time-reversed process. The rate to jump from a time-reversed coarse-grained state x_Θ to x'_Θ under the assumption of time-scale separation becomes

$$R^\Theta_{x'_\Theta,x_\Theta} = \frac{1}{\delta t}\frac{1}{V_{E,x}}\mathrm{tr}\{\Pi_\Theta(E,x')U_\Theta(\delta t)\Pi_\Theta(E,x)U^\dagger_\Theta(\delta t)\}. \tag{C.35}$$

Note that the number of microstates remains unchanged by the time-reversal operator: $V_{E,x_\Theta} = \mathrm{tr}\{\Theta\Pi_{E,x}\Theta^{-1}\} = \mathrm{tr}\{\Pi_{E,x}\}^* = V_{E,x}$. From eqn (C.33) it follows directly that

$$\frac{R_{x,x'}}{R^\Theta_{x'_\Theta,x_\Theta}} = \frac{V_{E,x}}{V_{E,x'}} = \exp\left[\frac{S_B(E,x) - S_B(E,x')}{k_B}\right], \tag{C.36}$$

which is the local detailed balance relation in full generality. The conventionally considered case from Section 2.3 assumes that the observable is even, $X = \Theta X\Theta^{-1}$, and the Hamiltonian has time-reversal invariance, $H = \Theta H\Theta^{-1}$. Under these circumstances $R^\Theta_{x'_\Theta,x_\Theta} = R_{x',x}$ and eqn (C.36) reduces to the conventional local detailed balance condition (2.43).

C.3 The Arrow of Time

After the mathematical treatment of time-reversal symmetry, we discuss some physical and philosophical implications. In fact, the world around us does not seem to obey time-reversal symmetry: ageing, the ability to remember the past but not the future and the fact that the rich get richer and the poor poorer are three simple examples that demonstrate our inability to reverse the **arrow of time** in our everyday life. Evolution, the second law of thermodynamics and the expansion of the universe are three further examples that seem to show a clear arrow of time. But how can such an arrow of time emerge from underlying equations of motion that obey time-reversal symmetry?

This apparent paradox was articulated by Loschmidt and Zermelo, who critically questioned Boltzmann's attempt in 1872 to derive the second law of thermodynamics on a purely mechanical and microscopic basis. Loschmidt in 1876 pointed out that a purely mechanical derivation to reach (and maintain) a stationary (i.e. equilibrium) state from some arbitrary initial state contradicts time-reversal symmetry. This argument is known as Loschmidt's *Umkehreinwand*, i.e. Loschmidt's *objection* based on (time-)*reversal*. In addition, Zermelo in 1896 pointed out that Poincaré's recurrence theorem forbids the derivation of a stationary state from an initial non-stationary state as the state of an isolated mechanical system must eventually return (very close) to its initial state. This argument is known as Zermelo's *Wiederkehreinwand*, i.e. Zermelo's *objection* based on *recurrences*.

However, Loschmidt's and Zermelo's arguments can be refuted by recalling how *unlikely* these arguments are for macroscopic systems. The vast majority of states of a macroscopic isolated mechanical system with a fixed energy closely resemble a maximum entropy state and the Poincaré recurrence time becomes immeasurably large (also compare with Section 1.3). After realizing this, Boltzmann emphasized the statistical character of the second law (as also Maxwell did) in his replies.

Nevertheless, even more than a hundred years after these objections, controversies about the origin of the arrow of time are still not settled. As the major purpose of this book is to clarify the foundations of quantum and classical stochastic thermodynamics, which necessarily also includes various arguments used to 'derive' the second law of thermodynamics, it seems worthwhile pointing out a few interesting observations—without any intention to provide a complete resolution.

A first question one could wonder about is whether the mathematical construct of time-reversal symmetry actually describes well the philosophical or human idea we associate with it. In fact, often time-reversal symmetry is epitomized by the transformation '$t \mapsto -t$', but as we saw above this very simple mapping does not correctly describe time-reversal symmetry, which is a much more complicated construction in general. Perhaps because of this reason some people prefer to speak about 'microreversibility' instead of 'time-reversal symmetry' and Wigner himself noted that it would be more appropriate to speak about 'reversal of the direction of motion'. Putting this question aside, let us proceed by assuming that the mathematical construct above correctly describes time-reversal symmetry.

Next, one could wonder whether the second law of (phenomenological) thermodynamics can actually be used to prove the arrow of time. In fact, the second law of thermodynamics is probably *the* outstanding physical law that does not possess time-reversal symmetry. However, remember that the laws of phenomenological thermodynamics are *postulates* or *axioms* in accordance with human experience (in the same way as Newton's or Schrödinger's equations are axioms in classical or quantum mechanics). Thus, saying that the second law proves the arrow of time is a tautology in so far as it simply shifts the problem from proving the arrow of time to proving the increase in entropy. At least, however, it gives us evidence that the arrow of time is linked to another physical concept, namely *entropy*. Thus, let us accept the idea that the second law of thermodynamics relates to the arrow of time.

But now, when we try to derive the second law from an underlying microscopic theory, we are back to Boltzmann's problem. Undeniably, our knowledge of how to 'derive' the second law of thermodynamics has improved since Boltzmann and many ways to do so are presented in this book. However, all the ways that the author is aware of share one common feature: the increase in entropy is only proven with respect to some special initial state. Thus, whereas we might nowadays be able to prove the second law under increasingly mild assumptions, we are *not* yet able to prove the arrow of time since all our derivations rely on a particularly chosen boundary condition. To put it differently: in this book we could replace the original Hamiltonian with its time-reversed counterpart, yet we would still be able to prove the second law in the same way as before. The second laws in this book do *not* explain the arrow of time.

This insight strongly suggests that the second law is *not a consequence of the microsopic equations of motion*, as already clearly stated by Boltzmann: 'The Second Law can never be proved mathematically by means of the equations of dynamics alone'. But of what is the second law then a consequence? One of Boltzmann's ideas, which he attributes to his assistant Schuetz, was the following. First, they assumed that the entire universe is in global equilibrium. Then, since they knew about the *statistical* character of thermodynamics, they concluded that there must be local fluctuations in the entropy of the universe. If the universe is sufficiently large, the chances of the existence of a very low entropy region somewhere in the universe could be sufficiently

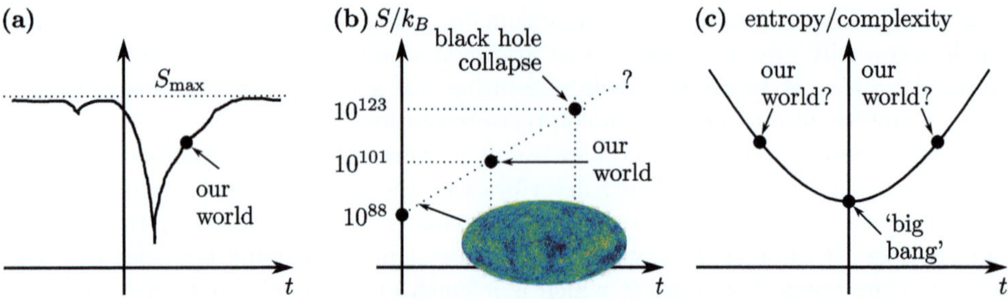

Fig. C.4 Three scenarios explaining the arrow of time in the cosmological universe. (a) An eternal and infinite universe showing a large spatiotemporal downward fluctuation in entropy away from its maximum value $S_{\max}$. (b) The big bang initialized the universe in a low entropy state of approximately $S_0 = 10^{88} k_B$ ('past hypothesis'). Shortly afterwards, the universe becomes transparent and the image of the cosmic microwave background (NASA/WMAP Science Team) shows radiation in almost perfect equilibrium (puzzle: why is this state not already close to maximum entropy in this scenario?). The entropy value $10^{123} k_B$ is computed by collapsing all matter and energy of the universe into a black hole. However, since black holes evaporate, this is not the end of the story. (c) Cosmological model that is time–reversal symmetric on large scales with a low entropy (or complexity) value at the centre (the 'big bang') and arrows of time emerging from there in both directions.

high. The fact that we experience a second law, and the fact that we even exist, is then a consequence of living in such a rare entropy fluctuation. This scenario is illustrated in part (a) of Fig. C.4.

This idea has been met with scepticism. One point of criticism is that a small local entropy fluctuation that creates only a *single* observer, whose brain gives the *impression* that we live in the current universe, is much *more likely* than a large local entropy fluctuation that creates a low entropy observer together with a low entropy Earth together with a low entropy solar system together with a low entropy galaxy, etc. The creation of such a single conscious observer by spontaneous thermal fluctuations is known as a **Boltzmann brain**.

Therefore, most researchers believe today that it is much more likely that the second law is *a consequence of the initial state of the universe right after the Big Bang* (the concept of the Big Bang was not known to Boltzmann and his contemporaries). In fact, if the thermodynamic entropy of the initial state of the universe were sufficiently low, an increase in thermodynamic entropy could be observed for a *very long* time, thus effectively explaining the second law of thermodynamics. This is also known as the **past hypothesis**. This scenario is illustrated in part (b) of Fig. C.4.

Notice that, even if the past hypothesis were true, it does not seem to explain why the *concept of time* (independent of any *arrow*) exists at all. Perhaps it turns out that the Hamiltonian of the universe is not time-reversal invariant. In fact, violations of time-reversal invariance *were observed* in systems described by electroweak interactions. Electroweak interactions could therefore explain the fundamental asymmetry between time and space. Furthermore, although the past hypothesis explains the *thermodynamic* arrow of time, it is not clear whether it also explains, for example, the biological arrow of time: Is evolution, ageing or the ability to remember the past but not the future a consequence of the second law?

We conclude by mentioning a third alternative to explain the emergence of an arrow of time on a cosmological scale, where a time–reversal symmetric universe has two branches, each locally defining its own arrow of time. This is illustrated in part (c) of Fig. C.4. In fact, it has been argued that this situation could be generic. Also doubts has been expressed about the applicability of the entropy concept to the universe as a whole because the general definition of gravitational entropy (except for the case of a black hole) remains unclear. Perhaps another concept (loosely called 'complexity' in Fig. C.4) is required to explain the arrow of time in the universe?

Further reading

Much of our knowledge about time-reversal symmetry, and symmetries in general in quantum mechanics, goes back to Wigner (1959). Finding pedagogically useful accounts of time-reversal symmetry in standard textbooks on quantum mechanics is, however, not so easy, but see, for instance, the book by Sakurai (1994) for an exception. The question of whether a Hamiltonian is invariant under time-reversal is also very important in the field of quantum chaos and part of my exposition was inspired by the book by Haake (2010).

The phrase 'arrow of time' was popularized in a book by Eddington (1928). Another historically important review about statistical mechanics, including many citations to the original references from Boltzmann, Loschmidt, Zermelo and others, can be found in the treatise by Ehrenfest and Ehrenfest (1911), which was translated into English by Moravcsik (Ehrenfest and Ehrenfest, 1959). I also directly quoted from Boltzmann (1895). Another modern and less technical account of many of the ideas I discussed above is given by Lebowitz (1993). I should also mention here the numerous research efforts showing why and how isolated quantum systems can effectively equilibrate, even if they are initialized out of equilibrium in a pure quantum state (Gemmer *et al.*, 2004; Borgonovi *et al.*, 2016; D'Alessio *et al.*, 2016; Gogolin and Eisert, 2016; Goold *et al.*, 2016; Deutsch, 2018; Mori *et al.*, 2018). To the best of my knowledge, however, this complementary line of research also cannot establish an asymmetry in time: isolated quantum systems initialized out of equilibrium equilibrate in both directions of time. Cosmological considerations about entropy and the arrow of time can be found at various places. The name 'past hypothesis' was coined by Albert (2000). The entropy estimates in part (b) of Fig. C.4 are due to Penrose (1989). A time–symmetric universe on large scales explaining an entropic arrow of time was suggested by Carroll and Chen (2004). Criticsim about the use of the entropy concept in cosmology can be found, e.g., in the article by Earman (2006). A generic explanation of a time–symmetric scenario with an arrow of time in terms of a complexity measure was found by Barbour *et al.* (2014). For further research on the question how the concept of time can emerge at all see Vaccaro (2016) and references therein.

References

Agarwalla, B. K. and Segal, D. (2018). Assessing the validity of the thermodynamic uncertainty relation in quantum systems. *Phys. Rev. B*, **98**, 155438.

Albert, D. Z. (2000). *Time and Chance.* Harvard University Press.

Alemany, A., Mossa, A., Junier, I., and Ritort, F. (2012). Experimental free-energy measurements of kinetic molecular states using fluctuation theorems. *Nat. Phys.*, **8**, 688–694.

Alicki, R. (1979). The quantum open system as a model of the heat engine. *J. Phys. A*, **12**, L103.

Allahverdyan, A. E. and Nieuwenhuizen, Th. M. (2005). Fluctuations of work from quantum subensembles: the case against quantum work-fluctuation theorems. *Phys. Rev. E*, **71**, 066102.

Allahverdyan, A. E., Balian, R., and Nieuwenhuizen, Th. M. (2004). Maximal work extraction from finite quantum systems. *Europhys. Lett.*, **67**, 565.

Allahverdyan, A. E., Hovhannisyan, K. V., Janzing, D., and Mahler, G. (2011). Thermodynamic limits of dynamic cooling. *Phys. Rev. E*, **84**, 041109.

Andrieux, D., Gaspard, P., Monnai, T., and Tasaki, S. (2009). The fluctuation theorem for currents in open quantum systems. *New J. Phys.*, **11**, 043014.

Barato, A. C. and Seifert, U. (2015). Thermodynamic uncertainty relation for biomolecular processes. *Phys. Rev. Lett.*, **114**, 158101.

Barbour, J., Koslowski, T., and Mercati, F. (2014). Identification of a Gravitational Arrow of Time. *Phys. Rev. Lett.*, **113**, 181101.

Barra, F. (2015). The thermodynamic cost of driving quantum systems by their boundaries. *Sci. Rep.*, **5**, 14873.

Bassett, I. M. (1978). Alternative derivation of the classical second law of thermodynamics. *Phys. Rev. A*, **18**, 2356–2360.

Benenti, G., Saito, K., and Casati, G. (2011). Thermodynamic bounds on efficiency for systems with broken time-reversal symmetry. *Phys. Rev. Lett.*, **106**, 230602.

Benenti, G., Casati, G., Saito, K., and Whitney, R. S. (2017). Fundamental aspects of steady-state conversion of heat to work at the nanoscale. *Phys. Rep.*, **694**, 1–124.

Bennett, C. H. (1982). The thermodynamics of computation—a review. *Int. J. Theor. Phys.*, **21**, 905.

Bergmann, P. G. and Lebowitz, J. L. (1955). New approach to nonequilibrium processes. *Phys. Rev.*, **99**, 578–587.

Bérut, A., Arakelyan, A., Petrosyan, A., Ciliberto, S., Dillenschneider, R., and Lutz, E. (2012). Experimental verification of Landauer's principle linking information and thermodynamics. *Nature (London)*, **483**, 187–189.

Binder, F., Correa, L. A., Gogolin, C., Anders, J., and Adesso, G. (eds) (2018). *Thermodynamics in the Quantum Regime: Fundamental Aspects and New Directions.* Springer Nature Switzerland, Cham.

Bochkov, G. N. and Kuzovlev, Yu. E. (1977). General theory of thermal fluctuations in nonlinear systems. *Sov. Phys. JETP*, **45**, 125–130.

Bochkov, G. N. and Kuzovlev, Yu. E. (2013). Fluctuation-dissipation relations. Achievements and misunderstandings. *Phys.-Usp.*, **56**, 590.

Boltzmann, L. (1895). On certain questions of the theory of gases. *Nature*, **51**, 413–415.

Borgonovi, F., Izrailev, F.M., Santos, L. F., and Zelevinsky, V. G. (2016). Quantum chaos and thermalization in isolated systems of interacting particles. *Phys. Rep.*, **626**, 1–58.

Breuer, H.-P. and Petruccione, F. (2002). *The Theory of Open Quantum Systems.* Oxford University Press, Oxford.

Bulnes Cuetara, G., Esposito, M., and Schaller, G. (2016). Quantum Thermodynamics with Degenerate Eigenstate Coherences, *Entropy*, **18**, 447.

Byrd, M. S. and Khaneja, N. (2003). Characterization of the positivity of the density matrix in terms of the coherence vector representation. *Phys. Rev. A*, **68**, 062322.

Cadney, J., Huber, M., Linden, N., and Winter, A. (2014). Inequalities for the ranks of multipartite quantum states. *Linear Algebra Appl.*, **452**, 153–171.

Campaioli, F., Pollock, F. A., and Vinjanampathy, S. (2018). Quantum batteries, in F. Binder, L. A. Correa, C. Gogolin, J. Anders, and G. Adesso, eds, *Thermodynamics in the Quantum Regime*. Springer, Cham.

Campisi, M., Talkner, P., and Hänggi, P. (2009). Fluctuation theorem for arbitrary open quantum systems. *Phys. Rev. Lett.*, **102**, 210401.

Campisi, M., and Fazio, R. (2016). The power of a critical heat engine. *Nat. Comm.*, **7**, 11895.

Carmichael, H. J. (1993). *An Open Systems Approach to Quantum Optics.* Lecture Notes, Springer, Berlin.

Carroll, S. M. and Chen, J. (2004). Spontaneous inflation and the origin of the arrow of time. *arXiv:hep-th/0410270.*

Chida, K., Desai, S., Nishiguchi, K., and Fujiwara, A. (2017). Power generator driven by Maxwell's demon. *Nat. Commun.*, **8**, 15310.

Chiribella, G., D'Ariano, G. M., and Perinotti, P. (2009). Theoretical framework for quantum networks. *Phys. Rev. A*, **80**, 022339.

Choi, M.-D. (1975). Completely positive linear maps on complex matrices. *Lin. Alg. Appl.*, **10**, 285–290.

Ciliberto, S. (2017). Experiments in stochastic thermodynamics: short history and perspectives. *Phys. Rev. X*, **7**, 021051.

Clausius, R. (1865). Ueber verschiedene für die Anwendung bequeme Formen der Hauptgleichungen der mechanischen Wärmetheorie. *Ann. Phys.*, **201**, 353–400.

Collin, D., Ritort, F., Jarzynski, C., Smith, S. B., Tinoco, I., and Bustamante, C. (2005). Verification of the Crooks fluctuation theorem and recovery of RNA folding free energies. *Nature (London)*, **437**, 231–234.

Cover, T. M. and Thomas, J. A. (1991). *Elements of Information Theory.* John Wiley & Sons, New York.

Cresser, J. (2019). Time-reversed quantum trajectory analysis of micromaser correlation properties and fluctuation relations. *Phys. Scr.*, **94**, 034005.

Crooks, G. E. (1998). Nonequilibrium measurements of free energy differences for microscopically reversible Markovian systems. *J. Stat. Phys.*, **90**, 1481–1487.

Crooks, G. E. (1999). Entropy production fluctuation theorem and the nonequilibrium work relation for free energy differences. *Phys. Rev. E*, **60**, 2721–2726.

D'Alessio, L., Kafri, Y., Polkovnikov, A., and Rigol, M. (2016). From quantum chaos and eigenstate thermalization to statistical mechanics and thermodynamics. *Adv. Phys.*, **65**, 239–362.

Davies, E. B. (1974). Markovian master equations. *Commun. Math. Phys.*, **39**, 91–110.

Davies, E. B. (1976). Markovian master equations. II. *Math. Ann.*, **219**, 147–158.

de Vega, I. and Alonso, D. (2017). Dynamics of non-Markovian open quantum systems. *Rev. Mod. Phys.*, **89**, 015001.

Deffner, S. and Campbell, S. (2017). Quantum speed limits: from Heisenberg's uncertainty principle to optimal quantum control. *J. Phys. A*, **50**(45), 453001.

Deffner, S. and Jarzynski, C. (2013). Information processing and the second law of thermodynamics: an inclusive, Hamiltonian approach. *Phys. Rev. X*, **3**, 041003.

Deutsch, J. M. (2018). Eigenstate thermalization hypothesis. *Rep. Prog. Phys.*, **81**, 082001.

Donald, M.J. (1987). Free Energy and the Relative Entropy, *J. Stat. Phys.* **49**, 81.

Earman, J. (2006). The 'past hypothesis': not even false. *Stud. Hist. Phil. Sci. B*, **37**, 399–430.

Eddington, A. (1928). *The Nature of the Physical World: The Gifford Lectures.* Cambridge University Press, Cambridge.

Ehrenfest, P. and Ehrenfest, T. (1911). *Begriffliche Grundlagen der statistischen Auffassung in der Mechanik*, pp. 3–90. Enzyklopädie der mathematischen Wissenschaften mit Einschluß ihrer Anwendungen. Teubner, Leipzig.

Ehrenfest, P. and Ehrenfest, T. (1959). *The Conceptual Foundations of the Statistical Approach in Mechanics.* Dover Pub., New York.

Eisert, J., Cramer, M., and Plenio, M. B. (2010). Colloquium: Area laws for the entanglement entropy. *Rev. Mod. Phys.*, **82**, 277–306.

Elouard, C., Herrera-Martií, D. A., Clusel, M., and Auffèves, A. (2017). The role of quantum measurement in stochastic thermodynamics. *npj Quantum Inf.*, **3**, 9.

Esposito, M. (2012). Stochastic thermodynamics under coarse graining. *Phys. Rev. E*, **85**, 041125.

Esposito, M., Harbola, U., and Mukamel, S. (2009). Nonequilibrium fluctuations, fluctuation theorems and counting statistics in quantum systems. *Rev. Mod. Phys.*, **81**, 1665.

Esposito, M., Lindenberg, K., and Van den Broeck, C. (2010). Entropy production as correlation between system and reservoir. *New J. Phys.*, **12**, 013013.

Flindt, C., Fricke, C., Hohls, F., Novotný, T., Netočný, K., Brandes, T., and Haug, R. J. (2009). Universal oscillations in counting statistics. *Proc. Natl. Acad. Sci. USA*, **106**, 10116–10119.

Gallavotti, G. (1999). Statistical Mechanics: A Short Treatise. Springer-Verlag, Berlin Heidelberg.

Gaveau, B. and Schulman, L. S. (1997). A general framework for non-equilibrium phenomena: the master equation and its formal consequences. *Phys. Lett. A*, **229**, 347–353.

Gavrilov, M., Chétrite, R., and Bechhoefer, J. (2017). Direct measurement of weakly nonequilibrium system entropy is consistent with Gibbs-Shannon form. *Proc. Natl. Acad. Sci. USA*, **114**, 11097–11102.

Gemmer, J., Michel, M., and Mahler, G. (2004). *Quantum Thermodynamics.* Lecture Notes in Physics, Springer, Heidelberg.

Gogolin, C. and Eisert, J. (2016). Equilibration, thermalisation, and the emergence of statistical mechanics in closed quantum systems. *Rep. Prog. Phys.*, **79**, 056001.

Gomez-Marin, A., Parrondo, J. M. R., and Van den Broeck, C. (2008). Lower bounds on dissipation upon coarse graining. *Phys. Rev. E* , **78**, 011107.

Goold, J., Huber, M., Riera, A., del Rio, L., and Skrzypzyk, P. (2016). The role of quantum information in thermodynamics—a topical review. *J. Phys. A*, **49**, 143001.

Gurvitz, S. A. and Prager, Ya. S. (1996). Microscopic derivation of rate equations for quantum transport. *Phys. Rev. B*, **53**, 15932.

Guryanova, Y., Friis, N., and Huber, M. (2020). Ideal Projective Measurements Have Infinite Resource Costs. *Quantum*, **4**, 222.

Haake, F. (2010). *Quantum Signatures of Chaos.* Springer-Verlag, Berlin Heidelberg.

Hartmann, F., Pfeffer, P., Höfling, S., Kamp, M., and Worschech, L. (2015). Voltage fluctuation to current converter with Coulomb-coupled quantum dots. *Phys. Rev. Lett.*, **114**, 146805.

Hartmann, M. (2006). Minimal length scales for the existence of local temperature. *Contemp. Phys.*, **47**, 89–102.

Hartmann, R. and Strunz, W. T. (2020). Accuracy assessment of perturbative master equations: embracing nonpositivity. *Phys. Rev. A*, **101**, 012103.

Hasegawa, Y. and Van Vu, T. (2019). Fluctuation theorem uncertainty relation. *Phys. Rev. Lett.*, **123**, 110602.

Holevo, A. S. (2001). *Statistical Structure of Quantum Theory.* Springer-Verlag, Berlin Heidelberg.

Holubec, V. and Ryabov, A. (2018). *Cycling tames power fluctuations near optimum efficiency. Phys. Rev. Lett.*, **121**, 120601.

Hong, J., Lambson, B., Dhuey, S., and Bokor, J. (2016). Experimental test of Landauer's principle in single-bit operations on nanomagnetic memory bits. *Sci. Adv.*, **2**, e1501492.

Horowitz, J. M. and Gingrich, T. R. (2020). Thermodynamic uncertainty relations constrain non-equilibrium fluctuations. *Nat. Phys.*, **16**, 15–20.

Hummer, G. and Szabo, A. (2001). Free energy reconstruction from nonequilibrium single-molecule pulling experiments. *Proc. Natl. Acad. Sci. USA*, **98**, 3658–3661.

Jacobs, K. (2009). Second law of thermodynamics and quantum feedback control: Maxwell's demon with weak measurements. *Phys. Rev. A*, **80**, 012322.

Jacobs, K. (2014). *Quantum Measurement Theory and its Applications.* Cambridge University Press, Cambridge.

Jamiołkowski, A. (1972). Linear transformations which preserve trace and positive semidefiniteness of operators. *Rep. Math. Phys.*, **3**, 275–278.

Jarzynski, C. (1997). Nonequilibrium equality for free energy differences. *Phys. Rev. Lett.*, **78**, 2690.

Jarzynski, C. (2004). Nonequilibrium work theorem for a system strongly coupled to a thermal environment. *J. Stat. Mech.*, P09005.

Jarzynski, C. (2007). Comparison of far-from-equilibrium work relations. *C. R. Phys.*, **8**, 495–506.

Jarzynski, C. and Wójcik, D. K. (2004). Classical and quantum fluctuation theorems for heat exchange. *Phys. Rev. Lett.*, **92**, 230602.

Josefsson, M., Svilans, A., Burke, A.M., Hoffmann, E.A., Fahlvik, S., Thelander, C., Leijnse, M., and Linke, H. (2018). A quantum-dot heat engine operating close to the thermodynamic efficiency limits. *Nat. Nanotechnol.*, **13**, 920–924.

Jun, Y., Gavrilov, M., and Bechhoefer, J. (2014). High-precision test of Landauer's principle in a feedback trap. *Phys. Rev. Lett.*, **113**, 190601.

Kemeny, J. G. and Snell, J. L. (1976). *Finite Markov Chains.* Springer-Verlag, New York.

Kliesch, M., Gogolin, C., Kastoryano, M. J., Riera, A., and Eisert, J. (2014). Locality of temperature. *Phys. Rev. X*, **4**, 031019.

Klimenko, A. Y. (2012). Teaching the third law of thermodynamics. *Open Thermodyn. J.*, **6**, 1–14.

Kohen, D., Marston, C. C., and Tannor, D. J. (1997). Phase space approach to theories of quantum dissipation. *J. Chem. Phys.*, **107**, 5236.

Kolmogorov, A. N. (2018). *Foundations of the Theory of Probability* (2nd English edn). Dover Publications, Mineola, New York.

Kondepudi, D. and Prigogine, I. (2007). *Modern Thermodynamics: From Heat Engines to Dissipative Structures.* John Wiley & Sons, Chichester.

Koski, J. V., Kutvonen, A., Khaymovich, I. M., Ala-Nissila, T., and Pekola, J. P. (2015). On-chip Maxwell's demon as an information-powered refrigerator. *Phys. Rev. Lett.*, **115**, 260602.

Kox, A. J. (2006). Confusion and clarification: Albert Einstein and Walther Nernst's heat theorem, 1911–1916. *Stud. Hist. Philos. Sci. B*, **37**, 101–114.

Kraus, K. (1983). *States, Effects and Operations: Fundamental Notions of Quantum Theory.* Springer-Verlag, Berlin Heidelberg.

Küng, B., Rössler, C., Beck, M., Marthaler, M., Golubev, D. S., Utsumi, Y., Ihn, T., and Ensslin, K. (2012). Irreversibility on the level of single-electron tunneling. *Phys. Rev. X*, **2**, 011001.

Kurchan, J. (2000). A quantum fluctuation theorem. *arXiv:cond-mat/0007360.*

Landauer, R. (1961). Irreversibility and heat generation in the computing process. *IBM J. Res. Dev.*, **5**, 183.

Landauer, R. (1998). The noise is the signal. *Nature*, **392**, 658–659.

Lebowitz, J. L. (1993). Boltzmann's entropy and time's arrow. *Phys. Today*, **46**, 32.

Lenard, A. (1978). Thermodynamical proof of the Gibbs formula for elementary quantum systems. *J. Stat. Phys.*, **19**, 575–586.

Li, L., Hall, M. J. W., and Wiseman, H. M. (2018). Concepts of quantum non-Markovianity: a hierarchy. *Phys. Rep.*, **759**, 1–51.

Lindblad, G. (1975). Completely positive maps and entropy inequalities. *Commun. Math. Phys.*, **40**, 147–151.

Lindblad, G. (1979). Non-Markovian quantum stochastic processes and their entropy. *Commun. Math. Phys.*, **65**, 281–294.

Lindblad, G. (1983). *Non-Equilibrium Entropy and Irreversibility.* D. Reidel Publishing, Dordrecht.

Liphardt, J., Dumont, S., Smith, S. B., Tinoco, I., and Bustamante, C. (2002). Equilibrium information from nonequilibrium measurements in an experimental test of Jarzynski's equality. *Science*, **296**, 1832–1835.

Liu, F. and Xi, J. (2016). Characteristic functions based on a quantum jump trajectory. *Phys. Rev. E*, **94**, 062133.

Lostaglio, M. (2019). An introductory review of the resource theory approach to thermodynamics. *Rep. Prog. Phys.*, **82**, 114001.

Macieszczak, K., Brandner, K., and Garrahan, J. P. (2018). Unified thermodynamic uncertainty relations in linear response. *Phys. Rev. Lett.*, **121**, 130601.

Maes, C. and Netočný, K. (2003). Time-reversal and entropy. *J. Stat. Phys.*, **110**, 269–310.

Martinazzo, R., Vacchini, B., Hughes, K. H., and Burghardt, I. (2011). Communication: universal Markovian reduction of Brownian particle dynamics. *J. Chem. Phys.*, **134**, 011101.

Maxwell, J. C. (1871). *Theory of Heat*. Longmans, Green, and Co., London.

McAdory, R. T. and Schieve, W. C. (1977). On entropy production in a stochastic model of open systems. *J. Chem. Phys.*, **67**(5), 1899–1903.

Merhav, N. and Kafri, Y. (2010). Statistical properties of entropy production derived from fluctuation theorems. *J. Stat. Mech.*, P12022.

Merkli, M. (2020). Quantum Markovian master equations: resonance theory shows validity for all time scales. *Ann. Phys. (N.Y.)*, **412**, 167996.

Miller, H. J. D. and Anders, J. (2017). Entropy production and time asymmetry in the presence of strong interactions. *Phys. Rev. E*, **95**, 062123.

Milz, S. and Modi, K. (2021). *Quantum Stochastic Processes and Quantum non-Markovian Phenomena*. PRX Quantum **2**, 030201.

Milz, S., Egloff, D., Taranto, P., Theurer, T., Plenio, M. B., Smirne, A., and Huelga, S. F. (2020*a*). When is a non-Markovian quantum process classical? *Phys. Rev. X*, **10**, 041049.

Milz, S., Sakuldee, F., Pollock, F. A., and Modi, K. (2020*b*). Kolmogorov extension theorem for (quantum) causal modelling and general probabilistic theories. *Quantum*, **4**, 255.

Mitchison, M. T. and Plenio, M. B. (2018). Non-additive dissipation in open quantum networks out of equilibrium. *New J. Phys.*, **20**(3), 033005.

Mohammady, M. H. and Romito, A. (2019). Efficiency of a cyclic quantum heat engine with finite-size baths. *Phys. Rev. E*, **100**, 012122.

Mori, T., Ikeda, T. N., Kaminishi, E., and Ueda, M. (2018). Thermalization and prethermalization in isolated quantum systems: a theoretical overview. *J. Phys. B*, **51**, 112001.

Muschik, W. (1977). Empirical foundation and axiomatic treatment of non-equilibrium temperature. *Arch. Ration. Mech. Anal.*, **66**, 379-401.

Nazarov, Y. V. and Blanter, Y. M. (2009). *Quantum Transport: Introduction to Nanoscience*. Cambridge University Press, Cambridge.

Nernst, W. (1906). Ueber die Berechnung chemischer Gleichgewichte aus thermischen Messungen. *Nachr. Kgl. Ges. Wiss. Gött.*, **1**, 1–40.

Nielsen, M. A. and Chuang, I. L. (2000). *Quantum Computation and Quantum Information*. Cambridge University Press, Cambridge.

Nitzan, A. (2006). *Chemical Dynamics in Condensed Phases: Relaxation, Transfer, and Reactions in Condensed Molecular Systems*. Oxford University Press, Oxford.

Nordsieck, Jr A., Lamb, W. E., and Uhlenbeck, G. E. (1940). On the theory of cosmic-ray showers I the Furry model and the fluctuation problem. *Physica*, **7**, 344.

Orlov, A. O., Lent, C. S., Thorpe, C. C., Boechler, G. P., and Snider, G. L. (2012). Experimental test of Landauer's principle at the sub-k_bT level. *Jpn. J. Appl. Phys.*, **51**, 06FE10.

Parrondo, J. M. R., Horowitz, J. M., and Sagawa, T. (2015). Thermodynamics of information. *Nat. Phys.*, **11**, 131–139.

Pearl, J. (2009). *Causality: Models, Reasoning and Inference.* Cambridge University Press, New York.

Penrose, O. (1970). *Foundations of Statistical Mechanics.* Pergamon Press, Oxford.

Penrose, R. (1989). *The Emperor's New Mind..* Oxford University Press, Oxford.

Perarnau-Llobet, M., Bäumer, E., Hovhannisyan, K. V., Huber, M., and Acin, A. (2017). No-go theorem for the characterization of work fluctuations in coherent quantum systems. *Phys. Rev. Lett.*, **118**, 070601.

Peterson, J. P. S., Sarthour, R. S., Souza, A. M., Oliveira, I. S., Goold, J., Modi, K., Soares-Pinto, D. O., and Céleri, L. C. (2016). Experimental demonstration of information to energy conversion in a quantum system at the Landauer limit. *Proc. R. Soc. A*, **472**, 20150813.

Piechocinska, B. (2000). Information erasure. *Phys. Rev. A*, **61**, 062314.

Pietzonka, P. and Seifert, U. (2018). Universal trade-off between power, efficiency, and constancy in steady-state heat engines. *Phys. Rev. Lett.*, **120**, 190602.

Planck, M. (1927). *Treatise on Thermodynamics* (3rd edn). Longmans, Green and Co., London.

Polettini, M. and Esposito, M. (2019). Effective fluctuation and response theory. *J. Stat. Phys.*, **176**, 94–168.

Pollock, F. A., Rodríguez-Rosario, C., Frauenheim, T., Paternostro, M., and Modi, K. (2018). Operational Markov condition for quantum processes. *Phys. Rev. Lett.*, **120**, 040405.

Pollock, F. A., Rodríguez-Rosario, C., Frauenheim, T., Paternostro, M., and Modi, K. (2018a). Non-Markovian quantum processes: Complete framework and efficient characterization. *Phys. Rev. A*, **97**, 012127.

Pottier, N. (2010). *Nonequilibrium Statistical Physics—Linear Irreversible Processes.* Oxford University Press, New York.

Potts, P. P., and Samuelsson, P. (2019). Thermodynamic uncertainty relations including measurement and feedback. *Phys. Rev. E*, **100**, 052137.

Proesmans, K., and Horowitz, J. M. (2019). Hysteretic thermodynamic uncertainty relation for systems with broken time-reversal symmetry. *J. Stat. Mech.*, 054005.

Purkayastha, A., Dhar, A., and Kulkarni, M. (2016). Out-of-equilibrium open quantum systems: a comparison of approximate quantum master equation approaches with exact results. *Phys. Rev. A*, **93**, 062114.

Pusz, W. and Woronowicz, S. L. (1978). Passive states and KMS states for general quantum systems. *Commun. Math. Phys.*, **58**(3), 273–290.

Reeb, D. and Wolf, M. M. (2015). Tight bound on relative entropy by entropy difference. *IEEE Trans. Inf. Theory*, **61**, 1458.

Reimann, P. (2008). Foundation of statistical mechanics under experimentally realistic conditions. *Phys. Rev. Lett.*, **101**, 190403.

Rempe, G., Schmidt-Kaler, F., and Walther, H. (1990). Observation of sub-Poissonian photon statistics in a micromaser. *Phys. Rev. Lett.*, **64**, 2783–2786.

Ribezzi-Crivellari, M. and Ritort, F. (2019). Large work extraction and the Landauer limit in a continuous Maxwell demon. *Nat. Phys.*, **15**, 660–664.

Rivas, Á. and Huelga, S. F. (2012). *Open Quantum Systems. An Introduction.* Springer, Berlin.

Roche, B., Roulleau, P., Jullien, T., Jompol, Y., Farrer, I., Ritchie, D.A., and Glattli, D.C. (2015). Harvesting dissipated energy with a mesoscopic ratchet. *Nat. Commun.*, **6**, 6738.

Rodríguez-Rosario, C. A., Modi, K., Kuah, A., Shaji, A., and Sudarshan, E. C. G. (2008). Completely positive maps and classical correlations. *J. Phys. A*, **41**(20), 205301.

Roncaglia, A. J., Cerisola, F., and Paz, J. P. (2014). Work measurement as a generalized quantum measurement. *Phys. Rev. Lett.*, **113**, 250601.

Roux, B. and Simonson, T. (1999). Implicit solvent models. *Biophys. Chem.*, **78**, 1.

Ruskai, M. B. (2002). Inequalities for quantum entropy: a review with conditions for equality. *J. Math. Phys.*, **43**(9), 4358–4375.

Šafránek, D., Deutsch, J. M., and Aguirre, A. (2019). Quantum coarse-grained entropy and thermalization in closed systems. *Phys. Rev. A*, **99**, 012103.

Šafránek, D., Deutsch, J. M., and Aguirre, A. (2019b). Quantum coarse-grained entropy and thermodynamics. *Phys. Rev. A*, **99**, 010101.

Saito, K. and Utsumi, Y. (2008). Symmetry in full counting statistics, fluctuation theorem, and relations among nonlinear transport coefficients in the presence of a magnetic field. *Phys. Rev. B*, **78**, 115429.

Sakurai, J. J. (1994). *Modern Quantum Mechanics* (revised edn). Addison-Wesley, Reading, MA.

Sánchez, R. and Büttiker, M. (2011). Optimal energy quanta to current conversion. *Phys. Rev. B*, **83**, 085428.

Sayrin, C., Dotsenko, I., Zhou, X., Peaudecerf, B., Rybarczyk, T., Gleyzes, S., Rouchon, P., Mirrahimi, M., Amini, H., Brune, M., Raimond, J.-M., and Haroche, S. (2011). Real-time quantum feedback prepares and stabilizes photon number states. *Nature*, **477**, 73–77.

Schaller, G. (2014). *Open Quantum Systems Far from Equilibrium.* Lecture Notes in Physics, Springer, Cham.

Schaller, G. and Nazir, A. (2018). The reaction coordinate mapping in quantum thermodynamics, in F. Binder, L. A. Correa, C. Gogolin, J. Anders, and G. Adesso, eds, *Thermodynamics in the Quantum Regime.* Springer, Cham.

Schaller, G., Emary, C., Kiesslich, G., and Brandes, T. (2011). Probing the power of an electronic Maxwell's demon: single-electron transistor monitored by a quantum point contact. *Phys. Rev. B*, **84**, 085418.

Schnakenberg, J. (1976). Network theory of microscopic and macroscopic behavior of master equation systems. *Rev. Mod. Phys.*, **48**, 571–585.

Schrödinger, E. (1952). Are there quantum jumps? *Br. J. Philos. Sci.*, **3**, 233.

Scully, M. O. and Zubairy, M. S. (1997). *Quantum Optics.* Cambridge University Press, Cambridge.

Seifert, U. (2005). Entropy production along a stochastic trajectory and an integral fluctuation theorem. *Phys. Rev. Lett.*, **95**, 040602.

Seifert, U. (2011). Stochastic thermodynamics of single enzymes and molecular motors. *Eur. Phys. J. E*, **34**, 26.

Seifert, U. (2016). First and second law of thermodynamics at strong coupling. *Phys. Rev. Lett.*, **116**, 020601.

Sekimoto, K. (1998). Langevin equation and thermodynamics. *Prog. Theor. Phys. Suppl.*, **130**, 17–27.

Sekimoto, K. (2010). *Stochastic Energetics.* Lecture Notes in Physics, vol. 799.

Springer, Berlin Heidelberg.
Skrzypczyk, P., Silva, R., and Brunner, N. (2015). Passivity, complete passivity, and virtual temperatures. *Phys. Rev. E*, **91**, 052133.
Sothmann, B., Sánchez, R., and Jordan, A. N. (2015). Thermoelectric energy harvesting with quantum dots. *Nanotechnology*, **26**, 032001.
Spohn, H. (1978). Entropy production for quantum dynamical semigroups. *J. Math. Phys.*, **19**, 1227–1230.
Spohn, H. and Lebowitz, J. L. (1979). Irreversible thermodynamics for quantum systems weakly coupled to thermal reservoirs. *Adv. Chem. Phys.*, **38**, 109–142.
Stokes, A. and Nazir, A. (2020). Implications of gauge-freedom for non-relativistic quantum electrodynamics. *arXiv:2009.10662*.
Stokes, A. and Nazir, A. (2022), Implications of gauge freedom for nonrelativistic quantum electrodynamics. *Rev. Mod. Phys.* **94**, 045003.
Strasberg, P. (2019*a*). Operational approach to quantum stochastic thermodynamics. *Phys. Rev. E*, **100**, 022127.
Strasberg, P. (2019*b*). Repeated interactions and quantum stochastic thermodynamics at strong coupling. *Phys. Rev. Lett.*, **123**, 180604.
Strasberg, P. (2020). Thermodynamics of quantum causal models: an inclusive, Hamiltonian approach. *Quantum*, **4**, 240.
Strasberg, P. and Díaz, M. G. (2019). Classical quantum stochastic processes. *Phys. Rev. A*, **100**, 022120.
Strasberg, P. and Esposito, M. (2017). Stochastic thermodynamics in the strong coupling regime: an unambiguous approach based on coarse graining. *Phys. Rev. E*, **95**, 062101.
Strasberg, P. and Esposito, M. (2019). Non-Markovianity and negative entropy production rates. *Phys. Rev. E*, **99**, 012120.
Strasberg, P. and Winter, A. (2021). First and Second Law of Quantum Thermodynamics: A Consistent Derivation Based on a Microscopic Definition of Entropy. PRX Quantum **2**, 030202.
Strasberg, P., Schaller, G., Brandes, T., and Esposito, M. (2013). Thermodynamics of a physical model implementing a Maxwell demon. *Phys. Rev. Lett.*, **110**, 040601.
Strasberg, P., Schaller, G., Brandes, T., and Esposito, M. (2017). Quantum and information thermodynamics: a unifying framework based on repeated interactions. *Phys. Rev. X*, **7**, 021003.
Strasberg, P., Díaz, M.G., and Riera-Campeny, A. (2021). *Clausius inequality for finite baths reveals universal efficiency improvements. Phys. Rev. E*, **104**, L022103.
Strasberg, P., Winter, A., Gemmer, J., and Wang, J. (2023). Classicality, Markovianity, and local detailed balance from pure state dynamics, *Phys. Rev. A* **108**, 012225.
Suárez, A., Silbey, R., and Oppenheim, I. (1992). Memory effects in the relaxation of quantum open systems. *J. Chem. Phys.*, **97**, 5101.
Tasaki, H. (2000). Jarzynski relations for quantum systems and some applications. *arXiv:cond-mat/0009244*.
Thierschmann, H., Sánchez, R., Sothmann, B., Arnold, F., Heyn, C., Hansen, W., Buhmann, H., and Molenkamp, L. W. (2015). Three-terminal energy harvester with coupled quantum dots. *Nat. Nanotechnol.*, **10**, 854–858.
Timpanaro, A. M., Guarnieri, G., Goold, J., and Landi, G. T. (2019). Thermodynamic uncertainty relations from exchange fluctuation theorems. *Phys. Rev. Lett.*, **123**, 090604.

Touchette, H. (2009). The large deviation approach to statistical mechanics. Phys. Rep. **478**, 1.
Touchette, H. (2015). Equivalence and nonequivalence of ensembles: thermodynamic, macrostate, and measure levels. *J. Stat. Phys.*, **159**, 987–1016.
Tuoriniemi, J. (2016). Physics at its coolest. *Nat. Phys.*, **12**, 11–14.
Trushechkin, A.S., Merkli, M., Cresser, J.D., and Anders, J. (2022), Open quantum system dynamics and the mean force Gibbs state, *AVS Quantum Sci.* **4**, 012301.
Utsumi, Y., Golubev, D. S., Marthaler, M., Saito, K., Fujisawa, T., and Schön, G. (2010). Bidirectional single-electron counting and the fluctuation theorem. *Phys. Rev. B*, **81**, 125331.
Uzdin, R. and Rahav, S. (2021). Passivity deformation approach for the thermodynamics of isolated quantum setups. *PRX Quantum*, **2**, 010336.
Vaccaro, J. A. (2016). Quantum asymmetry between time and space. *Proc. R. Soc. A*, **472**(2185), 20150670.
van Kampen, N. (1954). Quantum statistics of irreversible processes, *Physica* **20**, 603.
van Kampen, N. G. (2007). *Stochastic Processes in Physics and Chemistry* (3rd edn). North-Holland Publishing Company, Amsterdam.
Venuti, L. C. (2015). The recurrence time in quantum mechanics. *arXiv:1509.04352*.
von Neumann, J. (1929). Beweis des Ergodensatzes und des *H*-Theorems in der neuen Mechanik. *Z. Phys.*, **57**, 30–70.
von Neumann, J. (2010). Proof of the ergodic theorem and the H-theorem in quantum mechanics. *Eur. Phys. J. H*, **35**, 201–237.
Wallace, D. (2015). Recurrence theorems: a unified account. *J. Math. Phys.*, **56**(2), 022105.
Weiss, U. (2008). *Quantum Dissipative Systems* (3rd edn). World Scientific, Singapore.
Wigner, E. P. (1959). *Group Theory and Its Application to the Quantum Mechanics of Atomic Spectra*. Academic Press, New York.
Wilming, H., Gallego, R., and Eisert, J. (2016). Second law of thermodynamics under control restrictions. *Phys. Rev. E*, **93**, 042126.
Wiseman, H. M. and Milburn, G. J. (2010). *Quantum Measurement and Control*. Cambridge University Press, Cambridge.
Wolf, M. M., Eisert, J., Cubitt, T. S., and Cirac, J. I. (2008*a*). Assessing non-Markovian quantum dynamics. *Phys. Rev. Lett.*, **101**, 150402.
Wolf, M. M., Verstraete, F., Hastings, M. B., and Cirac, J. I. (2008*b*). Area laws in quantum systems: mutual information and correlations. *Phys. Rev. Lett.*, **100**, 070502.
Wu, L.-A., Segal, D., and Brumer, P. (2013). No-go theorem for ground state cooling given initial system-thermal bath factorization. *Sci. Rep.*, **3**, 1824.
Yamaguchi, M., Yuge, T., and Ogawa, T. (2017). Markovian quantum master equation beyond adiabatic regime. *Phys. Rev. E*, **95**, 012136.
Zhang, Y. (2019). Comment on 'Fluctuation theorem uncertainty relation' and 'Thermodynamic uncertainty relations from exchange fluctuation theorems'. *arXiv:1910.12862*.
Zhou, X., Dotsenko, I., Peaudecerf, B., Rybarczyk, T., Sayrin, C., Gleyzes, S., Raimond, J. M., Brune, M., and Haroche, S. (2012). Field locked to a Fock state by quantum feedback with single photon corrections. *Phys. Rev. Lett.*, **108**, 243602.

Index